T0339471

LEED v4 PRACTICES, CERTIFICATION, AND ACCREDITATION HANDBOOK

LEED v4 PRACTICES, CERTIFICATION, AND ACCREDITATION HANDBOOK

Second Edition

SAM KUBBA, Ph.D., LEED AP

Amsterdam • Boston • Heidelberg • London
New York • Oxford • Paris • San Diego
San Francisco • Singapore • Sydney • Tokyo
Butterworth-Heinemann is an imprint of Elsevier

Butterworth-Heinemann is an imprint of Elsevier
The Boulevard, Langford Lane, Kidlington, Oxford OX5 1GB, UK
225 Wyman Street, Waltham, MA 02451, USA

First edition 2010
Second edition 2016

Copyright © 2016, 2010 Elsevier Inc. All rights reserved.

No part of this publication may be reproduced or transmitted in any form or by any means, electronic
or mechanical, including photocopying, recording, or any information storage and retrieval system,
without permission in writing from the publisher. Details on how to seek permission, further
information about the Publisher's permissions policies and our arrangements with organizations such
as the Copyright Clearance Center and the Copyright Licensing Agency, can be found at our website:
www.elsevier.com/permissions.

This book and the individual contributions contained in it are protected under copyright by the
Publisher (other than as may be noted herein).

Notices
Knowledge and best practice in this field are constantly changing. As new research and experience
broaden our understanding, changes in research methods, professional practices, or medical treatment
may become necessary.

Practitioners and researchers must always rely on their own experience and knowledge in evaluating
and using any information, methods, compounds, or experiments described herein. In using such
information or methods they should be mindful of their own safety and the safety of others, including
parties for whom they have a professional responsibility.

To the fullest extent of the law, neither the Publisher nor the authors, contributors, or editors, assume
any liability for any injury and/or damage to persons or property as a matter of products liability,
negligence or otherwise, or from any use or operation of any methods, products, instructions, or ideas
contained in the material herein.

British Library Cataloguing-in-Publication Data
A catalogue record for this book is available from the British Library

Library of Congress Cataloging-in-Publication Data
A catalog record for this book is available from the Library of Congress

ISBN: 978-0-12-803830-7

For information on all Butterworth-Heinemann publications
visit our website at http://store.elsevier.com/

Working together
to grow libraries in
developing countries

ELSEVIER   Book Aid
            International

www.elsevier.com • www.bookaid.org

To my mother and father,
Who bestowed on me the gift of life ...
And to my wife and four children,
Whose love and affection inspired me on ...

CONTENTS

FOREWORD

Two decades ago, green building was not a part of the vernacular. It was a cutting-edge philosophy, championed by architects, engineers, designers, and other building industry professionals who dared to dream of a world far different than the one they lived in. We've come remarkably far in just two short decades. So far, in fact, that Dr. Kubba's *LEED v4 Practices, Certification and Accreditation* is now being issued as a second edition.

This book is the product of incredibly hard work, certainly on the part of Dr. Kubba, but also on the part of a larger movement. The green building movement is alive and well because of the willingness of thousands of people to admit what they didn't know, to seek answers and build consensus and create a new definition, to devise a new future.

When it comes to market transformation, few products or processes have had as much impact on an industry as LEED has had on the construction and building sector. Fifteen years into the lifetime of LEED, we see 1.85 million ft.^{2} of real estate certified every single day, all around the world. LEED has become synonymous with green building.

One of the most beautiful aspects of this movement and of LEED is that its development is iterative. We value the things that we learn along the way and we come together to improve upon what we have already built. LEED v4 is in many ways an outgrowth of LEED 2009; we took the best of what we already had and added to it based on the feedback we received from end users and the market at large. In this edition, Dr. Kubba addresses the changes, breaking down the credits and points, and outlines the process from start to finish.

In the First Edition Foreword, Rob Watson envisioned a day when the word "green" would no longer be necessary when referring to high-performance, sustainable buildings. I, too, look forward to the day when there is no question that a new building will meet LEED standards, a day when building renovation is done in consultation with green building standards, and there is no debate about whether it's worth it to build green.

I have every confidence that this second edition of the *LEED v4 Practices, Certification and Accreditation Handbook* will inform and inspire those seeking a better understanding of this movement. The more effectively we share information about green building design and construction, the more knowledgeable we will be as a community and an industry, and the more knowledgeable we are, the better equipped we will be to address the challenges of the future. This book is a critical piece of the puzzle, providing essential information and analysis to the green building professionals of today and tomorrow.

Scot Horst

PREFACE TO SECOND EDITION

Since its first printing in 2010, *LEED™ Practices, Certification, and Accreditation Handbook* has provided thousands of professionals and students with invaluable assistance in understanding the concept of sustainability and green construction. Furthermore, the book also assisted many students and professionals in taking the LEED™ exams, in addition to helping corporations to achieve LEED certification for proposed green building projects.

LEED™ (Leadership in Energy and Environmental Design) has emerged as one of the most important trends of "green" development and is revolutionizing the construction industry. It has gained tremendous momentum and has had a profound impact on our environment. The second edition of this reference handbook provides a solid foundation for the study of green building and LEED. It explains many of the important tenets of green building and sustainability, while providing strategies for implementation and specific case studies designed to broaden the reader's knowledge of sustainability and green building.

This second edition has been revised and fully updated to fill in many of the blanks as well as demystify LEED v4. Moreover, this book covers the topics and principles included in the LEED™ GA and LEED™ AP exams, providing all the information necessary to pass these exams. Likewise, this edition takes into account all the major changes that have taken place in the industry and LEED™ Accreditation since 2009. Of note, the USGBC released LEED v4 at the GreenBuild International Conference and Expo in November 2013. The Green Building Certification Institute (GBCI) has included the new LEED v4 content for all LEED exams in late spring 2014. We have incorporated all the new LEED v4 content in this second edition. Modifications of the credentialing process include a new three-tier system requiring applicants to first take the LEED™ Green Associate exam, followed by the LEED™ Professional Accreditation exam. The difference between the LEED Green Associate and LEED AP is clearly explained in the text. The requirements for taking these exams are also clearly explained in an easy-to-review format, pulling together all the critical points readers need to know about green building and sustainability to pass these exams. The USGBC has slightly relaxed the requirements for LEED v4, and project experience is no longer an eligibility requirement for those seeking LEED™ Accredited Professional (LEED AP) certification (although it is strongly recommended).

Finally, this edition will help the reader select a LEED v4 rating system to use for a particular project (e.g., LEED for Building Design and Construction, LEED for Operations and Maintenance, LEED for Interior Design and Construction, or LEED for Neighborhood Development).

Sam Kubba
PhD, LEED AP

ACKNOWLEDGMENTS

A book of this scope would not have been possible without the active and passive support of many friends and colleagues who have contributed greatly to my thinking and insight during the original writing of this book and who were in many ways instrumental in the crystallization and formulation of my thoughts on the subjects and issues discussed within. To them I am heavily indebted, as I am to the innumerable people and organizations that have contributed ideas, comments, photographs, illustrations, and other items that have helped make this book a reality instead of a pipe dream.

I must also unequivocally mention that without the unfailing fervor, encouragement, and wisdom of Mr. Kenneth McCombs, Senior Acquisitions Editor, Elsevier Science and Technology Books, this book would have remained on the drawing board, and this important second edition may never have become a reality. I must likewise acknowledge the wonderful work of Mr. Peter Jardim, Editorial Project Manager, Elsevier Science & Technology Books, and thank him for his unwavering commitment and support on the updating of this book. I also wish to acknowledge Mr. Gregory Harris for redesigning the second edition book cover. It is always a great pleasure working with them and the Elsevier family.

Likewise, I am particularly indebted to the US Green Building Council (USGBC) and its LEED Team, especially Ms. Ali Peterson for her assistance, continuous updates, and support on the new LEED™ version 4 Rating System and to Mr. Scot Horst, LEED™ Chief Product Officer for contributing the Foreword to this second edition. Likewise, my special thanks are extended to Mr. Paul Fair, LEED™ of Contracting Services, for reviewing many of the original chapters and for his expert comments and suggestions.

I would really be amiss if I failed to acknowledge my wife Ibtesam for her loving companionship and continuous support and for helping me prepare some of the original line illustrations. And last but not least, I wish to record my gratitude to all those who came to my rescue during the final stretch of this work – the many nameless colleagues – architects, engineers, and contractors who kept me motivated with their ardent enthusiasm, support, and technical expertise. To these wonderful professionals, I can only say, "Thank you." I relied upon them in so many ways, and while no words can reflect the depth of my gratitude to all of the above for their assistance and advice, in the final analysis, I alone must bear responsibility for any mistakes, omissions, or errors that may have found their way into the text.

INTRODUCTION – THE GREEN MOVEMENT YESTERDAY AND TODAY

The green building industry is a rapidly evolving field; indeed, the general perception of the green movement has been considerably transformed since its early formative days and is today a revolution sweeping across the United States, Europe, and much of the world. Jerry Yudelson, author of the *Green Building Revolution*, describes it as, "a revolution inspired by an awakened understanding of how buildings use resources, affect people, and harm the environment." But for a better understanding of the modern green movement, we need to try and trace its origins back to the beginning. However, it is almost impossible to determine precisely when a movement may have started. Long before the arrival of the industrial revolution and electrically powered heating and cooling systems, ancient and primitive populations were compelled to improvise using basic tools and natural materials to construct buildings that protected them from the harsh elements and extremes in temperature. As the ancients had very few other options at their disposal, ancient builders incorporated passive designs that took advantage of the resources provided by nature, namely, the sun and climate to heat, cool, and light their buildings. The Babylonians and Egyptians, for example, used adobe as their prime building material and built wind shafts (called *badger*) into their palaces and houses. These ancient palaces and houses were designed to take advantage of the courtyard principle. They also utilized narrow alleyways for shade and to overcome the many challenges of climate that confronted them.

More recently however, scholars like Mark Wilson expressed the belief that the concept of green building first appeared in America more than a century ago. Some historians believe that sustainability and environmentalism in America are actually rooted in the intellectual thought of the 1830s and 1840s.

Moving forward, the twentieth century witnessed the beginning of federal government action to preserve lands, and much of the credit for this is due to Teddy Roosevelt and John Muir for popularizing conservation. Roosevelt's visit to Yosemite in 1903 gained national publicity, and by 1916 the National Park Service had been established. Aldo Leopold (1887–1948) is one of many innovative philosophers whose theories have significantly influenced the North American green and environmental movements. Leopold recognized early on some of the benefits of building green, including environmental, economic, and social benefits. Unfortunately, however, the World Wars and the Great Depression pushed environmental concerns out of the mainstream and into the background of public consciousness. After World War II, environmental efforts continued to be focused on land conservation.

Following the 1930s, new building technologies began to fundamentally impact the urban landscape. The introduction of air conditioning, structural steel, vertical transport, and reflective glass suddenly made it possible to build multistory enclosed glass-and-steel-framed structures with controlled heating and cooling. The postwar economic boom was partly motivated by international architects such as Mies van der Rohe, whose International Style "glass box" provided a further catalyst in accelerating the pace of this phenomenon.

Other historians associate the first initial attempts to describe and spur Sustainable Development was with the publication of Rachel Carson's (1907–1964) book, *Silent Spring* (published in 1962), and the legislative fervor of the 1970s as a convenient marker, or with Henry David Thoreau who advocates in his book, *Marine Woods*, for the respecting of nature and also for an awakening to the need for conservation and federal preservation of virgin forests. Many believe that the green movement had its roots in the energy crises of the 1970s and the creative approaches to saving energy that emanated from it, such as smaller building envelopes and the use of active and passive solar design.

The 1973 OPEC oil crisis brought the cost of energy into sharp focus and effectively highlighted the need for diversified sources of energy and encouraged corporate and government investments in solar, wind, water, and geothermal sources of power. Additionally, the energy crises artificially created by the imposition of an oil embargo by OPEC in 1973 caused an upward spike in gasoline prices and caused long lines of vehicles at gas stations around the country. This had a dramatic effect on a small group of enlightened and forward thinking architects, ecologists, and environmentalists, who began questioning the wisdom of conventional building techniques and inspired them to seek new solutions to the problem of sustainability.

Also as a direct response to the crisis, the American Institute of Architects (AIA) formed an energy task force to study energy-efficient design strategies, which was later followed by an AIA Committee on Energy. The energy committee prepared several papers, including "A Nation of Energy Efficient Buildings," which became effective AIA tools for lobbying Capitol Hill. Among the more active committee members in the late 1970s were Donald Watson, FAIA, and Greg Franta, FAIA, when the AIA was also advocating building energy research. According to committee member Dan Williams, two groups were formed – one looked mainly at passive systems, such as reflective roofing materials and environmentally appropriate siting of buildings, to achieve energy savings, while the other looked primarily into solutions that involved the use of new technologies. The committee also collaborated with government and other organizations for more than a decade. In 1977, President Carter's administration founded what became the US Department of Energy; one of its principal tasks was to focus on energy usage and conservation.

Even as the immediate impact of the energy crisis began to ebb, pioneering efforts in energy conservation for buildings continued to advance. In England, Norman Foster's Willis Faber and Dumas Headquarters (1977) incorporated a grass roof, a daylighted atrium, and mirrored windows. But modern sustainability initiatives called for an integrated and synergistic design to both new construction and in the retrofitting of existing buildings.

In California numerous energy–efficient state office buildings were commissioned, including the Gregory Bateson Building (1978) in Sacramento (Figure I.1) which employed photovoltaics, underfloor rock–store cooling systems, and a computerized solar–tracking shading system. In 1977, a new

Figure I.1 Photo of the Gregory Batson Building in Sacramento, California (1978), which was a model energy-efficient state office building designed by Sim Van der Ryn and which employed several energy-saving tools, including an automated solar-tracking shading system with exterior ElectroShades®. *Source: MechoShade Systems, Inc.*

cabinet department, the Department of Energy, was created to address issues related to energy usage and conservation, and in the same year the Solar Energy Research Institute (later to be renamed the National Renewable Energy Laboratory) was founded to look at new energy technologies, such as photovoltaics.

In the United States, the 1980s and early 1990s saw significant efforts by sustainability proponents such as Robert Berkebile who received several awards, including the 2009 Heinz Award for his leadership and commitment to the environment, and Sandra Mendler, an architect specializing in the design of high-performance buildings, including office, mixed-use, and laboratory projects. Internationally, some enlightened architects were experimenting with prefabricated energy-efficient wall systems, water-reclamation systems, and modular construction units that were designed to minimize construction waste. These include University Professor Thomas Herzog of Germany, Malaysia's Kenneth Yeang, and England's Norman Foster (winner of the 1999 Pritzker Prize) and Richard Rogers, who is better known for his work on the Pompidou Center in Paris as well as awards including the RIBA Gold Medal, the Pritzker Prize, and the Thomas Jefferson Medal among other awards. At this time, Scandinavian governments set minimum standards for access to daylight and operable windows in workspaces. In 1987, the UN World Commission on Environment and Development, under Norwegian prime minister Gro Harlem Bruntland, was perhaps the first to define the term "sustainable development" as development that "meets the needs of the present without compromising the ability of future generations to meet their own needs."

In 1989, the AIA Energy Committee was transformed into a more broadly scaled AIA Committee on the Environment (COTE), and the following year, the AIA (through COTE) and the AIA Scientific Advisory Committee on the Environment obtained funding from the Environmental Protection Agency (EPA) to embark on the development of a building-products guide based on life-cycle analysis. This was the first such assessment of its kind to be undertaken in the United States. Individual product evaluations were ultimately compiled in the *AIA Environmental Resource Guide*, which was published in 1992.

In June 1992, Susan Maxman, who became the first female president of the national American Institute of Architects (AIA), represented the architectural profession at a UN Conference on Environment and Development in Rio de Janeiro. The conference drew delegations from 172 governments and 2400 representatives of nongovernmental organizations. This momentous event witnessed the passage of the Rio Declaration on Environment and Development, a blueprint for achieving global sustainability. Following on the heels of the Rio de Janeiro summit, Ms. Maxman chose sustainability as her theme for the June 1993 UIA/AIA World Congress of Architects. An estimated 6000 architects from around the world attended the Chicago event. Ms. Maxman also served as Chair of the Urban Land Institute's Environmental Council.

With the election of Bill Clinton to the presidency in November 1992, a number of proponents of sustainability began to circulate the idea to "green" the White House to decidedly improve the energy and environmental performance of 1600 Pennsylvania Avenue. And on Earth Day, April 21, 1993, President Clinton announced to the nation his plans to make the White House "a model for efficiency and waste reduction." The "Greening" of the White House consisted of a series of environmental upgrades of the White House complex, which consists of the Executive Residence, the 600,000-ft.[2] Old Executive Office Building across from the White House, and the White House Grounds. An energy audit was embarked on by the Department of Energy (DOE), as well as an environmental audit led by the Environmental Protection Agency (EPA), and a series of design charrettes in which nearly 100 design professionals, engineers, environmentalists, and government officials participated. They were required to formulate energy-conservation strategies using off-the-shelf technologies. Within three years, the results of these energy-conservation strategies produced numerous improvements to the nearly 200-year-old residence, such as an estimated $300,000 in annual energy and water savings, landscaping expenses, and solid-waste costs while at the same time reducing atmospheric emissions from the White House by an estimated 845 tons of carbon per year.

The flurry of federal greening projects was not the sole force propelling the sustainability movement in the 1990s. On September 14, 1998, President Clinton issued the first of three executive orders; it called upon the federal government to improve its use of recycled and "environmentally preferred" products (including building products). Another executive order in June 1999 was issued to encourage government agencies to improve energy management and reduce emissions in federal buildings through improved design, construction, and operation. A final report (under Chairman Ray C. Anderson) was issued recommending 140 specific actions to improve the nation's environment, many of which were related to building sustainability. The third executive order was issued in April 2000; it charged federal agencies to integrate environmental accountability into their daily decision-making and long-term planning.

The fruitful greening of the White House encouraged the greening of other properties in the massive federal portfolio. These included the Pentagon, the Presidio, and the US Department of Energy Headquarters as well as three national parks: Grand Canyon, Yellowstone, and Alaska's Denali. In 1996, the US Department of Energy signed a memorandum of understanding with AIA/COTE to work together on research and development, seeking to establish a program that contained a series of roadmaps for the construction of sustainable buildings in the twenty-first century. And on Earth Day 1998, the then-chair of AIA/COTE, Gail Lindsey, announced the first "Top 10 Green Projects," designed to increase the public's awareness of successful sustainable design, a program that has continued to the present day.

Individual federal departments were also making significant headway. For example, the Navy undertook eight pilot projects, including the Naval Facilities Engineering Command (NAVFAC) headquarters at the Washington Navy Yard. And in 1997, the Navy initiated development of an online resource, the *Whole Building Design Guide* (WBDG), with the object of incorporating sustainability requirements into mainstream specifications and guidelines. Seven other federal agencies have joined this project, which is now managed by the National Institute of Building Sciences (NIBS).

From the beginning, it became obvious that the green movement phenomena developed under the influenced of many people from all walks of life. These visionaries, both past and present, have recognized the need for serious changes in how we treat our environment. The championing of green issues by forward-thinking politicians, celebrities, and visionaries included people such as Al Gore, Ralph Nader, Robert Redford, and Leonardo DiCaprio. The movement was also impacted by the consequences of the 9/11 tragedy, gas shortages, visible climate change, the wars in Iraq and Afghanistan, and scientific consensus; these have all validated environmental concerns during the early years of this century. Former Vice President Al Gore's release in May 2006 of his acclaimed documentary film *An Inconvenient Truth* projected the climate crisis into the popular consciousness and raised public awareness of global warming, water contamination by toxic chemicals, and the exhaustion of our resources, and skyrocketing gas prices, among other consequences. To the surprise of many, the documentary was an instant hit and one of the many factors that facilitated the change of American attitudes toward the present climate predicament.

Lindsay McDuff, a green movement activist, says: "When politicians create or formulate policies, the business industries are consequently affected. With the rise in green policy, business executives from every arena are jumping on the green-movement bandwagon, basically out of the growing market demand. Being green has become a selling advantage in the business world, and eager companies are starting to jump at the chance to get ahead." The green movement has become an international conglomeration, seeking solutions to environmental concerns around the world.

Another important environmental activist is Robert Redford, who for over four decades has served as a trustee for the Natural Resources Defense Council and was the founder of the Institute for Resource Management. Redford also created the Sundance Institute in 1981, which recently developed and produced a television channel, "The Green," which is a broadband TV channel for environmentally themed films and documentaries. Like Redford, DiCaprio serves on the board of the Natural Resources Defense Council and is a member of Global Green USA. In 2007, DiCaprio also cowrote, *The 11th Hour*, a documentary that he created, produced, and narrated, on the state of the natural environment. Robert Redford and Leonardo DiCaprio are but two of the many celebrities who have taken advantage of their status to advance environmental issues and awareness.

THE MEANING OF GREEN BUILDING/SUSTAINABLE CONSTRUCTION

The phrases "green building" and "sustainable construction" are relatively new terms injected into our vocabulary. Their core message meant to improve current design and construction practices and standards so that the buildings we build today will last longer and be more efficient, cost less to operate, and contribute to healthier living and working environments for occupants, thereby helping to increase productivity. Green building/ sustainable construction is also about increasing the efficiency with which buildings and their sites utilize energy, water, and materials; protect natural resources; and improve the built environment so that ecosystems, people, and communities can thrive and prosper.

This handbook will help readers understand the fundamental concepts of sustainable design and green building practices, which have become common practice on projects worldwide. It is also intended as a practical study guide for the LEED Green Associate and LEED AP exams. This has become necessary because LEED accreditation and exam requirements have changed dramatically since the publication of the first edition. But to better understand these dramatic changes and today's modern green movement, we have to first define what is meant by "Green" and "Sustainability."

There are numerous definitions and concepts about *Green Building*; one such description of Green Building (also known as sustainable building) is as follows: "The practice of increasing the efficiency of buildings and their use of energy, water, and materials, and reducing building impacts on human health and the environment, through better siting, design, construction, operation, maintenance, and removal of the structure at the end of use, the complete building life cycle" ("Green Building" – Wikipedia, wikipedia.org).

Another definition by the EPA (Environmental Protection Agency) describes *green building* as "the practice of creating structures and using processes that are environmentally responsible and resource-efficient throughout a building's life-cycle from siting to design, construction, operation, maintenance, renovation, and deconstruction. This practice expands and complements the classical building design concerns of economy, utility, durability, and comfort. Green building is also known as a sustainable or high performance building." This essentially means that green building practices and strategies aim to reduce the environmental impact of building.

History tells us that over the years, the green-building phenomenon has significantly impacted both the US construction market and also the global marketplace as well. However, its precise impact on the building-construction industry and its suppliers has yet to be fully quantified. It is important to point out that we are witnessing a movement that has become truly prolific as state and local governments increasingly scramble to pass eco-friendly legislation or adopt codes and regulations that will promote sustainability and green building-construction practices.

Green building represents a model shift in the way we understand, design and construction today. Studies have consistently shown that buildings in the United States consume roughly one-third of all primary energy produced and close to two-thirds of electricity produced. Studies have also shown that roughly 30% of all new and renovated buildings in the United States have poor indoor environmental quality due to several factors, including noxious emissions, pathogens, and outgassing of harmful substances present in building materials. Some of the common building materials that release these gases (which are called volatile organic compounds (VOCs)) include caulks, sealants, and coatings, varnishes and/or stains, adhesives, paints, wall coverings, cleaning agents, fuels and combustion products, carpeting, vinyl flooring, fabric materials, and furnishings. These VOC sources are discussed in greater detail in Chapter 6 of this book. Persons that experience VOC exposure often report disagreeable odors; exacerbation of asthma; irritation to the eyes, nose, and throat; headaches and drowsiness. Health symptoms associated with VOC exposure can be minimized by choosing low VOC-emitting products. Efforts to address these environmental impacts and other issues are ongoing, and one of the solutions devised was the inclusion of sustainability practices in construction project objectives. The added sense of urgency caused by continually rising energy costs has inspired many in the construction industry to finally turn to the green movement for solutions and the momentum is now on the upswing.

But the principle of incorporating sustainable practices into conventional design and construction procedures will require a redefinition and reevaluation of the current roles of project participants to enable them to contribute effectively to sustainable project objectives. One of the primary characteristics of successful sustainable design is the implementation of a multidisciplinary and integrated team approach strategy, particularly during the early design phases. The incorporation of an integrated team approach, as well as early involvement, and greater participation of the various project members and stakeholders, will help ensure an end product that is more efficient and healthier for all concerned.

In 1993, Rick Fedrizzi, David Gottfried, and Mike Italiano founded the US Green Building Council (USGBC). Their mission was to promote sustainability in the building and construction industry. Outside the United States, the Building Research Establishment was developing its own building-assessment method, known as BREEAM, which was launched in 1990. The most ambitious international effort of the period was the Green Building Challenge (October 1998), held in Vancouver, British Columbia, Canada. It was a well-attended affair with representatives from 14 nations: Austria, Canada, Denmark, Finland, France, Germany, Japan, the Netherlands, Norway, Poland, Sweden, Switzerland, the United Kingdom, and the United States. Subsequent conferences took place in Maastricht, the Netherlands; Oslo, Norway; and other venues. The main goal of these conferences was to advance the state of the art in building environmental performance assessment methodologies and create an international assessment tool that takes into account regional and national environmental, economic, and social equity conditions. Another objective was to develop an internationally accepted generic framework that can be used to compare existing building environmental assessment methods and that can also be used by others to produce regionally based industry systems.

Even in the United States, the green building movement encouraged the creation of several green-building rating systems. At the forefront is the

Leadership in Energy and Environmental Design (LEED™) Green Building Rating System, developed by the USGBC. In 1998, the LEED™ 1.0 pilot program was released. By March 2000, 12 buildings had been certified under the pilot program. During the pilot period, there were extensive revisions and by March 2000 LEED™ 2.0 was released to the marketplace. As the LEED™ rating system evolved and developed, many building owners as well as the design and construction industry professionals gradually incorporated the LEED rating system into mainstream practice, as evidenced by the dramatic growth of LEED™ projects over the past decade. This upsurge in green building was fueled by a combination of forces including the General Services Administration requirement that all newly owned construction, build-to-suit leasing, and major renovations achieve a LEED® silver rating or higher.

Today, the LEED rating system continues to make inroads into the mainstream design and construction industry in the United States as reflected by the increase in the number of LEED™-certified green buildings and the increasing number of projects seeking LEED™ certification from the USGBC.

The use of LEED outside the United States continues to grow rapidly, and as of March 2015, the USGBC calculates that approximately 44% of all square footage pursuing LEED certification existed outside the United States. In fact, more than 150 countries and territories worldwide have resulted in LEED's adoption of the LEED™ NC rating system. On May 1, 2014, the USGBC released its ranking of the top 10 countries for LEED outside of the United States based on cumulative gross square meters (GSM) of space certified to LEED in each nation.

The full ranking★ is as follows:

Rank	Nation	GSM of LEED-certified space (million)	Total GSM of LEED-certified and registered space (millions)	Total number of LEED-certified and registered projects
1	Canada	17.74	58.66	4068
2	China	14.30	96.22	1638
3	India	11.64	66.22	1657
4	South Korea	3.84	16.61	242
5	Taiwan	2.98	6.97	114
6	Germany	2.90	7.32	365
7	Brazil	2.85	23.24	829
8	Singapore	2.16	3.86	91
9	UAE	1.82	47.16	850
10	Finland	1.45	3.56	148

★ Courtesy of USGBC.

The Indian Green Building Council (IGBC) was formed in 2001. By June 2014, more than 2 billion ft.^{2} area of green building projects had been registered. IGBC Chairman Prem C. Jain said that "in light of the increasing volumes of green building projects that IGBC is handling, IGBC feels it will be better managed if the LEED rating is handled by USGBC. Hence IGBC has agreed with USGBC for them to directly handle the LEED certification." And starting July 1, 2014, LEED projects in India are to be registered and certified directly by GBCI, the certification institute appointed by USGBC. The green-building goal as set by the IGBC envisages 10 billion ft.^{2} of green-building footprint to be registered for certification by 2022.

However, some organizations such as the North American Coalition on Green Building, which consists of 32 associations from various industries with a material interest in green building issues and is a strong supporter of green-building concepts, was created as a counter to the USGBC's LEED™ program as currently developed and implemented. The coalition believes that the present LEED rating system was created in the absence of a nationally recognized consensus process that allows for meaningful participation by all interested parties. The coalition was quick to voice its concerns with the LEED program to regulators as well as policy makers at the federal, state, and local levels of government and has started to explore alternatives. It is working with GBI on the possibility of endorsing the Green Globes program as a preferable alternative to LEED. Nevertheless, LEED continues to move forward and has taken important steps to address many of these concerns in the latest modifications (LEED™ V4).

In July 2006, an NPR news report by Chris Arnold highlighted the impact the green movement is having on the construction industry. One example given was a project in Boston where "in addition to nice-looking fixtures and appliances and wood floors, the bathroom faucets going in the project will save water, and the AC and heat systems were designed to save energy and boost indoor air quality." The 140-unit luxury condo development reportedly consumes 30% less electricity than a conventional building.

Federal research dedicated to green building practices and technologies has been disproportionately low when compared to the energy buildings consume and the greenhouse gases emitted, a recent study shows. Research on green building is estimated to constitute a mere 0.2% of all federally funded research, which comes to about $193 million per year. This is the approximate equivalent of a 0.02% of the estimated $1 trillion value of buildings constructed in the United States annually, even though the building-construction industry represents more than 10% of the US GDP.

Although federal funding for relevant research remains relatively low compared to funding for other research topics, Chris Pyke, director of research at the USGBC, states that there has been a significant increase in federal funding for green building research in recent years. To varying degrees, he says, the DOE and EPA have begun to recognize green building issues as legitimate topics of research. This is important especially after taking into account the major impact of buildings and the built environment on our quality of life, natural environment, and economy. The federal government and other relevant funding sources should be encouraged to provide appropriate support to these research programs and readily achievable strategies. Unless we move forward by significantly increasing and improving green-building practices, we are likely to experience a dramatic increase in the negative impacts of the built environment on human and environmental health in the years ahead.

A substantial increase in green-building research is vital if design and construction firms are to meet the many challenges pivotal to the nation's economy, health, and well being. These challenges include critical issues such as global warming, water shortages, indoor environmental-quality problems, and destruction of our ecosystem. Funding for green-building research should be increased to reflect the severity of the problems that buildings create. The current 0.2% of federal research dollars is totally inadequate to address these critical issues in a timely manner.

It has been consistently shown that buildings contribute substantially to environmental problems in the United States. For example, building operations account for some 38% of US carbon dioxide emissions, 71% of electricity use, and according to the Environmental Information Administration (2008), nearly 40% of total energy use; the latter number increases to an estimated 48% if the energy required to make building materials and construct buildings is included. Buildings consume 13.6% of the country's potable water, and EPA estimates of waste from demolition, construction, and remodeling amount to 136 million tons of landfill additions annually. Construction and remodeling of buildings account for 3 billion tons, or roughly 40%, of raw material use globally each year. They also negatively impact human health; up to 30% of new and remodeled buildings may experience acute indoor-air-quality problems in which indoor air pollutants are at concentrations typically between two and five times greater than those of outdoor air.

The impact of carbon emissions on global warming, which has become an issue of national concern, resulted in part in the collective adoption by the

AIA, ASHRAE, USGBC, Construction Specifications Institute (CSI), and the US Conference of Mayors' "the 2030 Challenge," which was developed by Architecture2030. Currently its goal is to design buildings that use 70% less annual energy than average for a building type. Thus, to satisfy the 2030 Challenge, a new hospital complex must emit 70% lower emissions than the current national average for hospitals. The reduction targets will decrease by 10% every 5 years until 2030 when all new construction will be required to have net zero carbon emissions.

A March 2007 UN report clearly reaffirms buildings' role in global warming; according to Achim Steiner, UN Under-Secretary General and UNEP Executive Director: "Energy efficiency, along with cleaner and renewable forms of energy generation, is one of the pillars upon which a decarbonized world will stand or fall. The savings that can be made right now are potentially huge and the costs to implement them relatively low if sufficient numbers of governments, industries, businesses, and consumers act." He goes on to say: "This report focuses on the building sector. By some conservative estimates, the building sector worldwide could deliver emission reductions of 1.8 billion tonnes [1 tonne = 1000 kilograms = 2025 pounds] of CO_2. A more aggressive energy-efficiency policy might deliver over two billion tonnes or close to three times the amount scheduled to be reduced under the Kyoto Protocol."

However, it will not be possible to meet the 2030 Challenge without a fundamental change in our present approach and improved knowledge of building energy issues. But besides the effect on climate, carbon emissions also cause direct damage to human health, according to Steiner. In November 2014, the UNEP Director said, Emissions of carbon and other pollutants are "responsible for approximately seven million premature deaths every year worldwide. That is more by far more than the combined premature deaths arising from HIV/AIDS and malaria combined."

A recently published report analyzed the global activity of green building from the perspective of early market adopters and construction-industry professionals in 45 countries; it confirms that green building is achieving an elevated adoption rate in most regions of the world and that the global marketplace is gradually but consistently undergoing a broad transformation to green. The green transformation can be attributed to a series of global pressures and trends, ranging from natural disasters to the dramatic increase and impact of green consumers. Together, these trends have motivated the industry to increasingly adopt sustainable design and construction methods in an effort to construct more efficient buildings designed to conserve energy and water, improve building operations, and enhance the health and well being of the global population. This is of particle significance since building construction is the largest source of carbon emissions in the United States.

The US market share of green construction continues to rise, particularly because state and local governments are using a variety of incentive-based techniques to encourage green-building practices. However, these efforts continue to encounter some obstacles and challenges along the way, including the cost of these new incentive programs, lack of available resources, and implementation difficulties. In order to help communities circumvent these obstacles, the American Institute of Architects (AIA) commissioned a report, *Local Leaders in Sustainability – Green Incentives*, which defines various incentive programs, examines the main barriers to success, and highlights best-practice examples from around the country, particularly improving design strategies, and improving the efficiency of established technologies and systems.

IMPACT OF GREEN BUILDING ON THE US BUILT ENVIRONMENT

According to the Department of Commerce (2008) the construction market accounts for 13.4% of the $13.2 trillion US GDP, of which the value of green-building construction is projected to increase to $60 billion by 2010, according to McGraw-Hill's "Key Trends in the European and U.S. Construction Marketplace" *SmartMarket Report*. Furthermore, it is estimated that 82% of corporate America will be greening at least 16% of real-estate portfolios in 2009; of these corporations, 18% are expected to be greening more than 60% of their real-estate portfolios.

Green building is increasingly seen as a business opportunity, and there are a number of sectors in the green-building industry that are expected to grow in the coming years, including office, education, healthcare, government, hospitality, and industrial. Of these, the three largest segments for nonresidential green-building construction – office, education, and healthcare – in 2008 accounted for more than 80% of total nonresidential green construction (Source: FMI 2008 US Construction Overview). According to the National Association of Realtors, "Office rents are projected to increase 2.3 percent in 2014 and 3.2 percent next year. Net absorption of office space in the U.S., which includes the leasing of new space coming on the market as well as space in existing properties, is likely to total 50.0 million in 2015."

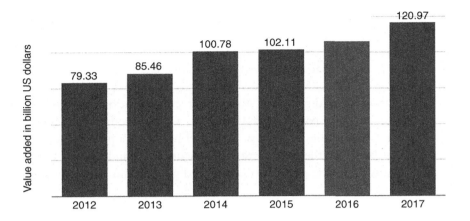

This statistic represents the projected value added of the US commercial building construction industry between 2012 and 2017. The graph shows that in 2015, the value added of this sector is expected to exceed US$102 billion (Source: Statista).

LEED™-registered and –certified projects represent a diverse cross-section of the industry. For example, roofing companies have turned their focus to technologies that will allow their customers to harness the energy a rooftop solution can provide. But these companies realize that to meet the demands and challenges that green technology brings requires a serious commitment financially and in terms of manpower, technology, training, and equipment.

One major effect that going green will have on the economy is the creation of new jobs and industries. Likewise, new green technologies are having a tremendous positive economic impact on the plumbing and electrical industries and are spurring economic growth for plumbing and electrical contractors around the country. We have recently witnessed an increased demand for new technologies in these industries, which have proven to be a positive economic stimulus to its members. Thus Plumbing contractors are in the forefront of advocating for and installing energy-efficient water systems, and promoting water conservation and energy efficiency through the installation and use of green technologies. Electrical contractors are also capitalizing on the green building movement, and firms are helping clients save on energy and other costs.

Studies on the costs and benefits of green buildings found that energy and water savings alone outweigh the initial cost premium in most green buildings and that green buildings may cost, on average, less than 2%

more to build than conventional buildings. This stands in contrast to the public's perception that green buildings would cost on average 17% more than conventionally designed buildings. In fact, one of the key findings in a landmark international study, *Greening Buildings and Communities: Costs and Benefits*, concludes, "Most green buildings cost 0 to 4 percent more than conventional buildings, with the largest concentration of reported 'green premiums' between 0 and 1 percent. Green premiums increase with the level of greenness, but most LEED™ buildings, up through gold level, can be built for the same cost as conventional buildings." A more detailed discussion of the green movement and green building is provided in Chapter 1.

CHAPTER 1

The Meaning of "Green Design" and "Sustainability"

1.1 DEFINING GREEN (SUSTAINABLE) BUILDING

Architectural/engineering professionals, people in the construction industry, and students often ask for a precise definition of green building and sustainable development. Although we know that green buildings are designed to reduce the overall impact of the built environment on human health and the natural environment, a precise definition is not easy since there are several definitions and concepts. For example, the EPA defines Green Building as "the practice of creating structures and using processes that are environmentally responsible and resource-efficient throughout a building's life-cycle from siting to design, construction, operation, maintenance, renovation, and deconstruction. This practice expands and complements the classical building design concerns of economy, utility, durability, and comfort. Green building is also known as a sustainable or high performance building." There are a number of reasons to build green, including potential environmental, economic, and social benefits, listed as follows (Source: EPA).

Environmental benefits
- Enhance and protect biodiversity and ecosystems
- Improve air and water quality
- Reduce waste streams
- Conserve and restore natural resources

Economic benefits
- Reduce operating costs
- Create, expand, and shape markets for green product and services
- Improve occupant productivity
- Optimize life-cycle economic performance

Social benefits
- Enhance occupant comfort and health
- Heighten aesthetic qualities
- Minimize strain on local infrastructure
- Improve overall quality of life

Copyright © 2016 Elsevier Inc.
All rights reserved.

1

Wikipedia describes Green Building as "the practice of increasing the efficiency of buildings and their use of energy, water, and materials, and reducing building impacts on human health and the environment, through better siting, design, construction, operation, maintenance, and removal of the structure at the end of use, the complete building life cycle" (wikipedia. org, "Green Building").

However, although sustainable or green building is a strategy for creating healthier and more resource-efficient models for construction, renovation, operation, maintenance, and demolition in which energy remains a pivotal component, green design needs to consider other environmental impacts. Recent research and experience have clearly shown that when buildings are designed and operated with their life-cycle impacts in mind, they usually provide substantially greater environmental, economic, and social benefits. And it is now apparent from the growing body of evidence that the most significant benefits can be achieved when the design and construction team adopts an integrated holistic approach from the earliest stages of a building project. Sustainable buildings use valuable resources more efficiently than buildings that are simply built to code. These green buildings are also kinder to the environment and provide indoor spaces that are generally more healthy, comfortable, and productive. It is inspiring to see the many changes that have taken place in the construction industry and the architectural/engineering professions over the last two decades in the promotion of environmentally responsible buildings.

Once the benefits of sustainable development were understood, architects, designers, builders, and building owners have become increasingly interested and involved in green building. Likewise, national and local programs encouraging green building are flourishing throughout the nation as well as globally. Thousands of projects have been built over the last two decades, providing tangible evidence of what green building can accomplish in terms of improved comfort levels, aesthetics, and energy and resource efficiency.

One of the major contributing factors to the rapid development of building efficiency is the demand from occupants and tenants who have to live and work inside these structures. There is a substantial body of evidence linking more efficient buildings with improved working conditions, leading to increased productivity, reduced turnover and absenteeism, and other benefits. The reason for these results is that in many cases building operating expenses represent less than 10% of an organization's cost structure, whereas the cost of personnel comprises the remaining 90%. This is a strong indicator that even small improvements in worker comfort can result in substantial dividends in performance and productivity.

This integrated whole-building approach is pivotal to a project's success and means that all aspects of a project, from site selection to structure to floor finish, are carefully considered. And focusing on a single component of a building can profoundly impact the project negatively, with unforeseen and unintended environmental, social, and/or economic consequences. For example, designing a building envelope that is not energy efficient can have a significantly adverse impact on indoor environmental quality. Likewise, exposure to materials such as asbestos, lead, and formaldehydes, which can have high volatile organic compound (VOC) emissions in a building, can precipitate significant health problems due to poor indoor-air quality, creating what is known as "sick building syndrome." An interdisciplinary team is thus considered a prerequisite to building green.

Although the related concepts of sustainable development and sustainability have become an integral feature of green building, it nevertheless is critical to consider any and all green options during the early design programming and planning phases. Furthermore, incorporating green strategies and materials during the early design phase is a great way to increase a project's potential market value. Continuing research has shown that sustainable developments can help lower operating costs over the life of the building by increasing productivity and utilizing less energy and water; green developments can also provide tenants and occupants with a healthier and more productive working environment. They can also significantly reduce environmental impacts by, for example, lessening storm-water runoff and the heat-island effect. Practitioners of green building often seek to achieve not only ecological but aesthetic harmony between a structure and its surrounding natural and built environment. The outward appearance and style of green buildings are not always immediately distinguishable from their less sustainable counterparts that are built to code.

History shows us that buildings have an enormous impact on the environment – both during the construction phase and throughout their operation. Building operation typically requires expending substantial amounts of energy, water, and materials while leaving behind large amounts of waste. Rob Watson, author of the *Green Building Impact Report*, issued in November 2008, states: "The construction and operation of buildings require more energy than any other human activity. The International Energy Agency (IEA) estimated in 2006 that buildings used 40% of primary energy consumed globally, accounting for roughly a quarter of the world's greenhouse-gas emissions. Commercial buildings comprise one-third of this total. Urbanization trends in developing countries are accelerating the growth of this sector relative to residential buildings, according to the World Business Council on Sustainable Development (WBCSD)."

The federal government is the nation's largest single landlord and has become one of the leaders in building green. In this respect, the General Services Administration (GSA) recently announced that it was applying stringent green building standards to its $12billion-construction portfolio of courthouses, post offices, border stations, and other buildings. In September 2014, a GSA task group recommended that at least 50% of the GSA's and the entire federal government's building area achieve net-zero energy (NZE) status by 2030 (GSA – RMIOUTLET). An NZE building is a one that produces as much energy through renewable resources as it consumes annually.

This role is echoed by Richard Fedrizzi, CEO and Founding Chairman of the US Green Building Council (USGBC), who comments: "The federal government has been at the forefront of the sustainable building movement since its inception, providing resources, pioneering best practices, and engaging multiple federal agencies in the mission of transforming the built environment." And on January 24 and 25, 2006, the first-ever conference took place; more than 150 federal facility managers and decision makers attended. In addition, 21 government agencies participated in formulating and witnessed the signing of the Federal Leadership in High Performance and Sustainable Buildings Memorandum of Understanding (MOU). With this MOU, signatory agencies committed to federal leadership in the design, construction, and operation of high-performance and sustainable buildings. The MOU represents a significant achievement by the federal government through its cumulative efforts to define common strategies and guiding principles of green building. The signatory agencies will also coordinate their efforts with others in the private and public sectors to achieve these goals.

Leah B. Garris, senior associate editor at *Buildings* magazine, says that "myth and misinformation surround the topic of sustainability, clouding its definition and purpose, and blurring the lines between green fact and fiction." But once you set aside the myths and misinformation relating to sustainability and green design, a number of pertinent strategies will become apparent that will help you achieve your green building objectives. Alan Scott, principal, Green Building Services, Portland, OR, says, "You can have a green building that doesn't really 'look' any different than any other building." Thus, a level of sustainability can easily be achieved by designing a green building that looks "normal." Ralph DiNola, principal at Green Building Services, says, "People don't really talk about the value of aesthetics in terms of the longevity of a building. A beautiful building will be preserved by a culture for a greater length of time than an ugly building." And, to be sustainable, it is important for a building to have a long-term, useful life. Aesthetics is a pivotal ingredient for longevity, and longevity is pivotal for sustainability.

In reality, sustainability is about conscious choices and not about spending more on superfluous options in hopes of earning an increased return on investment. It is about working with nature – not against it. Furthermore, it is not about buildings that are supposedly environmentally responsible but that ultimately sacrifice tenant/occupant comfort. This does not imply that purchasing green products or recycling assets at the end of their useful lives is not sustainable because it is. It is good for both the environment and for the health of a building's occupants. However, before making a final decision to go green, it is important to take the time to research what will work best for your particular project and offer the best possible return on investment instead of blindly buying into the myths that can defeat all your efforts to be sustainable. When there is little or no time set aside for sorting through the various sustainable technology options, a decision is often made at the last minute that turns out to be both costly and inappropriate.

1.1.1 The Need for Sustainable Design

The principal objective of sustainable design is to reduce, or completely avoid, running down of critical resources, including energy, water, and raw materials; prevent environmental dilapidation caused by facilities and infrastructure throughout their life cycle; and create built environments that are safe, livable, productive, and healthy. Ideally, building concepts and strategies should result in net positive benefits to all these areas. In addition to including sustainable design concepts in new construction, sustainable design advocates often prefer retrofitting existing buildings rather than building anew. This may partly be because retrofitting an existing building can often be more cost-effective, while minimizing environmental impact than by building a new facility.

Perhaps the main reason that green strategies are considered green is they typically work in harmony with the surrounding climatic and geographic conditions and not against them. This means that you need to fully understand the environment in which you are designing and building in order to utilize them to your advantage. Successful architects and designers fully understand that in order to succeed in sustainable design, they must be familiar with year-round weather conditions such as temperature, humidity, rainfall, site topography, prevailing winds, and indigenous plants. All impact sustainability in one way or another and depend to some extent on where you are. But in order to measure the degree of success of a sustainable design, the building's performance should be compared to a baseline condition. That baseline condition relates to where a building or project is located and the microclimate and environmental conditions of that specific location.

To attain sustainability, one cannot overemphasize the need to clearly identify and reduce a building's need for resources that are scarce or unavailable locally (e.g., water and energy) and increase the use of readily available resources, such as the sun, rain, wind, etc. A proficient understanding of the local climate is vital because it portrays a deep understanding of what is available: for example, the sun for heating and lighting, the wind for ventilation, and rain for irrigation and other water requirements (Figure 1.1a,b).

While the definition of *sustainable building design* is constantly changing, five areas immediately come to mind (this list is modeled after the Washington, DC–based USGBC's the Leadership in Energy and Environmental Design [LEED™] rating system):

- Sustainable Sites (SS)
- Water Efficiency (WE)
- Energy and Atmosphere (EA)
- Materials and Resources (MR)
- Indoor Environmental Quality (IEQ)

Site sustainability depends on climate and more specifically on the specifications and microelements that are particular to a given site (Figure 1.2). The materials employed in building a sustainable building will minimize life-cycle environmental impacts such as global warming, resource depletion, and human toxicity. And in order to avoid using inappropriate techniques that will invariably increase costs, it is important from the outset to understand the resources that are readily available and those that are not in the project area. For example, to get the optimum indoor-air quality, builders should use adhesives, paints, and sealants that do not release harmful chemicals into the air. Sustainability is facilitated by the use of high-performance insulation and installation of high-efficiency water, gas, and electrical systems. Passive cooling and "daylighting" are systems that can have a great impact on energy savings.

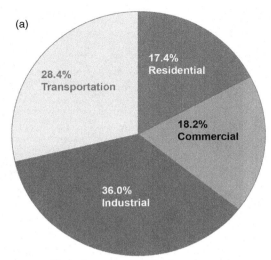

Greenhouse gas emissions by sector

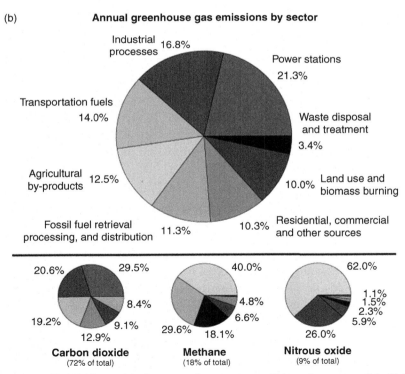

Figure 1.1 (a) Pie chart showing greenhouse-gas emissions by sector. (b) Global anthropogenic greenhouse-gas emissions broken down into eight sectors for the year 2000. *Sources: Part a, GreenerWorld Media, Inc. Part b, Robert Rohde, Wikipedia: Greenhouse Gas.*

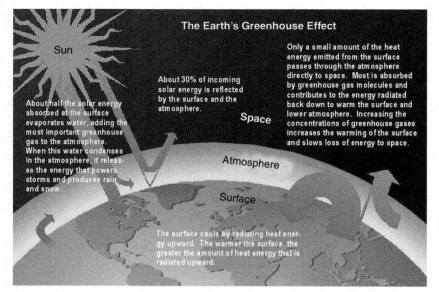

Figure 1.2 *Illustration Showing Greenhouse Gases in the Order of Relative Abundance.* These are water vapor, carbon dioxide, methane, nitrous oxide, and ozone. They come from natural sources and human activity. *Source: Darren Samuelsohn, Earth News, 2007.*

In promoting LEED, the USGBC emphasizes the simplicity of the program. But the uniqueness of the LEED™ certification system is that it typically mandates performance over process. In fact, over the past 15 years or so the USGBC, through the application of its widely circulated scoring system and other efforts, has dramatically impacted the way many contractors and their subcontractors operate.

1.2 GLOBAL UPSURGE IN THE GREEN BUILDING MOVEMENT

In April 2015, the WorldGBC announced the appointment of Terri Wills as its new Chief Executive Officer. Prior to joining the WorldGBC Ms Wills was Director of Global Initiatives with the C40 Cities Climate Leadership Group. WorldGBC Chairman, Bruce Kerswill, welcomed Ms. Terri as the new CEO and said, "The green building movement is growing apace – we have GBCs in 100 countries across the globe and the WorldGBC continues to grow in size and influence. Additionally, LEED rapidly became part of the mainstream and is currently the most popular and widely used green building rating system in the United States and globally. This is currently attested to by the more than 69,000 LEED building projects located in over 150 countries and territories (as of January 2015). Also, as of January 2015, more than 3.6 billion ft.^{2} of building space are LEED™-certified.

This increasing demand for building projects that use environmentally friendly and energy-efficient materials has spurred a strong green movement in the construction industry. An example of this is the expressed enthusiasm shown for green building as illustrated by the dramatic increase of nonresidential building starts in 2012 that were green (41%), as compared to all nonresidential building starts in 2005 (only 2%). It should be noted that LEED's remarkable strides in market transformation are tied to visible outcomes, including how buildings are oriented on their sites, the choice of cladding, size and location of windows, and exterior features such as green roofs, shading devices, solar PV panels or vertical-axis wind turbines. Storm water treatment ponds, bioswales, and a new trend toward indigenous plants are likewise transforming grounds and landscaping. This new generation of high-performance buildings with sustainability at the root of their design has taken on its own visual tone and is creating a recognizable style of high-performance sustainable architecture.

"Buildings should be treated as organic entities," Peter Busby, managing director at Perkins + Will, an architecture firm specializing in sustainable design, told attendees at the Canada Green Building Council's (CaGBC's) 2014 National Conference in Toronto, Canada. Busby also forecasts an imminent upsurge in the use of photovoltaic (PV) panels as the price (currently $2.50–$3-per-watt) continues to decline in step with dramatic improvements in efficiency, which is projected to double in watts-per-lumen output by the year 2017.

The quantity and quality of data relating to green business development and the environment remains inadequate. In this respect, government agencies, nonprofit groups, academic institutions, and corporations are beginning to improve data produced to allow quantifying, and assessing, simple measures of business environmental impact. The federal government in particular has until recently been deficient in its funding of research into green building. Figure 1.3 gives an indication of the amount of funding allocated to green building during the last decade as compared with other functions.

Mara Baum, Healthcare Sustainable Design Leader at HOK, says "other functions," within which sustainability research is allocated, make up 1.57% of total allocations. They "include education, training, employment, social services; income security; and commerce. Green building data was compiled

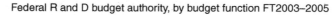

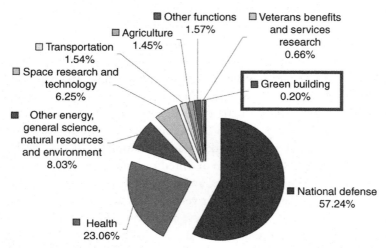

Figure   1.3 *Pie   Chart   Showing   Federal   Research-and-Development   Budget Appropriations According to Budget Function and the Minimal Amount Allocated to Green Building. Source: Mara Baum, USGBC.*

from agencies and the Office of Management and Budget (OMB); baseline federal R&D budget data comes from the National Science Foundation (NSF). The 0.2% funding toward green building does not include money from the Department of Defense." This is particularly alarming because, according to the American Institute of Architects (AIA), buildings are the leading source of greenhouse-gas emissions in the United States.

There are numerous options on how green building and LEED certification can save businesses money in the building process, but those options are not restricted to the construction and design process. With this in mind, Frank Hackett, an energy-conservation sales consultant for Mayer Electric Supply Co., Inc., says that one of the most basic things that a business can do to improve its efficiency is to update or retrofit its lighting system. One example according to Hackett is to modify and update existing lighting fixtures to use the more energy-efficient T-5 or T-8 fluorescent lamps as opposed to the T-12 models that are still being used in many parts of the country. Replacement of the magnetic ballasts in the lighting can also increase the system's energy efficiency. According to US Department of Energy (DOE) estimates, more than a modest percentage of a business's typical power bill is made up of lighting costs; reductions can therefore have significant economic benefits.

In recent years, many state and local governments and federal agencies are becoming increasingly eager to hop on the "green" bandwagon, and it is likely that updating and retrofitting existing lighting systems will gradually become mandatory because government regulations will leave businesses with no alternative. But other energy-saving options should be looked into as well, such as automatic controls that take advantage of natural light and automatically switch off the lights when no one is around. Sometimes tax deductions for these procedures are available, giving them an added advantage.

For residential projects, one of the more popular methods people are looking at to save money and energy over the long term is the application of so-called "green building" techniques to new homes or retrofitting existing homes. Green building materials can be anything from straw bale to low-flow toilets and blue-jean insulation made from recycled waste.

The US DOE recently issued a ruling that states must certify that their building codes meet the requirements in ASHRAE/IESNA's latest energy-efficiency standard. The DOE has determined that Standard 90.1-2013 (including earlier ASHRAE editions) will save more energy than the previously referenced Standards 90.1-1999 and 90.1-2004. This standard according to ASHRAE provides "the minimum requirements for energy-efficient design of most buildings, except low-rise residential buildings. It provides, in detail, the minimum energy-efficient requirements for design and construction of new buildings and their systems, new portions of buildings and their systems, and new systems and equipment in existing buildings, as well as criteria for determining compliance with these requirements." It is an indispensable reference for architects, engineers, and other professionals involved in design of buildings, building systems, and green building development.

As a US Act of Congress, President George W. Bush on October 3, 2008, signed into law H.R. 1424, which extended the Energy-Efficient Commercial Building Tax Deduction as part of the Emergency Economic Stabilization Act of 2008. It was enacted primarily to amend the Internal Revenue Code of 1986 to provide incentives for energy production and conservation, to extend certain expiring provisions, and to provide individual income tax relief. This tax deduction is not a tax credit – that is, the amount will not be directly subtracted from the tax owed, but an amount will be subtracted from the gross taxable income. Tax deductions offer benefits to the taxpayer; thus, this newly created program can be used as an incentive to assist in choosing energy-efficient building systems.

California Code of Regulations (CCR), Title 24, also known as the California Building Standards Code, is a compilation of three types of building criteria from three different origins (Source: California Building Standards Commission):
- *Building standards that have been adopted by state agencies without change from building standards contained in national model codes;*
- *Building standards that have been adopted and adapted from the national model code standards to meet California conditions; and*
- *Building standards, authorized by the California legislature, that constitute extensive additions not covered by the model codes that have been adopted to address particular California concerns.*

Soon after California passed its Revised Title 24 Code, most major cities throughout the United States began initiating some form of energy-efficiency standards for new construction and existing buildings. Additionally, a new law that took effect on January 1, 2009, states that owners of all nonresidential properties in California are mandated to make available to tenants, lenders, and potential buyers the energy consumption of their buildings as part of the state's participation in the federal ENERGY STAR® program. This information will then be given to the Environmental Protection Agency's ENERGY STAR portfolio manager, who will benchmark the data under its standards. Notwithstanding, the national model code standards adopted into Title 24 apply to all occupancies in California except for modifications adopted by state agencies and local governing bodies.

In April 2005 Washington State began requiring that all state-funded construction projects of more than 5000 ft.², including school district buildings, are to be built green. And in May 2006, Seattle approved a plan offering incentives to encourage site-appropriate packages of greening possibilities that would include green roofs, interior green walls, exterior vertical landscaping, air filtration, and storm-water runoff management. Seattle also became the first municipality in the United States to adopt the USGBC's LEED™ Silver rating for its own major construction projects. Likewise, Washington, DC's Green Building Act of 2006 went into effect in March 2007; it requires compliance with the LEED rating system for both private and public projects within the nation's capital. Washington, DC, has become the first major US city to require LEED™ compliance for private projects. These new green building standards became mandatory in the District in 2009 for privately owned, nonresidential construction projects with 50,000 ft.² or more; public projects will also have to comply with these new standards.

New York City's Local Law 86 (sometimes called "the LEED™ Law") took effect in January 2007. It basically requires that many of New York

City's new municipal buildings, as well as additions and renovations to its existing municipal buildings, achieve certain standards of sustainability that would meet various LEED™ criteria.

In Maryland, the Baltimore City Planning Commission voted in April 2007 to mandate developers to incorporate green building standards into their projects by 2010. Boston has taken it a step further, amending its zoning code to require green building for all public and private development projects in excess of 50,000 ft.[2].

Pennsylvania, which presently ranks second only to California in the number of LEED-certified buildings with 752 (as of May 2014), currently has four separate state program funding types, including a Sustainable Energy Fund, that provide grants and loans for energy-efficient and renewable-energy projects. According to the Pennsylvania Department of Community & Economic Development, these funding and financial incentives include:

Grants: A grant is a nonrepayable sum of money awarded for a specified purpose. Pennsylvania has a variety of grant programs available, from those that fund research and development projects and police regionalization to those that assist startup businesses and flood control projects.

Loans: A loan is a lent sum of money that is paid back with interest. You can find loans here for funding projects in distressed communities, jump-starting small businesses and more.

Loan guarantee: A loan guarantee is a promise by a lender to assume the debt obligations of a borrower if he or she defaults on a loan.

Tax credits: Tax credits are sums of money awarded to help offset specific taxes associated with various activities, such as expanding a business, investing in a particular industry or donating capital to scholarship programs (Source: Pennsylvania Department of Community & Economic Development).

Philadelphia has recently enacted a Green Roofs Tax Credit for costs incurred in installing a roof that supports living vegetation. Philadelphia has also proposed a sustainable zoning ordinance that mandates incorporating green roofs for buildings that occupy a minimum of 90,000 ft.[2] (see Figure 1.4, www.facilities.upen.edu).

Many other states – including Arizona, Arkansas, Colorado, Connecticut, Florida, Michigan, and Nevada – have followed suit, in addition to more than 60 cities and counties nationwide. In some states, green or sustainable building has ceased to be an option but rather a requirement. According to Stacey Richardson, a product specialist with the Tremco Roofing & Building

Figure 1.4 *One of Two Universities of Pennsylvania Building Projects That Received LEED Gold Certification by the USGBC.* Wharton School's Steinberg-Dietrich Hall West Tower Entrance addition was opened in early 2013. Designed by Kling Stubbins, it was built to achieve LEED Silver and exceeded that in its use of green features that have green roofs that manage storm water runoff and reduce cooling loads and heat island effects, utilize high-performance building materials, and have high-efficiency mechanical, lighting, and ventilation systems.

Maintenance division: "It is the way of the future, and industry developments in new green technology will provide building owners increasing access to energy-saving, environmentally friendly systems and materials. Everything from bio-based adhesives and sealants, low-VOC or recycled content building products, to the far-reaching capabilities of nanotechnology – the movement of building 'renewable' and 'energy-efficient' will only continue to strengthen."

While the United States is the uncontested global leader in green building construction, other countries around the world have also started to make substantial investments in sustainability. European Union (EU) leaders, for example, were able to agree on a new sustainable-development strategy that could potentially be pivotal in defining how the EU economy evolves in the coming decades. There are several green building assessment systems used around the world, such as Building Research Establishment's Environmental Assessment Method (BREEAM), Comprehensive Assessment System for Building Environmental Efficiency (CASBEE), Green Globes™ US, and Green Building (GB) Tool (Mago and Syal, 2007).

1.3 INCENTIVES FOR APPLYING GREEN PRINCIPLES

Green building tax benefits and incentives offered by governments are continually shifting, influenced by the economic and political climate. But architects/engineers, building contractors of all types, construction managers, owners, and developers can all find attractive opportunities by keeping up with them. But to maximize on a project's potential incentives and tax benefits, it is critical to make the decision to build green as early as possible in the design process.

1.3.1 Tax Deductions and Other Incentives

It is expected that 2015 will be a year of great change in green building standards, rating systems and codes. More than 65 local governments have already made a commitment to LEED™ standards in building construction, with some reducing the entitlement process by up to a year in addition to providing various tax credits. Annual energy costs have become a major office-building expense, but they can be reduced by as much as 30%, and these reductions will gradually increase with the development of new technologies.

Green Building research confirms the many benefits of developing and owning a green building. Beyond the health and environmental benefits of living and working in a green building, many local and state governments, utility companies, and other entities nationwide offer rebates, tax breaks, and other incentives to encourage the incorporation of ecofriendly elements in building projects. In fact, most major cities in the United States currently provide financial incentives for building green.

Through the Energy Policy Act of 2005, the US government now provides an Energy-Efficient Commercial Building Tax Deduction and other possible tax breaks, which initially applied largely to solar–electric systems, solar water heating systems and fuel cells, and which serve as an important incentive for commercial developers to construct energy efficient buildings as well as efficiency upgrades to homes. From January 1, 2008, The Energy Improvement and Extension Act of 2008 extended the tax credit to small wind-energy systems and geothermal heat pumps. One of the important revisions was an 8-year extension of the credit to December 31, 2016; other revisions included the ability to take the credit against the alternative minimum tax, and the removal of the $2000 credit limit for solar–electric systems beginning in 2009.

The US Environmental Protection Agency (EPA) website provides links to various funding sources for green building that are available to homeowners, industry, government organizations, and nonprofits in the form of grants, tax credits, loans, and other sources, both nationally and at the state and local levels.

The Database of State Incentives for Renewables and Efficiency (DSIRE), which is a nonprofit project funded by the US DOE through the North Carolina Solar Center and the Interstate Renewable Energy Council, also has information regarding local, state, federal, and utility incentives available for switching to renewable or efficient energy use. The US government's ENERGY STAR site can also light the path for consumers, homebuilders, and others to obtain federal tax credits for using energy-efficient products.

Property owners now generally agree that one of the more tangible benefits of attaining LEED™ certification for a building is the ability to use that achievement as a marketing tool. It has become apparent that certifiably green buildings are more likely to attract quality tenants and get higher rents. Likewise, designers and contractors with LEED-certified buildings in their portfolios often find that they have a distinct competitive edge. It is a good time therefore to be in the sustainable business.

1.3.2 Green Building Programs

A very effective and popular strategy to encourage green building is to incentivize the market through financial or structural incentives. The rewarding of building developers and homeowners who practice green building techniques spurs innovation and the pursuit for green building technologies. Throughout the United States and globally, cities are now promoting the use of various external green building programs. The City of Seattle is one such example, promoting a number of green building programs, including the following:

Built Green®: This is an environmentally friendly, nonprofit residential green building program developed by the Master Builders Association of King and Snohomish Counties in partnership with Seattle, King County, and a number of local environmental groups in Washington State. BUILT GREEN™ to set standards of excellence for residential developments, multifamily and single-family new construction and remodeling that can make a significant impact on housing, health, and the environment.

Energy Star Homes: This is a program for new homes that was created by the US EPA and DOE. The DOE posed a challenge to the homebuilding industry to build an increasing number of high-performance homes that achieve a 70 or better on the EnergySmart Home Scale (E-Scale). These homes will use at least 30% less energy than a typical new home built to code. The E-Scale allows homebuyers to understand – at a glance – how the energy performance of a particular home compares with others.

LEED™ for Homes: This is a consensus-developed, third party–verified, voluntary rating system that supports the design and construction of high-performance green homes, including affordable housing, mass-production homes, custom designs, stand-alone single-family homes, duplexes and townhouses, suburban and urban apartments and condominiums, lofts in historic buildings, and multifamily projects up to 8 stories. A LEED-certified home is designed and built in accordance with the rigorous guidelines of the LEED for Homes green building certification program. The benefits of a LEED home generally include lower energy and water bills and natural resources use; reduced greenhouse gas emissions; and less exposure to indoor toxins. In addition, a LEED rating can give homeowners confidence that their home is durable, healthy, environmentally friendly, and more comfortable for the occupants.

Seattle has also started implementing a Sustainable Building Policy that requires all new city-funded projects and renovations with more than 5000 ft.^{2} of occupied space to attain a LEED Silver rating. This policy affects all city departments that are involved with construction, such as the Department of Planning and Development (DPD), which monitors implementation of the policy. Seattle's green building program is now called CITY Green Building and resides within DPD.

1.3.3 Defining Sustainable Communities

The definition of sustainable communities remains somewhat elusive and continues to evolve over time as our knowledge about sustainability and new technologies increases. Some community planners are now trying to articulate a vision of how their community will grow in ways that will sustain its citizens' core values, which include:

- The community
- Environmental stewardship and responsibility
- Economic prosperity, opportunity, and security
- Social equity

Several cities in the United States have already started to adopt a comprehensive plan that includes goals and policies intended to help guide development toward a more sustainable future. This emerging group of sustainable communities, sometimes called green urbanism, seeks to apply leading-edge tools, models, strategies, and technologies to assist cities in meeting sustainability goals and policies. By applying an integrated, holistic, whole-systems approach to the planning of communities or neighborhoods, a city can achieve a still greater level of environmental protection.

Other compelling incentives for building owners and property developers to invest in green buildings and especially the LEED certification program include the financial benefit of operating a more efficient and less expensive facility. Adhering to LEED™ guidelines will help ensure that the facilities are designed, constructed, and operated more effectively, mainly because LEED focuses the design and construction team on operating life cycle costs, not initial construction costs.

Another compelling incentive involves the tax benefits that many states are now offering for green building and LEED compliance. An example is the State of New York, where former Governor Pataki established the New York State Green Building Tax Credit Program to facilitate the funding of concepts and ideas that will encourage green building. Through this program, the Hearst Building project, that is, the Hearst Tower, which is the Platinum LEED certified global headquarters of Hearst Corporation, in New York City was eligible for close to $5,000,000 in tax credits by achieving the required green building standards. The 46-story structure was designed by renowned architect Norman Foster and opened in 2006; it rises nearly 600 ft. above its landmark 6-story base. Hearst Tower is the first building to receive a Gold LEED rating for core and shell and interiors in New York City. In 2012, Hearst Tower also earned a Platinum LEED Rating for Existing Buildings, becoming the first building to receive both Gold and Platinum certifications.

It is surprising that until recently no single organization has taken a strong initiative to bring green construction to the American home market. Previously, various residential green building programs were in place and were sponsored for the most part by local homebuilder associations (HBAs), nonprofit organizations, and/or municipalities. Realizing that it was only a matter of time before the USGBC launched a LEED™-like program in this market, the National Association of Homebuilders (NAHB) and the NAHB Research Center (NAHB RC) took preemptive action and produced the Model Green Home Building Guidelines in addition to various utility programs.

Although the NAHB program provides many of the answers for the residential building market, LEED was for many years virtually the only game in town with respect to commercial construction. However, an alternative from Canada has been introduced that hopes to provide the US commercial construction industry with a simpler, less expensive approach to assessing and rating a building's environmental performance.

Green Globes is a web-based auditing tool that was developed by a Toronto-based environmental consultant, Energy and Environment Canada. Green Globes' strength is purported to be based on its reportedly rapid and economical method for assessing and rating the environmental performance of new and existing buildings. The Green Building Initiative (GBI) administers the building certification program in the United States, having brought the program in from Canada in 2006. The GBI purchased the rights to market the program in the United States, and GBI budgeted more than $800,000 as a first step to create national awareness of Green Globes as a viable alternative to the LEED program throughout the construction and development community and capture a viable percentage of its market share.

California and New York building codes are among the national leaders in sustainable development. California Governor Arnold Schwarzenegger's solar initiatives were instrumental in kick-starting the solar industry in his state, which just happen to be the nation's largest market. On December 14, 2004, Schwarzenegger signed Executive Order #S-20-04, requiring the design, construction, and operation of all new and renovated state-owned facilities to be LEED™ Silver-certified. The state is pursuing LEED Silver certification for new construction projects and LEED certification for existing buildings and facilities. The Executive Order sets a goal of reducing energy use in state-owned buildings by 20% by 2015 (from a 2003 baseline) and encourages the private commercial sector to set the same goal. California was followed by New Mexico, which passed a major green building tax credit in 2007, and Oregon, which passed a 35% tax credit for employing solar-energy systems.

History has shown that green building tenant attraction and retention continues to grow stronger, as major tenants increasingly favor healthier air quality over luxury amenities in premium properties, making a green building a sound investment and excellent long-term value. The main economic benefits of using energy more efficiently include the following:
- Saves operating costs on utility bills over the life of the building
- Reduces the unit cost of manufactured goods and services
- Increases resale and lease value of real estate

Among the main economic benefits of utilizing the environment more efficiently, thus reducing the environmental impact, are the following:

- Reduces water usage
- Reduces materials waste and disposal costs
- Reduces chemical use and disposal costs
- Encourages recycling and reuse of materials
- Encourages development of local markets for locally produced materials, saving on transportation costs

The main economic benefits of fostering social equity, improving indoor environment, and producing healthier places to work include the following:

- Increases productivity
- Reduces absenteeism
- Increases morale and corporate loyalty
- Reduces employee turnover

Sustainable construction and green buildings continue to enjoy an enviable reputation for excellence when it comes to a healthy work environment, and their developers and property owners continue to enjoy the well-deserved public perception of goodwill toward employees and the community in addition to the potential financial benefits that green buildings and sustainability bring to the table.

1.4 EMERGING DIRECTIONS: WHERE DO WE GO FROM HERE?

Over the last two decades the construction industry continues to undergo rapidly changing conditions, both in commercial and residential sectors. To meet the challenge of these demands and of government programs and increasingly stringent regulations regarding energy efficiency, alternative energy use, water conservation, and environmentally friendly building materials, working professionals including plumbers and electricians have found that they need to update their training on new equipment and technologies. Also, as a result of these new demands, entirely new jobs are emerging, such as solar panel installers, energy efficiency building auditors, and recycling specialists.

Greg Markvluwer of Herman Miller says, "The principles of green building, whether part of LEED certification process or independently undertaken, make good economic sense for all parties involved." The world of building design has been increasingly going green, so that today it has become an integral part of the mainstream and our global culture. Architects, designers, engineers, contractors, and manufacturers and federal, state, and local governments have all become willing participants in this emerging green phenomenon. Green building has not only become global but the trend toward sustainable design has shown that it is indeed now much more than a nominal trend.

In addition to the United States, LEED™ buildings can be found the world over in countries as diverse as Australia, Britain, Canada, China, India, Japan, Mexico, and Spain, to name but a few examples where the movement is well underway. Likewise, green building continues to impact and transform the building market, while revolutionizing our perception of how we design, inhabit, and operate our buildings. There are several factors that are accelerating this push toward building green. These include an increased demand for green construction, particularly in the residential sector; increasing levels of government initiatives; and improvements in the quality and availability of sustainable materials.

1.4.1 New Strategies

A recent report, "Greening Buildings and Communities: Costs and Benefits," is purported to be the largest international study of its kind; it is based on extensive financial and technical analyses of 150 green buildings, built from 1998 to 2008, in 33 US states and 10 countries. It provides the most detailed findings to date on the costs and financial benefits of building green. The conclusions of this report are outlined in Chapter 10.

Although green construction is booming, achieving sustainability on a large scale remains elusive, although as of January 2015, more than 3.6 billion ft.^{2} of building space is LEED certified. In 2014, 675.9 million ft.^{2} of real estate space became LEED certified, the largest area ever to become LEED certified in a single calendar year, and a 13.2% increase in total certified square-footage from 2013. In 2012, 41% of all nonresidential building starts were green, as compared to 2% of all nonresidential building starts in 2005. This trend is expected to intensify in the coming years, both in the United States and internationally. Table 1.1 shows some of the largest buildings certified by LEED to date.

Although the USGBC, a nonprofit membership organization, was founded in 1993, it was only in the last decade that it has become a potent driving force in the green building construction movement. It was able to do this largely through the development of its commercial-building rating system known as the Leadership in Energy and Design (LEED). Earning LEED certification commences in the early planning stage, where the

Table 1.1 Showing LEED's largest certified buildings in the United States

Owner/project	Location	Gross (ft.^{2})
Johnson Diversey/Global Headquarters	Sturtevant, WI	2,316,996
State of Illinois/McCormick Place West Expansion	Chicago, IL	2,226,000
EPA's National Headquarters Facilities	Washington, DC	1,878,453
State of California/Capitol Area East End Complex	Sacramento, CA	1,728,702
Silverstein Properties/7 World Trade Center	New York City, NY	1,682,000
Nitze-Stagen & Co./Starbucks Center	Seattle, WA	1,650,000
Goldman Sachs/Goldman Sachs Tower	Jersey City, NJ	1,556,915
General Motors/Lansing Assembly Plant	Lansing, MI	1,500,000
General Dynamics/Roosevelt C4 Facility	Scottsdale, AZ	1,500,000
Union Investment/111 South Wacker Drive	Chicago, IL	1,400,000
LaSalle Street Capital/ABN AMRO Plaza	Chicago, IL	1,375,058

Source: USGBC/GreenerWorld Media, Inc.

interested parties make the decision to pursue certification. Once this is accomplished, the next step is to register the project, which requires payment of an initial fee. Upon completion and commissioning of the project, with all the numbers and supporting documentation in, the project is submitted for evaluation and certification.

Today, USGBC has emerged as the leader in fostering and furthering green building efforts throughout the world. Although the LEED™ Green Building Rating System initially met some corporate resistance, and continues to face significant competition from other rating system programs such as Green Globes and BREEM, it nevertheless is overcoming these challenges and is increasingly becoming the national standard for green building in the United States; it is also recognized worldwide as an invaluable tool for the design and construction of high-performance, sustainable buildings.

In looking to the future, the USGBC has formulated a strategic plan in which it outlines the key strategic issues that face the green building community:

- Shift in emphasis from individual buildings toward the built environment and broader aspects of sustainability, including a more focused approach to social equity
- Need for strategies to reduce contribution of the built environment to climate change
- Rapidly increasing activity of government in the green building arena
- Increasing need for a focus on the greening of existing buildings;

- Lack of capacity in the building trades to meet the demand for green building
- Lack of data on green building performance
- Lack of education about how to manage, operate, and inhabit green buildings
- Increasing interest in and need for green building expertise internationally

One of the indicators reflecting international interest in the USGBC and LEED is evidenced each year by the large attendance and number of different countries represented at Greenbuild, the USGBC's International Conference and Expo. Nearly 17,500 people attended the Greenbuild 2114 conference in New Orleans, LA (source: USGBC Green Building Facts), compared to only 4200, who attended the same event in Austin, TX, in 2002. The 2014 Greenbuild attendees included representatives from all 50 states, over 85 countries, and 6 continents. There were also 552 exhibiting companies participating in 142,000 ft.^{2} of exhibit and display space on the trade show floor in New Orleans. There were more than 240 sessions, tours, summits, and workshops over the course of the week, and there were 37,250 h of continuing education credits earned at Greenbuild 2014. This venue is the world's largest conference dedicated to green building and has emerged as an important forum for international leaders in green building, one that will continue to facilitate the exchange of ideas and information on the subject.

Many project teams from different countries continue to successfully apply LEED as developed in the United States. However, the USGBC also recognizes that certain criteria, processes, or technologies may not be appropriate for all countries. To address this reality, the USGBC has agreed to sanction other countries to license LEED and to allow them to adapt the rating system to their specific needs while maintaining LEED's high standards. To date, many countries have expressed an interest in being LEED™-licensed, and several countries including Canada and India have already done so. In India, LEED has been adopted by the Confederation of Indian Industry. The USGBC also recognizes that successful strategies for encouraging and practicing green building will by necessity differ from one country to another, depending on local traditions and practices.

Internationally, the USGBC works through the World Green Building Council (WorldGBC), which was formed in 1999 by David Gottfried and officially launched at Greenbuild 2002 in Austin, TX. The World GBC is growing rapidly and represents large parts of the global construction sector. One of WorldGBC's goals is to help other countries to establish their own councils and find means to work effectively with local industry and policy makers.

The WorldGBC currently consists of a federation of more than 20 full-fledged member nations, including the United States and Canada and approximately 60 national green building councils under development. Their objective is devoted to transforming the global property industry to sustainability. Part of its mission is to support and promote individual green building councils, and it serves as a forum for knowledge transfer among councils. Its mission also includes encouraging the development of market-based environmental rating systems and recognizing global green building leadership. As one of the founding members of the WorldGBC, the USGBC is firmly committed to this mission.

The international importance of green building continues to be highlighted at the annual Greenbuild Conferences and Expos. Among the many international delegations was a recent high-level group from China that included the Vice Minister of the Ministry of Construction. Over the past decade, China's economy has been expanding and growing at an alarming rate and is on track to becoming the largest economy in the world by 2020. With such growth, however, comes a mixed bag of severe environmental problems, including a potential energy crisis. China has made impressive inroads to addressing these threats and reversing these environmental trends. To do so, China announced a new energy-efficiency strategy, of which green building is a primary component. Representatives of the Chinese Ministry of Construction and of the USGBC also signed a Memorandum of Understanding identifying points of dual interest for collaboration in the promotion of environmentally responsible buildings in both China and the United States. The combination of China's booming construction industry and its critical need for green buildings makes it a very exciting international opportunity for the USGBC.

These global developments reflect a few of the many ways the USGBC and its members are contributing to the international green building movement. And they represent but a small percentage of the opportunities that will appear in the future. But despite the enormous progress that has been achieved, there is still a great amount of work to be done to transform the built environment, both at home and aboard.

Following the attacks on the World Trade Center in New York and in Mumbai, India, global concern regarding the risk of terrorism has dramatically increased. This has created a greater demand for buildings that protect both the structure and the inhabitants. In many countries, architects and building owners are now demanding that their facilities be designed to have greater blast resistance and to better withstand the effects of violent storms such as tornadoes and hurricanes – for example, by the use of blast-resistant windows with protective glazing. The windows of federal buildings are now required to be designed to provide protection against potential threats. Also from the perspective of governments and the public alike, there are increasing demands for structures that are sustainable and meet environmental requirements. Taken together, these emerging trends require greater industry expertise, innovation, and solutions capable of addressing a wide range of issues.

Since the launching of LEED for New Construction & Major Renovations in 2000, the number of certified and registered projects has witnessed dramatic growth, with nearly 16,000 certified projects and 27,000 registered. It is a rating system that can be applied to commercial, institutional and residential buildings of four or more stories. The rating system has been applied to office buildings, manufacturing plants, hotels, laboratories, and many other building types.

Certified projects have been completed and verified through the USGBC's process, while registered projects are those that are still in the process of design or construction. The astonishing increase in LEED™ project registration for new construction is very significant, as it is a clear indicator of things to come.

This is echoed by Bob Schroeder, Industry Director (Americas) for Dow Corning's construction business, who says, "Today, sustainable design has been recognized by the industry and the public as a critical factor in achieving high-quality architecture and benefiting the building owners, the companies that occupy these structures, and the wider community." He goes on to say, "One very important industry trend has been the growing emphasis on sustainable, or green, construction. Years ago, structures that were built for World's Fairs were sometimes demolished almost soon after the event finished. It was a tremendous waste of limited resources and just what sustainable construction is meant to avoid." Architects and designers need to work more on bringing the outdoors inside with designs that increase natural daylight, giving a sense of spaciousness and harmony with nature. This can be facilitated by building siting and orientation.

1.4.2 Liability Issues

When taking on green/sustainable building, developers should be aware of the many potential risks involved. Ward Hubbell, previous president, Green Building Initiative, suggests that liability issues have become one of our most pressing problems. This is largely because some buildings that were designed to be green fail to live up to expectations. And often in the building industry, where there are failed expectations, lawsuits tend to follow. Hubbell says that "problems arise because building owners, designers, and builders often have very different ideas of what constitutes a successful green building, and they fail to explicitly communicate their thoughts at the outset of the project. As one might expect, these issues are compounded when the parties are relatively new to the green-building process." Hubbell further maintains that the two main areas for dissatisfaction are a building's failure to achieve a promised level of green building certification and its operational performance. Liability issues are discussed in greater detail in Chapter 10.

Decision Makers Table and LEED Credit Responsible Parties						
Credit	**Decision Makers**		**Credit**	**Decision Makers**	
Sustainable Sites			**Material and Resources**		
SS P1	Contractor	Civil engineer	MR P1	Owner		Architect
SS 1	Owner		MR 1.1	Owner	Contractor	Architect
SS 2	Owner	LEED AP	MR 1.2	Owner	Contractor	Architect
SS 3	Owner		MR 1.3		Contractor	Architect
SS 4.1	Owner		MR 2.1		Contractor	
SS 4.2		LEED AP and architect	MR 2.2		Contractor	
SS 4.3	Owner		MR 3.1		Contractor	Architect
SS 4.4	Owner	Civil engineer	MR 3.2		Contractor	Architect
SS 5.1	Owner	Contractor	Civil engineer	MR 4.1	Contractor	Architect
SS 5.2	Owner	Civil engineer	MR 4.2		Contractor	Architect
SS 6.1		Civil engineer	MR 5.1		Contractor	Architect
SS 6.2		Civil engineer	MR 5.2		Contractor	Architect
SS 7.1		Contractor	LEED AP/Land/Arch/Civil	MR 6	Contractor	Architect
SS 7.2		Contractor	LEED AP	MR 7	Contractor	Architect
SS8		LEED and lighting Dsgnr.	**Indoor Environmental Air Quality**		
Water Efficiency			EQ P1		Mechanical engineer
WE 1.1		Landscape architect	EQ P2	Owner	
WE 1.2	Owner	Landscape architect	EQ 1		Mechanical engineer
WE 2		Mechanical engineer	EQ 2		Mechanical engineer
WE 3.1		Mechanical engineer	EQ 3.1		Contractor	
WE 3.2		Mechanical engineer	EQ 3.2		Contractor	
Earth and Atmosphere			EQ 4.1		Contractor	Architect
EA P1	Owner	Contractor	Commissioning auth.	EQ 4.2	Contractor	Architect
EA P2		Mechanical engineer	EQ 4.3			Architect
EA P3	Owner	Mechanical engineer	EQ 4.4		Contractor	Architect
EA 1		Mechanical engineer	EQ 5			LEED AP
EA 2		Mechanical engineer	EQ 6.1		Mechanical engineer
EA 3	Owner	Contractor	Commissioning auth.	EQ 6.2		Mechanical engineer
EA 4		Mechanical engineer	EQ 7.1		Mechanical engineer
EA 5		Mech.Eng and facility eng.	EQ 7.2	Owner	Mechanical engineer
EA 6	Owner		EQ 8.1		LEED AP and architect
Look at each credit one by one and try and figure out why it makes sense that those are the responsible parties for that particular credit. You'll begin to notice some patterns for which trades work with which credits.			EQ 8.2		LEED AP and architect
Innovation in Design					
ID 1.1-1.4	Owner	Contractor	
ID 2	Owner	Contractor	LEED AP

Figure 1.5 *Highlighting Decision Makers and Parties Responsible for Achieving LEED™ Credits. Courtesy: GreenExam Academy.*

It suffices to say here that this is precisely why an integrated design process is needed, one that from the outset involves all the disciplines in a project working together, including the architect, engineers, contractor, building owner, suppliers, and others. A major benefit is that synergies are often achieved when the various parties contribute their advice and expertise from the earliest phases of the design, allowing them to perform their own functions with enhanced knowledge of the project in its entirety. The need for an integrated process continues from the building's design through construction, commissioning, and subsequent use. It further allows all of the relevant disciplines to agree on the scope of the project, its goals in terms of sustainability, cost, time, and accountability if the specified goals are not realized. Figure 1.5 highlights the responsibilities and decision-making parties for each credit in the LEED™ Certification Process.

1.4.3 Spectacular Landmarks

Another emerging trend is that of building national landmarks. Typical examples are the Sydney Opera House (Figure 1.6) and the Burj Khalifa

Figure 1.6 *Night Photo of the Sydney Opera House.* It is considered one of the most recognizable buildings in the world and has become the city's landmark with its iconic 14 triangular vaulted shells resembling white sails. In addition to representing Sydney, the opera house has also become a symbol for the country of Australia throughout the world. The original plan to build the opera house was won in competition in 1957 by the late Danish architect John Utzon, whose vision and design were too advanced for the architectural and engineering capabilities at the time. It was not until 1973 that the Opera House was finally opened by Queen Elizabeth II. *Source: beingaustralian.wordpress.com.*

Figure 1.7 *Illustration of Burj Khalifa.* Known as Burj Dubai before its inauguration, it is the tallest man-made structure in the world, and stands at 829.8 m (2722 ft.). The building, which was designed by Skidmore, Owings & Merrill and developed by Emaar Properties, officially opened on January 4, 2010. The Symbol of Dubai skyscraper has more than 160 habitable floors, 57 elevators, apartments, shops, swimming pools, spas, corporate suites, Italian fashion designer Giorgio Armani's first hotel, and an observation platform on the 124th floor. The total estimated cost for the project is estimated to be about US$1.5 billion. *Source: Skidmore, Owings & Merrill.*

(Figure 1.7). The latter is reportedly the tallest man–made structure in the world. Sometimes the driving factor is the desire for increased tourism, but more often cities are looking for symbols to foster local pride.

CHAPTER 2

Basic LEED™ Concepts

2.1 OVERVIEW – ESTABLISHING MEASURABLE GREEN CRITERIA

In December 1983, the World Commission on Environment and Development (WCED) was created and, by 1984, it was constituted as an independent body by the United Nations General Assembly. The creation of WCED by the United Nations was intended to address growing concerns "about the accelerating deterioration of the human environment and natural resources and the consequences of that deterioration for economic and social development." In establishing the commission, the UN General Assembly recognized that the environmental problems were global in nature and determined that it was in the common interest of all nations to establish policies for sustainable development (Report of the World Commission on Environment and Development: Our Common Future – http://www.un-documents.net/wced-ocf.htm). To rally countries to work and pursue sustainable development together, the UN decided to follow this by establishing the Brundtland Commission in 1987, which produced the Brundtland Report in August of the same year. It is in this document that the term "sustainable development" was first coined and defined. Some found the findings of this report particularly troubling, stating among other things,

> The 'greenhouse effect', one such threat to life support systems, springs directly from increased resource use. The burning of fossil fuels and the cutting and burning of forests release carbon dioxide (CO_2). The accumulation in the atmosphere of CO_2 and certain other gases traps solar radiation near the Earth's surface, causing global warming. This could cause sea level rises over the next 45 years large enough to inundate many low-lying coastal cities and river deltas. It could also drastically upset national and international agricultural production and trade systems.
>
> Another threat arises from the depletion of the atmospheric ozone layer by gases released during the production of foam and the use of refrigerants and aerosols. A substantial loss of such ozone could have catastrophic effects on human and livestock health and on some life forms at the base of the marine food chain. The 1986 discovery of a hole in the ozone layer above the Antarctic suggests the possibility of a more rapid depletion than previously suspected.

The report goes on to say, "A variety of air pollutants are killing trees and lakes and damaging buildings and cultural treasures, close to and sometimes thousands of miles from points of emission. The acidification of the environment threatens large areas of Europe and North America. Central Europe is currently receiving more than one gram of sulfur on every square meter of ground each year. The loss of forests could bring in its wake disastrous erosion, siltation, floods, and local climatic change. Air pollution damage is also becoming evident in some newly industrialized countries."

Today we are witnessing a substantial building boom that is sometimes supported by inferior design and construction strategies as well as highly inefficient HVAC systems, making buildings the largest contributors to global warming. Several federal and private organizations have tried (and continue to do so) to address these problems, and due partly to these efforts we are now seeing a surge of interest in green concepts and sustainability to the extent that "green" has become an integral part of the mainstream of the construction industry. Many project owners have become aware of the numerous benefits of incorporating green strategies and are increasingly aspiring to achieve Leadership in Energy and Environmental Design (LEED) certification for their buildings. The green building rating systems serve two principal functions – promoting high-performance buildings and helping create demand for sustainable construction. Green buildings are economically viable and ecologically benign, and their operation is sustainable over the long term.

Responding to this, the Partnership for Achieving Construction Excellence and the Pentagon Renovation and Construction Program Office put out the *Field Guide for Sustainable Construction*, which has been developed to assist and educate field workers, supervisors, and managers in making decisions that help the project team meet sustainable project goals. The following is a summary of this Field Guide's key points:

1. *Procurement* – Specific procurement strategies to ensure sustainable construction requirements are addressed.
2. *Site/environment* – Methods to reduce the environmental impact of construction on the project site and surrounding environment are identified.
3. *Material selection* – Identifies environmentally friendly building materials as well as harmful and toxic materials that should be avoided.
4. *Waste prevention* – Methods to reduce and eliminate waste on construction projects are identified.
5. *Recycling* – Identifies materials to recycle at each phase of construction and methods to support the onsite recycling effort.

6. *Energy* – Methods to ensure and improve the building's energy performance, reduce energy consumed during construction, and identify opportunities to use renewable energy sources.

7. *Building and material reuse* – Identifies reusable materials and methods to facilitate the future reuse of a facility, systems, equipment, products, and materials.

8. *Construction technologies* – Identifies technologies that can be used during construction to improve efficiency and reduce waste (especially paper).

9. *Health and safety* – Methods to improve the quality of life for construction workers are identified.

10. *Indoor Environmental Quality (IEQ)* – Methods to ensure IEQ measures during construction are managed and executed properly.

The Department of Energy (DOE) has also come out with an Environmental Protection Program, the goals and objectives of which are "to implement sound stewardship practices that are protective of the air, water, land, and other natural and cultural resources impacted by DOE operations and by which DOE cost effectively meets or exceeds compliance with applicable environmental; public health; and resource protection laws, regulations, and DOE requirements." To achieve these objectives, DOE's requirements must be accomplished by implementing environmental management systems (EMSs) at DOE sites. An EMS is a continuing cycle of planning, implementing, evaluating, and improving processes and actions undertaken to achieve environmental goals. Some of these goals and objectives are outlined as follows:

1. *Goal*: Protect the environment and enhance mission accomplishment through waste prevention.

   Objective: Reduce environmental hazards, protect environmental resources, minimize life-cycle cost and liability of DOE programs, and maximize operational capability by eliminating or minimizing the generation of wastes that would otherwise require storage, treatment, disposal, and long-term monitoring and surveillance (i.e., future environmental legacies).

2. *Goal*: Protect the environment and enhance mission accomplishment through reduction of environmental releases.

   Objective: Reduce environmental hazards, protect environmental resources, minimize life-cycle cost and liability of DOE programs, and maximize operational capability by eliminating or minimizing the use of toxic chemicals and associated releases of pollutants to the environment that would otherwise require control, treatment, monitoring, and reporting.

3. *Goal*: Protect the environment and enhance mission accomplishment through environmental preferable purchasing.

Objective: Reduce environmental hazards, conserve environmental resources, minimize life-cycle cost and liability of DOE programs, and maximize operational capability through the procurement of recycled content, bio-based content, and other environmentally preferable products, thereby minimizing the economic and environmental impacts of managing toxic by-products and hazardous wastes generated in the conduct of site activities.

4. *Goal*: Protect the environment and enhance mission accomplishment through incorporation of environmental stewardship in program planning and operational design.

Objective: Reduce environmental hazards, conserve environmental and energy resources, minimize life-cycle cost and liability of DOE programs, and maximize operational capability by incorporating sustainable environmental stewardship in the commissioning of site operations and facilities.

5. *Goal*: Protect the environment and enhance mission accomplishment through postconsumer material recycling.

Objective: Protect environmental resources, minimize life-cycle cost of DOE programs, and maximize operational capability by diverting materials suitable for reuse and recycling from landfills, thereby minimizing the economic and environmental impacts of waste disposal and long-term monitoring and surveillance.

To implement the aforementioned, it is important for readers to have a clear understanding of LEED™ certification and why this will help property owners remain competitive in an increasingly green market. Certification gives independent verification that a building has met accepted guidelines in these areas, as required in the LEED™ Green Building Rating System. LEED certification of a project provides recognition of its quality and environmental stewardship. The LEED Green Building Rating System is now widely accepted by public and private owners, both in the United States and internationally, further fueling the demand for achieving green building certification.

The LEED rating system rapidly made inroads into the mainstream design and construction industry, and soon contractors and property developers realized that they can contribute towards a project's success in achieving green objectives and goals by first, understanding the LEED process and the specific role and methodology in achieving LEED credits, and then, through early involvement and participation throughout the various project phases by incorporating an integrated team approach in the design process. It also goes without saying that measureable benchmarks are required in order to accomplish verification and confirm a building's acceptable performance. It is interesting to note that ASHRAE puts this responsibility squarely on the shoulders of the owner to define design intent requirements. But to be in a position to correctly evaluate a building or project, certain information must be made available regarding the criteria on which the project's design and execution was based. So unless the project's plans and specifications, etc. are prepared in a manner that has measurable results, it would be difficult to make a meaningful evaluation of a project to ascertain if it has met the required objectives and original design intent. Furthermore, in terms of sustainability, before measureable green criteria can be established, it is necessary to first agree on a finite definition of green construction and to articulate exactly what is required to achieve this.

The National Association of Home Builders (NAHB) has put forward a set of green home building guidelines that "should be viewed as a dynamic document that will change and evolve as new information becomes available, improvements are made to existing techniques and technologies, and new research tools are developed." The NAHB says that their Model Green Home Building Guidelines were written to help move environmentally friendly home building concepts further into the mainstream marketplace, and is one of two rating systems that make up NAHBGreen, the National Green Building Program. The NAHB point system consists of three different levels of green building – Bronze, Silver, and Gold – which is available to builders wishing to use these guidelines to rate their projects. "At all levels, there are a minimum number of points required for each of the seven guiding principles to ensure that all aspects of green building are addressed and that there is a balanced, whole-systems approach. After reaching the thresholds, an additional 100 points must be achieved by implementing any of the remaining line items." Table 2.1 outlines the points required for the three different levels of green building.

Although the appearance of a Green Building may be similar to other building forms, the conceptual approach in their design is different in that it revolves round a concern for the environment by extending the life span of natural resources, provide human comfort and well-being, security, productivity, and energy efficiency. In turn, this approach will result in reduced operating costs including energy, and water, as well as other intangible benefits. According to the Indian Green Building Council (IGBC) which administers

Table 2.1 The NAHB point system that is available to builders wishing to use these guidelines to rate their projects

Guidelines	Bronze	Silver	Gold
Lot design, preparation, and development	8	10	12
Resource efficiency	44	60	77
Energy efficiency	37	62	100
Water efficiency	6	13	19
Indoor environmental quality	32	54	72
Operation, maintenance, and homeowner education	7	7	9
Global impact	3	5	6
Additional points from sections of your choice	100	100	100

Points required for the three different levels of Green Building– Bronze, Silver, and Gold – are shown.

the LEED India rating system, India currently has 1419 IGBC-accredited professionals, 607 certified green buildings, 3106 registered projects, and a 2.74 billion-ft.² green building footprint. The salient attributes of a green building as highlighted by the IGBC are as follows:

- Minimal disturbance to landscape and site condition
- Use of recycled and environmental friendly building materials
- Use of nontoxic and recycled/recyclable materials
- Efficient use of water and water recycling
- Use of energy efficient and ecofriendly equipment
- Use of renewable energy
- Indoor air quality for human safety and comfort
- Effective controls and building management systems

The *Whole Building Design Guide* (WBDG) also sets out certain objectives and principles of sustainable design:

Objectives:

1. Avoid resource depletion of energy, water, and raw material.
2. Prevent environmental degradation caused by facilities and infrastructure throughout their life cycles.
3. Create built environments that are livable, comfortable, safe, and productive.

Principles:

1. Optimize site potential
2. Optimize energy use
3. Protect and conserve water
4. Use environmentally preferred products
5. Enhance Indoor Environmental Quality
6. Optimize operations and maintenance procedures

James Woods, executive director of the Building Diagnostics Research Institute notes,

Building performance is a set of facts and not just promises. If the promises are achieved and verified through measurement, beneficial consequences will result and risks will be managed. However, if the promises are not achieved, adverse consequences are likely to lead to increased risks to the occupants and tenants, building owners, designers and contractors; and to the larger interests of national security and climate change.

Needless to say, different sustainability experts will offer sometimes conflicting views. For example, Alan Bilka, a sustainable design expert with ICC Technical Services, says,

Over time, more and more 'green' materials and methods will appear in the coders and/or have an effect on current code text. But the implications of green and sustainable building are so wide and far reaching that their effects will most certainly not be limited to one single code or standard. On the contrary, they will affect virtually all codes and will spill beyond the codes. Some green building concepts may become hotly contested political issues in the future, possibly requiring the creation of new legislation and/or entirely new government agencies.

2.2 USGBC LEED GREEN BUILDING RATING SYSTEM

The most appropriate manner to be able to contribute to the success of a LEED project is to become familiar with the requirements and opportunities offered by the new program. Comprehensive documentation can be found on the US Green Building Council (USGBC) website (www.leedbuilding.org), from accreditation requirements to careers and e-newsletters. The USGBC has transformed the way architects/engineers and developers design, build, maintain, and operate our buildings, homes, and communities. To be successful in earning LEED certification, the design process must begin in the very initial planning stage, where the stakeholders involved make the decision to pursue certification. The official first step must be to register the project and payment of an initial fee. Upon completion of the project, and after all the numbers are in, including preparation of all supporting documentation, the project is submitted for evaluation and certification. Once this has been determined, the project is listed on the LEED project list. The summary sheet showing the tally of credits earned becomes available for most certified projects.

2.2.1 The Process Overview

The LEED Green Building Rating System consists of a set of performance standards used in the certification of commercial, institutional, and other building types in both the public and private sectors with the intention of promoting healthy, durable, and environmentally sound practices. An LEED certification provides independent, third-party verification that a building project has achieved the highest green building and performance measures according to the level of certification achieved. Setting up an integrated project team to include the major stakeholders of the project such as the developer/owner, architect, engineer, landscape architect, contractor, and asset and property management staff will help jump start the process. This implementation of an integrated, systems-oriented approach to green project design, development, and operations can yield significant synergies while enhancing the overall performance of a building. During the initial project team meetings, the project's goals will be articulated and the LEED certification level sought will be determined.

LEED v4 is the newest version of the world's premier benchmark for high-performance green buildings. This version is bolder, more specialized, and designed for an improved user experience. It builds on the fundamentals of previous versions while offering a new system that prepares all LEED projects in a portfolio to perform at a higher level.

All projects applying for LEED certification must meet a set of minimum program requirements (MPRs). Should a project not meet the MPRs, it becomes ineligible for LEED certification. The USGBC states that "minimum program requirements define the types of buildings that LEED was designed to evaluate. Taken together, they serve three goals: to give clear guidance to customers, to reduce complications that occur during the LEED certification process and to protect the integrity of the LEED program." MPRs describe the eligibility for each system and are intended to "evolve over time in tandem with the LEED rating systems."

LEED v4 Minimum program requirement #1: Must be in a permanent location on existing land

> *Intent*: The LEED rating system is designed to evaluate buildings, spaces, and neighborhoods in the context of their surroundings. A significant portion of LEED requirements is dependent on the project's location; therefore, it is important that LEED projects are evaluated as permanent structures. Locating projects on existing land is important to avoid artificial landmasses that have the potential to displace and disrupt ecosystems.
>
> *Requirements*: All LEED projects must be constructed and operated on a permanent location on existing land. No project that is designed to move at any point in its lifetime may pursue LEED certification. This requirement applies to all land within the LEED project.

Minimum program requirement #2: Must use reasonable LEED boundaries

Intent: The LEED rating system is designed to evaluate buildings, spaces, or neighborhoods and all environmental impacts associated with those projects. Defining a reasonable LEED boundary ensures that a project is accurately evaluated.

Requirements: The LEED project boundary must include all contiguous land that is associated with the project and supports its typical operations. This includes land altered as a result of construction and features used primarily by the project's occupants, such as hardscape (parking and sidewalks), septic or stormwater treatment equipment, and landscaping. The LEED boundary may not unreasonably exclude portions of the building, space, or site to give the project an advantage in complying with credit requirements. The LEED project must accurately communicate the scope of the certifying project in all promotional and descriptive materials and distinguish it from any noncertifying space.

Minimum program requirement #3: Must comply with project size requirements

Intent: The LEED rating system is designed to evaluate buildings, spaces, or neighborhoods of a certain size. The LEED requirements do not accurately assess the performance of projects outside of these size requirements.

Requirements: All LEED projects must meet the size requirements listed as follows.

1. LEED BD + C and LEED O + M Rating Systems
2. The LEED project must include a minimum of 1000 ft.^{2} (93 m^2) of gross floor area.
3. LEED ID + C Rating Systems
4. The LEED project must include a minimum of 250 ft.^{2} (22 m^2) of gross floor area.
5. LEED for Neighborhood Development Rating Systems
6. The LEED project should contain at least two habitable buildings and be no larger than 1500 acres.
7. LEED for Homes Rating Systems
8. The LEED project must be defined as a "dwelling unit" by all applicable codes. This requirement includes, but is not limited to, the International Residential Code stipulation that a dwelling unit must include "permanent provisions for living, sleeping, eating, cooking, and sanitation."

2.2.2 How LEED Works

Projects apply for LEED certification under a particular rating system (e.g., LEED for Building Design & Construction, LEED for Interior Design & Construction, LEED for Building Operations & Maintenance, Homes, etc.). Projects achieve points by satisfying various requirements, which are geared toward different methods of green building practice. If a project earns enough points, it gets a ranking (Certified, Silver, Gold or Platinum) and is then dubbed a "LEED certified" project.

The LEED Green Building Rating System inspires and instigates global adoption of sustainable green building practices through the adoption and execution of universally understood and accepted tools and performance criteria. The LEED system is based on awarding points relative to performance, including sustainable site development, energy and atmosphere, water efficiency, IEQ, materials and resources, regional priority, and innovation in design. Designers can select the points that are most appropriate to their projects to achieve a LEED rating. A total of 100 + 10 points are possible. Depending on the points, platinum, gold, silver, or certified ratings are awarded.

When the USGBC first introduced the Leadership in Energy and Environmental Design (LEED) Green Building Rating System, version 1.0, in December 1998, it was a pioneering effort. Today, LEED has become the predominant means for certifying green buildings, and has recently released a new version, LEED v4, which succeeds LEED 2009 (formerly known as LEED v3), and which is the first major LEED overhaul since LEED 2009 came out. The most notable change from LEED v3 to v4 is an expansion that includes rating systems for different building types and renovation types. The following are the LEED v4 categories:

LEED for building design and construction
- BD + C: New Construction
- BD + C: Core and Shell
- BD + C: Data Centers
- BD + C: Healthcare
- BD + C: Hospitality
- BD + C: Retail
- BD + C: Schools
- BD + C: Warehouses and Distribution Centers

LEED for operations and maintenance
- O + M: Existing Buildings
- O + M: Data Centers
- O + M: Hospitality
- O + M: Retail

- O + M: Schools
- O + M: Warehouses and Distribution Centers

LEED for interior design and construction

- ID + C: Commercial Interiors
- ID + C: Hospitality
- ID + C: Retail

LEED for building design and construction

- BD + C: Homes
- BD + C: Multifamily Midrise

LEED for neighborhood development

- ND: Plan
- ND: Built Project

As can be seen from the aforementioned, in addition to increased requirements, LEED v4 will now also include Location and Linkage (transportation), and Awareness and Education. The development of LEED v4 took more than 3 years to be approved.

According to USGBC.org, LEED users will find that a number of incorporated changes are geared toward the user experience, including

- The new LEED™ Online portal is easier to use and simplifies the process and requirements for credit submittal
- The customer experience now includes customer account management, the LEED Coach service and Proven Provider
- A LEED Dynamic Plaque, allowing you to track your LEED performance in real time
- Restructured reference guides, incorporating a new interactive web-based version, which offers video tutorials and downloadable templates and presentations
- Three new webinar suites dedicated to v4
- More accessible forms and calculators that are easier to use and that can be seen without a registered project

One can see from the aforementioned that LEED v4 has been significantly transformed by the many changes, both major and minor, to the rating system. The major LEED systems will be affected, and the overlap between them will increase substantially. The v4 version is bolder, more specialized, and provides an improved user experience.

2.2.3 The LEED Points System

LEED is continually evolving simple point-based structure that has set the green building standard and has made it the most widely accepted green program in the United States.

The various LEED categories differ in their scoring systems based on a set of required "prerequisites" and a variety of "credits" in the seven major categories listed later. In LEED v2.2 for new construction and major renovations for commercial buildings, there were 69 possible points, and buildings were able to qualify for four levels of certification.

With LEED v4, it has become much less complicated to figure out how many points a building receives and where that places it in the continuum of green building achievement. The new USGBG LEED green building certification levels for all systems are more consistent and are shown later (previous rating system is shown in brackets). There are four levels of certification — the number of points a project earns determines the level of LEED certification that the project will receive. Typical certification thresholds are as follows:

Certified: 40–49 (26–32) points

Silver: 50–59 (33–38) points

Gold: 60–79 (39–51) points

Platinum: 80 plus (52–69) points

The number of points available per LEED system has been increased so that all LEED systems have 100 base points as well as 10 possible innovation and regional bonus points, which bring the possible total points achievable for each category to 110. Figures 2.1–2.4 depict buildings that have received various levels of LEED certification.

As can be seen from the aforementioned, the maximum achievement in LEED–NC was increased from 69 points to 100 points in LEED 2009, which has remained the same for LEED v4. Unfortunately, it remains unclear sometimes how the added 31 points are distributed. Aurora Sharrard research manager at Green Building Alliance (GBA) says,

> *The determination of which credits achieve more than 1 point (and how many points they achieve) is actually the most complex part of LEED 2009 (LEED v3). LEED has always implicitly weighted buildings' impacts by offering more credits in certain sections. However, in an effort to drive greater (and more focused) reduction of building impact, the USGBC is now applying explicit weightings to all LEED credits. The existing weighting scheme was developed by the National Institute of Standards and Technology (NIST). The USGBC hopes to have its own weighting system for future LEED revisions, but currently, LEED credits are proposed to be weighted based on the following categories, which are in order of weighted importance:*

- Greenhouse gas emissions
- Indoor environmental quality fossil fuel depletion

- Particulates
- Water use
- Human health (cancer)
- Ecotoxicity
- Land use
- Eutrophication
- Smog formation
- Human health (noncancer)
- Acidification
- Ozone depletion

As will be seen, the weighting preferences in the LEED 2009 system puts much greater emphasis on energy, which is very appropriate and addresses some of the criticism levied against the earlier version of the LEED Rating System. For updates to the latest version of the LEED Rating System

Figure 2.1 *The West Building of the Phoenix Convention Center in Phoenix, Arizona.* The first phase of a $600-million expansion to be completed, it has been awarded a LEED Silver certification by the US Green Building Council (USGBC). The West Building, which opened in June 2006, was designed by Leo A Daly and HOK and achieved the LEED certification for incorporating energy use, lighting, water, and material use as well as incorporating a variety of other sustainable strategies. *Source: AV Concepts.*

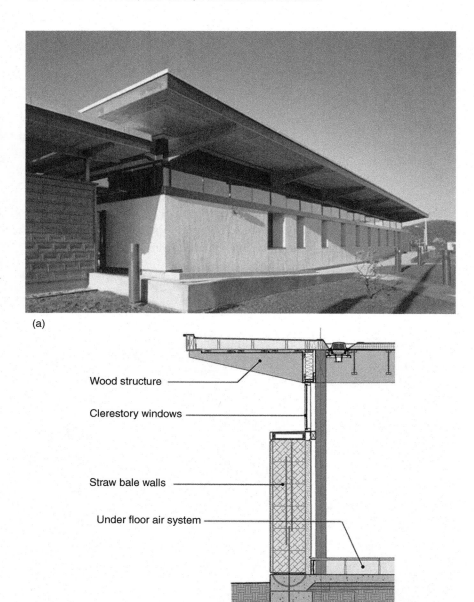

(a)

(b)

Figure 2.2 (a) The Santa Clarita Transit Maintenance is one of the first LEED Gold-certified straw-bale buildings in the world. The resource- and energy-efficient transit facility was designed by HOK and exceeds California energy efficiency standards by more than 40%, securing a new standard for straw-bale in high performance building design. (b) Diagram shows section taken through the exterior wall of the transit facility. The designers reportedly opted for a solar photovoltaic canopy to shade buses and provide nearly half of the building's annual energy needs. An electronic monitoring system is in place to track thermal comfort, energy efficiency, and moisture levels. *Source: HOK Architects.*

Figure 2.3 (a) Bird's eye-view of Donald Bren School of Environmental Science & Management at UCSB, which is a LEED Gold Pilot project. It utilized silt fencing, straw bale catch basins, and scheduled grading activities in accordance with the project's erosion control plan. (b) Bren Hall – faculty and students on the third-floor terrace. *Source: Bren School website: http://www.esm.ucsb.edu/*

(LEED v4), visit www.USGBC.org. There has also been an increase in the Innovation and Design (ID) credits from 4 points to 5 points. An additional point can be achieved for having a LEED–accredited professional (LEED AP) on the project team (bringing the total ID points to 6). The introduction of the new category of Regional Priority adds another potential 4 bonus points (bringing the total points possible to 110).

Figure 2.4 *Interior of BP America's New Government Affairs Office in Washington, DC.* Designed by Fox Architects, the 22,000-ft.² building achieved a LEED platinum level. *Source: Fox Architects.*

Mike Opitz, a senior professional mechanical engineer specializing in sustainability, energy efficiency, and third-party certification compliance for commercial and public sector real estate was Vice President of LEED Implementation for the USGBC until 2011. He highlights many of the changes incorporated (by category) in the LEED v3. These include the following:

Sustainable Sites

SSp1. Construction Activity Pollution Prevention (NC, CS, Schools) – Compliance verification is now required throug h photo documentation, inspection logs, or reports.

SSc4.1. Alternative Transportation, Public Transportation Access (NC, CS, Schools) – Exemplary performance option regarding "doubling transit ridership" has been formally incorporated into the reference guide.

SSc4.3. Alternative Transportation: Low-Emitting and Fuel Efficient Vehicles (NC, CS, Schools) – Defines preferred parking through discounts for spaces as at least 20% below the normal price. For NC projects, defines a compliance path option for a low–emitting vehicle sharing program. *SSc5.1. Maximize Open Space (NC, CS, Schools)* – In the implementation section, there is a clause that states "if a 10-acre site contains 5 acres of greenfield and 5 acres of previously developed land, site disturbance must be limited in the greenfield area, and native and adapted vegetation must be protected or restored for at least 50% (excluding the building footprint) of the previously developed site area."

SSc7.1. Heat Island Effect: Non-Roof (NC, CS, Schools) – Allows area shaded by solar panels or other renewable energy sources (e.g., over parking) to be added to qualifying area calculation.

Water Efficiency

WEp1 and WEc2. Water Use Reduction (NC, CS, Schools) – The 20% requirement that was formerly a point is now a prerequisite. Points begin at a 30% reduction and rise up to 45% for exemplary performance. There is some helpful information about how the EPA WaterSense program requirements relate to the LEED requirements.

Energy and Atmosphere

EAp1 (EAc3). Fundamental (Enhanced) Commissioning of Building Energy Systems (NC, CS, and Schools) – Provides much greater detail of step-by-step process for moving through owner's project requirements, basis of design, and commissioning coordination as well as greater detail about who is and is not allowed to act as the commissioning authority.

EAp2 (EAc1). Minimum (Optimize) Energy Performance (NC, CS, and Schools) – Calculations are now based on ASHRAE 90.1-2007 instead of the less stringent 2004 edition, and the required benchmarks have been lowered as well (the v3 prerequisite calls for a 10% reduction over 2007 standard, whereas in v2.2 a 14% reduction over the 2004 standard is required). Additional ASHRAE Advanced Energy Design Guide compliance paths have been added for retail, small warehouse buildings, and schools.

EAc2. On-Site Renewable Energy (NC, CS, and Schools) – This credit now offers regular points generating as little as 1% and as much as 13% of the power needed by the building (exemplary performance is set at 15%). There is also a clause that clearly forbids including energy that is generated but not used or energy that is to be sold back to the grid (presumably via net-metering). Furthermore, the owner cannot sell Renewable Energy Credits (RECs) for the power they are claiming as part of calculations. There is also a documentation requirement in regard to "any incentives that were provided to support the installation of on-site renewable energy systems."

EAc5.2. Measurement and Verification: Tenant Submetering (CS only) – In addition to v2 requirements, the project team must "provide a process for corrective action if the results of the M&V plan indicate that energy savings are not being achieved."

Materials and Resources

This edition of the reference guide references MasterFormat 2004 in its definitions of what is and is not included in the qualified materials figures for MRc3-7. Instead of the 1997 divisions 2-10 definition, they now cite 2004 divisions 3-10, 31.6, 32.1, 32.3, and 32.9. Division 12 (furniture and furnishings) is still optional.

MRp1. Storage and Collection of Recyclables (NC, CS, and Schools) – Clearly indicates that "it may be possible to create a central collection area that is outside of the building footprint or project site boundary" if you can "document how the recyclable materials will be transported to the separate collection area."

MRc1.1. Building Reuse: Maintain Existing Walls, Floors, and Roof (NC, CS, and Schools) – Because additional points are assigned to these credits, the thresholds for achievement have changed: 55%–95% for NC, 25%–75% for CS, and 75%–95% for Schools. Only CS has an exemplary performance option at 95%.

MRc7. Certified Wood – Goes into much greater detail about FSC chain-of-custody reporting requirements and how to report them for LEED purposes.

Indoor Environmental Quality

EQc4.3. Low-Emitting Materials: Flooring Systems (NC, CS, and Schools) – This was formerly the "carpet systems" standard, now expanded to included hard surface flooring. The standard carpets remain the same. FloorScore is used for other surfaces, and any stains or sealants on wood or other materials must comply with EQc4.1 requirements. An alternative option is to use products meeting CHiPS requirements.

EQc5. Indoor Chemical and Pollutant Source Control (NC, CS, and Schools) – Metal grates by main entries must now be 10 ft. long instead of just 6 ft. Also, there is a new requirement to "provide containment for appropriate disposal of hazardous liquid wastes in places where water and chemical concentrate mixing occurs." There are also additional requirements for separating "battery banks used to provide temporary back-up power."

EQc7.2. Thermal Comfort: Verification (NC and Schools) – There is now a clause that states that earning this credit "is contingent on achieving EQc7.1, Thermal Comfort – Design."

EQc8.1. Daylight and Views: Daylight (NC, CS, and Schools) – This credit has abandoned the 2% daylight factor calculation in favor of one that uses a more simple multiplication of glazing visible transmittance and window-to-floor area ratio. There are also now maximum footcandle levels (500 fc) if a simulation is being used. There is a clause that rids you of the maximum levels if you "incorporate view-preserving automated shades for glare control." Finally, there is a fourth "combination" option that allows you to use any of the aforementioned to document individual compliant spaces.

2.2.4 Building Certification Model

During LEED v3, the project certification process had moved to the Green Building Certification Institute (GBCI), which is a third-party organization that provides independent oversight of professional credentialing and project certification programs to determine if they have met the standards set forth by the LEED rating system. It was established in 2008 with the support of USGBC. The revised LEED v3 process is an improved ISO-compliant certification process that is designed to grow with the green building movement. In April 2014, the International WELL Building Institute and the GBCI announced a formal collaboration that will streamline how LEED and WELL work together and demonstrates that green building and health and wellness go hand in hand. However, the USGBC will continue to administer the development and ongoing improvement of the LEED rating system, and will remain the primary source for LEED and green building education.

The LEED v3 building certification infrastructure that is based on ISO standards, and that is administered by the GBCI has, in addition, 10 new certification bodies assigned to the certification process. These new LEED certification bodies include

- ABS Quality Evaluations, Inc. (http://www.abs-qe.com)
- BSI Management Systems America, Inc. (http://www.bsigroup.com)
- Bureau Veritas North America, Inc. (http://www.us.bureauveritas.com)
- DNV Certification (http://www.dnv.us/certification/managementsystems/index.asp)

- Intertek (http://www.intertek-sc.com)
- KEMA-Registered Quality, Inc. (http://www.kema.com)
- Lloyd's Register Quality Assurance Inc. (http://www.lrqausa.com)
- NSF-International Strategic Registrations (http://www.nsf.org)
- SRI Quality System Registrar, Inc. (http://www.sri-i.com)
- Underwriters Laboratories-DQS Inc. (http://www.ul.com/mss)

2.2.5 What's New?

After 3 years of review and collaboration, the changes from LEED v3 to LEED v4 are nuanced and intricate. According to the USGBC, LEED v4 is a further improvement on LEED 2009 (LEED v3) because it

- includes a focus on materials that goes beyond how much is used to get a better understanding of what's in the materials we specify for our buildings and the effect those components have on human health and the environment,
- takes a more performance-based approach to IEQ to ensure improved occupant comfort,
- brings the benefits of smart grid thinking to the forefront with a credit that rewards projects for participating in demand response programs, and
- provides a clearer picture of water efficiency by evaluating total building water use.

Likewise, LEED Online v3 is a greatly improved system compared to its predecessor, by providing enhanced functionality. For example, the USGBC notes some of the new project management improvement tools incorporated into v3, such as

- *Project organization* – the ability to sort, view, and group LEED projects according to a number of project traits, such as location, design or management firm, etc.
- *Team member administration* – increased functionality and flexibility in making credit assignments, adding team roles, and assigning them to team members. For example, credits are now assigned by team member name rather than by project role.
- *Status indicators and timeline* – clearer explanation of the review and certification process and highlights steps as they are completed in specific projects. The system now displays specific dates related to each phase and step, including target dates that each review is to be returned to the customer.
- *LEED Support for Certification Review and Submittals*

- LEED Online v3 offers many other enabling features to support the LEED certification review process, as well as enhancements to the functionality of submittal documentation and certification forms:
  - *End-to-end process support* – the new system will guide project teams through the certification process, from initial project registration through the various review phases. Furthermore, it will provide assistance to beginners during the registration phase to help them determine the type of LEED rating system that is best suited for their project.
  - *Improved midstream communication* – a midreview clarification page allows a LEED reviewer to contact the project team through the system when minor clarifications are required to complete the review.
  - *Data linkages* – LEED Online v3 automatically fills out fields in all appropriate forms after user inputs data the first time, which saves time and helps ensure project-wide consistency. Override options are available when required.
  - *Automatic data checks* – new system alerts users when incomplete or required data are missing, thus allowing user to correct error before application submission, thus avoiding delays.
  - *Progressive, context-based disclosure of relevant content* – upon selection of an option, the new system will simplify the process of completing forms by only showing data fields that are relevant to the customer's situation and hiding all extraneous content.

2.3 LEED VARIANTS AND OTHER SYSTEMS USED WORLDWIDE

Rating and certification systems are required to facilitate defining green buildings in the market. They tell us how environmentally sound a building is and clarify the extent to which green components have been incorporated and identify the sustainable principles and practices that have been employed. There are many different green building rating systems being used around the world, and each with its pros and cons depending on the type of certification targeted for a specific building. While many people may agree with the definition of a "green" building as being in the eye of the beholder, rating or certifying a green building helps remove some of that subjectivity. Moreover, rating a green building makes a property more marketable by informing tenants and the public about the environmental benefits of a property and also discloses the additional innovation and effort that the owner has invested to achieve a high performance building.

Energy and resource conservation are the general logic behind the design and construction of a green building project. A holistic approach to design means strategically integrating mechanical, electrical, and materials systems, which often substantially increases the efficiency – the complexity of which is not always apparent. Rating a green building identifies those differences objectively, and quantifies their contribution to energy and resource efficiency. Rating buildings can reduce implied risks because rating systems usually require independent third-party testing of the various elements which means there is less risk that these systems will not perform as predicted. Likewise, when a building is formally rated or certified, the risk of the project being marketed under the perception that it is green when in fact it is not is much less.

Several examples of major rating systems used in the United States are outlined as follows:

LEED®: This rating system is a product of the USGBC, and is the most widely applied rating system in the United States for commercial buildings. The LEED framework consists of several rating categories, applicable to different points in a building's lifecycle as discussed in other parts of this chapter. Numerous municipalities and government departments, including the General Services Administration (GSA) as well as an increasing number of private investors and owners have instituted policies requiring LEED certification for new construction projects.

Green Globes: This rating system is discussed in the next section. The Green Globes website says it "delivers an online assessment protocol, rating system and guidance for green building design, operation and management. It is interactive, flexible and affordable, and provides market recognition of a building's environmental attributes through third-party verification." It is basically an interactive, web-based commercial green building assessment protocol offered by the Green Buildings Institute (GBI).

ENERGY STAR®: This is a joint program of the Environmental Protection Agency (EPA) and the US Department of Energy (www. energystar.gov). The program is initiated to reduce greenhouse gas emissions and other pollutants caused by the inefficient use of energy and make it easy for consumers to identify and purchase energy-efficient products that offer savings on energy bills without sacrificing performance, features, and comfort. The program is designed for existing buildings, consisting of an Energy Performance Rating System that is free, with an online tool that focuses on energy performance. The impacts of other factors such as materials, indoor air quality, or recycling are not taken into consideration. The system essentially compares the energy performance of a particular building to that of a national stock of similar buildings. Data entered into the ENERGY STAR Portfolio Manager tool will model energy consumption based on a building's size, occupancy, climate, and space type. A minimum of 1 year of utility information input is required, after which the property is assigned a rating from 1 to 100. Buildings that acquire a score of 75 or more can apply and receive the ENERGY STAR label.

Other Green Building Rating Systems Worldwide: Many countries have developed their own standards of energy efficiency for buildings. Only one or two of these systems are currently available in the United States, but they may still prove influential in the emerging green building industry. Examples include BREEAM in the UK, Green Star in Australia, and BOMA Go Green Plus in Canada. The following are some examples of building environmental assessment tools currently in use by different countries around the world.

Australia: The Green Building Council of Australia (GBCA), established in 2002, is a not-for-profit industry association that promotes sustainability in the built environment. It has developed a green building standard known as Green Star. The Green Star environmental rating system is accepted as the Australian industry standard for green buildings. In three states, it has been mandated as a minimum for Government office accommodation. The Green Star environmental rating tools for buildings benchmark the potential of buildings based on nine environmental impact categories. Other standards used include EER (Energy Efficiency Rating) and NABERS (National Australian Built Environment Rating System), which is a government initiative to measure and compare the environmental performance of Australian buildings. Of note, the GBCA has recently forged a new partnership with the Australian Supply Chain Sustainability School to fasttrack the industry's green skills.

Canada: Green Globes and LEED are the two main rating systems used in Canada. Established in December 2002, the Canada Green Building Council acquired an exclusive license in 2003 from the USGBC to adapt the LEED rating system to Canadian circumstances. The Canadian LEED for Homes rating system was released on March 3, 2009. In 1982, Canada implemented the "R-2000" to promote construction that exceeds their building code to increase energy efficiency and promote sustainability. The R-2000 Standard is an industry-endorsed technical performance standard for energy efficiency, indoor air tightness quality, and environmental responsibility in home construction. An optional feature of the R-2000 home program is the EnerGuide rating service, which is available across Canada and which allows homebuilders and homebuyers to measure and rate the performance of their homes, and confirm that those specifications have been met.

Regional initiatives based on R-2000 include Energy Star for New Homes, Built Green, Novoclimat, GreenHome, Power Smart for New Homes, and GreenHouse. In March 2006, Canada's first green building point of service, Light House Sustainable Building Centre, opened in Vancouver, BC, which is funded by Canadian government departments and businesses to help implement green building practices and to recognize the economic value of green building as a new regional economy.

China: In China, there are two sets of national building energy standards (one for public buildings and another for residential buildings). Although China has developed mandatory building energy standards, they are narrow in their scope and lack a strong regulatory framework to incorporate energy-efficient standards in construction. Moreover, Ministry of Construction (MoC) enforcement remains problematic, and in 2005, the central government put in place a building inspection program to monitor the implementation of building energy efficiency. Under this program, design institutions, developers, and construction companies will lose their licenses or certificates if they do not comply with the regulations.

Standards and Ratings. The MoC recently introduced the "Evaluation Standard for Green Building" (GB/T 50378-2006), which is similar in structure and rating process to the USGBC's LEED (which itself is also being used). The building energy consumption data will be collected by MoC, and will be used to assess building performance; a three-star green building certificate will be awarded to qualified buildings. Market demand in China has also demonstrated a keen interest in LEED. This is verified by the fact that China is the third largest market for LEED in the world, with over 1500 LEED registered and certified projects. Green Olympic Building Assessment System (GOBAS) is another green building rating system, which was developed from Japan's Comprehensive Assessment System for Building Environment Efficiency (CASBEE). High-performance building projects are being supported both by the government and business. WBGC (an SBCI Member) has assisted the Ministry of Construction in China to establish the China Green Building Council. China's new green building standard, which was recently launched, is meant to complement better known labels like BREEAM (UK) and LEED, which are currently only used in office buildings for multinationals or upscale apartments.

France: The French government formed six working groups to find ways to redefine France's environment policy. The proposed recommendations were then put to public consultation, leading to a set of recommendations released at the end of October 2007. This process was named "Le Grenelle de l'Environnement," and recommendations were put to the French parliament in early 2008. The six working groups addressed climate change, biodiversity and natural resources, health and the environment, production and consumption, democracy and governance, and competitiveness and employment. Since January 1, 2013, all new buildings or parts must meet the requirements of the RT 2012. It is the aim to limit the consumption of primary energy housing 50 kW h/m² on an average through bioclimatic design and energy-efficient buildings. These developments are intended to match with European and international regulations and frameworks.

Germany: In January 2009, the first German standard for the new certificates was developed for sustainable buildings by the DGNB (Deutsche Gesellschaft für nachhaltiges Bauen e.V. – German Society for Sustainable Construction) and the BMVBS (Bundesministeriums für Verkehr, Bau und Stadtentwicklung – Federal Ministry of Transport, Building and Urban Affairs).

There are a number of German organizations that employ green building techniques, such as

- The Solarsiedlung (Solar Village) in Freiburg, Germany, which features energy-plus houses.
- The Sonnenschiff (Sun Ship) in Freiburg, Germany, which is also built according to German solar architect Rolf Disch PlusEnergy standards.
- The Vauban development, also in Freiburg.
- Houses designed by Baufritz, incorporating passive solar design, heavily insulated walls, triple-glaze doors and windows, nontoxic paints and finishes, summer shading, heat recovery ventilation, and graywater treatment systems.
- The new Reichstag building in Berlin, which produces its own energy. (Source: Wikipedia, the free encyclopedia)

India: The Energy and Resource Institute of India plays a key role in developing green building awareness and strategies in the country. A rating system called GRIHA was adopted by the Govt. of India as the National Green Building Rating System for the country, and measures are being taken to spread awareness about this rating system. GRIHA aims at ensuring that all types of buildings become green buildings. One of the strengths of GRIHA is that it puts great emphasis on local and traditional construction knowledge and even rates nonair-conditioned buildings as green.

Another organization that is playing an active role in promoting sustainability in the Indian construction sector is the Confederation of Indian Industry (CII). The CII is the central pillar of the Indian Green Building Council or IGBC. The IGBC is licensed by the LEED Green Building Standard from the USGBC and is currently responsible for certifying LEED New Construction and LEED Core and Shell buildings in India. All other projects are certified through the USGBC. There are many energy-efficient buildings in India, situated in a variety of climatic zones (Figure 2.5).

In February 2007, the Indian Bureau of Energy Efficiency (BEE) launched the Energy Conservation Building Code (ECBC), which is

Figure 2.5 *The Sohrabji Godrej Green Business Centre in Hyderabad, India.* In 2001, this building encouraged the development of green building in the country. This was the first platinum-rated green building under the LEED rating system, outside the USA, boasting energy savings of 63%. *Source: Confederation of Indian Industry.*

set for energy efficiency standards for design and construction with any building of minimum conditioned area of 1000 m² and a connected demand of power of 500 kW or 600 kVA. On February 25, 2009, the BEE launched a five-star rating scheme for office buildings operated only in the daytime in three climatic zones: composite, hot and dry, and warm and humid.

Israel: Israel has recently implemented a voluntary new standard for "Buildings with Reduced Environmental Impact" SI-5281; this standard is based on a point rating system (55 = certified and 75 = excellence) and together with complementary standards 5282-1, 5282-2 for energy analysis and 1738 for sustainable products provides a system for evaluating environmental sustainability of buildings. USGBC LEED rating system has also been implemented on several building projects in Israel, and there is a strong industry drive to introduce an Israeli version of LEED in the near future. In Figure 2.6 is a photo of Israel's first Platinum Building.

Figure 2.6 *The Porter School of Environmental Studies is Israel's First LEED Platinum Building.* It is conveniently located within walking distance of public transportation – Tel Aviv University Railway Station and a bus terminal. The building's sustainable features include solar PV panels, wastewater recycling, a green roof, and an "EcoWall" with solar tubes that will provide heat during the winter while during the summer the same wall will provide shading to keep the interior cool.

Japan: In 2001, a joint industrial/government/academic project was created with the support of the Housing Bureau, under the Ministry of Land, Infrastructure, Transport and Tourism (MLIT). This led to the creation of the Japan GreenBuild Council (JaGBC) and the Japan Sustainable Building Consortium (JSBC), which in turn created the CASBEE system. JaGBC/JSBC is continuously developing and updating the CASBEE system, and currently the CASBEE certification is available for new construction, existing building, renovation, market promotion, heat island, urban development, urban area (cities), plus buildings and detailed home.

CASBEE is composed of four assessment tools, CASBEE for predesign, for new construction, for existing buildings, and for renovations, which with the other expanded tools for specific purposes, are called the "CASBEE Family" and are designed to serve at each stage of the design process. Each tool is intended for a separate purpose and target user, and is designed to accommodate a wide range of uses (offices, schools, apartments, etc.) in the evaluated buildings. Likewise, the process of obtaining CASBEE certification differs from LEED. The LEED certification process starts at the beginning of the design process, with review and comments taking place throughout the design and construction of a project and which correspond to the building lifecycle, according to the following policies: (1) The system should be structured to award high assessments to superior buildings, thereby enhancing incentives to designers and others. (2) The assessment system should be as simple as possible. (3) The system should be applicable to buildings in a wide range of building types. (4) The system should take into consideration issues and problems peculiar to Japan and Asia. Of note, the 2014 editions of CASBEE tools were released in February 2015.

Malaysia: The main organization promoting green practices and building techniques is SIRIM Berhad, formerly known as the Standards and Industrial Research Institute of Malaysia (SIRIM). This is a corporate organization owned wholly by the Malaysian Government.

Mexico: The Mexico Green Building Council (MexicoGBC) is the principal organization promoting sustainable building technology, policy, and best practice. It formally announced its international launch during the Greenbuild conference on November 11, 2004. As a national, nonprofit organization, the Council's mission is "to promote sustainable development through the realization and construction of a superior built environment."

New Zealand: The New Zealand Green Building Council (NZGBC) was founded in 2005 and is a not-for-profit, industry organization dedicated to accelerating the development and adoption of market-based green building practices. In 2006–2007, several major milestones were achieved including becoming a member of the World GBC and the launch of the Green Star NZ (Office Design Tool, and welcoming of member companies).

South Africa: The Green Building Council of South Africa (GBCSA) is an independent, nonprofit, membership-based organization that was formed in 2007 by leaders from all sectors of the commercial property industry. It is a full member of the World Green Building Council and the official certification body of buildings under the Green Star SA Rating System. Its objective is to ensure that all buildings are built and operated in an environmentally sustainable way so that all South Africans work and live in healthy, effective, and productive environments. Its development of a Green Star SA rating tool is based on the Green Building Council of Australia tools, to provide the property industry with an objective measurement tool for green/sustainable buildings and to recognize and reward environmental leadership. Each Green Star SA rating tool reflects a different market sector (e.g., office, retail, multiunit residential, etc.). Green Star SA – Office was the first tool developed and which was released in final form (version 1) at the Green Building Council of South Africa Convention & Exhibition '08 in November 2008.

South Africa is in the process of incorporating an energy standard SANS 204, which aims to provide energy-saving practices as a basic standard in the South African context. Green Building Media which was launched 2007 has also played an instrumental role in green building in South Africa. On September 2014, the GBCSA announced the launch of the "EDGE" rating system, to be utilized for homes in South Africa. IFC's EDGE Green Buildings Certification System assesses the financial viability of green building projects at the early design stage.

United Kingdom: The Association for Environment Conscious Building (AECB) has promoted sustainable building in the United Kingdom since 1989. Now under the Energy Performance of Building Directive (EPBD) Europe has made a mandatory energy certification since January 4, 2009. A mandatory certificate called the Building Energy Rating system (BER) and a certification Energy Performance Certificate (EPC) is needed by all buildings that measure more than 1000 m^2 (approximately 10,765 ft.^{2}) in all the European nations.

In March 2009, the UK Green Building Council (UK–GBC) also called for the introduction of a Code for Sustainable Buildings to cover all nondomestic buildings, both new and existing. Of note, the Code for Sustainable Buildings is owned by the government but developed, managed, and implemented by industry and covers refurbishment as well as new construction. Also, the Welsh Assembly Government planning policy sets a national standard for sustainability for most new buildings proposed in Wales from September 1, 2009.

The Building Research Establishment's Environmental Assessment Method (BREEAM) is a leading and widely used environmental assessment method for buildings, setting the standard for best practice in sustainable design and a measure used to describe a building's environmental performance. According to the BREEAM,

[A] certificated BREEAM assessment is delivered by a licensed organization, using assessors trained under a UKAS accredited competent person scheme, at various stages in a buildings life cycle. This provides clients, developers, designers and others with:

- *market recognition for low environmental impact buildings,*
- *confidence that tried and tested environmental practice is incorporated in the building,*
- *inspiration to find innovative solutions that minimize the environmental impact,*
- *a benchmark that is higher than regulation,*
- *a system to help reduce running costs, improve working and living environments,*
- *a standard that demonstrates progress towards corporate and organizational environmental objectives.'*

There have been many significant changes to BREEAM, mainly in response to an evolving and changing construction industry and public agenda. Recent BREEAM changes include the following:

- The setting of minimum requirements for energy and water consumption.
- The introduction of an "outstanding" rating. Currently, an "excellent" rating is achieved when a building scores 70% of the potentially available score. "Outstanding" designs will need to achieve 85%.
- Postconstruction certification. This has previously been an option but it will now be compulsory for every rating.
- Innovation. Design teams and clients will be invited to put forward project innovations that will earn them extra credits. The ideas will be vetted at the design stage by a committee to see if it will achieve an environmental benefit. Clients must commit to monitoring the

performance of whatever they are innovating and make this public, whether it works or not.

- Materials selection. Minimum requirements will be put in place. For example, four of the seven key building materials must be rated.
- A+ to D in the, as yet, unpublished fourth edition of the *Green Guide to Building Materials*.

ESOS has recently been introduced to facilitate the UK meeting its requirements under the EU Energy Efficiency Directive and is expected to affect more than 9000 of the largest companies in the United Kingdom. The scheme mandates that these companies undertake obligatory assessments looking at energy use and energy efficiency opportunities at least once every 4 years. ESOS-compliant energy audits are employed to assess total energy consumption (using verifiable energy data) over a consecutive 12-month period known as the reference period. What many miss in the detail is that the reference period must overlap with the qualification date and end before December 5, 2015.

BREEAM is the preferred scheme for a number of the national Green Building Councils across Europe, including the Netherlands, Norway, and others.

United States: The United States has in place several sustainable design organizations and programs. The most prevalent is the USGBC, which is a nonprofit trade organization that promotes sustainability in how buildings are designed, built, and operated. The USGBC is best known for the development of the LEED rating system and Greenbuild, a green building conference that promotes the green building industry.

Today, LEED is adopted in more than 150 countries and territories worldwide and has 76 chapters and is 197,000 LEED professionals strong. It also has 12,870 member organizations from all sectors of the building industry; it works to promote buildings that are environmentally responsible, profitable and healthy places to live and work. As discussed in this and Chapter 4, the USGBC through its GBCI offers industry professionals the chance to develop expertise in the field of green building and to receive accreditation as green building professionals.

The National Association of Home Builders, which is a trade association representing home builders, remodelers, and suppliers to the industry, has formed a voluntary residential green building program called NAHBGreen (www.nahbgreen.org). This program incorporates an online scoring tool, national certification, industry education, and training for local verifiers. The online scoring tool is free to both builders and homeowners.

The Green Building Initiative (GBI) is a nonprofit network of building industry leaders working to mainstream building approaches that are environmentally progressive but also practical and affordable for builders to implement. The GBI has introduced a web-based rating tool called Green Globes, and which is discussed in the following section.

The United States Environmental Protection Agency's Energy Star program rates commercial buildings for energy efficiency and provides Energy Star qualifications for new homes that meet its standards for energy-efficient building design.

2.4 THE CHALLENGE OF GREEN GLOBES

Although there are a number of green building rating systems in the United States, the two national systems most widely used for commercial structures are LEED (Leadership in Energy and Environmental Design) and Green Globes (Go Green Plus). LEED Green Building Rating System® is focused on assessing *new* construction high-performance, sustainable buildings. Go Green targets *existing* buildings owners who want to have a more environmentally friendly building. In this section, we will compare Green Globes with the LEED Rating System. Although Green Globes currently has a minute share of Green Globe–certified buildings in the United States (about 55 buildings) that pales in comparison with LEED certified buildings, it is moving energetically forward to try and increase its market share within the United States. In March 2009, the GBI and the American Institute of Architects (AIA) signed a memorandum of understanding, pledging to work together to promote the design and construction of energy-efficient and environmentally responsible buildings. Likewise, a new memorandum of understanding was signed between ASHRAE and the GBI to work together to accelerate the adoption of sustainability principles in the built environment. Green Globes® rating system is also on track to become the first American National Standard for commercial green buildings.

2.4.1 Green Globes Emerges to Challenge LEED

History and Background: The birth of the Green Globes system lies in the BREEAM, which was brought to Canada in 1996 in cooperation with ECD Energy and Environment, and was initially developed as a rating and assessment system to monitor and assess green buildings in Canada.

The Green Globes environmental assessment and rating system represents more than a decade of research and refinement by a wide range of prominent international organizations and experts. Among the persons that were instrumental to the formation of this project include Jiri Skopek, John Doggart and Roger Baldwin, who were the principal authors of the BREEAM Canada document. The Canadian Standards Association that same year, published BREEAM Canada for Existing Buildings. This was followed in 1999 by the formation of the BREEAM Green Leaf ecorating program, and in 2000, BREEAM Green Leaf took another giant step forward in its development by becoming an online assessment and rating tool under the name Green Globes for Existing Buildings. In the same year, BREEAM Green Leaf for the Design of New Buildings was developed for the Canadian Department of National Defense and Public Works and Government Services Canada. The product underwent a further iteration in 2002 by a panel of experts including representatives from Arizona State University, the Athena Institute, BOMA, and several federal departments in Canada.

During 2002, Green Globes for Existing Buildings went online in the United Kingdom as the Global Environmental Method (GEM), and efforts began to incorporate BREEAM Green Leaf for the Design of New Buildings into the online Green Globes for New Buildings. In Canada, Go Green (for Existing Buildings) is owned and operated by the Canada Building Owners and Manufacturers Association (BOMA). All other Green Globes products in Canada are owned and operated by ECD Energy and Environment Canada. BOMA Canada adopted the system in 2004 under the name *Go Green Comprehensive* (now known as *Go Green Plus*). The Canadian federal government also later announced plans to adopt *Go Green Plus* for its entire real estate portfolio.

Additionally, in 2004, the GBI acquired the US rights to the Green Globes building assessment and certification program and adapted it for the US market as an alternative to the LEED building rating system. The GBI committed itself to refining the system to ensure that it reflects changing opinions and ongoing advances in research and technology, and, in so doing, to involve multiple stakeholders in an open and transparent process. To that end, GBI in 2005 became the first green building organization to be accredited as a standards developer by the ANSI and started the process to establish Green Globes as an official ANSI standard.

Defining Green Globes: It is an environmental assessment, education, and rating system that is promoted in the United States under the auspices of the GBI, a Portland, Oregon–based nonprofit organization. Canada's federal government has been using the Green Globes suite of tools for several years under the Green Globes name, and it has been the basis for the Building

Owners and Manufacturer's Association of Canada's Go Green Plus program. Adopted by BOMA Canada in 2004, Go Green Plus was chosen by Canada's Department of Public Works and Government Services, which has an estimated 300 buildings in its existing portfolio. The Green Globes system has also been used by the Continental Association for Building Automation (CABA) to power a building intelligence tool called Building Intelligence Quotient (BiQ).

2.4.2 Green Globes – An Alternative to LEED

Green Globes (Canada) is a product of the BREEAM, which was developed in the UK. One of BREEAM's creators, ECD Consultants, Ltd., used it as the basis for a Canadian assessment method called BREEAM Green Leaf. The original intention of creating BREEAM Green Leaf was to allow building owners and managers to self-assess the performance of their existing buildings. This was followed by ECD creating Green Globes as a Web-based application of Green Leaf.

The Green Globes™ system is a Web-based green building performance interactive software tool that came from Canada, and competes with the better-known, and more expensive, LEED system from the USGBC. Green Globes was introduced to the US market as an alternative to the USGBC's LEED® Rating System. The GBI was established to promote the use of the National Association of Homebuilders' (NAHB's) Model Green Home Building Guidelines and has expanded into the nonresidential building market by licensing Green Globes for use in the United States. GBI is supported by a number of industry groups including the Wood Promotion Network that object to some provisions in LEED and, as trade associations, are not allowed to join the USGBC.

Green Globes was released in Canada in January 2002 and consists of a series of questionnaires, customized by project phase and the role of the user in the design team (e.g., architect, mechanical engineer, or landscape architect). A total of eight design phases are supported. A separate Green Globes model, for assessing the performance of *existing* buildings, has not been licensed into the United States yet. The questionnaires produce design guidance appropriate to each team member and project phase. Green Globes users can order a Green Globes third-party assessment at any time, upon purchasing a subscription, during the completion of the questionnaire, or after completion of the questionnaire. After an online self-assessment is completed and payment is made, a GBI representative contacts the project manager or owner to schedule the assessment and provide the assessor name and contact information. Completion of the preassessment checklist helps prepare for the assessment process; this can be downloaded from the Green Globes customer training area.

Formal rating and certification programs are necessary to provide a mechanism to ensure that new construction project teams or facilities management staff is fully aware of the environmental impact of design and/or operating management decisions. It also offers a visible means to quantify/measure their performance and allows recognition for their achievements and hard work at the end of the process (Figure 2.7). Although investment in environmental/sustainability ratings and certification is currently a

Figure 2.7 New 356,000-ft.² Newell Rubbermaid Corporate Headquarters in Atlanta, GA. This building achieved a two-globe rating using the Green Globes New Construction module. *Courtesy: Green Building Initiative.*

choice for the leading proponents of ecological responsibility, emerging standards such as GBI Proposed ANS Standard 01-200 XP and legislative recognition of "green building and operations" practices may signify and facilitate broader adoption in the future.

Green Globes is designed to be cost effective in that through its value-added online system and a comprehensive yet streamlined in person third-party review process, significant savings on consulting fees are made that were normally associated with green certification. There is an annual per building license fee for use of the online tool as well as a third-party assessment fee. Rates are based on a number of factors including size of project (hectares/acres), number of integrated developments and location (environmentally sensitive areas). Users can register/subscribe for both the annual license for the online Green Globes tools and choose to purchase a third-party assessment (required for certification). Third-party assessor travel expenses are billed separately. The following are some of the Green Globes costs:

Green Globes® existing building rating/certification package: $5270* per building

Green Globes® new construction rating/certification package: $7270* per building

Software subscription costs	Price
Green Globes CIEB existing building 1-year subscription	$1000
Green Globes new construction	
One-project subscription	$500
Three-project subscription	$1500
Ten-project subscription	$2500

Note: A Green Globes subscription is required for third-party assessment/certification.

Third-party assessments/Green Globes certification	Price
Green Globes CIEB assessment/rating	$3500*
Green Globes NC Stage I assessment	$2000
Green Globes NC Stage II assessment/rating	$4000*
Green Globes NC Stage I and II assessment/rating	$6000*

Note: *Pricing for buildings more than 250,000 ft.² in size or departing significantly from standard commercial building complexity will be custom quoted prior to assessment services being performed.

Travel for GBI assessor to/from building location: Invoice actual expenses + 20% after assessment *or* pay a flat fee of $1000 upfront.

As a LEED alternative, the Green Globes building rating system provides an assessment tool for characterizing a building's energy efficiency and environmental performance. The system also provides guidance for green building design, operation, and management; when compared to LEED, some feel that Green Globes' appeal may be enhanced by the flexibility and affordability the system may provide while simultaneously providing market recognition of a building's environmental attributes through recognized third-party verification. And from a practical and marketing perspective, it should not be necessary to pursue LEED certification in order to demonstrate to tenants, their customers, clients, and building visitors that a building's owners and management are taking steps to be more environmentally responsible.

The GBI currently oversees Green Globes in the United States. GBI has also become an accredited standards developer under the ANSI and is in the process of establishing Green Globes as an official ANSI standard.

The ANSI process has always been a consensus-based process, involving a balanced committee of varying interests including users, producers, interested parties and NGOs who basically conduct a thorough technical review through an ANSI-approved, open and transparent process. The standard continues to be monitored by this committee and will continue to follow ANSI approved rules and procedures for updating the standard.

Surprisingly, neither LEED nor Green Globes (or Energy Star for that matter) provides continuous, longitudinal monitoring of energy efficiency or building performance. This indicates that building measurements and ratings are concluded on a one-off basis that must then be reverified later on. This is a significant shortcoming in terms of practicality of greening existing real estate, because buildings are dynamic and rarely perform in an identical manner week after week.

Green Globes generates numerical assessment scores at two of the eight project phases; these are the schematic design phase and construction documents phase. These scores can either be used as self-assessments internally, or they can be verified by third-party certifiers. Projects that have had their scores independently verified can use the Green Globes logo and brand to promote their environmental performance. The Green Globes questionnaire corresponds to a checklist with a total of 1000 points listed in seven categories (for New Construction) as opposed to LEED's 110 points in eight categories (Figure 2.8).

Green Globes however differs from LEED by not holding projects accountable for strategies that are not applicable, which is why the actual number of points available varies by project. For example, points are available

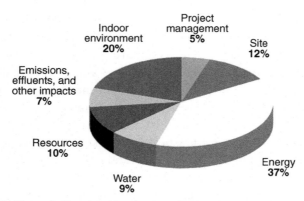

Figure 2.8 *Pie Diagram Showing Distribution of Points in Green Globes Rating System for New Construction.* Source: Building Green LLC.

for designing exterior lighting to avoid glare and skyglow, but for a project with no exterior lighting, a user can select N/A, which removes those points from the total number available so as not to penalize the project. A rating of one or more Green Globes is applied to projects based on the percentage of applicable points they have achieved. In Canada, the ratings range from one to five Green Globes, whereas in the United States, the lowest rating was eliminated and the rest adjusted so that the highest rating is four Globes. Ward Hubbell, executive director of GBI, says that the objective of this was to have something that people are accustomed to, a four-stage system, which is roughly comparable with the four levels of LEED.

Green Globes is reportedly broader than LEED in terms of technical content, including points for topics such as optimized use of space, acoustical comfort, and an integrated design process. It is difficult to compare the levels of achievement required to claim points in the two systems because they are organized differently and also because the precise requirements within Green Globes are not publicly available. One of Green Globes' main attractions for industry groups supporting GBI in the United States is that it recognizes all the mainstream forest certification systems, whereas LEED previously referenced only the Forest Stewardship Council's program. Green Globes also awards points for the use of life-cycle assessment methods in product selection, although it does not specify how those methods should be applied.

Ward Hubbell also claims that Green Globes is on a par with LEED with respect to overall achievement levels, and notes, "We did carry out a harmonization exercise with LEED – not credit-by-credit; we compared objectives." The actual development of the Green Globes system in Canada, as well as its subsequent adaptation for the United States has involved many iterations and participation by a wide range of organizations and individuals. Changes originally made to adapt Green Globes for the US market do not appear to be substantive, for example, converting units of measurement, referencing US rather than Canadian standards and regulations, and incorporation of US programs such as the EPA's Target Finder.

Supporters of Green Globes tried to block the introduction of LEED into Canada but lost a close vote in a committee of the Royal Architectural Institute of Canada that led the creation of the Canada Green Building Council (CaGBC) in 2003. Not surprisingly, Alex Zimmerman, president of CaGBC, has some criticisms of Green Globes and notes that in Canada Jiri Skopek, president of ECD Energy and Environment Canada, has been the primary developer of Green Globes and in the past was its sole certifier. Zimmerman also says, "While there are more certifiers now, it is not clear who they are, how they were chosen, or who they are answerable to." GBI responded to this criticism in the United States by training a network of independent certifiers to verify Green Globes ratings and who have access to the report generated by the Green Globes website, as well as other relevant information such as the project drawings, results of an energy simulation, specifications, and commissioning plan.

Green Globes has been advanced as a green certification system in areas with strong timber-industry lobbying presence because it opposes favoring FSC over SFI forest certification. The consensus is that legislation to encourage green building in states like Virginia and Arkansas, is likely to include Green Globes in addition to LEED. Furthermore, a number of federal agencies such as the Department of the Interior are also reportedly considering an endorsement of Green Globes. It is possible that Green Globes presence on the American scene has had a beneficial impact on LEED, perhaps prompting it to improve its rating system and release LEED 2009 Version 3. It is also important to recognize that Green Globes can attract a significant following that for various reasons are alienated by LEED certifications costs and complexity. This must be good for the green building industry and the environment.

2.4.3 Comprehensive Environmental Assessment and Rating Protocol

The Green Globes assessment protocol for Existing Buildings covers six different areas, with each area having an assigned number of points that are

Table 2.2 Depicting the six different assessment categories of Green Globes for existing buildings

Assessment category	Points	Description
Energy	350	Performance, efficiency, management, CO_2, transportation
Indoor environment	185	Air quality, lighting, noise
Emissions and effluents	175	Boilers, water effluents, hazmat
Resources	110	Waste reduction, recycling
Environmental management	100	EMS documentation, purchasing, environmental awareness
Water	80	Performance, conservation, management
Total	1000	

For new construction, also include site analysis. Table 2.2 shows a clear emphasis on energy, which takes up more than a third of the total points. There are no prerequisites.

utilized to quantify overall building performance. For New Construction, there are seven categories (includes a Site category), and the points will differ slightly. The categories for Existing Buildings are shown in Table 2.2.

The scoring for the six Green Globe categories is based on approximately 150 questions that are completed via the online questionnaire within the Green Globes tool. There are pop-up "tool tips" embedded within the questionnaire to address frequently asked questions and add clarifications regarding the input data requirements that will appear during the survey. The time normally required to input data and complete the survey is roughly 2–3 h per building; this does not include time required to research and gather required information for the survey.

Green Globes Ratings: In order to earn a formal Green Globes rating/certification, a building must be evaluated by an independent third party that is recognized, trained, and affiliated with the GBI. Both new construction and existing buildings can be formally rated or certified within the Green Globes system. Buildings that achieve 35% or more of the 1000 points possible in the rating system are eligible candidates for one or more Green Globes rating. A summary of rating levels and how they relate to environmental achievement can be seen in Figure 2.9.

Steps to Green Globes rating/certification: GBI states that the process for obtaining formal Green Globes rating/certification is fairly straightforward and essentially consists of implementing the following steps:

Step 1: Purchase a subscription to either Green Globes NC or CIEB

Step 2: Login to Green Globes at the GBI website with your username and password

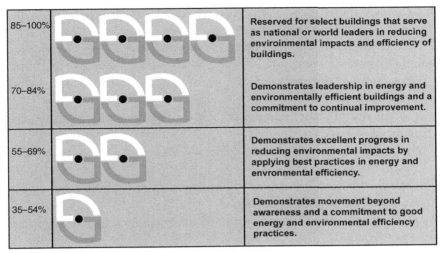

85–100%		Reserved for select buildings that serve as national or world leaders in reducing enviroinmental impacts and efficiency of buildings.
70–84%		Demonstrates leadership in energy and environmentally efficient buildings and a commitment to continual improvement.
55–69%		Demonstrates excellent progress in reducing environmental impacts by applying best practices in energy and envronmental efficiency.
35–54%		Demonstrates movement beyond awareness and a commitment to good energy and environmental efficiency practices.

Figure 2.9 *Green Globes Rating Levels in the United States.* Source: *Green Building Initiative.*

Step 3: Select the tool you have purchased (NC or CIEB) to go to Green Globes

Step 4: Add a building and enter basic building information

Step 5: Use step-through navigation and building dashboard to complete the survey

Step 6: Print your report to see your building projected rating and get feedback using automatic reports

Step 7: Order a third-party assessment and Green Globes rating/certification (if automated report indicates a predicted rating of at least 35% of 1000 points)

Step 8: Schedule and complete a third-party building assessment. Third-party assessment for Green Globes-NC occurs in two comprehensive stages: The first stage includes a review of the Construction Documents developed through the design and delivery process. The second stage includes a walk-through of the building postconstruction.

Step 9: Receive the Green Globes rating/certification

Advantages of the Green Globes rating/certification system: Green Globes is marketed as an economical, practical, and convenient means for obtaining comprehensive environmental and sustainability certification for new or existing commercial buildings. It provides a complete, integrated system that has been developed to enable design teams and property managers to focus their resources on the processes of actual environmental improvement of facilities and operations, rather than on costly, cumbersome, and lengthy certification and rating processes.

Other Advantages of Green Globes relative to other Rating/Certification Systems include the following:

- Projects can be evaluated and informally self-assessed for a low registration fee.
- Consultants are not required for the certification process – thereby reducing costs.
- Submission requirements are generally less complicated than other rating/certification systems.
- Online web tools provide a convenient, proven, and effective way to complete the assessment process.
- Upfront commitment to a lengthy and costly rating/certification process is not required.
- The entire process is fairly rapid, with minimal waiting for final rating/certification.
- The estimated rating number of Green Globes that will be achieved is largely known in advance of the decision to pursue certification because the self-assessed score is always available to users.

Many states have now formally recognized the GBI's Green Globes environmental assessment and rating system in legislation. These include New Jersey, Arkansas, Connecticut, Hawaii, Maryland, Minnesota, North Carolina, Oklahoma, Pennsylvania, South Carolina, Kentucky, Illinois, and Wisconsin.

CHAPTER 3

LEED™ Documentation Process and Technical Requirements

3.1 GENERAL OVERVIEW

The Leadership in Energy and Environmental Design (LEED™) Green Building Rating System reflects the US Green Building Council's (USGBC's) continuing effort to provide a global standard for what constitutes a "green building." It is also intended to be used as a design guideline and third-party certification tool to improve contractor performance and occupant well-being, as well as improved environmental performance and greater economic returns of buildings using established green strategies and innovative practices, standards, and technologies. Each new version of LEED improves on the previous version and embraces new opportunities and trends in the building industry. One of the key areas of focus for LEED v4 is greater information transparency and recording of building product materials and components. The LEED Green Building Rating System is essentially a voluntary, consensus-based national standard that aims at developing high-performance, sustainable buildings. It is important to stay updated on the latest LEED™ programs and requirements. The latest version of LEED is LEED v4. The majority of the LEED programs can now be downloaded at no cost from the USGBC website at www.usgbc.org. It is important to note that the updated LEED v4 version of the rating system is now incorporated into the credential exams, so contractors and architects, engineers, and students interested in taking the new exam will need to get up to speed on LEED v4.

LEED™-Online: For the design professional or building contractor who has never undertaken a LEED project before, the idea of completing all the required documentation can appear daunting. We have come a long way over the last two decades. According to the USGBC, the entire LEED documentation and certification process has been greatly streamlined and has also gone online (i.e., become paperless) in an effort to simplify the process and reduce the time and cost of achieving LEED certification. Project teams are no longer required to submit binders of documentation

LEED v4 Practices, Certification, and Accreditation Handbook
http://dx.doi.org/10.1016/B978-0-12-803830-7.00003-7

Copyright © 2016 Elsevier Inc.
All rights reserved.

71

that often take weeks to prepare, but now have the option of submitting 100% of their documentation online. LEED v4 projects became the first to experience the new LEED-Online platform with streamlined documentation and process. This allows you to spend your time achieving credits, not documenting them. LEED-Online is a user-friendly interface that enables project team members to upload credit templates, track credit interpretation requests (CIRs), manage key project details, contact customer service, and communicate with reviewers throughout the design and construction reviews. With LEED-Online, all LEED™ information, resources, and support are accessible in a centralized location (see http:// www.usgbc.org/). The first requirement, however, to have the project certified by the USGBC under the LEED™ system, is it must be registered with the USGBC.

This latest version of LEED-Online has incorporated significant improvements in browser compatibility. For example, it supports all new versions of Internet Explorer, Safari, and Firefox. Moreover, to get assistance, there is a link on the login page that says "check system requirements." This will run a check to ensure that everything is compatible and lets users know should anything needs adjusting.

Streamlining the certification process is but one of a series of recently implemented LEED™ process modifications that will simplify the documentation and certification process, making it more user-friendly and easier for project teams to manage, without negatively impacting the technical rigor and quality of the LEED process. Recent process refinements are the resultant harvest of continuous market surveys, discussions with various organizations and individuals who apply LEED™ practices in addition to the recently formed technical partnership between USGBC and Adobe. Of note, project teams will be able to choose to use LEED v2009, rather than the new rating system, until October 2016. After that deadline, LEED v4 will become mandatory for new projects. The LEED™ credit requirements will remain essentially unchanged. Project teams are still required to verify their achievements through third party validation and ensure that the building is built as designed.

The changes to the accreditation process announced by USGBC toward the end of 2008, going forward includes the replacement of the LEED NC designation with the LEED Building Design and Construction (BD+C) designation and the LEED CI designation with LEED for Interior Design and Construction (ID+C). This is in order to maintain alignment with the latest LEED 2009 Rating System. Details of the changes to the LEED AP program can be found at the Green Building Certification Institute website: www.gbci.org. As of 2009, LEED™ eligibility for certification is available for New Construction, Existing Buildings: Operations and Maintenance, Commercial Interiors, Core and Shell Development, Homes, Schools, Healthcare (pilot stage), Retail (pilot stage), and Neighborhood Development (pilot stage). These categories include subcategories that cover specific building types. LEED v4 will address new market sectors by introducing rating systems for a broader selection of building types that now include warehouses, data centers, and hospitality, distribution centers, existing schools and retail, and midrise residential projects.

3.2 CREDIT CATEGORIES

There are several ways to denote a building's "greenness." In the United States, and now in many countries around the world, the most widely recognized method for certifying green buildings is through an independent third party, the USGBC's program LEED, the latest version being LEED v4.

As stated earlier, 2009 heralded the implementation of a much-anticipated and urgently needed modification of the USGBC LEED green building rating system. The USGBC's latest revamped rating system includes a series of major technical advancements focused on improving energy efficiency, reducing carbon emissions, and addressing additional environmental and human health concerns. This overhaul is the outcome of 8 years of user feedback including 7000 comments and credit interpretation rulings from USGBC members and stakeholders as well as years of committee meetings. Brendan Owens, vice president for LEED™ technical development, sums it up nicely: "The conclusion of the balloting process marks the culmination of tireless work done by representatives from all corners of the building industry."

The LEED™ version 2009 modifications came at a very appropriate time as numerous criticisms were being raised pointing to the inadequacy of earlier versions of the rating system. For example, Charles Crawford, a San Diego architect and professor at NewSchool of Architecture and Design, while agreeing to the LEED program's importance, believes that the program has significant shortcomings. He says, "The points program they administer is heavily tilted toward rewarding greener technology over basic 'passive' principles. So, for example, you get more points for an energy-efficient air conditioner than you do for a naturally ventilated AC-free building." Another example cited by Crawford is LEED's apparent lack of consideration for the vast amounts of fuel expended in overseas transportation. He goes on to say that "A green product manufactured by a multinational overseas and shipped to the U.S. is often worth more points than a product manufactured only a few miles away."

The revised rating systems address many of these criticisms, which are now more appropriately formulated. The new system aligns the prerequisite/credit structure of each category (Figure 3.1) into a common denominator so that the same set of credits is typically offered under each rating system (Figure 3.2a,b). Other significant changes to LEED 2009 are the incorporation of Regional Priority credits and revision of its credit weightings. These are discussed further and in Chapter 4. The intent of the USGBC is for LEED to transition into what the Council terms a "predictable development cycle" to help drive continuous improvements and allow the market to effectively participate in LEED's growth and development.

Under the LEED® 2009 certification program, green building design focuses on seven basic categories as shown in Figure 3.3a. The total point

Credit Category	New Prerequisite	Revised Prerequisite
Sustainable Sites	Environmental site assessment now applicable to schools and healthcare; site management policy	Updated reference standard for construction activity
Water Efficiency	Building level water metering	Water use reduction now split into outdoor and indoor water use reduction
Energy and Atmosphere	Building level energy metering	Fundamental commissioning of building energy systems now called fundamental commissioning and verification; updated reference standard for minimum energy performance
Materials and Resources	Construction and demolition waste management; PBT source reduction – mercury; ongoing purchasing and waste policy; facility alterations and additions policy	Storage and collection of recyclables
Indoor Environmental Quality	Green cleaning policy	Minimum indoor air quality performance; environmental tobacco smoke control; minimum acoustical performance

Figure 3.1 Some updates to the prerequisites for LEED v4 credit categories.

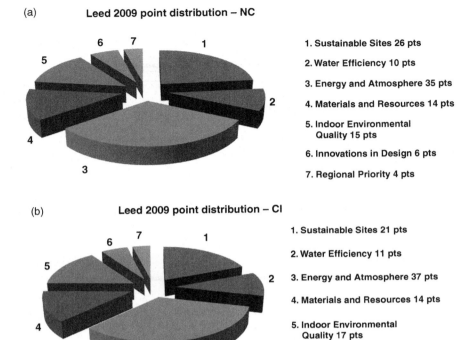

(a) **Leed 2009 point distribution – NC**

1. Sustainable Sites 26 pts
2. Water Efficiency 10 pts
3. Energy and Atmosphere 35 pts
4. Materials and Resources 14 pts
5. Indoor Environmental Quality 15 pts
6. Innovations in Design 6 pts
7. Regional Priority 4 pts

(b) **Leed 2009 point distribution – CI**

1. Sustainable Sites 21 pts
2. Water Efficiency 11 pts
3. Energy and Atmosphere 37 pts
4. Materials and Resources 14 pts
5. Indoor Environmental Quality 17 pts
6. Innovations in Design 6 pts
7. Regional Priority 4 pts

Figure 3.2 (a,b) Pie diagrams showing the LEED™ NC v2009 and LEED™ CI v2009 point distribution system, which incorporates a number of major technical advancements focused on improving energy efficiency, reducing carbon emissions, and addressing additional environmental and human health concerns.

distribution for the seven categories is 100 points plus 6 possible points for Innovations in Design and 4 additional possible points for Regional Priority. In Figure 3.3b, we see the latest distribution system as portrayed in LEED v4. Notice that there are two new categories: Location and Transportation and Integrative Process.

3.2.1 Location and Transportation

The LEED Location and Transportation credits (LT) encourage project teams to take advantage of the infrastructure elements in existing communities that provide environmental and human health benefits. LEED says that the Location and Transportation (LT) category "rewards

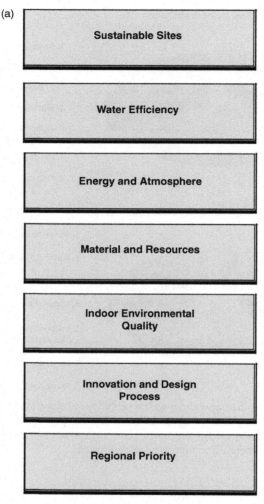

Leed 2009 credit categories

Figure 3.3 (a) The seven LEED 2009 credit categories: (1) Sustainable Sites, (2) Water Efficiency, (3) Energy and Atmosphere, (4) Materials and Resources, (5) Indoor Environmental Quality, (6) Innovations in Design, (7) Regional Priority.

thoughtful decisions about building location, with credits that encourage compact development, alternative transportation, and connection with amenities, such as restaurants and parks. The LT category is an outgrowth of the Sustainable Sites category, which formerly covered location-related topics. Whereas the SS category now specifically addresses on-site ecosystem

(b)   **Leed v4 points by category**

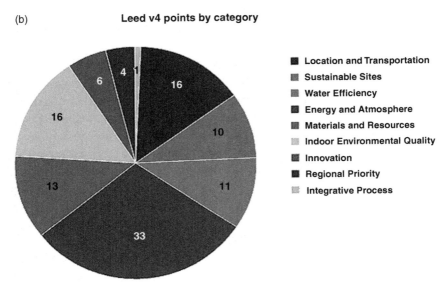

Figure 3.3 *(cont.)* (b) The nine credit categories in LEED v4 now include Location and Transportation and Integrative Process.

services, the LT category considers the existing features of the surrounding community and how this infrastructure affects occupants' behavior and environmental performance."

Location and Transportation credits are designed to focus on the integration of smart growth principles such as revitalization, reuse of existing infrastructure, mixed–use development, and pedestrian-friendly design to further diminish carbon emissions thereby improving energy efficiency and advancing human health. These are not entirely new concepts, as many of the credits are currently listed under the Sustainable Sites (SS) category, but it will be important for the reader to note which credits will be making the transfer to LT.

Best Practices for Transportation:

- Encourage transportation alternatives to driving to reduce the adverse environmental impact of automobile use and look for a site that is well suited to take advantage of mass transit, that is, near public transportation.
- Provide preferred parking for alternative-fuel vehicles.
- Reduce parking capacity and encourage carpooling.
- Include adequate secure bicycle storage, showers, and changing rooms.

3.2.2 Sustainable Sites Credit Category

The planning and design of green buildings requires a conscious decision from the start to build in an environmentally friendly manner. The intent of the *Sustainable Sites* category is to encourage good stewardship of the land, in addition to ensuring that any adverse project impacts on surrounding areas during and after construction are minimized. An excellent example of a contemporary building designed to achieve a gold rating in LEED's certification program is the San Elijo Lagoon Nature Center in San Diego County, California (Figure 3.4). This nature center reportedly features state-of-the-art, museum-quality exhibits interpreting the natural and cultural resources of the San Elijo Lagoon as well as the sustainable design features of the building. The two-story 5525-ft.^{2} building is constructed from recycled and environmentally friendly building materials including block, steel, rebar, and decorative glass. Additionally, in excess of 90% of the debris from demolition and construction was recycled, and the building's insulation was made from recycled denim. To ensure good indoor air quality, the structure was built using low or no volatile organic compounds (VOCs), materials, paints, and sealers. The center also boasts a green roof that filters pollutants

Figure 3.4 *The San Elijo Lagoon Nature Center Building in San Diego County, California.* It incorporates a number of green building elements including radiant floor heating, green planted roof, recycled cotton insulation, certified renewable lumber, onsite renewable energy in the form of photovoltaics that is designed to provide 52% of energy requirements, natural daylighting and ventilation, stormwater filtering, native vegetation, and recycled water used for both irrigation and to supply the toilets.

from storm water. High-efficiency lighting is incorporated into the design, and more than half of the building's energy is provided by solar panels. Native, drought-tolerant plants are used for landscaping around the center to reduce water usage, and the building makes use of recycled water for irrigation and restrooms. The nature center also provides bike racks in its parking lot.

To achieve maximum credits under the LEED rating system, a building owner should, whenever possible, consider choosing an appropriate site in terms of urban redevelopment, and Brownfield development. It is important that on identifying a site, the building design proceeds to utilize the site conditions to maximum advantage and to minimize site disturbances to the surrounding ecosystem. Careful consideration should be given to the positioning of buildings on the site and their relationship to the site's topography, landscape, and proximity to amenities as these are key factors that influence a building's impact on the environment and the community. The Sustainable Sites category also encourages the use of alternative transportation options, reducing site disturbance, and storm water management. Credit is also given to projects that achieve a reduced heat island effect, reduce light pollution, and incorporate onsite renewable energy.

3.2.2.1 Summary of Best Practices

Project selection/siting:

- Identify and develop on appropriate sites to minimize the environmental impact of construction.
- Select an urban infill site or a brownfield site to build on as opposed to an undisturbed greenfield site, farmland, or wetland so that building footprints can be minimized and land used efficiently.
- Buildings should preferably be located in dense urban areas to take advantage of existing infrastructure.
- Disruption of existing habitats should be avoided; ample open space is to be provided.
- Protect and retain existing landscaping and natural features. Plants should be selected that have low water and pesticide needs and which generate minimum plant trimmings. The use of compost and mulches should be encouraged to save water and time.
- Recycled content paving materials and furnishings to be used whenever possible.

Reduce heat island effect:

- Reduce heat islands (the heating up of a site based on the heat captured in dark-colored surfaces) as they can disturb local microclimates and increase overall summer cooling loads, leading to increased levels of greenhouse gas and air pollution.
- Use cool roofing strategies like the use of light colored or living "green" roofs.
- Parking should be located in garages, preferably underground.
- Utilize light-colored pavement and shade trees whenever possible.

Reduce light pollution:

- Design lighting to prevent excessive emissions to the night sky that would negatively affect the comfort of neighbors and the habits of migratory birds. This can be achieved by using low-intensity, shielded fixtures with proper cutoffs in addition to ensuring lights are turned off or dimmed during nonbusiness hours.

3.2.3 Water Efficiency Credit Category

The Water Efficiency (WE) credits focus on reducing water use inside and outside the building as well as the importance of metering for better water management. Water efficiency can be defined as achieving a desired result or level of service or the accomplishment of a function, task, or process, with the least amount of necessary water. Water efficiency reflects the relationship between the amount of water needed for a particular purpose or function and the amount of water actually used or delivered.

There is a distinct difference between water conservation and water efficiency although the two terms are frequently used interchangeably. Water conservation can be considered to be water management practices that improve the use of water resources to benefit people or the environment. It basically constitutes a beneficial reduction in water use, loss, or waste.

Water efficiency differs from water conservation in that it focuses on reducing waste. It also means that the key for water efficiency is reducing waste, not restricting use. Consumers have a major role to play in water efficiency by making small behavioral changes to reduce water wastage and through the choice of more water-efficient products. Water efficiency can be enhanced by various means such as fixing leaking taps, taking showers rather than baths, installing displacements devices inside toilet cisterns, and using dishwashers and washing machines with full loads.

The intent of the Water Efficiency category is to encourage the thoughtful use of water. LEED credits are given to building and landscape designs that reduce the use of potable water for irrigation and wastewater such as the use of a gray water system that recovers rainwater or other nonpotable water for site irrigation. Credits are also given when there is a total reduction in potable water use in the building through various water conservation strategies.

Because potable water is becoming a limited resource and is pivotal to sustaining life, economic development, and the environment, water efficiency is an essential element of green building practices. Water efficiency and stormwater management practices can be developed and implemented through a collaborative process in many areas and often complement site-related strategies to improve multiple building systems.

Other methods for reducing wastewater by increasing water efficiency (and save on utility bills) is through the implementation of water-efficient plumbing strategies such as ultra-low-flush or dual-flush toilets, low-flow automatic faucets, low-flow shower heads, and other water-saving plumbing fixtures such as low-flow urinals or water-free urinals (Figure 3.5).

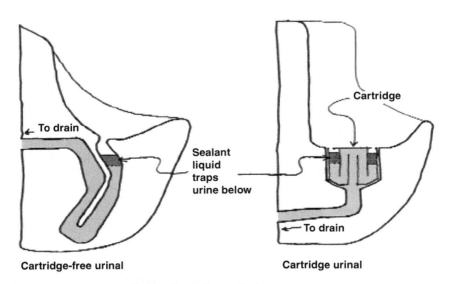

Figure 3.5 *Two Types of Waterless Urinals Currently on the Market.* Waterless urinals do not use water and have no flush valves, which can save considerably on water and maintenance. However, local code requirements should be checked prior to installation, as some jurisdictions do not allow waterless urinals.

For water conservation, "Practice Greenhealth" lists the following steps to establish a water conservation program:

1. Audit current water use.
   a. Install water meters at strategic locations in the facility.
   b. Read/record water readings weekly and analyze the data. Look for high water use areas, trends, and unusual occurrences. Identify water conservation opportunities.
2. Identify water conservation opportunities.
   a. Fix drips, leaks, and unnecessary flows.
   b. Implement changes to improve practices in cleaning, laundry, and kitchens.
   c. List opportunities requiring engineering/equipment solutions (toilets, sterilizers, boiler, chillers, etc.). Determine cost of opportunities and potential return on investment.
3. Determine cost of opportunities and potential return on investment.
4. Prioritize water conservation opportunities.
5. Develop a phased plan that fits your budget.
6. Obtain funding (revise plan, if necessary).
7. Implement plan.
8. Measure and document success.

3.2.3.2 Summary of Best Practices

- Employ stormwater management techniques to prevent/minimize pollution, sedimentation, and flooding of receiving waters.
- Employ recirculating systems for centralized hot water distribution and install point-of-use hot water heating systems for more distant locations.
- Capture and reuse stormwater for nonpotable applications such as landscape irrigation and toilet flushing.
- Employ a water budget approach when possible that schedules irrigation for landscaping.
- Encourage use of bioretention systems such as rain gardens and bioswales into landscaping strategies to store and treat stormwater.
- Landscaped areas and buildings should be metered separately. Microirrigation (which excludes sprinklers and high-pressure sprayers) should be used to supply water in nonturf areas.
- Use irrigation controllers and self-closing nozzles on hoses.
- Eliminate or minimize the need for irrigation through use of native and adaptive species. Where irrigation is unavoidable, techniques should be used that have proven to be more efficient and use less water, such as

drip irrigation systems, low-volume/low-angle sprinklers, and nighttime watering.

- Install a vegetated green roof to filter stormwater and reduce runoff.
- Incorporate water-efficient plumbing technologies and strategies into the design; for instance, composting toilets and waterless urinals use no water and ultrahigh-efficiency plumbing fixtures such as low-flow lavatory faucets with automatic controls are very useful.

3.2.4 Energy and Atmosphere Credit Category

The first prerequisite in the *Energy and Atmosphere* (EA) category is that the building undergoes fundamental commissioning to ensure that its building systems are operating in the manner they were designed to operate. Other prerequisites for this category include attaining a minimum energy performance (in compliance with ASHRAE 90.1-2004) and eliminating the use of chlorinated fluorocarbon (CFC)–based refrigerants in new building HVAC&R equipment. The EA category in LEED v4 now requires an additional new prerequisite – building-level energy metering. To meet the requirements of this prerequisite, the building must install a meter (or submeters) that track the total building energy consumption at least monthly. The project is also obligated to providing that data to USGBC for at least 5 years. A project can also earn an additional point for more rigorous metering and tracking of its energy usage. This is in line with USGBC's increased emphasis on building performance, and not just design. The most significant change, however, may be that the minimum energy standard has become the 2010 edition of ASHRAE 90.1. The change from the 2007 to the 2010 version represents a nearly 20% savings in overall building energy use according to the US Department of Energy. EA credit 1: Optimizing Energy Performance has been adjusted to reflect the more stringent base standard. It also has two new credits, one for advanced metering and one for demand response programs. It is important for architects, engineers, and contractors involved with wall systems to understand these changes in order to benefit from them. LEED credits are given for projects that further optimize energy performance, employ renewable energy generated onsite, and purchase green power from a Green-e utility program, or other acceptable supply source such as through Green-e Tradable Renewable Certificates. Additional credits are given for more enhanced commissioning and long-term continuous measurement and verification of building performance.

Green buildings employ high performance systems and strategies to achieve increased energy efficiency. There are numerous aspects of a building design that can influence its energy performance, and many of the most significant utility cost savings can be realized through implementing energy efficiency techniques.

ASHRAE 90.1-2004 is the baseline criteria that registered projects are generally required to meet to satisfy the prerequisite requirements. Where local codes are more stringent, these must be followed (which should include an explanation with the submittal for LEED certification). Of particular relevance is the California Title 24 standard, which is more stringent than the new Standard 90.1-2007 and which is accepted without further evaluation. Attaining and surpassing the energy efficiency levels of the California Title 24 standard is not difficult, yet most projects strive only to meet the standard than surpass it. With regard to this standard, the Energy Commission agreed to a California Building Standards Commission request that all parts of the California Building Standards Code (Title 24), including the Building Energy Efficiency Standards, be effective on August 1, 2009, so that the entire code update will have the same effective date. The 2008 Energy Efficiency Standards have incorporated new measures to reduce energy use and greenhouse gas emissions. California's Building Energy Efficiency Standards were updated in 2013 on an approximate 3-year cycle. The 2013 Standards, which went into effect on July 1, 2014, are a significant improvement upon the 2008 Standards for new construction of, and additions and alterations to, residential and nonresidential buildings.

Implementation of passive design strategies can have a dramatic effect on building energy performance. Such measures would typically include building shape and orientation, passive solar design, and the appropriate use of natural lighting. Studies have consistently shown that providing natural lighting has a positive impact on productivity and well-being.

ENERGY STAR Portfolio Manager. This is an online tool for assessing a building's energy and water consumption. It can be used for a single building or for an entire portfolio. It is a useful benchmarking tool that can help building owners identify areas for efficiency improvements, track the performance of these improvements over time, and compare their building's performance with national averages. It is discussed further in Chapter 4.

3.2.4.3 Summary of Best Practices
- Correct sizing of HVAC equipment is imperative for the building demand. Effective strategies such as demand-control ventilation, variable speed pumping, heat recovery, and economizer cycles should be incorporated whenever possible.

- Eliminate all use of CFC and HCFC refrigerants, in order to reduce the project's potential adverse impact on ozone depletion and global warming.
- Employ high-efficiency ENERGY STAR rated appliances, computers, and equipment whenever possible.
- Employ an appropriately sized and energy-efficient heat/cooling system in conjunction with a thermally efficient building shell.
- Minimize the electric loads from lighting, equipment, and appliances. Thus, for example, great savings can be achieved by switching to energy-efficient light bulbs, which use up to 75% less energy and which reportedly have a life cycle 10 times longer than incandescent bulbs.
- Coordinate energy efficient lighting with daylight strategies, coupled with installation of high-efficiency lighting systems with advanced lighting controls. Motion sensors and occupancy sensors tied to dimmable lighting controls should be included. Task lighting also helps reduce general overhead light levels.
- Maximize the use of light colors for roofing and wall finish materials. Wall and ceiling insulation should have high *R*-values. Minimal glass to be used on east and west exposures (because of low angle of sun) to reduce heat gain.
- Ensure proper building systems commissioning is conducted to verify that the project's energy-related systems are designed, installed, calibrated, and operating properly and as intended.
- Check that the building envelope and systems are designed to maximize energy performance by addressing insulation, glazing ratios, and glass efficiency.
- Consider the use of alternative energy sources such as on-site power generation technologies like solar thermal, photovoltaic panels, and wind turbines. Renewable energy sources are symbolic of emerging technologies for the future. The use of green power should be encouraged by purchasing electricity generated from renewable sources, such as wind and solar.
- Employ measurement and verification systems to monitor energy used by the various energy-consuming systems within the building and that provides useful data about the building systems performance while identifying potential maintenance requirements for performance optimization.
- Computer modeling has great potential and is an invaluable tool in optimizing the design of electrical and mechanical systems and the building shell.

3.2.5 Materials and Resources Credit Category

The Materials and Resources (MR) category has arguably seen the most significant changes in the transition to LEED v4 (from LEED 2009) since the USGBC first released it in 1999. The prerequisites and credits that have not witnessed significant change from the LEED 2009 version are Storage and Collection of Recyclables and Construction and Demolition Waste Management Planning. One of the significant changes in LEED v4 that reaches across several of the credits relates to regional materials. LEED 2009 (LEED v3) included a credit that awarded points for materials and products that were extracted, harvested, and manufactured within a 500-mile radius of the project site. This credit has been eliminated as a standalone credit in LEED v4. In LEED v4, regional materials must be extracted, manufactured, and installed all within 100 miles for the project team to gain any benefit. Also, LEED v3 was focused on single product attributes and materials costs, whereas LEED v4 pursues life-cycle thinking, ingredient transparency, number of products weighted by criteria, and material cost.

The new MR credits include the following:

1. Building Life-Cycle Impact Reduction: This would impact the project team on major renovations to existing buildings. The Building Life-Cycle Impact Reduction contains four options: historic building reuse, renovation of abandoned or blighted buildings, building and material reuse, or a whole-building life-cycle assessment. It is intended that this credit encourages reuse and reduces the building's environmental impact. Building Product Disclosure and Optimization for Environmental Product Declarations (EPD), is more product-specific and addresses Environmental Product Declarations; generally, the credit provides direction on the requirements and standards a manufacturer needs to follow to prepare an EPD. The aim for the EPD credit is to encourage the use of products with restricted impacts throughout their lifetimes, and from manufacturers that provide transparency about the product's ingredients and manufacturing procedures.

   In LEED 2009, these credits focused mainly on green attributes of products (e.g., recycled content, bio-based content, certified wood and regional sourcing), On the other hand, LEED v4 focuses on product disclosure. Individual credits for recycled content, regional materials, and certified wood have been eliminated in LEED v4. Six of the 13 available points in the Materials and Resources (MR) category now relate to building product disclosure in the form of environmental product declarations (EPD), supply chain reporting, material ingredient reporting, etc.

These new credits attempt to capture more of a complete view of the material's sustainability throughout its life cycle. They not only motivate project teams to employ more sustainable materials but also encourage product manufacturers to provide better, detailed information about where their products originated from, how they were produced, and what they comprise. Of note, the Building Product Disclosure and Optimization credits also have options to achieve points based on the cost or number of compliant products.

The Materials and Resources category generally requires that a designated area be provided for the collection and storage of recyclables on the site, preferably during design and operations. The category also gives credit and encourages the use of recycled materials, rapidly renewable materials (that are harvested within a 10-year or shorter cycle), locally manufactured environmentally responsible materials, and certified wood. Credit is also given to projects that reuse parts of the existing building on the site and to projects that implement a construction waste management plan, thus reducing waste going to landfills.

The CI Reference Guide quite rightly states that "building materials choices are important in sustainable design because of the extensive network of extraction, processing and transportation steps required to process them." Indeed, the energy and resources required extend a project's impact far beyond the building itself. This is why careful selection and disposal of materials can immensely benefit a building's environmental impact and promote wellness (Figure 3.6). Toward this end, Penny Bonda and Katie Sosnowchik state in their book *Sustainable Commercial Interiors* that "first and foremost, designers must remember that sustainable materials must have certain attributes: they must be healthy, durable, appropriate, and easily maintained with minimal environmental impacts throughout their life cycle."

3.2.5.4 Summary of Best Practices

- Locate in an existing building and reuse as much of the building materials as possible.
- Provide easily accessible collection and storage points for recyclable materials.
- Applying construction waste management operations in building out a tenant space to reduce debris by recycling these materials. Property management firms like Cassidy & Pinkard Colliers, for example, require contractors to recycle a minimum of 50% of the waste generated on-site.

Figure 3.6 *An Interior Space in Jupiter, Florida.* It illustrates the impact that a careful selection of furnishings and finishes can have on a project. Diligent choices were made, such as cork for a sustainable material for flooring and walls, low-volatile organic compound paints, natural lighting, reuse of furnishings, green cleaning, and maintenance of the interior. *Source: Denise Robinette – www.healthylivinginteriors.com*

- In building out new office spaces, preference should be given to materials that are harvested and manufactured locally or regionally.
- Preference should be given to materials that contain a high percentage of recycled content.
- Priority should be given to rapidly renewable materials and materials that are reclaimed, salvaged, or refurbished.
- Whenever possible, use wood that has been certified by the Forest Stewardship Council (FSC) or other acceptable organization.

Materials Efficiency: When selecting sustainable building materials and products for a project, start by evaluating some of the important inherent characteristics that could adversely impact the environment. These include zero or low off-gassing of harmful air emissions and zero or low toxicity in addition to reused and recycled content, high recyclability, durability, longevity, and local production. Such products also promote resource conservation and efficiency. The use of recycled-content products also helps develop markets for recycled materials that are being diverted from landfills. It is particularly important to achieve LEED credits. Thus, the ability to maintain 40% or 60% of a building's Interior Non-Structural Components, for example, can earn LEED™ MR Credits.

The use of dimensional planning and other material efficiency strategies helps reduce the amount of building materials needed and cut construction costs. For example, designing rooms on 4-ft. modules minimizes waste by conforming to standard-sized wallboard and plywood sheets. A further example is the reuse and recycling of construction and demolition materials (say as a base course for building foundation) reduces the materials designated for landfills and cuts costs. Plans should be prepared for managing materials through deconstruction, demolition, and construction.

3.2.6 Indoor Environmental Quality Credit Category

The *Indoor Environmental Quality* (IEQ) category seeks to ensure that green buildings have among other things, optimal lighting, thermal comfort and healthy indoor air quality for their occupants. Indoor Environmental Quality in LEED v4 has essentially remained very similar to the original credit category from 2009. It continues to address minimum indoor air quality performance and environmental tobacco smoke control in its prerequisites (thus contributing to the comfort and well-being of the occupants). Additionally, it addresses issues like daylight, views, thermal comfort, and low-emitting materials in its credits. Credits are also given for implementing outdoor air delivery monitoring, installation of CO_2 sensors, increased ventilation effectiveness, indoor air quality management during construction and prior to occupancy, the use of low-emitting materials, adhesives, paints, and finishes and allowing occupants to control the systems in their personal workspace.

With LEED v4, the USGBC attempts to reward design teams that provide a superior indoor environment across the spectrum of human health and comfort. Minimum IEQ performance necessitates outdoor air intake observation in addition to minimum outdoor air compliance, based on the measured-performance theme noted throughout the rating system. Low-emitting materials credits earned via product selections are now pooled into a single LEED v4 credit, and projects must show compliance for at least two product categories to earn points. Designers and contractors should update their LEED 2009 template specifications to reflect the new LEED v4 VOC content and product emissions standards, particularly since the credit in LEED v4 for low-emitting materials has been adjusted to focus on VOC emissions rather than content.

New in LEED v4 for New Construction is Indoor Environmental Quality (EQ) Credit 9: Acoustic Performance, which receives one point. This credit was previously only found in the LEED for Schools rating system. It should be noted that this credit contains sound transmission class (STC) requirements for interior walls ranging from 45 to 60 contingents on the occupancy of adjacent rooms.

Another new credit is Interior Lighting Quality, which combines criteria for fixtures, light sources, and interior surfaces with the previous Lighting Controls credit. Indoor environmental quality is a pivotal component of green buildings. Numerous studies have confirmed the effect of the indoor environment on the health and productivity of building occupants as outlined in Chapter 7 of this book. Ventilation, thermal comfort, air quality, and access to daylight and views are all cardinal factors that play a critical role in determining indoor environmental quality.

Occupant Health and Safety: Recent research leaves no doubt that buildings with good overall environmental quality can reduce the rate of respiratory disease, allergy, asthma, sick building symptoms, in addition to enhancing worker performance. The potential financial benefits of improving indoor environments exceed costs by a factor of 8 and 14 (Fisk and Rosenfeld, 1998).

Many building materials and cleaning/maintenance products emit toxic gases, including VOCs and formaldehyde. These gases can have an adverse impact on occupants' health and productivity.

Minimize the potential of indoor microbial contamination through selection of materials resistant to microbial growth and by providing effective drainage from the roof and surrounding landscape. In addition, adequate ventilation is required for all bathrooms, proper drainage of air-conditioning coils, and installation of other building systems to control humidity. "Enhanced Indoor Air Quality Strategies" is a new credit in the IEQ category that counts on increased ventilation from the LEED v3 rating system. In addition to including requirements for increased ventilation, it also embraces requirements for carbon dioxide monitoring, cross-contamination prevention, and air contamination prevention and monitoring. Determining which elements of the credit need to be met depends on whether the building uses mechanical, natural, or mixed-mode ventilation.

3.2.6.5 Summary of Best Practices
- Efficient ventilation systems and a high-efficiency, in-duct filtration system helps to prevent the development of indoor air quality problems and contribute to the comfort and well being of building occupants. Heating and cooling systems that maintain adequate ventilation and proper filtration can have a significant and positive impact on indoor air quality.
- Facilitate environmental tobacco smoke (ETS) control by prohibiting smoking within buildings or near building entrances. Outdoor smoking areas should be designated at least 25 ft. from openings serving occupied spaces and air intakes.
- Install carbon dioxide and airflow sensors in order to provide occupants with adequate fresh air when required.
- Use of zero emissions or low-emitting construction materials and interior finish products (that contain minimal or no VOCs) will improve indoor air quality. Such materials include adhesives, sealants, paints, carpet and flooring, furniture, and composite wood products and insulation.
- On-site stored or installed absorptive materials should be protected during construction from moisture damage and control particulates through the use of air filters.
- If air handlers are to be used during construction, filtration media with a minimum efficiency reporting value (MERV) of 8 are to be installed at each return grill as determined by ASHRAE 52.2-1999.
- All filtration media are to be replaced immediately before occupancy. Conduct, when possible, a minimum 2-week flush-out with new filtration media with 100% outside air after construction ends and before occupancy of the affected space. Where building occupants may be exposed to potentially hazardous particulates, biological contaminants, and chemical pollutants that adversely affect air and water quality, new air filtration media is to be provided with an MERV of 13 or better.
- Install permanent entryway systems such as grilles or grates to prevent occupant-borne contaminants from entering the building.
- Incorporate design strategies that maximize daylight and views for building occupants.
- Occupants' thermal comfort can be maintained by incorporating adjustable features such as thermostats or operable windows.

- HVAC systems and building envelope should be designed to meet requirements of ASHRAE 55-2004, Thermal Comfort Conditions for Human Occupancy.
- During construction, meet or exceed the recommended design approaches of the Sheet Metal and Air Conditioning Contractors' National Association (SMACNA) IAQ Guidelines for Occupied Buildings Under Construction, 1995, Chapter 3.
- All adhesives used are to meet or exceed the requirements of the South Coast Air Quality Management District (SCAQMD) Rule #1168. Aerosol adhesives should meet requirements of Green Seal Standard GC-36.
- Paints and coatings used on interior to comply with VOC limits set forth by the following types: architectural paints and primers for walls and ceilings to comply with Green Seal Standard GS-11; anticorrosive and antirust paints for interior ferrous surfaces to comply with Green Seal Standard GC-03; clear wood finishes, coatings, stains, sealers, and shellacs to comply with SCAWMD Rule 1113.
- All carpet systems must meet or exceed requirements of Carpet & Rug Institute's (CRI's) Green Label Indoor Air Quality Test Program. All carpet cushion installed to meet requirements of CRI Green Label Plus Program.
- Composite wood and agrifiber products must be void of added urea-formaldehyde resins (Figure 3.7).

3.2.7 Innovation and Design Process Credit Category

LEED v4 states that the intent of the Innovation and Design Process category is "to provide design teams and projects the opportunity to be awarded points for exceptional performance above requirements set by the LEED Green Building Rating System™ and/or innovative performance in Green Building categories not specifically addressed by the LEED Green Building Rating System™." This category therefore provides an opportunity to receive additional points for performance that exceeds LEED requirements. A credit may also be achieved by having a LEED™ AP as a principal participant of the project team.

LEED v4 Requirements:

Credit can be achieved through any combination of the Innovation in Operations and Exemplary Performance paths as described as follows:

   Path 1. Innovation in design (1–5 points for NC and CS, 1–4 points for Schools)

Figure 3.7 *Kitchen by EcoFriendly™ Cabinetry Using Composite Woods.* The manufacturer guarantees cabinetry to emit 0% volatile organic compounds. Avoid use of solvent-based stains or coatings, which are known for emitting formaldehyde. *Source: Executive Kitchens, Inc.*

Achieve significant, measurable environmental performance using a strategy not addressed in the LEED 2009 for New Construction and Major Renovations, LEED 2009 for Core and Shell Development, or LEED 2009 for Schools Rating Systems.

One point is awarded for each innovation achieved. No more than 5 points (for NC and CS) and 4 points (for Schools) under IDc1 may be earned through Path 1 – Innovation in design.

Identify the following in writing:
- *The intent of the proposed innovation credit.*
- *The proposed requirement for compliance.*
- *The proposed submittals to demonstrate compliance.*
- *The design approach (strategies) used to meet the requirements.*

Path 2. Exemplary performance (1–3 points)

Achieve exemplary performance in an existing LEED 2009 for New Construction, Schools and Core and Shell prerequisite or credit that allows exemplary performance as specified in the LEED Reference Guide for Green Building Design & Construction, *2009 Edition. An exemplary performance point may be earned for achieving double the credit requirements and/or achieving the next incremental percentage threshold of an existing credit in LEED.*

One point is awarded for each exemplary performance achieved. No more than 3 points under IDc1 may be earned through Path 2 – Exemplary performance.

Path 3. Pilot credit (1–4 points)

Attempt a pilot credit available in the Pilot Credit Library at www.usgbc.org/ pilotcreditlibrary. Register as a pilot credit participant and complete the required documentation. Projects may pursue up to 4 Pilot Credits total.

3.2.8 Regional Priorities Credit Category

Regional Priority is essentially identical to the credit in LEED 2009. Projects can earn up to four out of six points available for using strategies identified by that region's USGBC council or chapter. One point is awarded for each Regional Priority credit achieved, up to a maximum of four. Although the USGBC has always strongly stressed integrated project design, or "IPD," it now awards 1 point for using a collaborative design process from the predesign phase through the design phases. The project team must identify potential synergies across credit categories and document how their early analyses informed their project requirements and basis of design.

The intent is "to provide an incentive for the achievement of credits that address geographically specific environmental, social equity, and public health priorities."

Requirements: You can "Earn up to four of the six Regional Priority credits. These credits have been identified by the USGBC regional councils and chapters as having additional regional importance for the project's region. A database of Regional Priority credits and their geographic applicability is available on the USGBC website, www.usgbc.org/rpc."

3.2.9 Integrative Process Credit Category

The new LEED v4 Integrative Process (IP) category includes a prerequisite for healthcare projects called Integrative Project Planning and Design. This prerequisite requires projects to utilize an ongoing analysis approach and forces the project team to understand the sustainable goals of the owner from the beginning of the planning and design process. When using this category, the focus should be on achieving "synergies across disciplines and building systems" through collaborative efforts across disciplines. This credit requires preliminary analysis in the areas of energy and water use, including site conditions, massing and orientation of the building, and the building envelope – areas where the architect and structural engineer can provide valuable input. This requires the team to evaluate synergies and opportunities that sometimes fail to be analyzed in the traditional design-bid-build delivery approach. Likewise, this credit category enables projects to earn a point for analyzing synergies between various basic "passive" approaches (e.g., building orientation, site placement, and massing) and systems that impact energy use and occupant comfort, including building envelope considerations, glazing, and energy-related systems.

As outlined in the NC and CI reference guides, "Sustainable design strategies and measures are constantly evolving and improving. New technologies are continually introduced to the marketplace and up-to-date scientific research influences building design strategies. The purpose of this LEED™ category is to recognize projects for innovative building features and sustainable building knowledge." It also helps foster creative thinking and research into new areas that have yet to be explored. Check the LEED website for constant updates.

Building Operation and Maintenance (O+M): Green building measures cannot attain their goals unless they work as intended. Building commissioning includes testing and adjusting the mechanical, electrical, and plumbing systems to ensure that all equipment meets the design criteria set out by the owner. It also includes instructing the staff on the operation and maintenance of equipment. Over time, building performance can be assured through measurement, adjustment, and upgrading. Proper maintenance is necessary to ensure that a building continues to perform as designed and commissioned.

The LEED for Building Operations and Maintenance rating system includes market-sector adaptations for existing buildings, data centers, warehouses and distribution centers, hospitality, schools, and retail. It contains the addition of new credits, as well as the combination of various LEED 2009 credits into condensed, clearer versions.

3.3 PROJECT DOCUMENTATION, SUBMITTALS, AND CERTIFICATIONS

The intent of LEED certification is to provide independent, third-party verification that a building project meets the highest green building and performance measures. All LEED™-certified projects receive a LEED™ plaque, which is a nationally recognized symbol stating that a building is environmentally friendly, responsible, profitable, and a healthy place to live and work, which is why LEED certification is sought after by many building owners.

Before LEED-Online was available, the registration of projects was via a paper certification process. However, applications and documentation for LEED project certification can now be submitted 100% online in an easy-to-use interface featuring Adobe LiveCycle technology (see https://leedonline .usgbc.org/). The new facility of LEED-Online means that project team members are now able to upload credit templates, track CIRs, manage key project details, contact customer service, and communicate with reviewers throughout the design and construction reviews.

The LEED™ certification process consists of the following five basic stages:

1. Register the project.
2. Integrate LEED requirements.
3. Obtain technical support.
4. Document the project for certification.
5. Receive certification.

3.3.1 Step 1: Register the Project

Registering a LEED project is the first step toward earning LEED certification and should be completed as early as possible, preferably during schematic design. "Once you submit the registration form and payment, your project will be accessible in LEED Online, where you'll have access to a variety of tools and resources necessary to apply for LEED certification." From here, you can assemble your project team, and the documentation process begins. Have questions? Registration can be completed by submitting an online registration form via the USGBC website, which also provides clear instructions on what is required. It essentially consists of inputting project contact information and other relevant information such as project name, location, square footage, site area, and building type; a brief narrative description; and a preliminary LEED credit scorecard. Images can be uploaded for inclusion with the project posting on the USGCB website list of all LEED™-registered projects. Architects and owners can choose to withhold registration data from public view if they have any confidentiality concerns. Upon registering, project teams receive information, tools, and communication that will help guide through the certification process.

An important advantage of early registration, that is, during early phases of project design, is that it ensures maximum potential for achieving the targeted certification. Another advantage of registration is that it establishes point of contact with USGBC and provides access to essential information, software tools, and communications. Furthermore, registration allows access to a database of existing CIRs and rulings as well as to the four sections of online Credit Templates:

1. Template status
2. Manage template
3. Required documents
4. Documentation status

Registration fees: The LEED™ registration fee depends on member status. It is currently a flat fee of $900.00 for USGBC members and $1,200.00 for nonmembers and is paid upfront at the time of registration. ND registration is $1,500, and Homes registration varies depending on number and type of homes in the assessment. Certification: Total certification cost depends on system and project size.

Certification fees: Fees for LEED certification vary and are based on the rating system that the project is certifying under (e.g., NC, CI, EB, CS, etc.) and the project type, project size, member status, and level of review required; review can include construction-phase drawings, design-phase drawings, or both. This fee is paid when the project team submits documentation for review. Certification fees are waived for projects that receive Platinum LEED certification and will receive a rebate for all certification fees paid at the completion of the certification process. Table 3.1 provided by the USGBC outlines current rates:

3.3.2 Step 2: Integrate LEED Requirements

To earn LEED certification, a project must satisfy all LEED™ prerequisites and earn a minimum number of points outlined in the LEED Rating System under which it is registered. Architects here use the LEED™ scorecard early on to reveal the project's potential and to explore possibilities for integrated solutions. When a proposed design solution can be used to achieve, or contribute to achieve, more than one credit that usually suggests that it is an effective solution. Integrated design typically mandates that appropriate time and effort be exerted to carry out the quantifiable performance and cost analysis during the earliest design phase because this will save time, money, and effort in later phases. It should be noted that the LEED™ v4 rating systems differs in several aspects (such as credit weightings and the introduction of regional priority) from LEED 2009 and earlier rating system versions. This will be clarified throughout the book. It is always wise to check with the USGBC website for the latest updates.

3.3.3 Step 3: Obtain Technical Support

Obtaining technical support during the design and certification process for LEED registered projects are facilitated by the allocation of two free requests for a *credit interpretation ruling* on any technical or administrative questions that may arise during the design phase and which may require clarification. The CIRs provide a database of Requests or Rulings that are available on the USGBC website to assist CIR customers in the understanding of LEED

Table 3.1 Outline of current LEED certification rates

	Less than 50,000 ft.²	50,000–500,000 ft.²	More than 500,000 ft.²	Appeals (if applicable)
LEED for: New construction, commercial interiors, schools, and core and shell full certification	**Fixed rate**	**Based on square-footage**	**Fixed rate**	**Per credit**
Design review				
Members	$1,250.00	$0.025/ft.²	$12,500.00	$500.00
Nonmembers	$1,500.00	$0.030/ft.²	$15,000.00	$500.00
Expedited fee*	$5,000.00 regardless of square footage			$500.00
Construction review				
Members	$500.00	$0.010/ft.²	$5,000.00	$500.00
Nonmembers	$750.00	$0.015/ft.²	$7,500.00	$500.00
Expedited fee*	$5,000.00 regardless of square footage			$500.00
Combined design and construction review				
Members	$1,750.00	$0.035/ft.²	$17,500.00	$500.00
Nonmembers	$2,250.00	$0.045/ft.²	$22,500.00	$500.00
Expedited fee*	$10,000.00 regardless of square-footage			$500.00
LEED for existing buildings	**Fixed rate**	**Based on square-footage**	**Fixed rate**	**Per credit**
Initial certification review				
Members	$1,250.00	$0.025/ft.²	$12,500.00	$500.00
Nonmembers	$1,500.00	$0.030/ft.²	$15,000.00	$500.00
Expedited fee*	$10,000.00 regardless of square-footage			$500.00
Recertification review*				
Members	$625.00	$0.0125/ft.²	$6,250.00	$500.00
Nonmembers	$750.00	$0.015/ft.²	$7,500.00	$500.00
Expedited fee*	$10,000.00 regardless of square-footage			$500.00

LEED for core and shell: precertification	**Fixed rate for all projects**	**Per credit**
Members	$2,500.00	$500.00
Nonmembers	$3,500.00	$500.00
Expedited fee*	$5,000.00	$500.00

* In addition to regular review fee.
** The Existing Building Recertification Review fee is due upon the customer submitting an application for recertification review. Before submitting, customer service (leedinfo@usgbc.org) should be contacted to receive a promotion code.
Source: USGBC.

credits and how they may apply to their specific projects and how technologies and strategies may be successfully applied to earn points. This can greatly assist project teams during the certification process, especially when it is unclear whether or not a particular strategy applies to a given credit; a CIR can be submitted, and the resultant ruling will determine the suitability of the approach. It should be emphasized that CIR rulings will not guarantee or award any credits – it merely provides specific information guidance regarding applicability. The fee for each CIR credit submittal is $220.

The USGBC strongly advises that before submitting a CIR request, project teams should:

1. Review the intent of the prerequisite credit in question to ensure that the project satisfies that intent.
2. Check the appropriate LEED™ *Reference Guide* to see if it contains the required answers.
3. Review the large online CIR database for previous CIRs logged by other projects on relevant credits to see if any listed CIRs address the raised issues.
4. If a satisfactory answer cannot be found, consider contacting LEED™ customer service to look into it and confirm whether it warrants a new CIR.

If a new CIR is warranted, this should be submitted via the LEED-Online form. When submitting a new CIR, there are several points to remember:

1. Do not include the credit name of your contact information as the database automatically tracks this information.
2. Confidential information should not be included as the submitted text will be posted on the USGBC website.
3. The CIR should not be formatted as a letter. Only the inquiry and essential background information should be included.
4. Each CIR should refer to only one LEED credit or prerequisite (unless there is technical justification to do otherwise). The CIR should preferably be presented in the context of the credit intent.
5. The CIR inquiry should only include essential project strategy and relevant background and/or supporting information.
6. Your CIR submission text should not exceed 600 words or 4000 characters including spaces.
7. Submissions of plans, drawings, cut-sheets, or other attachments are NOT permitted.
8. Note that CIRs can be viewed by all USGBC members, nonmembers with registered projects, and workshop attendees.
9. CIRs can only be requested by LEED™ Registered Project Team Members.

It is advisable to proofread the CIR text for clarity, readability, and spelling before submission. Credit language or achievement thresholds cannot be modified through the CIR process. In cases where the Technical Advisory Group cannot satisfactorily address the inquiry in hand, the USGBC states that "the Council reserves the right to circulate the interpretation request to the LEED™ Steering Committee and/or relevant LEED™ Committees as required." Should the project team that submitted the CIR disagree with the credit interpretation ruling, the submitter has recourse to appeal. Of note, the CIR and ruling must be submitted with the LEED™ application to ensure an effective credit review.

Furthermore, once registered, project team members are given access to the electronic LEED™ "letter template" files that is now in place and which is designed to help streamline the certification process. These templates provide the primary means for documenting key LEED™ credit certification data and represent the core of a LEED™ certification submittal.

3.3.4 Step 4: Documenting the Project to be Certified

The USGBC has transformed this process into a paperless one using online submissions. This has made it possible for teams to upload and access documents on a Web location designated for their project.

The USGBC has established a two-phase application process allowing LEED™ certification documentation to be executed in two stages: (1) a preliminary submission, which may be submitted for comment, after which it is revised and then resubmitted, and (2) a final submission. Preliminary certification submissions should include two copies of the letter template file on CD-ROM and two three-ring binders containing the project's LEED scorecard, project narrative, letter templates, illustrative drawings and photographs, and any other relevant backup documentation required for all targeted credits, tabbed by credit. The USGBC also allows complete submittal in one phase (design and construction submittals together).

LEED Letter Templates: These consist of preformatted submittal sheets that are required for documentation of each LEED prerequisite and credit in the USGBC certification process. The Letter Templates outline the specific project data needed to demonstrate achievement of the LEED™ performance requirements and include calculation formulas where applicable. Sample

Letter Templates are available for download from the USGBC website for review purposes only in a nonexecutable format. Participants in projects registered with the USGBC may access fully executable templates through LEED-Online.

The following is pertinent information required to be submitted online with the application for LEED certification so that the application process can proceed:

1. The LEED rating system for which you are submitting (NC, CI, EB, CS, etc.).
2. Project contact information, project type, project size, number of occupants, and anticipated date of construction completion (see USGBC website to download forms).
3. Provide a project narrative including descriptions of at least three project highlights.
4. LEED™ Project Checklist, which should include project prerequisites, credits, and total anticipated score.
5. Copies of latest LEED Letter Templates and supporting documentation.
6. Complete list of all CIRs used, including dates of applied rulings.
7. Drawings, photos, and diagrams (8.5 × 11 in. or 11 × 17 in.) that illustrate and explain the project, including
    a. site plan,
    b. typical floor plan,
    c. typical building section,
    d. typical or primary elevation, and
    e. photo or rendering of project as required.
8. Payment of the appropriate certification fees. These are made in different stages (i.e., design and construction phase). The application will only be reviewed by the USGBC once payment is fully processed.

Upon completing the design phase and once the USGBC receives the preliminary project submittal, it performs a technical review of the documentation and issues a "Preliminary LEED™ Review." This review essentially assesses the initial status of the targeted credits and includes relevant comments. Credits are not actually awarded after the design phase. However, the USGBC will mark each credit and place it into one of four categories: likely to be earned, denied, audited, or pending. The pending category includes technical advice to project teams regarding clarifications that will be required in the final submission. The audited category will indicate additional documentation that will be required.

Project teams are then required to respond by submitting a final clarified and updated compilation of the project's documentation.

3.3.5 Step 5: Receipt of Certification

Following this second submission, the USGBC issues a "Final LEED™ Review" report that outlines the final status of all targeted credits (i.e., *credit achieved* or *credit denied*), along with the project's level of certification. Project teams have the right to appeal if they disagree with the final assessment.

The new LEED v4 process innovations will facilitate the documentation and certification process and render it more user-friendly without diminishing the technical rigor and quality of LEED that the community has come to expect. The LEED credit requirements themselves have not changed, and project teams are still required to verify their achievements through third-party validation and through documentation ensure that the building is built as it was designed.

Other LEED process refinements recently introduced, such as the two-part documentation submittals, reflect the manner in which project teams work. It gives design teams an interim opportunity to modify design documents prior to commencing construction and ensure that the project is on track for its certification goals. Likewise, with the revised instruments of service procedure, documentation requirements have been aligned with existing instruments of service (to reduce additional project documentation). The USGBC has also introduced a "building in a feedback loop" that provides improved customer service throughout the LEED process, in addition to USGBC's continued implementation of procedures to make customer feedback and interaction an integral part of that process.

Appeals: A project team has the option to appeal a ruling (within 25 days after Final LEED™ Review) if in their opinion sufficient grounds exist where a credit has been denied in the Final LEED Review. The cost of an appeal is $500 per credit. Appeal submittals are all done via LEED-Online.

Because a different review team will undertake reviewing the denied appeal, the following is required to be submitted with the appeal:

- LEED registration information, including project contact, project type, project size, number of occupants, date of construction completion, etc.
- An overall project narrative including at least three project highlights.
- The LEED Project Checklist scorecard indicating project prerequisites and credits and the total score for the project.

- Drawing and photos illustrating the project, including
  - site plan,
  - typical floor plan,
  - typical building section,
  - typical or primary elevation,
  - photo or rendering of project, and
  - complete list of all CIRs used.
- Original, resubmittal, and appeal submittal documentation for only those credits that are being appealed. Narratives for each to be included as well.

3.4 GREENING YOUR SPECIFICATIONS

Construction documents typically consist of working drawings and specifications and are essential to convey the building design concept to the contractor. They provide the contractor with the necessary information to bid and build a project. The more accurately a concept is conveyed, the more likely it will be realized. It is important therefore that the building specifications be an integral part of the written documents and go hand in hand with the drawings; they describe the materials to be used as well as the methods of installation. They also prescribe the quality and standards of construction required to be achieved on the project.

Thus, in order to facilitate communication of the building design concept, the construction industry has standardized the format for construction documents. The working drawings describe the location, size, and quantity of materials, whereas the specifications (the written documents that accompany the working drawings) describe the quality of construction. For example, if a working drawing shows a plaster wall, the specifications would include the description of the plaster mix, lath and paper backing, and finishing techniques. To do this more effectively, several standard formats were developed. However, the most widely used today is the standard organizational format for specifications developed by the Construction Specifications Institute (CSI), which is now used by manufacturers, architects, engineers, interior designers, contractors, and building officials throughout the United States to format construction specifications in building contracts. The obvious purpose of this format is to assist the user in locating specific types of information.

Likewise, when specifying green building materials, it is best to follow the CSI's *MasterFormat*™ as most specifications are organized according to this format. Moreover, Green specifications can be used to benchmark the efficacy of other environmental specifications. Likewise, environmental goals for a specific project can easily be implemented into the CSI *MasterFormat*. There is also a wealth of information on greening your specifications on the Internet. The EPA format, in particular, is designed to help supplement project specifications.

Furthermore, the EPA team will reportedly assist in the development and modification of project specifications to meet LEED credit requirements. But in order to achieve this, a clear understanding is necessary of how the specifications can best be used as a proactive mechanism to assist in procuring materials that are environmentally friendly and for collecting required LEED information from subcontractors and suppliers (Figure 3.8).

BuildingGreen.com has extensive information on sustainable design based on the CSI MasterFormat (2014–2004) hierarchy. The specifications are basically general guidelines as to product selection and installation and may not be appropriate for a specific project, which is why before using the Guideline Specifications the reader should read their disclaimer. Further information regarding GreenSpec® can be found on the BuildingGreen.com website: http://www.buildinggreen.com/menus/divisions.cfm.

According to the CSI,

MasterFormat is the specifications-writing standard for most commercial building design and construction projects in North America. It lists titles and section numbers for organizing data about construction requirements, products, and activities.

Figure 3.8 The roots of Integrated Project Delivery (IPD) can be traced back to large healthcare projects, like the Sutter Medical Center Extension in Sacramento, California. *Photo: KMD Architects.*

By standardizing such information, MasterFormat facilitates communication among architects, specifiers, contractors and suppliers, which helps them meet building owners' requirements, timelines and budgets.

Although *GreenSpec* is organized according to the CSI *MasterFormat* (2014–2004) structure for organizing products, the seventh edition of *GreenSpec* introduced an entirely new approach to guideline specifications. The previous guideline specifications for a range of sections throughout the various divisions have been replaced with a much more comprehensive set of guideline specifications for four sections in Division 1 only. The CSI and CSC recently released a MasterFormat® 2014 Update.

There are several significant modifications included in the LEED v4 edition update. These include:

- The addition of selective demolition sections to most divisions of the format.
- Providing guidance for companies and project managers who need to incorporate demolition into their plans.
- Renaming Division 40 from "Process Equipment" to "Process Interconnections" and overhauling its content.
- Reorganizing "Process Liquid Pumps" in Division 43, as a result of input from members of the Hydraulic Institute.
- Moving "Manufactured Planters" and "Site Seating and Tables" to Division 32 to better align with other site-based work and renaming their former Division 12 location to "Interior Public Space Furnishings" to address similar types of furnishings used in indoor settings.
- Revising "Equipment" in Division 11 to update, simplify, and rationalize its content and organization.
- Expanding "Agreement Forms" in Section 00 52 00 to improve alignment with standard forms of agreement (Source: CSI).

The MasterFormat Maintenance Task Team conducts a biennial revision cycle process, publishing updates to the format every 2 years. This committee of volunteers develops the changes to the format based on input from industry supporters and proposals from individual users that were submitted through www.masterformat.com. The Task Team comprises appointees from CSI, CSC, ARCAT, ARCOM, Building Systems Design, Inc. (BSD), Specification Consultants in Independent Practice (SCIP), Digicon, and Canadian National Master Specifications. MasterFormat is a master list of numbers and titles classified by work results for construction practices. It is used to organize project manuals, detail cost information, and relate drawing notations to specifications. By fostering fuller and more detailed construction specifications, MasterFormat is designed to reduce costly changes and delays in projects due to incomplete, misplaced or missing information. "Before 2004, MasterFormat consisted of 16 Divisions. The standard is the most widely used standard for organizing specifications and other written information for commercial and institutional building projects in the US and Canada. It provides a master list of divisions, and section numbers and titles within each division, to follow in organizing information about a facility's construction requirements and associated activities. Standardizing the presentation of such information improves communication among all parties involved in construction projects." (Wikipedia). For more information on the 2014 updates, please visit www.masterformat.com.

The following is the latest GreenSpec® Guideline Specifications based on the current CSI MasterFormat Divisions (April 2014):

Procurement and Contracting Requirements Group
- *Division 00 – Procurement and Contracting Requirements*
  Specifications Group
General Requirements Subgroup
- *Division 01 – General Requirements*
  Facility Construction Subgroup
- *Division 02 – Existing Conditions (Ex. Alterations to existing natural conditions)*
- *Division 03 – Concrete (Ex. Footings)*
- *Division 04 – Masonry (Ex. Concrete block and brick work)*
- *Division 05 – Metals (Ex. Steel framing)*
- *Division 06 – Wood, Plastics, and Composites (Ex. House framing)*
- *Division 07 – Thermal and Moisture Protection (Ex. Insulation and water barriers)*
- *Division 08 – Openings (Ex. Doors, windows, and louvers)*
- *Division 09 – Finishes*
- *Division 10 – Specialties*
- *Division 11 – Equipment*
- *Division 12 – Furnishings*
- *Division 13 – Special Construction*
- *Division 14 – Conveying Equipment*
- *Division 15 – RESERVED FOR FUTURE EXPANSION*
- *Division 16 – RESERVED FOR FUTURE EXPANSION*
- *Division 17 – RESERVED FOR FUTURE EXPANSION*
- *Division 18 – RESERVED FOR FUTURE EXPANSION*
- *Division 19 – RESERVED FOR FUTURE EXPANSION*
Facility Services Subgroup:
- *Division 20 – RESERVED FOR FUTURE EXPANSION*
- *Division 21 – Fire Suppression*
- *Division 22 – Plumbing*
- *Division 23 – Heating Ventilating and Air Conditioning*

- *Division 24 – RESERVED FOR FUTURE EXPANSION*
- *Division 25 – Integrated Automation*
- *Division 26 – Electrical*
- *Division 27 – Communications*
- *Division 28 – Electronic Safety and Security*
- *Division 29 – RESERVED FOR FUTURE EXPANSION*

Site and Infrastructure Subgroup:
- *Division 30 – RESERVED FOR FUTURE EXPANSION*
- *Division 31 – Earthwork*
- *Division 32 – Exterior Improvements*
- *Division 33 – Utilities*
- *Division 34 – Transportation*
- *Division 35 – Waterway and Marine works*
- *Division 36 – RESERVED FOR FUTURE EXPANSION*
- *Division 37 – RESERVED FOR FUTURE EXPANSION*
- *Division 38 – RESERVED FOR FUTURE EXPANSION*
- *Division 39 – RESERVED FOR FUTURE EXPANSION*

Process Equipment Subgroup:
- *Division 40 – Process Interconnections*
- *Division 41 – Material Processing and Handling Equipment*
- *Division 42 – Process Heating, Cooling, and Drying Equipment*
- *Division 43 – Process Gas and Liquid Handling, Purification and Storage Equipment*
- *Division 44 – Pollution and Waste Control Equipment*
- *Division 45 – Industry-Specific Manufacturing Equipment*
- *Division 46 – Water and Wastewater Equipment*
- *Division 47 – RESERVED FOR FUTURE EXPANSION*
- *Division 48 – Electrical Power Generation*
- *Division 49 – RESERVED FOR FUTURE EXPANSION*

You can get more detailed information by simply clicking on one of the above divisions on the BuildingGreen website. For example by clicking Division 22: Plumbing, you would get additional information as show below:
CSI Division 22: Plumbing
  22 01 40: Operation and Maintenance of Plumbing Fixtures (4 products)
  22 07 00: Plumbing Insulation (1 product)
  22 11 16: Domestic Water Piping (4 articles, 1 product)
  22 13 00: Sanitary Sewerage, Facilities (1 article, 10 products)
  22 13 16: Sanitary Waste and Vent Piping (2 articles, 8 products)
  22 14 00: Facility Storm Drainage (1 product)
  22 16 00: Greywater Systems (3 articles, 9 products)
  22 32 00: Domestic Water Filtration Equipment (6 products)
  22 34 01: Fuel-Fired Residential Water Heaters (12 articles, 27 products)
  22 34 02: Fuel-Fired Commercial Water Heaters (1 article, 34 products)
  22 35 00: Domestic Water Heat Exchangers (5 articles, 10 products)
  22 41 13: Residential Toilets (12 articles, 49 products)

22 41 14: Composting Toilet Systems (7 articles, 8 products)
22 41 15: Residential Urinals (17 articles, 1 product)
22 41 16: Residential Lavatories and Sinks (2 articles, 1 product)
22 41 39: Residential Faucets, Supplies and Trim (12 articles, 15 products)
22 41 42: Residential Showerheads (12 products)
22 42 00: Commercial Plumbing Fixtures (8 articles)
22 42 13: Commercial Toilets (25 articles, 22 products)
22 42 14: Commercial Composting Toilet Systems (7 articles, 4 products)
22 42 15: Commercial Urinals (19 articles, 11 products)
22 42 39: Commercial Faucets, Supplies and Trim (13 articles, 21 products)
22 42 42: Commercial Showerheads (1 article, 11 products)
22 42 43: Flushometers (9 articles, 5 products)
22 47 16: Pressure Water Coolers (1 article, 2 products)
22 51 00: Swimming Pool Plumbing Systems (5 products)

For projects seeking LEED certification, or that wish to track their project's performance against LEED, the specifications include details on how LEED requirements relate to the expressed requirements. Many LEED credits may not be addressed directly in the Guideline Specifications primarily because attaining those credits are determined by choices made in site selection or design and are not affected by product choices or other activities governed by these sections. The responsibility lies with the designer to ensure that any such credits have been satisfactorily addressed in the design and construction process.

Earlier versions of LEED required the submittal of extensive documentation from contractors and subcontractors to verify compliance with credit requirements. However, with the shift to online submissions, documentation requirements have been dramatically reduced.

For projects pursuing LEED certification, the contractor should be provided with a "LEED Submittal Form" for each LEED credit that the contractor is to provide documentation. The contractor would then complete the form and attach any additional documentation to it. Project managers sometimes link receipt of the completed forms to payment requests from the contractor at appropriate points in the construction process. In addition, there may be submittals required for LEED or for the client that are not typically within the scope of the specifications document.

The following is a guidance document example that is based on the *Whole Building Design Guide – Federal Green Construction Guide for Specifiers*. It consists of sample specification language intended to be inserted into project specifications on this subject as appropriate to "greening" your specifications. Certain provisions, where indicated, are required for US federal agency projects. Sample specification language is numbered to clearly distinguish it from advisory or discussion material. Each sample is preceded by identification of the typical location in a specification section, where it would appear using the SectionFormat™ of the CSI; the six-digit section number cited is per CSI Masterformat™ 2014. For a more complete set, visit the Whole Building Design Guide web site at: http://fedgreenspecs.wbdg.org.

3.4.1 Section 03 40 00 – Precast Concrete

SPECIFIER NOTE:

Resource Management: Plant fabrication handles raw materials and by-products at a single location that typically allows greater efficiency and better pollution prevention than job site fabrication.

Aggregates for use in concrete include normal sand and gravel, crushed stone, expanded clay, expanded shale, expanded slate, pelletized or extruded fly ash, expanded slag, perlite, vermiculite, expanded polystyrene beads, or processed clay, diatomite, pumice, scoria, or tuff.

Architectural items (e.g., planters, lintels, bollards) fabricated from lightweight and recycled content aggregates are available. The quantity and type of recycled materials vary from manufacturer to manufacturer and include cellulose, fiberglass, polystyrene, and rubber.

Autoclaved aerated concrete (AAC) is a type of lightweight precast concrete prevalent in Europe, Asia, and the Middle East and recently available through manufacturing facilities in the United States. It is made with Portland cement, silica sand or fly ash, lime, water, and aluminum powder or paste. The aluminum reacts with the products of hydration to release millions of tiny hydrogen gas bubbles that expand the mix to approximately five times the normal volume. When set, the AAC is cut into blocks or slabs and steam-cured in an autoclave.

Toxicity/IEQ: Refer to Section 03 30 00 - Cast-In-Place Concrete. Precast concrete generally requires less Portland cement per volume of concrete for similar performance due to better quality control.

Performance: Performance is more predictable in precast operations since more exact dimensions, placement of reinforcing, and surface finishing can be obtained. Precast concrete can be fabricated with continuous insulation. AAC is significantly lighter (about one-fifth the weight of traditional concrete) than normal concrete and can be formed into blocks or panels. Lighter-weight concretes generally have greater fire and thermal resistance but less strength than traditional normal-weight concrete. A full range of lightweight concretes are available and their strength and weight is determined by the aggregates used.

PART 1 – GENERAL

1.1 SUMMARY

A. This Section includes:

  1. Autoclaved Aerated Concrete (AAC).

B. Related Sections:

  1. 03 30 00 – Cast-In-Place Concrete.

1.2 SUBMITTALS

A. Product data. Unless otherwise indicated, submit the following for each type of product provided under work of this Section:

SPECIFIER NOTE:

Green building rating systems often include credit for materials of recycled content. USGBC-LEED™ v3, for example, includes credit for materials with recycled content, calculated on the basis of preconsumer and postconsumer percentage content, and it includes credit for use of salvaged/recovered materials. However, in LEED v4, the credit for materials of recycled content requires materials to meet both regional and recycled content criteria to maximize points. (Attention should be given to the latest LEED™ requirements and recommendations.)

Green Globes US also provides points for reused building materials and components and for building materials with recycled content.

  1. Recycled Content:

   a. Indicate recycled content; indicate percentage of preconsumer and postconsumer recycled content per unit of product.

   b. Indicate relative dollar value of recycled content product to total dollar value of product included in project.

   c. If recycled content product is part of an assembly, indicate the percentage of recycled content product in the assembly by weight.

   d. If recycled content product is part of an assembly, indicate relative dollar value of recycled content product to total dollar value of assembly.

SPECIFIER NOTE:

Specifying local materials may help minimize transportation impacts; however it may not have a significant impact on reducing the overall embodied energy of a building material because of efficiencies of scale in some modes of transportation.

Green building rating systems frequently include credit for local materials. Transportation impacts include: fossil fuel consumption, air pollution, and labor.

USGBC-LEED™ 2009 includes credits for materials extracted/harvested and manufactured within a 500-mile radius from the project site. However, this OPTION within the new v4 version encourages local sourcing. The radius from the construction site for extraction, manufacturing, and purchasing of the product has now been reduced to within a 100-mile radius of the construction site. Green Globes US also provides points for materials that are locally manufactured.

    2. Local/Regional Materials:

        a. Sourcing location(s): Indicate location of extraction, harvesting, and recovery; indicate distance between extraction, harvesting, and recovery and the project site.

        b. Manufacturing location(s): Indicate location of manufacturing facility; indicate distance between manufacturing facility and the project site.

        c. Product Value: Indicate dollar value of product containing local/regional materials; include materials cost only.

        d. Product Component(s) Value: Where product components are sourced or manufactured in separate locations, provide location information for each component. Indicate the percentage by weight of each component per unit of product.

B. Submit environmental data in accordance with Table 1 of ASTM E2129 for products provided under work of this Section.

C. Documentation of manufacturer's take-back program for *(units, full and partial) (packaging) (xxxx)*. Include the following:

    1. Appropriate contact information.

    2. Overview of procedures.

        a. Indicate manufacturer's commitment to reclaim materials for recycling and/or reuse.

    3. Limitations and conditions, if any, applicable to the project.

PART 2 - PRODUCTS

SPECIFIER NOTE:

EO 13423 includes requirements for Federal Agencies to use "sustainable environmental practices, including acquisition of biobased, environmentally preferable, energy-efficient, water-efficient, and recycled-content products."

Specifically, under the Sustainable Building requirements per Guiding Principle #5 Reduce Environmental Impact of Materials, EO 13423 directs Federal agencies to "use products meeting or exceeding EPA's recycled content recommendations" for EPA-designated products and for other products to "use materials with recycled content such that the sum of post-consumer recycled content plus one-half of the pre-consumer content constitutes at least 10% (based on cost) of the total value of the materials in the project."

2.1 MATERIALS

A. Load-bearing and non-load-bearing AAC elements: Comply with ASTM C1386.

PART 3 - EXECUTION

3. X SITE ENVIRONMENTAL PROCEDURES

A. Waste Management: As specified in Section 01 74 19 – Construction Waste Management and as follows:

  1. Broken, waste AAC units: May be used as nonstructural fill *(if approved by Architect/Engineer)*.

  2. Coordinate with manufacturer for *take-back program*. Set aside *(scrap) (packaging) (xxxx)* to be returned to manufacturer for recycling into new product.

CHAPTER 4

LEED™ Professional Accreditation, Standards, and Codes

4.1 OVERVIEW – LEED™ V4 NEW FEATURES AND IMPROVEMENTS; CERTIFICATION TIER 1, TIER 2, AND TIER 3

From the outset, LEED professional accreditation was designed to distinguish building professionals with the knowledge and skills to successfully steward the LEED™ certification process and sustainable design in general. LEED has indeed made some dramatic advances in recent years; toward this end, the LEED Professional Accreditation program was in 2008 transitioned to the Green Building Certification Institute (GBCI) to manage. The GBCI with the support of the USGBC® now handles exam development and delivery to ensure objective, balanced management of the credentialing program. The GBCI states that "the specific credits in the rating system provide guidance for the design and construction of buildings of all sizes in both private and public sectors." Jerry Yudelson, author of *The Green Building Revolution*, comments that "the essence of LEED, and its particular genius, is that it is a point–based rating system that allows vastly different green buildings to be compared in the aggregate." Yudelson goes on to say, "The LEED rating is a form of "eco-label" that describes the environmental attributes of a project." (See Figure 4.1.)

In an effort to capitalize on the existing market momentum, the USGBC has significantly revamped and reorganized the existing LEED™ Rating Systems. The previous LEED™ version 3 (v3) green building certification system was launched on April 27, 2009, and built on the fundamental structure and familiarity of the earlier rating system, while providing a new structure to ensure that the LEED™ v3 rating system incorporated new technology and addressed the most urgent priorities. According to the USGBC, LEED 2009 was a significant improvement, particularly with the inclusion of three major enhancements to the LEED rating system: harmonization, credit weightings, and regionalization.

In November 2013, The US Green Building Council (USGBC) announced the launching of LEED v4, the newest version of the LEED green building program, at the annual Greenbuild International Conference and

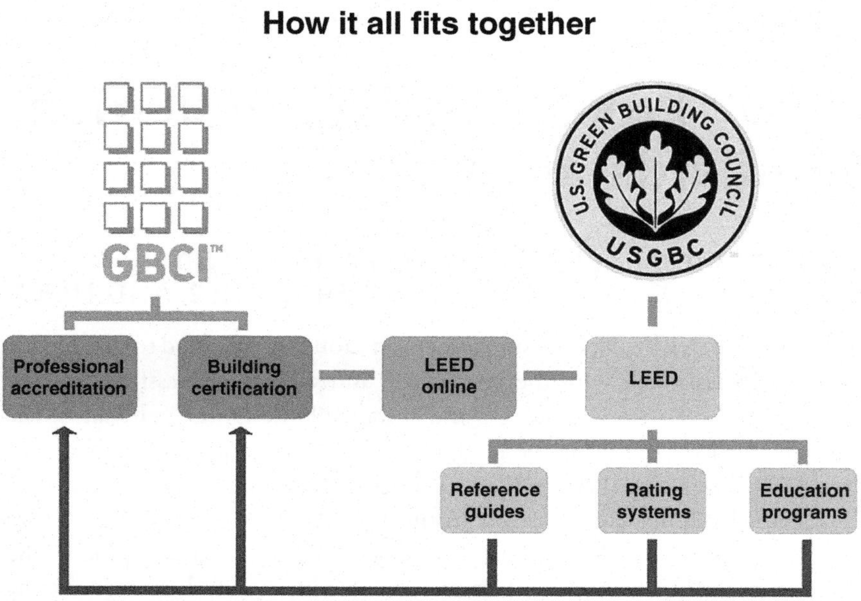

How it all fits together

Figure 4.1 *Diagram Showing the Relationship Between USGBC and GBCI and How They Relate to Project Certification and LEED Accreditation. Source: USGBC.*

Expo in Philadelphia. LEED has played a critical role in revolutionizing the green building marketplace since 1998 as the world's premier benchmark for the design, construction, and operation of high-performance green buildings.

4.1.1 Harmonization and Alignment

This means that LEED™ credits across the rating systems will bring the core elements of LEED into one elegant rating system. And according to USGBC, this is intended to synchronize the development and deployment of LEED rating systems while creating capacity to respond to previously underserved markets. Credits and prerequisites from all LEED rating systems have now been updated, consolidated and aligned, drawing on their most effective common denominators, so that credits and prerequisites should be consistent across all LEED 2009 and LEED v4 rating systems. Necessary precedent-setting and clarifying information from Credit Interpretation Rulings (CIRs) were incorporated into the rating systems. LEED™ for Homes and LEED™ for Neighborhood Development have not changed under LEED 2009. USGBC also announced that "a scrub of the existing Credit Interpretation Rulings (CIRs) was conducted and necessary precedent-setting and clarifying language has been incorporated into the prerequisites/credits." The reader must update for the LEED four requirements.

4.1.2 Credit Weightings

The second major advancement that comes with LEED 2009 is that credits now consist of different weightings that rely on their ability to impact different environmental and human health concerns. Penny Bonda, coauthor of Sustainable Commercial Interiors, describes the new credit weightings as "scientifically grounded re-evaluations that place an increased emphasis on energy use and carbon emission reductions." The recently revised LEED credit weightings therefore awards more points to green strategies and practices that reflect the most positive impact on critical issues such as energy efficiency, CO_2 reductions, and life cycle performance of assemblies.

For the revised weightings, credits were evaluated individually against a list of 13 environmental impact categories that include climate change, indoor environmental quality, water intake, resource depletion, and others (Figure 4.2a,b). LEED prioritized the impact categories, and credits were assigned a value based on the size of their contribution to mitigating each impact. This resulted in giving the most value to credits that offer the greatest potential for making the biggest change. And while the credits are all intact, they nevertheless are now worth different amounts. In LEED 2009, the evaluation of each credit against the set of impact categories produces a point total for each credit; the point totals for all credits come to 100, and these final point totals are reflected in the rating system scorecard. In previous LEED versions, for example, the total points for the New Construction category came to 69 points and for the Commercial Interiors category to 57 points.

LEED v4 boasts nine distinct categories in which projects can earn points toward certification. A new category, Integrative Process, was added. In addition, the credits related to the building site and location are split into two separate categories in LEED v4: Sustainable Sites (SS) and a new category, Location and Transportation (LT). Those credits related to site development, rainwater management, and heat island effect are found in the SS category, and those related to site selection, density, and transit land in the LT category. However, the total number of available points has remained constant at 110 points, and the certification levels remained unchanged. The Figure 3.3b shows the points allocated to each category with the new categories in green.

The new category, Integrative Process, focuses on achieving "synergies across disciplines and building systems" through collaborative efforts across

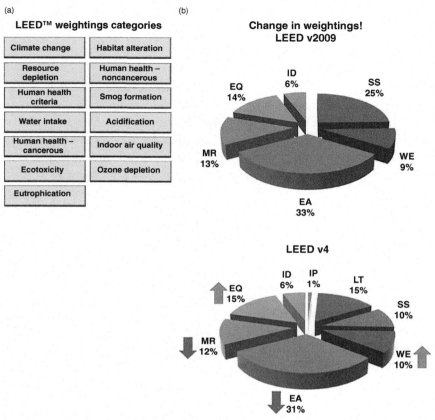

Figure 4.2 (a) Diagram illustrating the 13 LEED™ weightings categories on which much of the LEED 2009 and LEED v4 rating systems are based. (b) Pie charts showing difference in weightings between LEED 2009 and LEED v4. *Source: Part a, USGBC.*

disciplines. This credit requires preliminary analysis in the areas of energy and water use, including site conditions, massing and orientation of the building, and the building envelope.

4.1.3 Regionalization

One of the most significant changes in the LEED 2009 and LEED v4 Rating Systems are the incorporation of Regional Priority Credits. This was to enhance the flexibility of LEED and provide a more effective method of addressing the need for regional adaptation. Moreover, USGBC's introduction of Regional Bonus Credits is intended to increase the value of pursuing credits that address environmental areas of concern in a project's region. This is in contrast to the rating systems used prior to LEED 2009, which was applied uniformly across the United States and point values were equal for achievability of credits across various regions.

The LEED v4 Regional priority is basically identical to the credit in LEED 2009. There are 6 regional priority credit options and teams can be awarded for up to 4 of the 6 options in each rating system, which will be identified as regional credits similar to current exemplary performance points and are meant to address regionally prioritized environmental issues that have been identified as regionally critical within a project's environmental zone by the regional USGBC Board. A maximum of four points are available for project teams to pursue. The implication of this is that in the new rating system, a project can earn up to four extra points, or one point for each of four such Regional Priority Credits. USGBC Chapters and Regional Councils played a crucial role in this effort, based on their knowledge of issues of concern in their locales.

LEED for Homes was launched in 2008 and was rolled out on a national basis. LEED for Homes is a voluntary initiative to actively promote more sustainable building practices in the home building industry. The LEED for Homes rating system is targeting the top 25% of homes with best practice environmental features. It should also be noted that LEED for Homes format differs somewhat from the other certification categories such as New Construction and Commercial Interiors. Additional LEED for Homes credit categories in v4 include Location and Linkage credits. These encourage construction on previously developed or infill sites and promote walkable neighborhoods with access to efficient transportation options and open space.

Additionally, there are the Awareness and Education credits. The intent is to encourage home builders and real estate professionals to provide homeowners, tenants, and building managers with the appropriate education and tools they require to understand and make the most of the green building features of their homes. Readers are strongly advised to check the USGBC website for the latest updates.

4.2 TIER 1: LEED CERTIFICATION GREEN ASSOCIATE

The following sections list out the main categories for LEED certification and upon which much of the LEED™ v4 exams are based. The LEED Green Associate and LEED AP exams have recently changed to reflect the improvements and adjustments to the new LEED v4 rating system. The earlier exams, which were put in place until May 31, 2014, focused on the 2009 version of the rating system. However, project teams will still be able to choose to use LEED v2009, rather than the new v4 rating system, until October 2016. After that deadline, LEED v4 will become mandatory for all new projects. LEED Green Associate exam candidates should download the *LEED Green Associate Candidate Handbook* for the latest information and updates about the exam and LEED policies.

4.2.1 General Overview

To assist professionals who want to demonstrate green building expertise in nontechnical fields of practice, GBCI has created the LEED Green Associate credential. This credential basically denotes basic knowledge of green design, building construction, and operations and denotes that an individual has an up-to-date understanding of the most current green building principles and practices, and is committed to its professional future. And although the USGBC and GBCI do not release information about how they score the exams, the organization has released a "Cheat Sheet," for the new LEED v4 exams that basically outlines content areas in the form of Task Domains and Knowledge Domains, which are covered in the exam. These are discussed in Section 4.2.5.

The LEED Green Associate exam consists of one part containing 100 randomly delivered multiple-choice questions in which the candidate has 2 h to complete the exam (no additional time will be provided). Additionally, there is also an optional 10-min tutorial and an optional 10-min exit survey. The content of the exam is designed to focus on the LEED project process (including integrated design), core sustainability concepts, green building terminology, and several aspects of the LEED rating systems. The exam is computer-based with questions and answers displayed on screen. The computer records answers and times the exam. You are able to change your answers, skip questions, and flag questions for later review. Should a candidate exit the exam session prior to completion, the exam cannot be restarted and the exam session and fee are forfeited.

The fees associated with the LEED Green Associate exam are as follows:
- $50 application fee,
- $200 exam fee (per exam appointment) for USGBC national members and full-time students or
- $250 exam fee (per exam appointment) for all others, and
- $50 biennial CMP renewal fee.

4.2.2 Registration

To register for the LEED Associate exam, a candidate must

1. Log in to their Credentials account by using their existing USGBC® site user account or creating a new account if the candidate does not have one.
2. Verify that the name entered matches the name on the ID that is presented at the test center. If it does not match, the candidate must update his/her name in the site user account "settings." Contact the GBCI if any issues are experienced relating to updating your name.
3. The credential exam applied for should be selected and the instructions on the screen are to be followed to complete the application.
4. The candidate will be redirected to prometric.com/gbci to schedule the exam date and location.
5. When the exam appointment is scheduled, the candidate will receive a confirmation number onscreen and from Prometric through an email.
6. The confirmation number should be recorded, as it will be needed to confirm, cancel, or reschedule an appointment through the Prometric website, prometric.com/gbci (Prometric's website is available for scheduling, rescheduling, canceling, and confirming exam appointments 24 h/day. To schedule an exam, the candidate will need to have first applied and registered with GBCI. To reschedule, cancel, or confirm an exam appointment, the Prometric-issued 16-digit confirmation number will be needed).
7. Once an exam has been scheduled, the confirmation notice from Prometric should be printed and saved. It may be used for any communication with Prometric about the proposed exam. Also, once candidates register and pay for their exam, they have 1 year to schedule their exam session. They may request one 6-month extension of this 1-year period. After three unsuccessful attempts, a candidate must wait 90 days before submitting a new registration to GBCI. For information on GBCI's Exam refund/rescheduling policy please visit usgbc.org. To schedule five or more candidates at one time, customer service should be contacted (Source:GBCI).

4.2.3 Eligibility Requirements

The GBCI recommends that candidates planning to take the LEED® Green Associate™ Exam familiarize themselves with LEED and green building concepts and requirements. This can be done through educational courses, volunteering, or work experience (prior to testing). All candidates must also agree to the Disciplinary and Exam Appeals Policy and credentialing maintenance requirements. Additionally, if audited, they must be willing to provide requested information and be 18 years of age or older.

Of note, the eligibility requirements for the LEED Green Associate exam do not require candidates to have experience in the form of documented involvement on an LEED-registered project, employment, or previous employment in a sustainable field of work, engagement in or completion of an education program that addresses green building principles. Candidates however, are still required to agree to the Disciplinary and Exam Appeals Policy and Credential Maintenance Program and submit to an application audit.

4.2.4 Are You Ready?

As a proposed candidate for the LEED Green Associate exam, ask yourself: Are You Ready? One month before the proposed exam date, make sure that the name given (first name) and surname (last name) in the usgbc.org account matches the given name and surname on the identification you will present at the test center. If the candidate's names do not match, he/she will not be allowed to test and the exam fee will be forfeited. Furthermore, the candidate should check and update the address listed in the usgbc.org profile so that the certificate is mailed to the most current address. One week prior to the exam date, check that the date, time, and location of the proposed exam is correct. If inaccuracies exist, go to usgbc.org to make the necessary corrections and for information about rescheduling or missing your exam.

4.2.5 LEED Green Associate Exam Specifications

Specifications: The following provides a general description of exam content areas for the LEED Green Associate exam. The USGBC and GBCI do not release information about how the exams are scored. However, the organization has released a "Cheat Sheet," for the new LEED v4 exams. This "Cheat Sheet" outlines content areas in the form of Task Domains and Knowledge Domains that are covered on the exam. Task Domains reflect the tasks necessary to perform LEED safely and effectively, and include concepts such as LEED™ Project and Team Coordination, LEED Certification Process, Analyses Required for LEED Credits, and Advocacy and Education for Adoption for LEED Rating System. Knowledge Domains reflect the rating systems' credit categories and what a candidate is required to know. These include concepts such as LEED™ Process, Integrative Strategies, LEED™ credit categories, and Project Surroundings and Public Outreach.

Exam part 1: LEED™ Green Associate Exam – According to the USGBC, the LEED Green Associate tests your general knowledge of green building practices for both commercial and residential spaces and both new construction and existing buildings as well as how to support other professionals working on LEED projects.

Task Domains (Green Associate) LEED™ Green Associate Tasks (Source: USGBC/GBCI)

- Communicate broad and basic green building concepts to team or colleagues.
- Research and create a library of sustainable building materials.
- Assist others with sustainability goals.
- Create project profiles/case studies/press release.
- Serve as a green advocate to clients, team members, and the general public.
- Stay current on any updates to LEED and green strategies in general.
- Navigate in LEED™ Online.
- Assist project leader with LEED correspondence to all project team members.
- Assist in managing documentation process.
- Assist in managing timeline of LEED™ certification.

Knowledge Domains: LEED® Process (16 questions)

- Organization fundamentals (e.g., mission/vision; nonprofit; role of USGBC/GBCI)
- Structure of LEED rating systems (e.g., credit categories, prerequisites, credits, and/or Minimum Program Requirements for LEED Certification)
- Scope of each LEED rating system (e.g., rating system selection; rating system families (BD+C, ID+C, O+M, ND, Homes))
- LEED™ development process (e.g., consensus based; stakeholder and volunteer involvement; rating system updates/evolution)
- Credit categories (e.g., goals and objectives of each (LT, SS, WE, EA, MR, EQ, IN, RP); synergies)
- Impact categories (e.g., what should a LEED project accomplish?)
- LEED certification process (e.g., certification levels) Components of LEED Online and Project Registration
- Other rating systems

Integrative Strategies (8 questions)
- Integrative process (e.g., early analysis of the interrelationships among systems; systems thinking; charrettes)
- Integrative project team members (e.g., architect, engineer, landscape architect, civil engineer, contractor, facility manager, etc.)
- Standards that support LEED (e.g., breadth not depth of American Society of Heating, Refrigeration and Air-conditioning Engineers (ASHRAE); Sheet Metal and Air Conditioning Contractors National Association (SMACNA) guidelines; Green Seal, ENERGY STAR®, HERs, Reference Standards listed in ACPs, etc.)

Location and Transportation (7 questions)
- Site selection (e.g., targeting sites in previously developed and brownfields/high-priority designation area, avoiding sensitive habitat, located in areas with existing infrastructure and nearby uses, reduction in parking footprint)
- Alternative transportation (e.g., type, access, and quality; infrastructure and design)

Sustainable Sites (7 questions)
- Site assessment (e.g., environmental assessment, human impact)
- Site design and development (e.g., construction activity pollution prevention; habitat conservation and restoration; exterior open space; rainwater management; exterior lighting; heat island reduction)

Water Efficiency (9 questions)
- Outdoor water use (e.g., use of graywater/rainwater in irrigation; use of native and adaptive species)
- Indoor water use (e.g., concepts of low-flow/waterless fixtures; water-efficient appliances; types and quality)
- Water performance management (e.g., measurement and monitoring)

Energy and Atmosphere (10 questions)
- Building loads (e.g., building components, space usage (private office; individual space; shared multioccupant spaces))
- Energy efficiency (e.g., basic concepts of design, operational energy efficiency, energy auditing, commissioning)
- Alternative and renewable energy practices (e.g., demand response, renewable energy, green power, carbon offsets)
- Energy performance management (e.g., energy use measurement and monitoring; building automation controls/advanced energy metering; operations and management; benchmarking; ENERGY STAR)
- Environmental concerns (e.g., sources and energy resources; greenhouse gases; global warming potential; resource depletion; ozone depletion)

Materials and Resources (9 questions)
- Reuse (e.g., building reuse, material reuse, interior reuse, furniture reuse)
- Life-cycle impacts (e.g., concept of life-cycle assessment; material attributes; human and ecological health impacts; design for flexibility)
- Waste (e.g., construction and demolition; maintenance and renovation; operations and ongoing; waste management plan)
- Purchasing and declarations (e.g., purchasing policies and plans; environmental preferable purchasing (EPP); building product disclosure and optimization (i.e., raw materials sourcing; material ingredients; environmental product disclosure))

Indoor Environmental Quality (8 questions)
- Indoor air quality (e.g., ventilation levels; tobacco smoke control; management of and improvements to indoor air quality; low-emitting materials; green cleaning)
- Lighting (e.g., electric lighting quality, daylight)
- Sound (e.g., acoustics)
- Occupant comfort, health, and satisfaction (e.g., controllability of systems, thermal comfort design, quality of views, assessment/survey)

Project Surroundings and Public Outreach (11 questions)
- Environmental impacts of the built environment (e.g., energy and resource use in conventional buildings; necessity of green buildings; environmental externalities; triple bottom line)
- Codes (e.g., relationship between LEED and codes (building, plumbing, electrical, mechanical, fire protection); green building codes)
- Values of sustainable design (e.g., energy savings over time; money-saving incentives; healthier occupants; costs (hard costs, soft costs); life cycle)
- Regional design (e.g., regional green design and construction measures as appropriate, regional emphasis should be placed in Sustainable Sites and Materials and Resources)

The exam also contains 15 pretest questions.

The primary sources for the development of the LEED® Professional Exams are the LEED Rating Systems. The following list of references is not meant to be comprehensive. When combined with the test specifications, the candidate has the material from which the exam is based. LEED Green Associate Exam This exam is designed to test the general knowledge of green building practices and how to support other professionals working on LEED projects.

References:
- USGBC. Green Building and LEED™ Core Concepts Guide. 3rd Edition. USGBC, 2011. (Print and digital versions available.)

- USGBC. Introductory and Overview Sections. LEED™ Building Design + Construction Reference Guide. v4 Edition. USGBC, 2013. Web. (Note that the introductory and overview sections are available to download separately from purchasing the full reference guide.)
- USGBC. LEED™ v4 Impact Category and Point Allocation Process Overview. USGBC, 2013. Web.
- USGBC. LEED™ v4 User Guide. USGBC, 2013. Web.
- USGBC. Guide to LEED™ Certification: Commercial. USGBC, 2014. Web.
- "LEED™ Certification Fees." USGBC, 2014. Web.
- "Rating System Selection Guidance." USGBC, 2014. Web.
- "Addenda Database." USGBC. Web.

4.3 TIER 2: LEED PROFESSIONAL ACCREDITATION

4.3.1 General Overview

Practitioner experience is critical to the LEED AP designation and, as such, project proficiency will be tested objectively within the LEED AP with specialty exam itself.

The following qualifications are strongly advised to be considered for the LEED Professional Accreditation Exam, but are not strictly required:

- Attendance at a USGBC-sponsored LEED™ Training Workshop.
- Formal training as an architect, engineer, interior designer, facilities manager, or contractor.
- Tenure in the building design, green building, and construction industry, or as a building management or operations professional, facilities staff, or executive.

4.3.2 LEED v4 BD+C: New Construction and Major Renovations Project Checklist (LEED NC – Courtesy USGBC)

Integrative Process (IP)	1 possible point
Location and Transportation (LT)	16 possible points
LT Credit – Sensitive land protection	1
LT Credit – High-priority site	2
LT Credit – Surrounding density and diverse uses	5
LT Credit – Access to quality transit	5
LT Credit – Bicycle facilities	1
LT Credit – Reduced parking footprint	1
LT Credit – Green vehicles	1

Sustainable Sites (SS)	10 possible points
SS Prerequisite – Construction activity pollution prevention	Required
SS Credit – Site assessment	1
SS Credit – Site development – protect or restore habitat	2
SS Credit – Open space	1
SS Credit – Rainwater management	3
SS Credit – Heat island reduction	2
SS Credit – Light pollution reduction	1

Water Efficiency (WE)	11 possible points
WE Prerequisite – Outdoor water use reduction	Required
WE Prerequisite – Indoor water use reduction	Required
WE Prerequisite – Building-level water metering	Required
WE Credit – Outdoor water use reduction	2
WE Credit – Indoor water use reduction	6
WE Credit – Cooling tower water use	2
WE Credit – Water metering	1

Energy and Atmosphere (EA)	33 possible points
EA Prerequisite – Fundamental commissioning and verification	Required
EA Prerequisite – Minimum energy performance	Required
EA Prerequisite – Building-level energy metering	Required
EA Prerequisite – Fundamental refrigerant management	Required
EA Credit – Enhanced commissioning	6
EA Credit – Optimize energy performance	18
EA Credit – Advanced energy metering	1
EA Credit – Demand response	2
EA Credit – Renewable energy production	3
EA Credit – Enhanced refrigerant management	1
EA Credit – Green power and carbon offsets	2

Materials and Resources (MR)	13 possible points
MR Prerequisite – Storage and collection of recyclables	Required
MR Prerequisite – Construction and demolition waste management planning	Required
MR Credit – Building life-cycle impact reduction	5
MR Credit – Building product disclosure and optimization – EPD	2
MR Credit – Building product disclosure and optimization – raw materials	2
MR Credit – Building product disclosure and optimization – material ingredients	2
MR Credit – Construction and demolition waste management	2

Indoor Environmental Quality (EQ)	16 possible points
EQ Prerequisite – Minimum indoor air quality performance	Required
EQ Prerequisite – Environmental tobacco smoke (ETS) control	Required
EQ Credit – Enhanced indoor air quality strategies	2
EQ Credit – Low-emitting materials	3
EQ Credit – Construction indoor air quality management plan	1
EQ Credit – Indoor air quality assessment	2
EQ Credit – Thermal comfort	1
EQ Credit – Interior lighting	2
EQ Credit – Daylight	3
EQ Credit – Quality views	1
EQ Credit – Acoustic performance	1
Innovation in Design (ID)	**6 possible points**
ID Credit – 1 Innovation in design	5
ID Credit – 2 LEED Accredited Professional	1
Regional Priority (RP)	**4 possible points**
RP Credit – 1 regional priority	1
RP Credit – 2 regional priority	1
RP Credit – 3 regional priority	1
RP Credit – 4 regional priority	1
Totals	**Possible points 110**

4.3.3 LEED v4 BD+C: Core and Shell Development Project Checklist (LEED CS)

LEED BD+C: Core and Shell is the appropriate rating system to use if the gross floor area is more than 40% incomplete at the time of certification. The following is the LEED™ v4 Core and Shell Development Checklist and Scorecard (Courtesy USGBC).

Integrative Process (IP)	1 possible point
Location and Transportation (LT)	**20 possible points**
LT Credit – LEED for neighborhood development location	16
LT Credit – Sensitive land protection	2
LT Credit – High-priority site	3
LT Credit – Surrounding density and diverse uses	6
LT Credit – Access to quality transit	6
LT Credit – Bicycle facilities	1
LT Credit – Reduced parking footprint	1
LT Credit – Green vehicles	1

Sustainable Sites (SS)	11 possible points
SS Prerequisite – Construction activity pollution prevention	Required
SS Credit – Site assessment	1
SS Credit – Site development – protect or restore habitat	2
SS Credit – Open space	1
SS Credit – Rainwater management	3
SS Credit – Heat island reduction	2
SS Credit – Light pollution reduction	1
SS Credit – Tenant design and construction guidelines	1
Water Efficiency (WE)	**11 possible points**
WE Prerequisite – Outdoor water use reduction	Required
WE Prerequisite – Indoor water use reduction	Required
WE Prerequisite – Building-level water metering	Required
WE Credit – Outdoor water use reduction	2
WE Credit – Indoor water use reduction	6
WE Credit – Cooling tower water use	2
WE Credit – Water metering	1
Energy and Atmosphere (EA)	**33 possible points**
EA Prerequisite – Fundamental commissioning and verification	Required
EA Prerequisite – Minimum energy performance	Required
EA Prerequisite – Building-level energy metering	Required
EA Prerequisite – Fundamental refrigerant management	Required
EA Credit – Enhanced commissioning	6
EA Credit – Optimize energy performance	18
EA Credit – Advanced energy metering	1
EA Credit – Demand response	2
EA Credit – Renewable energy production	3
EA Credit – Enhanced refrigerant management	1
EA Credit – Green power and carbon offsets	2
Materials and Resources (MR)	**14 possible points**
MA Prerequisite – Storage and collection of recyclables	Required
MA Prerequisite – Construction and demolition waste management planning	Required
MA Credit – Building life-cycle impact reduction	6
MA Credit – Building product disclosure and optimization – environmental product declarations	2
MA Credit – Building product disclosure and optimization – sourcing of raw materials	2
MA Credit – Building product disclosure and optimization – material ingredients	2
MA Credit – Construction and demolition waste management	2

Indoor Environmental Quality (EQ)	10 possible points
EQ Prerequisite – Minimum indoor air quality performance	Required
EQ Prerequisite – Environmental tobacco smoke (ETS) control	Required
EQ Credit – Enhanced IAQ strategies	2
EQ Credit – Low-emitting materials	3
EQ Credit – Construction IAQ management plan	1
EQ Credit – Daylight	3
EQ Credit – Quality views	1
Innovation in Design (ID)	**6 possible points**
ID Credit – Innovation	5
ID Credit – LEED Accredited Professional	1
Regional Priority (RP)	**4 possible points**
RP Credit – 1 regional priority	4
Totals	**Possible points 110**

4.3.4 LEED v4 Professional Accreditation Requirements for Commercial Interiors

LEED for Commercial Interiors is a green building rating system for tenant space in office, retail, and institutional buildings developed by the USGBC (Courtesy USGBC).

Integrative Process (IP)	2 possible points
Location and Transportation (LT)	**18 possible points**
LT Credit – LEED for neighborhood development location	18
LT Credit – Surrounding density and diverse uses	8
LT Credit – Access to quality transit	7
LT Credit – Bicycle facilities	1
LT Credit – Reduced parking footprint	2
Water Efficiency (WE)	**12 possible points**
WE Prerequisite – Indoor water use reduction	Required
WE Credit – Indoor water use reduction	12
Energy and Atmosphere (EA)	**38 possible points**
EA Prerequisite – Fundamental commissioning and verification	Required
EA Prerequisite – Minimum energy performance	Required
EA Prerequisite – Fundamental refrigerant management	Required
EA Credit – Enhanced commissioning	5
EA Credit – Optimize energy performance	25

EA Credit – Advanced energy metering	2
EA Credit – Renewable energy metering	3
EA Credit – Enhanced refrigerant management	1
EA Credit – Green power and carbon offsets	2
Materials and Resources (MR)	**13 possible points**
MR Prerequisite – Storage and collection of recyclables	Required
MR Prerequisite – Construction and demolition waste management planning	Required
MR Credit – Long-term commitment	1
MR Credit – Interiors life-cycle impact reduction	4
MR Credit – Building product disclosure and optimization – environmental product declarations	2
MR Credit – Building product disclosure and optimization – sourcing of raw materials	2
MR Credit – Building product disclosure and optimization – material ingredients	2
MR Credit – Construction and demolition waste management	2
Indoor Environmental Quality (EQ)	**17 possible points**
EQ Prerequisite – Minimum indoor air quality performance	Required
EQ Prerequisite – Environmental tobacco smoke (ETS) control	Required
EQ Credit – Enhanced indoor air quality strategies	2
EQ Credit – Low-emitting materials	3
EQ Credit – Construction indoor air quality management plan	1
EQ Credit – Indoor air quality assessment	2
EQ Credit – Thermal comfort	1
EQ Credit – Interior lighting	2
EQ Credit – Daylight	3
EQ Credit – Quality views	1
EQ Credit – Acoustic performance	2
Innovation (I)	**6 possible points**
ID Credit – Innovation	5
ID Credit – LEED Accredited Professional	1
Regional Priority (RP)	**4 possible points**
RP Credit – Regional priority: specific credit	1
RP Credit – Regional priority: specific credit	1
RP Credit – Regional priority: specific credit	1
RP Credit – Regional priority: specific credit	1
Totals	**Possible points 100**

4.3.5 LEED v4 Professional Accreditation Requirements for Existing Buildings: Operations and Maintenance (Courtesy USGBC)

Location and Transportation (LT)	15 possible points
LT Alternative Transportation	15
Sustainable Sites (SS)	10 possible points
SS Prerequisite – Site management policy	Required
SS Credit – Site development – protect or restore habitat	2
SS Credit – Rainwater management	3
SS Credit – Heat island reduction	2
SS Credit – Light pollution reduction	1
SS Credit – Site management	1
SS Credit – Site improvement plan	1
Water Efficiency (WE)	12 possible points
WE Prerequisite – Indoor water use reduction	Required
WE Prerequisite – Building-level water metering	Required
WE Credit – Outdoor water use reduction	2
WE Credit – Indoor water use reduction	5
WE Credit – Cooling tower water use	3
WE Credit – Water metering	2
Energy and Atmosphere (EA)	38 possible points
EA Prerequisite – Energy efficiency best management practices	Required
EA Prerequisite – Minimum energy performance	Required
EA Prerequisite – Building-level energy metering	Required
EA Prerequisite – Fundamental refrigerant management	Required
EA Credit – Existing building commissioning – analysis	2
EA Credit – Existing building commissioning – implementation	2
EA Credit – Ongoing commissioning	3
EA Credit – Optimize energy performance	20
EA Credit – Advanced energy metering	2
EA Credit – Demand response	3
EA Credit – Renewable energy and carbon offsets	5
EA Credit – Enhanced refrigerant management	1
Materials and Resources (MR)	8 possible points
MR Prerequisite – Ongoing purchasing and waste policy	Required
MR Prerequisite – Facility maintenance and renovation policy	Required
MR Credit – Purchasing – ongoing	1
MR Credit – Purchasing – lamps	1
MR Credit – Purchasing – facility management and renovation	2

MR Credit – Solid waste management – ongoing	2
MR Credit – Solid waste management – facility management and renovation	2
Indoor Environmental Quality (EQ)	**17 possible points**
EQ Prerequisite – Minimum indoor air quality performance	**Required**
EQ Prerequisite – Environmental tobacco smoke (ETS) control	**Required**
EQ Prerequisite – Green cleaning policy	Required
EQ Credit – Indoor air quality management program	2
EQ Credit – Enhanced indoor air quality strategies	2
EQ Credit – Thermal comfort	1
EQ Credit – Interior lighting	2
EQ Credit – Daylight and quality views	4
EQ Credit – Green cleaning – custodial effectiveness assessment	1
EQ Credit – Green cleaning – products and materials	1
EQ Credit – Green cleaning – equipment	1
EQ Credit – Integrated pest management	2
EQ Credit – Occupant comfort survey	1
Innovation (I)	**6 possible points**
ID Credit – Innovation	5
ID Credit – LEED Accredited Professional	1
Regional Priority (RP)	**4 possible points**
RP Credit – Regional priority specific credit	1
RP Credit – Regional priority specific credit	1
RP Credit – Regional priority specific credit	1
RP Credit – Regional priority specific credit	1
Totals	**Possible points 110**

4.3.6 LEED v4 Professional Accreditation Requirements for New Construction in Schools (Courtesy USGBC)

IP Credit Integrative Process	1 possible point
Location and Transportation (LT)	15 possible points
LT Credit – Neighborhood development location	15
LT Credit – Sensitive land protection	1
LT Credit – High-priority site	2
LT Credit – Surrounding density and diverse uses	5
LT Credit – Access to quality transit	4
LT Credit – Bicycle facilities	1
LT Credit – Reduced parking footprint	1
LT Credit – Green vehicles	1

Sustainable Sites (SS)	12 possible points
SS Prerequisite – Construction activity pollution prevention	Required
SS Prerequisite – Environmental site assessment	Required
SS Credit – Site assessment	1
SS Credit – Site development – protect or restore habitat	2
SS Credit – Open space	1
SS Credit – Rainwater management	3
SS Credit – Heat island reduction	2
SS Credit – Light pollution reduction	1
SS Credit – Site master plan	1
SS Credit – Joint use of facilities	1
Water Efficiency (WE)	12 possible points
WE Prerequisite – Outdoor water use reduction	Required
WE Prerequisite – Indoor water use reduction	Required
WE Prerequisite – Building-level water metering	Required
WE Credit – Outdoor water use reduction	2
WE Credit – Indoor water use reduction	7
WE Credit – Cooling tower water use	2
WE Credit – Water metering	1
Energy and Atmosphere (EA)	31 possible points
EA Prerequisite – Fundamental commissioning and verification	Required
EA Prerequisite – Minimum energy performance	Required
EA Prerequisite – Building-level energy metering	Required
EA Prerequisite – Fundamental refrigerant management	Required
EA Credit – Enhanced commissioning	6
EA Credit – Optimize energy performance	16
EA Credit – Advanced energy metering	1
EA Credit – Demand response	2
EA Credit – Renewable energy production	3
EA Credit – Enhanced refrigerant management	1
EA Credit – Green power and carbon offsets	2
Materials and Resources (MR)	13 possible points
MR Prerequisite – Storage and collection of recyclables	Required
MR Prerequisite – Construction and demolition waste management planning	Required
MR Credit – Building life-cycle impact reduction	5
MR Credit – Building product disclosure and optimization – environmental product declarations	2

MR Credit – Building product disclosure and optimization – sourcing of raw materials	2
MR Credit – Building product disclosure and optimization – material ingredients	2
MR Credit – Construction and demolition waste management	2
Indoor Environmental Quality (EQ)	16 possible points
EQ Prerequisite – Minimum IAQ performance	Required
EQ Prerequisite – Environmental tobacco smoke control	Required
EQ Prerequisite – Minimum acoustic performance	Required
EQ Credit – Enhanced IAQ strategies	2
EQ Credit – Low-emitting materials	3
EQ Credit – Construction IAQ management plan	1
EQ Credit – IAQ assessment	2
EQ Credit – Thermal comfort	1
EQ Credit – Interior lighting	2
EQ Credit – Daylight	3
EQ Credit – Quality views	1
EQ Credit – Acoustic performance	1
Innovation in Design (ID)	6 possible points
ID Credit – Innovation	5
ID Credit – LEED Accredited Professional	1
Regional Priority (RP)	4 possible points
RP Credit – Regional priority	4
Totals	Possible points 110

4.3.7 LEED v4 Professional Accreditation Requirements for New Construction in Retail (Courtesy USGBC)

IP Credit Integrative Process	1 possible point
Location and Transportation (LT)	16 possible points
LT Credit – Neighborhood Development Location	16
LT Credit – Sensitive land protection	1
LT Credit – High-priority site	2
LT Credit – Surrounding density and diverse uses	5
LT Credit – Access to quality transit	5
LT Credit – Bicycle facilities	1
LT Credit – Reduced parking footprint	1
LT Credit – Green vehicles	1

Sustainable Sites (SS)	10 possible points
SS Prerequisite – Construction activity pollution prevention	Required
SS Credit – 1 Site assessment	1
SS Credit – Site development – protect or restore habitat	2
SS Credit – Open space	1
SS Credit – Rainwater management	3
SS Credit – Heat island reduction	2
SS Credit – Light pollution reduction	1
Water Efficiency (WE)	12 possible points
WE Prerequisite – Outdoor water use reduction	Required
WE Prerequisite – Indoor water use reduction	Required
WE Prerequisite – Building-level water metering	Required
WE Credit – Outdoor water use reduction	2
WE Credit – Indoor water use reduction	7
WE Credit – Cooling tower water use	2
WE Credit – Water metering	1
Energy and Atmosphere (EA)	33 possible points
EA Prerequisite – Fundamental commissioning and verification	Required
EA Prerequisite – Minimum energy performance	Required
EA Prerequisite – Building-level energy metering	Required
EA Prerequisite – Fundamental refrigerant management	Required
EA Credit – Enhanced commissioning	6
EA Credit – Optimize energy performance	1–18
EA Credit – Advanced energy metering	1
EA Credit – Demand response	2
EA Credit – Renewable energy production	2
EA Credit – Enhanced refrigerant management	1
EA Credit – Green power and carbon offsets	2
Materials and Resources (MR)	13 possible points
MR Prerequisite – Storage and collection of recyclables	Required
MR Prerequisite – Construction and demolition waste management planning	Required
MR Credit – Building life-cycle impact reduction	5
MR Credit – Building product disclosure and optimization – environmental product declarations	2
MR Credit – Building product disclosure and optimization – sourcing of raw materials	2

MR Credit – Building product disclosure and optimization – material ingredients	2
MR Credit – Construction and demolition waste management	2
Indoor Environmental Quality (EQ)	**15 possible points**
EQ Prerequisite – Minimum indoor air quality performance	**Required**
EQ Prerequisite – Environmental tobacco smoke (ETS) control	**Required**
EQ Credit – Enhanced indoor air quality (IAQ) strategies	2
EQ Credit – Low-emitting materials	3
EQ Credit – Construction indoor air quality management plan	1
EQ Credit – Indoor air quality (IAQ) assessment	2
EQ Credit – Thermal comfort	1
EQ Credit – Interior lighting	2
EQ Credit – Daylight	3
EQ Credit – Quality views	1
Innovation in Design (ID)	**6 possible points**
ID Credit – Innovation	5
ID Credit – LEED Accredited Professional	1
Regional Priority (RP)	**4 possible points**
RP Credit – 1 Regional priority	4
Totals	**Possible points 110**

4.3.8 LEED v4 Professional Accreditation Requirements for Homes Design and Construction

The USGBC designed the LEED for Homes rating system to promote the design and construction of high-performance green homes because green homes use less energy, water, and natural resources; create less waste; and are healthier and more comfortable for the occupants. One example of a LEED-certified home is in Port St. Lucie, Florida, which boasts hurricane-resistant wall panels that provide superior energy efficiency, ENERGY STAR appliances, and a drought-tolerant and low-maintenance landscape (Figure 4.3).

The LEED for Homes Rating System has 35 topic areas, each with a specific intent or goal. The LEED for Homes was not officially launched until February 2008. LEED for Homes is available for building design and construction projects for single-family homes and multifamily projects up to 8 stories.

Figure 4.3 Photo of a Port St. Lucie (Florida) home that became the first in the state to earn a certification through the USGBC's LEED for Homes pilot program after several inspections and performance tests were completed by Florida Solar Energy Center. The home completed the certification process on March 19, 2007. *Courtesy: Florida Solar Energy Center.*

LEED BD+C: Homes and multifamily low-rise: Designed for single-family homes and multifamily buildings between 1 and 3 stories. Projects 3–5 stories may choose the Homes rating system that corresponds to the ENERGY STAR program in which they are participating.

LEED BD+C: Multifamily midrise: Designed for midrise multifamily buildings between 4 and 8 stories.

4.3.8.1 LEED v4 Building Design and Construction: Multifamily Lowrise (Courtesy USGBC)

IP Credit Integrative Process	2 possible points
Location and Transportation (LT)	15 possible points
LT Prerequisite – Floodplain avoidance	Required
Credit – LEED for neighborhood development location	15
Credit – Site selection	8
Credit – Compact development	3
Credit – Community resources	2
Credit – Access to transit	2
Sustainable Sites (SS)	7 possible points
SS Prerequisite – Construction activity pollution prevention	Required
SS Prerequisite – No invasive plants	Required

SS Credit – Heat island reduction	2
SS Credit – Rainwater management	3
SS Credit – Nontoxic pest control	2
Water Efficiency (WE)	**12 possible points**
WE Prerequisite – Water metering	**Required**
Credit – Total water use	12
Credit – Indoor water use	6
Credit – Outdoor water use	4
Energy and Atmosphere (EA)	**38 possible points**
EA Prerequisite – Minimum energy performance	**Required**
EA Prerequisite – Energy metering	**Required**
EA Prerequisite – Education of the homeowner, tenant, or building manager	**Required**
EA Prerequisite – Home size	**Required**
EA Credit – Annual energy use	29
EA Credit – Efficient hot water distribution	5
EA Credit – Advanced utility tracking	2
EA Credit – Active solar-energy design	1
EA Credit – HVAC start-up credentialing	1
EA Credit – Building orientation for passive solar	3
EA Credit – Air infiltration	2
EA Credit – Envelope insulation	2
EA Credit – Windows	3
EA Credit – Space heating and cooling equipment	4
EA Credit – Heating and cooling distribution systems	3
EA Credit – Efficient domestic hot water equipment	3
EA Credit – Lighting	2
EA Credit – High efficiency appliances	2
EA Credit – Renewable energy	4
Materials and Resources (MR)	**10 possible points**
MR Prerequisite – Certified tropical wood	**Required**
MR Prerequisite – Durability management	**Required**
MR Credit – Durability management verification	1
MR Credit – Environmentally preferable products	4
MR Credit – Construction waste management	3
MR Credit – Material efficient framing	2
Indoor Environmental Quality (EQ)	**16 possible points**
EQ Prerequisite – Ventilation	**Required**
EQ Prerequisite – Combustion venting	**Required**
EQ Prerequisite – Garage pollutant protection	**Required**
EQ Prerequisite – Radon-resistant construction	**Required**
EQ Prerequisite – Air filtering	**Required**
EQ Prerequisite – Environmental tobacco smoke	**Required**
EQ Prerequisite – Compartmentalization	**Required**

EQ Credit – Enhanced ventilation	3
EQ Credit – Contaminant control	2
EQ Credit – Balancing of Heating and cooling distribution systems	3
EQ Credit – Enhanced compartmentalization	1
EQ Credit – Enhanced combustion venting	2
EQ Credit – Enhanced garage pollutant protection	2
EQ Credit – Low-emitting products	3
Innovation in Design (ID)	**6 possible points**
ID Preliminary rating	**Required**
ID Credit – Innovation	5
ID Credit – LEED Accredited Professional	1
Regional Priority (RP)	**4 possible points**
RP Credit – Regional priority: specific credit	4
Totals	**Possible points 110**

4.3.8.2 LEED v4 Building Design and Construction: Multifamily Midrise (Courtesy USGBC)

IP Credit Integrative Process	**2 possible points**
Location and Transportation (LT)	**15 possible points**
LT Prerequisite – Floodplain avoidance	Required
Performance Path	
Credit – LEED for neighborhood development location	15
Prescriptive Path	
Credit – Site selection	8
Credit – Compact development	3
Credit – Community resources	2
Credit – Access to transit	2
Sustainable Sites (SS)	**7 possible points**
SS Prerequisite – Construction activity pollution prevention	Required
SS Prerequisite – No invasive plants	Required
SS Credit – Heat island reduction	2
SS Credit – Rainwater management	3
SS Credit – Nontoxic pest control	2
Water Efficiency (WE)	**12 possible points**
WE Prerequisite – Water metering	Required
Performance Path	
Credit – Total water use	12

Prescriptive Path

Credit – Indoor water use	6
Credit – Outdoor water use	4

Energy and Atmosphere (EA) — 37 possible points

EA Prerequisite – Energy metering	Required
EA Prerequisite – Minimum energy performance	Required
EA Prerequisite – Education of the homeowner, tenant, or building manager	Required
EA Credit – Annual energy use	30
EA Credit – Efficiency hot water distribution	5
EA Credit – Advanced utility tracking	2

Materials and Resources (MR) — 9 possible points

MR Prerequisite – Certified tropical wood	Required
MR Prerequisite – Durability management	Required
MR Credit – Durability management verification	1
MR Credit – Environmentally preferable products	5
MR Credit – Construction waste management	3

Indoor Environmental Quality (EQ) — 18 possible points

EQ Prerequisite – Ventilation	Required
EQ Prerequisite – Combustion venting	Required
EQ Prerequisite – Garage pollutant protection	Required
EQ Prerequisite – Radon-resistant construction	Required
EQ Prerequisite – Air filtering	Required
EQ Prerequisite – Environmental tobacco smoke	Required
EQ Prerequisite – Compartmentalization	Required
EQ Credit – Enhanced ventilation	3
EQ Credit – Contaminant control	2
EQ Credit – Balancing of heating and cooling distribution systems	3
EQ Credit – Enhanced compartmentalization	3
EQ Credit – Enhanced combustion venting	2
EQ Credit – Enhanced garage pollutant protection	1
EQ Credit – Low-emitting products	3
EQ Credit – No environmental tobacco smoke	1

Innovation in Design (ID) — 6 possible points

ID preliminary rating	Required
ID Credit – Innovation	5
ID Credit – LEED AP homes	1

Regional Priority (RP) — 4 possible points

RP Credit – Regional priority: specific credit	1
RP Credit – Regional priority: specific credit	1
RP Credit – Regional priority: specific credit	1
RP Credit – Regional priority: specific credit	1

Totals — Possible points 110

4.3.9 LEED v4 Professional Accreditation Requirements for Neighborhood Development (Courtesy USGBC)

This applies to individuals participating in the planning, design, and development of walkable neighborhoods and communities. According to the USGBC, "The LEED™ for Neighborhood Development program (LEED-ND) integrates the principles of smart growth, New Urbanism and green building into the first national rating system for neighborhood design. The rating system was developed by the U.S. Green Building Council (USGBC) in partnership with the Congress for the New Urbanism (CNU) and the Natural Resources Defense Council (NRDC)."

Smart Location and Linkages (LL)	28 possible points
LL Prerequisite – Smart location	Required
LL Prerequisite – Imperiled species and ecological communities	Required
LL Prerequisite – Wetland and water body conservation	Required
LL Prerequisite – Agricultural land conservation	Required
LL Prerequisite – Floodplain avoidance	Required
LL Credit – Preferred locations	10
LL Credit – Brownfield remediation	2
LL Credit – Access to quality transit	7
LL Credit – Bicycle facilities	2
LL Credit – Housing and jobs proximity	3
LL Credit – Steep slope protection	1
LL Credit – Site design for habitat or wetland and water body conservation	1
LL Credit – Restoration of habitat or wetlands and water bodies	1
LL Credit – Long-term conservation management of habitat or wetlands and water bodies	1
Neighborhood Pattern and Design (ND)	41 possible points
ND Prerequisite – Walkable streets	Required
ND Prerequisite – Compact development	Required
ND Prerequisite – Connected and open community	Required
ND Credit – Walkable streets	9
ND Credit – Compact development	6
ND Credit – Mixed-use neighborhoods	4
ND Credit – Housing types and affordability	7
ND Credit – Reduced parking footprint	1
ND Credit – Connected and open community	2

ND Credit – Transit facilities	1
ND Credit – Transportation demand management	2
ND Credit – Access to civic and public space	1
ND Credit – Access to recreation facilities	1
ND Credit – Visitability and universal design	1
ND Credit – Community outreach and involvement	2
ND Credit – Local food production	1
ND Credit – Tree-lined and shaded streetscapes	2
ND Credit – Neighborhood schools	1
Green Infrastructure and Buildings	**31 possible points**
Prerequisite – Certified green building	Required
Prerequisite – Minimum building energy performance	Required
Prerequisite – Indoor water use reduction	Required
Prerequisite – Construction activity pollution prevention	Required
Credit – Certified green buildings	5
Credit – Optimize building energy performance	2
Credit – Indoor water use reduction	1
Credit – Outdoor water use reduction	2
Credit – Building reuse	1
Credit – Historic resource preservation and adaptive reuse	2
Credit – Minimized site disturbance	1
Credit – Rainwater management	4
Credit – Heat island reduction	1
Credit – Solar orientation	1
Credit – Renewable energy production	3
Credit – District heating and cooling	2
Credit – Infrastructure energy efficiency	1
Credit – Wastewater management	2
Credit – Recycled and reused infrastructure	1
Credit – Solid waste management	1
Credit – Light pollution reduction	1
Innovation and Design Process	**6 possible points**
Credit – Innovation	5
Credit – LEED Accredited Professional	1
Innovation and Design Process	**4 possible points**
Credit – Regional priority credit: region defined	1
Credit – Regional priority credit: region defined	1
Credit – Regional priority credit: region defined	1
Credit – Regional priority credit: region defined	1
Totals	**Possible points 110**

4.4 TIER 3: LEED AP FELLOW CATEGORY

According to the USGBC,

The LEED Fellow designates the most exceptional professionals in the green building industry, and it is the most prestigious designation awarded by the Green Building Certification Institute (GBCI).

The LEED Fellow Program was established to recognize outstanding LEED APs who have demonstrated exceptional achievement in key mastery elements related to:
- *Technical knowledge and skill*
- *A history of exemplary leadership in green building*
- *Significant contributions in teaching, mentoring, or research with proven outcomes*
- *A history of highly impactful commitment, service, and advocacy for green building and sustainability*

The program was developed to honor and recognize LEED APs who have made a significant contribution to green building and sustainability at a regional, national or global level.

Evaluation

The criteria for assessing LEED Fellow nominees are based on five major dimensions of green building and sustainability that have been identified as mastery elements. The application process allows nominees to demonstrate that they have made exceptional and quantifiable contributions to the world of green building using their skills and abilities in four of the five mastery elements. One of the elements must be Technical Proficiency, and the nominee may choose the other three elements from Education and Mentoring, Leadership, Commitment and Service, and Advocacy.

1. Technical Proficiency: *A LEED Fellow demonstrates expert-level technical proficiency. He or she is experienced and knowledgeable in the application of multiple LEED rating systems and has provided significant contributions to LEED projects. A LEED Fellow is adept at identifying technical or procedural solutions to green building challenges and has demonstrated a sustained level of accomplishment for at least ten years.*
2. Education and Mentoring: *A LEED Fellow provides education, training and mentoring – sharing knowledge about LEED, sustainability and green building with others, both inside and outside of his or her own organization. The Fellow demonstrates this through measurable outcomes.*
3. Leadership: *A LEED Fellow is a leader in his or her own organization and in the field of green building. He or she plays an important role in instituting and applying sustainability practices and procedures within his or her own organization as well as within clients' projects and the community, and can demonstrate the impact of this leadership.*
4. Commitment and Service: *A LEED Fellow demonstrates a history of commitment and service to green building. This can include contribution*

and service to GBCI and USGBC or other Green Building Councils, as well as community service that have furthered green building or sustainability.

5. *Advocacy: A LEED Fellow is a longstanding advocate for sustainable ideas, concepts and technologies related to or promoting green building or sustainability. He or she delivers presentations and speeches and writes articles or books explaining green building and sustainability. A LEED Fellow also proactively encourages the adoption and use of LEED rating systems with clients, communities or government entities.*

Eligibility
To be nominated as a LEED Fellow, nominees must have:
- *Held a LEED Professional credential for eight or more years*
- *A LEED AP with Specialty credential in good standing*
- *Ten years of professional green building experience*
- *Agreed to be nominated*

To nominate a LEED Fellow, nominators must have:
- *A LEED AP with Specialty credential in good standing*
- *Ten years of professional green building experience*
- *Nominated no one else for the current LEED Fellow class*
- *Not been nominated for the current LEED Fellow class*
- *Agreed to be a nominator*

Source: US Green Building Council – published on January 1, 2012.

According to Beth Holst of GBCI, the GBCI Board of Directors has approved the creation of this credential but has not "framed out" in detail what it means. The evaluation process is supervised by the LEED Fellow Evaluation Committee and supported by the GBCI.

4.5 IDENTIFYING STANDARDS THAT SUPPORT LEED V4 CREDITS

The number of organizations today writing and maintaining standards continue to increase. The majority of these standards are developed by trade associations, government agencies, or standards writing organizations. Additionally, a long-standing relationship exists between construction codes and standards that address design, installation, testing, and materials related to the construction industry. The fundamental role standards play (including "green-related standards") in the building regulatory process is that they denote an extension of the code requirements and are therefore equally enforceable. However standards only have legal standing when specified by a specific code that is accepted by a jurisdiction.

Building standards function as a valuable design guideline to architects and engineers while establishing a framework of acceptable practices from which codes are frequently taken. When a standard is stipulated, the acronym of the standard organization and a standard number is called out. National and international standards are a critical aspect of supporting and achieving LEED credits. The more relevant organizations and standards relating to green issues include the following.

4.5.1 Organizations and Agencies

American Institute of Architects (AIA): Based in Washington, DC, the AIA has been the leading professional membership association for licensed architects, emerging professionals, and allied partners since 1857. The goal of the AIA is to serve as the voice of the architecture profession and the resource for our members in service to society. The AIA carries out its goal through advocacy, information, and community.

American National Standards Institute (ANSI): The ANSI approves standards as American National Standards, and provides information and access to the world's standards. It is also the official US representative to the world's leading standards bodies, including the International Organization for standardization (ISO). It provides and administers the only recognized system in the United States for establishing standards.

American Society of Heating, Refrigerating and Air-Conditioning Engineers (ASHRAE): The ASHRAE an international organization whose purpose is to advance the arts and sciences of heating, ventilation, air conditioning, and refrigeration for the public's benefit. ASHRAE's stated purpose is to write "standards and guidelines in its fields of expertise to guide industry in the delivery of goods and services to the public." The HVAC&R industry is making great strides in the areas of indoor air quality, protection of the ozone layer, and energy conservation. ASHRAE leads this effort through its standards and guidelines. There are currently 120 Standards and 22 Guidelines published by ASHRAE.

ASTM International (previously known as *American Society of Testing and Materials*): Is one of the largest voluntary standards development organizations in the world and provides a global forum for the development and publication of voluntary consensus standards for materials, products, systems, and services having internationally recognized quality and applicability.

American Wind Energy Association (AWEA): The AWEA is a national trade association representing wind power project developers, equipment suppliers, services providers, parts manufacturers, utilities, researchers, and others involved in the wind industry – one of the world's fastest growing energy industries. In addition, AWEA represents hundreds of wind energy advocates from around the world. The mission of the American Wind Energy Association is to promote wind power growth through advocacy, communication, and education.

BIFMA International (formerly known as the Business and Institutional Furniture Manufacturers Association): The BIFMA is a nonprofit trade association of furniture manufacturers and suppliers addressing issues of common concern. The association was established in 1973 and serves as the industry voice for workplace solutions by providing standards development, statistical data generation, government relations, industry promotion, and education.

California Energy Commission (CEC): Formally the Energy Resources Conservation and Development Commission, it is California's primary energy policy and planning agency and was created by the Legislature in 1974. It is located in Sacramento and has responsibilities that include forecasting future energy needs and keeping historical energy data; licensing thermal power plants 50 MW or larger; promoting energy efficiency by setting the state's appliance and building efficiency standards and working with local government to enforce those standards; supporting public interest energy research that advances energy science and technology through research, development, and demonstration programs; supporting renewable energy by providing market support to existing, new, and emerging renewable technologies; providing incentives for small wind and fuel cell electricity systems; and providing incentives for solar electricity systems in new home construction; implementing the state's alternative and renewable fuel and vehicle technology program; and planning for and directing state response to energy emergencies.

Code of Federal Regulation (CFR): The CFR is the codification (arrangement of) of the general and permanent rules published in the Federal Register by the executive departments and agencies of the Federal Government. It is divided into 50 titles that represent broad areas subject to federal regulation. Each volume of the CFR is updated once each calendar year and is issued on a quarterly basis.

Chartered Institution of Building Services Engineers (CIBSE): The CIBSE is the professional body that received its Royal Charter in 1976 and exists to support the science, art, and practice of building services engineering, by providing its members and the public with first-class information and education services and promoting the spirit of fellowship which guides its work.

Carpet and Rug Institute (CRI): The CRI is a nonprofit trade association representing the manufacturers of more than 95% of all carpet made in the United States, as well as their suppliers and service providers. The CRI coordinates with other segments of the industry, such as distributors, retailers, and installers, to help increase consumers' satisfaction with carpet and to show how carpet creates a better environment.

Center for Resource Solutions (CRS): The CRS is a national nonprofit with global impact. CRS brings forth expert responses to climate change issues with the speed and effectiveness necessary to provide real-time solutions. Our leadership through collaboration and environmental innovation builds policies and consumer-protection mechanisms in renewable energy, greenhouse gas reductions, and energy efficiency that foster healthy and sustained growth in national and international markets.

US Department of Energy (DOE): The primary mission of the DOE is to advance the national, economic, and energy security of the United States; to promote scientific and technological innovation in support of that mission; and to ensure the environmental cleanup of the national nuclear weapons complex. In addition, part of its mission is protecting the environment by providing a responsible resolution to the environmental legacy of nuclear weapons production.

US Energy Information Administration (EIA): The EIA was created by Congress in 1977 and is an independent statistical agency within the US Department of Energy. The EIA's mission is to provide policy-independent data, forecasts, and analyses to promote sound policy making, efficient markets, and public understanding regarding energy and its interaction with the economy and the environment. The agency collects data on energy reserves, production, consumption, distribution, prices, technology, and related international, economic, and financial matters. This information is disseminated as policy-independent data, forecasts, and analyses. EIA programs cover data on coal, petroleum, natural gas, electric, renewable, and nuclear energy.

US Environmental Protection Agency (EPA): The EPA was founded in July 1970 by the White House and Congress working together in response to the growing public demand for cleaner water, air, and land. Prior to the establishment of the EPA, the federal government was not structured to make a coordinated attack on the pollutants that harm human health and degrade the environment. The EPA was assigned the daunting task of repairing the damage already done to the natural environment and to establish new criteria to guide Americans in making a cleaner environment a reality.

US Federal Emergency Management Agency (FEMA): FEMA is an independent agency of the federal government, reporting to the president. Since its founding in 1979, FEMA's mission has been clear: to reduce loss of life and property and protect our nation's critical infrastructure from all types of hazards through a comprehensive, risk-based, emergency management program of mitigation, preparedness, response, and recovery.

Forest Stewardship Council (FSC): The FSC was created in 1993 to establish international forest management standards (known as the FSC Principles and Criteria) to ensure that forestry practices are environmentally responsible, socially beneficial, and economically viable. Certification is a "seal of approval" awarded to forest managers who adopt environmentally and socially responsible forest management practices and to companies that manufacture and sell products made from certified wood. Certification enables consumers, including architects, designers, and specifiers, to identify and procure wood products from well-managed sources and thereby use their purchasing power to influence and reward improved forest management activities around the world. LEED accepts certification established by the internationally recognized FSC. Of note, the FSC itself does not issue certifications but rather accredits certification bodies in order to maintain its independence from those seeking certification. The FSC offers two types of certificates. The holder of a Forest Management Certificate can maintain that its operations comply with FSC standards. However, before selling products as FSC-certified, a Chain of Custody Certificate must also be obtained.

Green Seal (GS): The GS is a Washington, DC-based independent nonprofit, standard-setting organization certifying a range of products and services. GS (www.greenseal.org) is an independent organization that provides third-party "green" certification to various products and services. Approved products carry a GS logo that is well recognized throughout industry and government as a leading environmental standard. Manufacturers pay GS a fee for each product that is reviewed for certification. In addition to complying with GS standards, manufacturers of approved products are subject to ongoing factory inspections (at the manufacturer's expense), product testing, and annual maintenance fees.

GBCI: The GBCI was established as a separately incorporated entity with the support of the US Green Building Council. GBCI administers the LEED certification program, performing third-party technical reviews and verification of LEED-registered projects to determine if they have met the standards set forth by the LEED rating system. These programs support the application of proven strategies for increasing and measuring the performance of buildings and communities as defined by industry systems, including the Leadership in Energy and Environmental Design (LEED) Green Building Rating Systems. Dedicated GBCI technical experts ensure the building certification process meets the highest levels of quality and integrity.

GREENGUARD Environmental Institute: It is a nonprofit organization that oversees the GREENGUARD Certification Program, a program that seeks to establish acceptable indoor air standards for indoor products, environments, and buildings. This third-party testing program evaluates low-emitting products and materials. The GREENGUARD Indoor Air Quality Certification Program certifies products based on their measured chemical emissions levels. This incorporates the Standard for Cleaning Products and Systems and the Standard for Children and Schools. (The latter standard requires reduced chemical levels compared to the former.) Chemical emissions are measured and reviewed across a broad range of exposure levels established by the US EPA and the Agency for Toxic Substances and Disease Registry (ATSDR).

Greenguard Certification Program: The GREENGUARD Certification Program is an industry-independent, third-party testing program for low-emitting products and materials. The program is implemented by the Greenguard Environmental Institute.

Illuminating Engineering Society of North America (IESNA): The IESNA is the recognized technical authority on illumination. Its principal objective is to communicate information on all aspects of good lighting practice to its members, to the lighting community, and to consumers, through a variety of programs, publications, and services. IESNA is a forum for the exchange of ideas and information and a vehicle for its members' professional development and recognition. Through technical committees, with hundreds of qualified individuals from the lighting and user communities, the IESNA correlates research, investigations, and discussions to guide lighting professionals and laypersons via consensus-based lighting recommendations.

Industrial Material Exchange (IMEX): IMEX is a regional program of local governments working together to protect public health and environmental quality by reducing the threat posed by the production, use, storage, and disposal of hazardous materials.

International Organization for Standardization (ISO): ISO is the world's largest developer and publisher of International Standards. It is a network of the national standards institutes of more than 160 countries, one member per country, with a Central Secretariat in Geneva, Switzerland, that coordinates the system. It is a nongovernmental organization that forms a bridge between the public and private sectors. On the one hand, many of its member institutes are part of the governmental structure of their countries, or are mandated by their government. On the other hand, other members have their roots uniquely in the private sector, having been set up by national partnerships of industry associations.

New Building Institute (NBI): The NBI is incorporated as a 501(c)(3) nonprofit in 1997. NBI works with national, regional, state, and utility groups to promote improved energy performance in commercial new construction, managing projects involving building research, design guidelines, and code activities. Additionally, NBI serves as a carrier of ideas between states and regions, researchers, and the market.

National Fenestration Rating Council (NFRC): The NFRC is a nonprofit organization that administers the only uniform, independent rating and labeling system for the energy performance of windows, doors, skylights, and attachment products. Its goal is to provide fair, accurate, and reliable energy performance ratings.

US EPA Office of Solid Waste & Emergency Response (OSWER): The OSWER provides policy, guidance, and direction for the Agency's solid waste and emergency response programs. It develops guidelines for the land disposal of hazardous waste and underground storage tanks and provides technical assistance to all levels of government to establish safe practices in waste management. The Office administers the Brownfields program, which supports state and local governments in redeveloping and reusing potentially contaminated sites. The Office also manages the Superfund program to respond to abandoned and active hazardous waste sites and accidental oil and chemical releases as well as to encourage innovative technologies to address contaminated soil and groundwater.

Sustainable Building Industry Council (SBIC): The SBIC is an independent, nonprofit trade association that seeks to dramatically improve the long-term performance and value of buildings through our outreach, advocacy, and education programs.

The SBIC was founded in 1980 as the Passive Solar Industries Council (PSIC) and served as a clearinghouse on passive solar design for the major building trade groups, including those representing architects, home builders, the glass industry, and building owners. In the 1990s, the PSIC's scope expanded to include the other major aspects of sustainable design and construction, including optimizing site potential, minimizing energy use, using renewable energy sources, conserving and protecting water, using environmentally preferable products, enhancing indoor environmental quality, and optimizing operations and maintenance practices.

The South Coast Air Quality Management District (SCAQMD): A governmental organization in Southern California with the mission to maintain healthful air quality for its residents. The organization established source specific standards to reduce air quality impacts, including South Coast Rule #1168.

Sheet Metal and Air Conditioning Contractors' National Association (SMACNA): This standard provides an overview of air pollutants associated with construction, control measures, construction process management, quality control, communications with occupants, and case studies. Consult the referenced standard for measures to protect the building HVAC system and maintain acceptable indoor air quality during construction and demolition activities.

US Green Building Council (USGBC): The USGBC is a nonprofit organization that was founded in 1993 to transform the way buildings and communities are designed, built, and operated, enabling an environmentally and socially responsible, healthy, and prosperous environment that improves the quality of life. A steering committee of the USGBC developed the Leadership in Energy and Environmental Design (LEED) Green Building Rating System to provide universally understood and accepted tools and performance criteria that encourage and accelerate global adoption of sustainable green building and development practices. LEED encourages construction practices that meet specified standards, resolving much of the negative impact of buildings on their occupants and on the environment. Green buildings in the United States are certified with this voluntary, consensus-based rating system. The latest version of LEED is version 4.

4.5.2 Referenced Standards and Legislation

ASHRAE90.1 (2013 edition): Building energy standard covering design, construction, operation, and maintenance.

ASHRAE 52.2–2012 (addenda a, b, and d): Standardized method of testing building ventilation filters for removal efficiency by particle size.

ASHRAE 55–2013: Standard describing thermal and humidity conditions for human occupancy of buildings.

ASHRAE 62.1–2007: Standard that defines minimum levels of ventilation performance for acceptable indoor air quality.

ASHRAE 192: Standard for measuring air-change effectiveness.

ASTM E408: Standard of inspection-meter test methods for normal emittance of surfaces.

ASTM E903: Standard of integrated-sphered test method for solar absorptance, reflectance, and transmittance.

Comprehensive Environmental Response, Compensation, and Liability Act (CERCLA): Also known as Superfund, the CERCLA provides a Federal "Superfund" to clean up uncontrolled or abandoned hazardous-waste sites as well as accidents, spills, and other emergency releases of pollutants and contaminants into the environment. Through CERCLA, EPA was given the power to seek out those parties responsible for any release and ensure their cooperation in the cleanup.

EcoLogo: A standards-based labeling system in the Environmental Choice Program (ECP), Environment Canada's ecolabeling program, this program certifies a range of products, including hand cleaners, window and glass cleaners, boat and bilge cleaners, vehicle cleaners, degreasers, cooking appliance cleaners, cleaning products with low potential for environmental illness and endocrine disruption, bathroom cleaners, dish cleaners, carpet cleaners, and disinfectants.

Advanced Buildings: Energy Benchmark for High Performance Buildings (E-Benchmark): For Option A, install HVAC systems that comply with the efficiency requirements outlined in the New Buildings Institute, Inc.'s publication "Advanced Buildings: Energy Benchmark for High Performance Buildings (EBenchmark)" prescriptive criteria for mechanical equipment efficiency requirements, Sections 2.4 (less ASHRAE Standard 55), 2.5, and 2.6.

ANSI/ASHRAE 52.2–1999: Method of Testing General Ventilation Air-Cleaning Devices for Removal Efficiency by Particle Size. This standard presents methods for testing air cleaners for two performance characteristics: the ability of the device to remove particles from the air stream and the device's resistance to airflow. The minimum efficiency reporting value (MERV) is based on three composite average particle size removal efficiency (PSE) points. Consult the standard for a complete explanation of MERV value calculations.

ANSI/ASHRAE Standard 55–2004: Thermal Environmental Conditions for Human Occupancy. This standard specifies the combinations of indoor thermal environmental factors and personal factors that produce thermal environmental conditions acceptable to predicted percentage of the occupants within a defined space and provide methodology to be used for most applications, including naturally ventilated spaces.

ANSI/ASHRAE Standard 62.1–2004: Ventilation for Acceptable Indoor Air Quality American Society of Heating, Refrigerating and Air-Conditioning Engineers. This standard specifies minimum ventilation rates and indoor air quality (IAQ) levels to reduce the potential for adverse health effects. The standard specifies that mechanical or natural ventilation systems be designed to prevent uptake of contaminants, minimize the opportunity for growth and dissemination of microorganisms, and filter particulates, if necessary. Makeup air inlets should be located away from contaminant sources such as cooling towers; sanitary vents; and vehicular exhaust from parking garages, loading docks, and street traffic.

ANSI/ASHRAE/IESNA 90.1–1999: Energy Standard For Buildings Except Low-Rise Residential. On-site renewable or site-recovered energy that might be used to capture EA Credit 2 is handled as a special case in the modeling process. If either renewable or recovered energy is produced at the site, the ECB Method considers it free energy and it is not included in the Design Energy Cost. See the Calculation section for details.

ANSI/ASHRAE/IESNA 90.1-2004: Energy Standard for Buildings Except Low-Rise Residential Buildings. Standard 90.1-2004 was formulated by the American Society of Heating, Refrigerating and Air-Conditioning Engineers, Inc. (ASHRAE), under an American National Standards Institute (ANSI) consensus process. Standard 90.1-2004 establishes minimum requirements for the energy-efficient design of buildings, except low-rise residential buildings. Of note, the LEED v3 references ASHRAE 90.1-2007 was updated by *ASHRAE 90.1-2010* which is the latest updated version. The provisions of this standard do not apply to single-family houses, multifamily structures of three habitable stories or fewer above grade, manufactured houses (mobile and modular homes), buildings that do not use either electricity or fossil fuel, or equipment and portions of buildings systems that use energy primarily for industrial, manufacturing, or commercial processes. Building envelope requirements are provided for semiheated spaces, such as warehouses.

ANSI/ASTM-779-03: Standard Test Method for Determining Air Leakage Rate by Fan Pressurization. Acceptable sealing of residential units shall be demonstrated by blower door tests conducted in accordance with ANSI/ASTM-779-03, Standard Test Method for Determining Air Leakage Rate By Fan Pressurization, and using the progressive sampling methodology defined in Chapter 7, "Home Energy Rating Systems (HERS) Required

Verification And Diagnostic Testing," of *California Low Rise Residential Alternative Calculation Method Approval Manual.*

ASTM E1903-97 Phase II Environmental Site Assessment: This guide covers a framework for employing good commercial and customary practices in conducting a Phase II environmental site assessment of a parcel of commercial property. It covers the potential presence of a range of contaminants that are within the scope of CERCLA, as well as petroleum products.

ASTM Standard E1980-01: Standard Practice for Calculating Solar Reflectance Index of Horizontal and Low-Sloped Opaque Surfaces. This standard describes how surface reflectivity and emissivity are combined to calculate a Solar Reflectance Index (SRI) for a roofing material or other surface. The standard also describes a laboratory and field-testing protocol that can be used to determine SRI.

ASTM E408-71 (1996)e1: Standard Test Methods for Total Normal Emttance of Surfaces Using Inspection-Meter Techniques. This standard describes how to measure total normal emittance of surfaces using a portable inspection-meter instrument. The test methods are intended for large surfaces where nondestructive testing is required. See the standard for testing steps and a discussion of thermal emittance theory.

ASTM E903-96: Standard Test Method for Solar Absorptance, Reflectance, and Transmittance of Materials Using Integrating Spheres. Referenced in the ENERGY STAR roofing standard, this test method uses spectrophotometers and need only be applied for initial reflectance measurement. Methods of computing solar-weighted properties from the measured spectral values are specified. This test method is applicable to materials having both specular and diffuse optical properties. Except for transmitting sheet materials that are inhomogeneous, patterned, or corrugated, this test method is preferred over Test Method E1084. The ENERGY STAR roofing standard also allows the use of reflectometers to measure solar reflectance of roofing materials. See the roofing standard for more details.

California Title 24 2001: Though Title 24 is recognized as being more stringent for EA Prerequisite 2, for consistency, and fairness, projects in California must use Standard 90.1- 2007 in determining performance in EA Credit 1.1 (for the latest version, see http://www.bsc.ca.gov/Rulemaking/emergency/Emerg.aspx).

CIBSE Applications Manual 10–2005: Natural Ventilation in Non-Domestic Buildings. This manual sets out the various approaches to ventilation and cooling of buildings, summarizes the relative advantages and disadvantages of those approaches, and gives guidance on the overall approach to design. The

AM 10 (2005) provides detailed information on how to implement a decision to adopt natural ventilation, either as the sole servicing strategy for a building or as an element in a mixed-mode design.

Energy Policy Act (EPAct) of 1992: This Act was promulgated by the US government and addresses energy and water use in commercial, institutional, and residential facilities.

ENERGY STAR is a government-backed program administered by the US EPA and the DOE. Its goal is to protect the environment by identifying and promoting the use of energy efficient products and services. The program labels more than 35 product categories, including residential and light commercial HVAC equipment, as well as new homes and commercial buildings. For labeling commercial buildings, ENERGY STAR evaluates conformance to energy efficiency and indoor environmental standards, primarily ASHRAE and IES standards. It uses a statistical analysis data set (the DOE's Commercial Buildings Energy Consumption Survey) to compare energy intensity of similar buildings across the country. The blue ENERGY STAR label can make it easier to choose energy-efficient appliances, because it can only be placed on products that meet their strict energy efficiency requirements. ENERGY STAR appliances often cost more to buy than their less efficient counterparts, but cost less to operate because of the lower energy requirements, which also benefits the environment. Some utilities and state governments even offer financial incentives to use these energy-efficient appliances.

EPA ENERGY STAR Roofing Guidelines: In addition to several other building product categories, the ENERGY STAR program identifies roofing products that reduce the amount of air conditioning needed in buildings and can reduce energy bills. Roofing products with the ENERGY STAR logo meet the EPA criteria for reflectivity and reliability. Roofing products that meet ENERGY STAR criteria are a good starting point for achievement of this credit, but note that ENERGY STAR requirements are not as stringent as LEED credit requirements; LEED also accounts for good emissivity in the SRI calculation. An ENERGY STAR rating alone does not necessarily meet LEED credit requirements.

FTC Guides for the Use of Environmental Marketing Claims, 16 CFR 260.7(e): According to the guide, "A recycled content claim may be made only for materials that have been recovered or otherwise diverted from the solid waste stream, either during the manufacturing process (preconsumer), or after consumer use (postconsumer)." To the extent the source of recycled content includes preconsumer materials, the manufacturer or advertiser must be able to substantiate that the preconsumer material would otherwise have entered the solid waste stream. In asserting a recycled content claim, distinctions may be made between preconsumer and postconsumer materials. In such cases, any express or implied claim about the specific preconsumer or postconsumer content of a product or package must be substantiated.

Green Label: The CRI is a trade organization representing the carpet and rug industry. Green Label Plus is an independent testing program that identifies carpets with very low emissions of volatile organic compounds (VOCs). Carpet pad must meet or exceed CRI Green Label testing and product requirements.

Green Label Plus: The CRI is a trade organization representing the carpet and rug industry. Green Label Plus is an independent testing program that identifies carpets with very low emissions of VOCs. Carpet must meet or exceed CRI's Green Label Plus testing and product requirements. (Green Label Plus does not address backer or adhesive.)

Green-e Renewable Electricity Certification Program: The Green-e Program is a voluntary certification and verification program for green electricity products. Where the Green-e logo is exhibited on a product, it means that it is greener and cleaner than the average retail electricity product sold in that particular region. To be eligible for the Green-e logo, companies must meet certain threshold criteria for their products. Criteria include qualified sources of renewable energy content such as solar electric, wind, geothermal, biomass, and small or certified low-impact hydro facilities; "new" renewable energy content (to support new generation capacity); emissions criteria for the nonrenewable portion of the energy product; absence of nuclear power; and other criteria regarding renewable portfolio standards and block products. Criteria are often specific per state or region of the United States. Refer to the standard for further details.

Green Seal Standard GS-03: This is an independent nonprofit organization that promotes the manufacture and sale of environmentally friendly consumer products. GS-03 is a standard that sets VOC limits for anticorrosive and antirust paints.

Green Seal Standard GS-11: This is an independent nonprofit organization that promotes the manufacture and sale of environmentally friendly consumer products. GS-11 is a standard that sets VOC limits for commercial flat and nonflat paints.

Green Seal Standard 36 (GS-36): This is an independent nonprofit organization that promotes the manufacture and sale of environmentally friendly consumer products. GS-36 is a standard that sets VOC limits for commercial adhesives.

HERS: "Home Energy Rating Systems (HERS) Required Verification and Diagnostic Testing." California Low Rise Residential Alternative Calculation Method Approval Manual Acceptable sealing of residential units shall be demonstrated by blower door tests conducted in accordance with ANSI/ASTM-779-03, Standard Test Method for Determining Air Leakage Rate By Fan Pressurization, *and* using the progressive sampling methodology defined in Chapter 7 "Home Energy Rating Systems (HERS) Required Verification And Diagnostic Testing" of the *California Low Rise Residential Alternative Calculation Method Approval Manual.*

IESNA Recommended Practice Manual: Lighting for Exterior Environments (IESNA RP-33-99) Illuminating Engineering Society of North America. This standard provides general exterior lighting design guidance and acts as a link to other IESNA outdoor lighting Recommended Practices (RPs). IESNA RP documents address the lighting of different types of environments. RP-33 was developed to augment other RPs with subjects not otherwise covered and is especially helpful in the establishment of community lighting themes and SS WE EA MR EQ ID in defining appropriate light trespass limitations based on environmental area classifications. Light level recommendations in RP-33 are lower than in many other RPs, mainly because RP-33 was written to address environmentally sensitive lighting.

Management Measures for Sources of Non-Point Pollution in Coastal Waters, January 1993: This document discusses a variety of management practices that can be incorporated to remove pollutants from stormwater volumes. Chapter 4, Part II, addresses urban runoff and suggests a variety of strategies for treating and infiltrating stormwater volumes after construction is completed.

Natural Ventilation in Non-Domestic Buildings, A Guide for Designers, Developers and Owners (Good Practice Guide G237): The Good Practice Guide 237 is available for no charge, but registration (also free) is required to get access to the guide. Under the Energy section of the website, search for "natural ventilation" to find the Guide. The Good Practice Guide 237 is based on an earlier version of the CIBSE AM10.

Resource Conservation and Recovery Act (RCRA): Congress enacted RCRA to address the increasing problems the nation faced from its growing volume of municipal and industrial waste.

South Coast Rule #1113: South Coast Rule #1113 sets VOC limits for architectural coatings.

South Coast Rule #1168: The South Coast Rule #1168 VOC sets limits for adhesives.

US Energy Policy Act of 1992 (EPACT): Covers efficiency levels of general purpose industrial motors. The real intent of the law is to reduce the rate of energy consumption in the United States by requiring the use of energy-efficient products. To accomplish this goal, EPAct mandates that most industrial AC motors imported or produced for sale in the United States must meet energy-efficient requirements as defined by Table 12.10 from NEMA Standard MG 1.

US EPA "Compendium of Methods for the Determination of Air Pollutants in Indoor Air": Conduct baseline IAQ testing, after construction ends and prior to occupancy, using testing protocols consistent with the US EPA "Compendium of Methods for the Determination of Air Pollutants in Indoor Air," which is further detailed in the CI *Reference Guide*.

US EPA ETV: US EPA's Environmental Technology Verification (ETV). The protocol uses a climatically controlled test chamber in which the seating product or furniture assembly being tested is placed. A controlled quantity of conditioned air is drawn through the chamber with emission concentrations measured at set intervals over a 4-day period.

4.6 LEED EDUCATION AND TAKING THE LEED AP EXAM

The latest version of the LEED exam is LEED Version 4. LEED APs work in all sectors of the building industry and are recognized experts in green building practices and principles and the LEED Rating System. Eligibility requirements to take the LEED v4 exam include agreeing to disciplinary policy and credential maintenance guidelines. Demonstrate or document involvement in support of LEED projects, be employed in a sustainable field of work or engaged in an education program in green building principles and LEED, and submit to an application audit. Project experience will no longer be an eligibility requirement for those seeking LEED Accredited Professional (LEED AP) certification.

4.6.1 Preparing for the LEED AP Exam

The LEED AP exam, though not particularly difficult, does require a solid comprehension of the integrated approach to Green Building, in addition to the ability to recall a host of facts related to the LEED rating system. Eligibility standards, exam content, exam standards, fees, and guidelines are subject to change.

The issue of whether a project is able to receive the Innovation Credit for utilizing a LEED AP on a project using the LEED Green Associate credential requires clarification. The assumption is that this benefit will remain limited to Legacy LEED APs and LEED AP+'s only.

The LEED Professional Accreditation Exam has several sections. Each question on the exam will typically be from one of these sections. The exam sections include

1. Green Building Design and Construction Industry Knowledge
2. Knowledge of the LEED Rating System
3. LEED™ Resources and Processes
4. Understanding Integrative Green Design Strategies

It should be mentioned that LEED AP exam subject areas overlap somewhat, and some questions will fit into more than one area. There are several areas of competency to fully comprehend for the LEED AP Exam. These include

- Implement the LEED Process,
- Coordinate the Project and the Team,
- Knowledge of LEED™ Credit Intents and Requirements, and
- Able to Verify, Participate in, and Perform technical Analyses Required for LEED Credits.

4.6.2 New LEED Reference Guides

The new LEED reference guides have been consolidated into five books that address buildings by type and phase:

1. *LEED™ BD+C* spans three separate LEED rating systems including LEED™ for New Construction, LEED™ for Schools, and LEED for Core and Shell. It is well suited for professionals interested in the design and construction phases of sustainable building. This specialty covers not only new commercial and residential construction but also education and healthcare. "A revision was made to the Global Alternative Compliance Path supplement to the LEED 2009 Reference Guide for Green Building Design and Construction. For credits SSc7.1 Heat Island Effect–Non-Roof and SSc7.2 Heat Island Effect–Roof, this revision provides more options for credit compliance in locations where product-manufacturing data are not readily available. The guidance introduces new strategies to achieve the credit requirements by including more guidance related to in-place testing, lab testing, and using data from a previous project to achieve the credit. The guidance also includes an expanded list of resources for global teams to reference when pursuing these credits." (Green Building Design and Construction).

The USGBC says that The LEED 2009 and the October 2014 update of the Reference Guide for Green Building Design and

Construction is a user's manual that guides a LEED project from registration to certification of the design and construction of new or substantially renovated commercial or institutional buildings and high-rise residential buildings of all sizes, both public and private. The reference guide is a unique tool for all new construction projects, as it incorporates guidance for the three design and construction based LEED rating systems: LEED™ 2014 update for New Construction and Major Renovations, LEED 2014 update for Schools New Construction and Major Renovations and LEED 2014 update for Core and Shell Development. The USGBC says, "For each credit or prerequisite, the guide provides: intent, requirements, point values, environmental and economic issues, related credits, summary of reference standards, credit implementation discussion, timeline and team recommendations, calculation methods and formulas, documentation guidance, examples, operations and maintenance considerations, regional variations, resources, and definitions."

2. *LEED™ AP ID+C*: Green Interior Design and Construction credential suits candidates interested in interior design, remodeling, or tenant fit-out. The LEED ID+C focuses on design, construction, and improvement of interior and tenant spaces including retail and office. The LEED for Retail Commercial Interiors rating system is part of LEED ID+C, with emphasis on retail projects.

According to the USGBC, the LEED™ *Reference Guide for Green Interior Design and Construction* is a user's manual that guides a LEED project from registration to certification of high-performance building interiors. This guide is really designed to provide the necessary tools to promote sustainable choices to be made by tenants and designers, who may not have control over whole building operations. The LEED Reference Guide for Green Interior Design and Construction also includes detailed information on the process for achieving LEED certification, as well as detailed credit and prerequisite information, resources and standards for the LEED™ for Commercial Interiors Green Building Rating System. "For each credit or prerequisite, the guide provides: intent, requirements, point values, environmental and economic issues, related credits, summary of reference standards, credit implementation discussion, timeline and team recommendations, calculation methods and formulas, documentation guidance, examples, operations and maintenance considerations, regional variations, resources, and definitions."

3. *LEED™ AP O+M*: Green Building Operations and Maintenance (for rating systems that address LEED™ for Existing Buildings; LEED™ for Existing Schools). This is a fast growing area of opportunity reflecting the existing buildings sector. The increasing cost of energy combined with incentives for green improvements is fueling the shift to sustainable operations, management, and maintenance. The operations and maintenance category focuses on performance and is the only LEED rating system that requires recertification every 5 years. If you are a property manager/developer, on-site engineer, retrofitter, or sustainability consultant, the LEED AP O+M may prove to be appealing.

The latest LEED Reference Guide for Green Building Operations and Maintenance includes detailed information on the process for achieving LEED certification, detailed credit and prerequisite information, resources and standards for the LEED v4 for Existing Buildings: Operations and Maintenance Rating System. For each credit or prerequisite, the guide provides intent, requirements, point values, environmental and economic issues, related credits, summary of reference standards, credit implementation discussion, timeline and team recommendations, calculation methods, and independently of the hard copy.

The USGBC describes the Reference Guide for Green Building Operations and Maintenance as "a user's manual that guides a LEED™ project from registration to certification of a high-performance existing building. This guide is specifically designed to provide the tools necessary for building owners, managers and practitioners to maximize the operational efficiency of their buildings, while minimizing environmental impacts."

4. *LEED AP Homes*: If you are interested in residential construction, LEED for Homes may be the answer. The only LEED system to focus on highly efficient, healthy, and durable residential properties, LEED for Homes works with Green Raters to ensure that homes are built to the highest standards.

5. *LEED AP ND*: Urban planning is more important than ever. With increasing focus on reducing sprawl, the LEED for Neighborhood Development rating system emphasizes integrated planning, design, and development of sustainable, walkable communities connected to the surrounding habitats.

The new LEED v4 Reference Guides can be purchased from the USGBC website. It should be checked for prices and updated information.

4.6.3 Procedure for Taking the LEED AP Exam

All LEED exams are administered at a Prometric testing center. The following are the main points that the USGBC suggests you follow to taking the LEED AP exam:

- Register online to take the test through www.gbci.org (GBCI will ask you to pay the required fee) and schedule your test date and local test center location online utilizing the Prometric website (www.prometric. com/gbci).

- Arrive at the test center at least 30 min before your scheduled appointment to complete the required check-in process before testing begins. Make sure you have correct identification ready when you check in. Ensure that the name on your ID exactly matches the name you provided when you registered for the exam.

- You will not be permitted to take anything into the testing room, and you will be required to put all personal belongings (books, notes, etc.) in a locker provided at the test center.

- You will be given paper and pencil before you enter the exam for notes and calculations. Use these to write down important details that are fresh in your mind just before you start the computer-based exam.

- Upon being seated, you will have the option of being given 10 min to do a quick tutorial on how to use the testing software and program. Once you complete this tutorial, you can begin taking the exam.

- You will have exactly 2 h to complete the exam.

- The exam will consist of 100 multiple-choice questions, the majority of which will have multiple answers. The system allows you to skip by questions and/or to mark questions you want to come back to later. You can review ALL of your answers at the end, or you can do so at any time by pressing the review answers button. Unanswered questions will show an "I" for incomplete. "Marked" questions will be flagged with a red flag to make it easier to find them. You can call up any question to review your answer and make any changes you wish prior to completing the exam.

- The maximum passing score is 200 points. Minimum passing score is 170 points. All unanswered or incomplete exam questions will be scored as incorrect when time expires.

- Upon completing the exam, click "End Exam" and your score will appear on the computer screen. If you have not completed the exam and you are reviewing questions, do *not* click "Finish" as this will exit you from the exam and you will not be able to get back in.

- Unsuccessful candidates can schedule to retake the exam by repeating the previously described process. There is no limit to the number of times a failing candidate may retake the exam, but the full examination fee must be paid each time.
- Candidates who successfully pass the Tier 1 exam will be awarded LEED Green Associate certificates through US mail and listed as LEED Green Associate Professionals on the USGBC Web site. Candidates who successfully pass the Tier 2 exam will be awarded LEED Accredited Professional certificates through US mail and listed as LEED Accredited Professionals on the USGBC Web site.

4.6.3.1 Exam Review: The Three Tiers of LEED Credentials – LEED v4 (Figure 4.4)

Tier 1: LEED Green Associate – The exam for the LEED Green Associates covers *basic green building knowledge* and will not require the technical details of the individual rating systems (e.g., you are not required to know the difference between ASHRAE 55 and ASHRAE 62.1) nor the in-depth knowledge that it takes to actually get LEED buildings built. It will be one exam that is the same for all of the different LEED tracks (Figure 4.4).

In terms of the LEED Green Associate exam, Beth Holst, Vice President of Credentialing for the GBCI, states that the exam will focus on the green

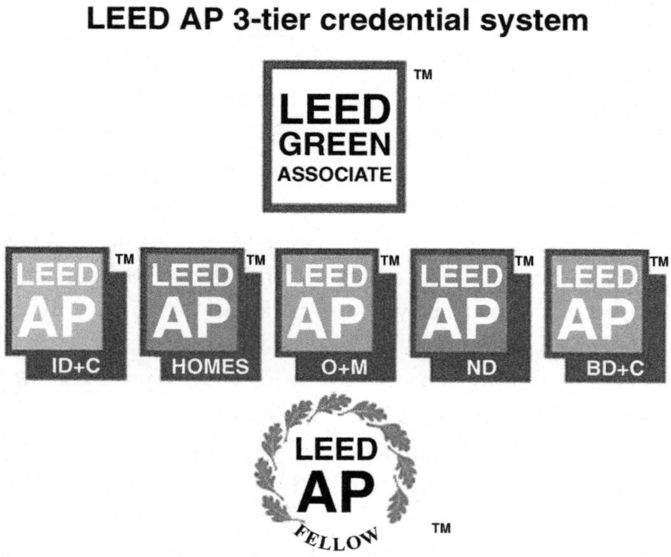

LEED AP 3-tier credential system

Figure 4.4 *Diagram Showing the New Three-Tier LEED™ Credential System.*

building principles and information that are consistent and present across all of the various LEED specialties, such as New Construction, Commercial Interiors, Existing Buildings Operations and Maintenance, etc. Although the LEED Green Associate exam is similar in principle and length to the existing LEED AP exam, it will reportedly better test a candidate's general knowledge of good environmental practices and his or her understanding of green building design, construction and operations.

Ms. Holst indicated that the *LEED Green Associate* credential will provide a level of distinction for those who are looking to establish basic green building knowledge while working to implement the processes and principles of the LEED certification system and is ideal for students, teachers, marketers, manufacturers, program managers, and anyone who has ever wanted to understand what LEED is and why it is important to the green building industry without actually needing to know all the technical details related to its implementation.

Also, to be eligible, a candidate is required to agree to disciplinary policy and credential maintenance guidelines.

Tier 2: Specialized Credential System – The LEED Accredited Professional. Possessing a LEED Accredited professional credential signifies that you have an in-depth knowledge of sustainability and sustainable development and an active participant in today's green building movement.

This tier builds on Green Associate exam and tests credit knowledge of specific credit details and regulations. It gets into the in-depth knowledge of the rating system for the area that you work in most. The LEED AP test is more appropriate for practitioners actively working on LEED projects at the certification level.

Here is a picture of the LEED rating systems specialties under LEED™ 2015:

- LEED™ AP BD+C: Building Design + Construction
- LEED™ AP ID+C: Interior Design + Construction
- LEED™ AP Homes: Homes
- LEED™ AP O+M: Building Operations + Maintenance
- LEED™ AP ND: Neighborhood Development

Tier 3: LEED™ AP Fellow – The LEED AP Fellow distinction is expected to encompass an "elite class" of leading professionals who are distinguished by their years of experience and who have demonstrated a great depth of work in green building. Nominees for Fellow must be/have:

- A LEED AP with specialty credential in good standing
- Ten years of professional green building experience

- Held a LEED Professional credential for 8 or more years
- Can demonstrate significant contributions to advancing the movement; Made an exceptional impact in key mastery areas

Existing LEED APs, (Legacy LEED Aps) may opt-in or upgrade to the new LEED AP+ by enrolling in the new-tiered system. A current LEED AP also has the option to do nothing in which case he/she will continue to be designated a LEED AP without a specialty title (such as O&M, BD&C, ID&C) in the LEED Professional Directory. Of note, the "Legacy LEED AP" title has recently been replaced by the "LEED AP without Specialty" title for this group of professionals.

4.6.4 Tips on Passing the Exam

Success really does require a detailed study of the reference manual and the USGBC/GBCI website in addition to this Handbook. Pay particular attention to the introductory sections on green building design/construction and the introductions to each individual credit area (Introductory pages to ID, LT, SS, WE, EA, MR, EQ, IN, RP). Many questions are drawn from these areas.

- For the exam, the candidate should understand:
  - How to register a project
  - LEED implementation process
  - The steps that are involved in the certification process
  - The Credit Interpretation Process
  - The Appeal Process
  - Stakeholder involvement in innovation – who is (and is not) part of a project team?
- Know and memorize which items are prerequisites and which are credits. There are usually several questions that will ask you to differentiate.
- Know the various industry publications (Green Spec, Sustainable Industries, Building Green, Environmental Design and Construction, etc.)
- Fully understand each of the opportunities for exemplary performance, innovation and design, and regional priority credit thresholds and percentages.
- Know which credits have calculations and ensure that you know how to perform them – especially credits dealing with FTE or number of plumbing fixtures.
- Know all of the SMACNA Standards and what LEED points they relate to.
- Practice tests and, then, more practice tests! Sample practice tests can be found on various Internet sites. A sample practice test can be found in the exhibits of this handbook.

- Know all of the sample questions on USGBC's study guide (website). There are often a number of these on the test nearly word for word.
- Many questions focus on the Energy and Atmosphere credits, especially strategies to optimize energy performance.
- Understand the 3 types of renewable energy discussed in the reference guide.
- Understand the differences between the Fundamental Commissioning prerequisite and the Enhanced Commissioning credit.
- Understand the differences between the IAQ prerequisite and the IAQ credit.
- Toilets use the most water of any fixture in commercial buildings.
- Know who the commissioning agent should be for Fundamental Commissioning and for Enhanced Commissioning.
- Know that for Brownfields, the letter must be from the local regulatory agency or the regional EPA office.
- There are 2 free credit interpretations. Know how to apply for an CIR interpretation.
- Green Seal is the standard by which paints are measured in VOCs. Know the difference between GS-36 (commercial adhesives), GS-11(paints and coatings), and GS-3 (anticorrosive paints).
- Make sure you scan through all of the "Summary or Referenced Standards." There are a couple of questions that ask which of these is not a reference standard.
- Know the definitions for emmissivity and albedo as well as how they are used within the different credits (heat islands, energy efficiency).
- Know the definition and uses of the composting toilet, waterless urinals, low flow fixtures, etc.
- Rapidly Renewable Resources: know definition and the various materials.
- Understand the difference between Stormwater Management, Erosion and Sedimentation Control, and Wastewater Technologies.
- ASHRAE 55 and ASHRAE 62: Be very familiar with where they apply.
- Try and memorize each prerequisite and credit number, name, intent, requirements, and any referenced standards, strategies, exemplary performance possibilities, point value, submittals, and the decision makers.
- Some of the test questions may unsettle you. Don't let them; just skip the ones you don't think you know and come back to them later. Concentrate first on completing the questions you know. Mark any questions you are not sure of and return to them toward the end. There should be plenty of time to review the questions.

CHAPTER 5

Design Strategies and the Green Design Process

5.1 OVERVIEW: CONVENTIONAL VERSUS GREEN DELIVERY SYSTEMS

Over the last few decades, *global* climate change has become increasingly apparent, and most climate experts have come to the conclusion that humans are the major contributors to this development. Although most of our time is spent inside buildings, we nevertheless seem to take for granted the shelter, protection, comfort, air, and light that buildings provide, and rarely do we give much thought to the systems that allow us to enjoy these services unless there's a power interruption or other problem. Moreover, few Americans understand the environmental consequences of maintaining indoor comfort levels. This may be partly because buildings are often deceptively complex. While connecting us with the past, they reflect our greatest legacy for the future. Buildings also provide us with shelter to live and work, embody our culture, and play a pivotal part in our daily life on this earth. But the role of buildings continues to change as they become increasingly costly to build and maintain, as well as requiring constant adjustment to function effectively over their life cycle. President Obama clearly understood this and the important role of the Federal Government when he said, "As the largest consumer of energy in the US economy, the federal government can and should lead by example when it comes to creating innovative ways to reduce greenhouse gas emissions, increase energy efficiency, conserve water, reduce waste, and use environmentally responsible products and technologies."

When public facilities are designed, built, and operated using green building principles, we find that this encourages designers and contractors in the private sector to also design, build green, and implement sustainable building practices. Additionally, the US military was an early innovator in green building and today reportedly has more square feet of green building than anyone else. The US Air Force has recently made it a requirement that all construction be capable of LEED™ (Leadership in Energy and Environmental Design) Silver certification.

LEED v4 Practices, Certification, and Accreditation Handbook
http://dx.doi.org/10.1016/B978-0-12-803830-7.00005-0
Copyright © 2016 Elsevier Inc.
All rights reserved.

Sustainable or green building probably means different things to different people. Generally, it means smart construction and design; it is smart for the consumer and smart for Earth. This is particularly relevant today with the supply of the world's fossil fuel rapidly dwindling, the concerns for energy security increasing, and the impact of greenhouse gases on our climate rising. There is a dire need to find ways to reduce energy loads, increase efficiency, and employ renewable energy resources in our facilities. Green construction is environmentally friendly because it uses sustainable, location-appropriate building materials and employs building techniques that reduce energy consumption. Indeed, the primary objectives of sustainable design are to avoid resource depletion of energy, water, and raw materials and prevent environmental degradation caused by facilities and infrastructure throughout their life cycle. Likewise, it places a high priority on health, which is why green buildings are also designed and operated to create healthier and more productive working and learning environment. And through the use of natural light and other green features, green buildings also help create safer, more comfortable, and improved living environments. Because sustainable buildings use resources like energy, water, materials, and land much more efficiently than typical buildings, sustainable buildings are more cost-effective, saving taxpayers' money by reducing operations and maintenance costs.

In addition to the classic building design concerns of economy, utility, durability, and delight, green design strategies underline emerging concerns regarding occupant health, the environment, and resource depletion. There are many green design strategies and measures that can be employed to help address these concerns, including the following:

- Reduce human exposure to hazardous materials. Choose nontoxic, nonregulated materials. Where toxic materials might be required, have them generated on-site in preference to having them made elsewhere and shipped.
- Maximize the use of renewable energy and materials that are sustainably harvested.
- When possible, choose natural materials rather than synthetic.
- Conserve nonrenewable energy and scarce materials.
- Use water efficiently and minimize wastewater and runoff.
- Use energy as efficiently as possible and minimize the ecological impact of energy and materials used.
- Protect and restore local air, water, soils, flora, and fauna.

- Encourage occupants to use bicycles, mass transit, and other alternatives to fossil-fueled vehicles.
- Optimize site selection in order to preserve green space and minimize transportation impacts.
- Building orientation should take maximum advantage of sunlight and microclimate.
- Minimize materials impacts by employing green products.

The majority of conventionally designed and constructed buildings existing today are having negative impacts on the environment as well as on occupant health and productivity. Moreover, these buildings are expensive to operate and maintain, and they contribute to excessive resource consumption, waste generation, and pollution. Reducing negative impacts on our environment and establishing new environmentally friendly goals and adopting guidelines that facilitate the development of green/sustainable buildings such as outlined by the US Green Building Council (USGBC) and similar organizations must be a priority for our generation. USGBC's LEED™ guidelines include required and recommended practices that are intended to reduce life-cycle environmental impacts associated with the construction and operation of both commercial and municipal developments and major remodel projects.

Green/sustainable construction is rapidly entering the mainstream in the United States. Moreover, the adoption of building green is an important factor in helping to reduce the carbon footprint. Facilitating this is the rising costs of energy and building materials, coupled with warnings from the EPA about the toxicity of today's treated and human-made materials, which have prompted engineers and architects to take a new look at building techniques that use native resources as construction materials and utilize natural phenomena in the heating and cooling process. The majority of green buildings are efficient and high-quality; they last longer and cost less to operate and maintain. They also provide greater occupant satisfaction than conventional developments, which is why most sophisticated buyers and lessors prefer them and would consider paying a premium for these benefits.

A true green building takes advantage of the Sun's seasonal position to heat a building's interior in winter and often incorporates design features such as light shelves, overhanging eaves, or landscaping to mitigate the Sun's heat in summer. Room orientation should be considered to assist in improving ventilation. Green building construction takes into account renewable, recycled, and native building materials to minimize the adverse impact on the environment and to harmonize with its surroundings. Naturally insulative materials such as earth and straw are often employed in the construct of walls. Some green buildings also incorporate solar, wind, or other alternative energy into the heating, ventilation, and air-conditioning (HVAC) system to reduce operational costs and minimize the use of fossil fuels. Green buildings also generally avoid using potentially toxic materials such as treated woods, plastics, and petroleum-based adhesives, which can degrade air and water quality and cause health problems.

Most of today's buildings use mechanical equipment powered by electricity or fossil fuels for heating, cooling, lighting, and maintaining indoor air quality. According to the US Department of Energy (DOE), buildings in the United States annually consume more than one-third of the nation's energy and contribute roughly 36% of the carbon dioxide (CO_2) emissions released into the atmosphere. The fossil fuels used to generate electricity and condition buildings pay a heavy toll on the environment; they emit a plethora of hazardous pollutants (e.g., volatile organic compounds) that cost building occupants and insurance companies millions of dollars annually in health care costs. In fact, nearly 85% of greenhouse gas emissions in the United States today comprise CO_2, much of which is instigated by the buildings and land use patterns that local regulations have in place. Mining and extraction of fossil fuels adds to the environmental impacts, in addition to the instability in pricing, which causes concern among both businesspeople and homeowners. The need for creating buildings that use less energy both reduces and stabilizes costs, as well as having a positive impact on the environment. The incorporation of operating and maintenance factors into the design process of a building project will contribute significantly to healthy working environments, higher productivity, and reduced energy and resource costs. Whenever possible, designers should always specify "green" materials and systems that simplify and reduce maintenance and life-cycle costs, use less water and energy, and are cost-effective. It is no secret that green/sustainable buildings provide much better value when compared with conventional buildings and systems.

The DOE had the foresight to realize early on the need for buildings that were more energy efficient, and in 1998 took the initiative and decided to work with the commercial buildings industry to develop a 20-year plan for research and development on energy-efficient commercial buildings. The primary mission of DOE's High-Performance Buildings Program is to help create better buildings that save energy and provide a quality, comfortable environment for workers. The principal target of the program is the building community, especially building owners/developers, engineers, and architects. Today, we have the knowledge and technologies required to reduce energy use in our homes and workplaces without compromising comfort and aesthetics. Unfortunately, however, the building industry is not taking full advantage of these advances, and to this day, many buildings are still being designed and operated without considering all the environmental impacts.

High Performance: An especially cost-effective approach to high-performance design is to equip designers with the right tools – particularly architects and designers working at the early stages of design. Of particular importance is careful consideration of building performance, which is a valuable prerequisite for designers. High-performance design techniques can be employed to create fundamentally better buildings and at the same time reduce the capital needed to build them. Once performance measures are selected, it is necessary to follow up and establish performance targets and the metrics to be used for each measure. Minimum expectations, or baselines, are generally defined by codes and standards. Otherwise, performance baselines can be set to exceed the average performance of a building type. Alternatively, performance can be measured against the most recent building built or against the performance of the best building of a particular type that can be found documented.

Several important green building rating systems have been developed over the years to set standards for the evaluation of high performance. The most widely recognized system for rating building performance in the United States is LEED, which provides various consensus-based criteria to measure performance, along with useful reference to baseline standards and performance criteria. The latest LEED version is LEED™ v4. Nevertheless, LEED™ certification, by itself, does not ensure high performance in terms of energy efficiency as certification may have been achieved by acquiring other nonenergy-related categories such as Materials and Resources or Sustainable Sites. For this reason, specific energy-related goals must still be set.

High-performance building projects are more effectively initiated with a green design "charrette" or multidisciplinary kick-off meeting to formulate a clear road map for the project team to follow. A green design *charrette* is useful because it allows team professionals, often with the assistance of green design experts and facilitators, to brainstorm design goals and alternatives. This goal-setting approach helps identify green strategies for members of the design team and helps facilitate the group's ability to reach a consensus on performance targets for the project.

Integrated design is enhanced by the use of computer energy modeling tools such as the Department of Energy's DOE 2.1E and other computer programs that can inform the building team of energy-use implications very early in the design process by factoring in relevant information, such as climate data, seasonal changes, building massing and orientation, and daylighting. It facilitates prompt exploration of cost-effective design alternatives for the building envelope and mechanical systems by forecasting energy use of various alternatives in combination.

Building Information Modeling (BIM): As the world of buildings continues to change and grow in complexity, additional programs and information are having a major impact on the entire design, planning and construction community. Among them is the introduction of BIM software, which is a recent trend in computer-aided design and which is being touted by many industry professionals as a lifesaver for complicated projects because of its ability to correct errors at the design stage and accurately schedule construction. BIM consists mainly of 3D modeling concepts in addition to information database technology and interoperable software in a desktop computer environment that architects, engineers, and contractors can use to design a facility and simulate construction. With BIM technology, therefore, an accurate virtual model of a building can be digitally constructed and allows sharing that information with each other. It also helps architects, engineers, and developers visualize what is to be built in a simulated environment and to detect any potential design, construction, or operational issues and to resolve them in a virtual environment well before the construction phase in the real world. BIM is gradually changing the role of drawings for the construction process, improving architectural productivity, and making it easier to consider and evaluate design alternatives. BIM, therefore, represents a new paradigm within AEC, one that encourages integration of the roles of all stakeholders on a project. Moreover, some industry professionals forecast that buildings will soon be built directly from the electronic models that BIM creates, and that the design role of architects and engineers will dramatically change (Figure 5.1a,b). BIM is also helping to facilitate the process of integrating the various design teams' work, thus adding to the urgency of utilizing an integrated team process.

BIM consists of 3D modeling concepts, information database technology, and interoperable software in a computer application environment that architects, engineers, and contractors can use to design a facility and simulate construction. This modeling technology enables members of the project team to create a virtual model of the structure and all of its systems

Figure 5.1 (a) Highlands Lodge Resort and Spa project, a joint venture of Q&D Construction and Swinerton Builders, Inc. where Vico, a BIM software package was used. The five-star hotel and high-end luxury condominiums have a total gross floor area of 406,500 ft.² (37,720 ft.²) and are built on a roughly 20-acre site at the Northstar-at-Tahoe Ski Resort in Northern California. (b) Miami Science Museum, designed by designed by Grimshaw Architects uses Building Information Modeling (BIM). The Miami Science Museum carefully integrates the direction and location of the sun into its design – engaging in both passive and active solar design strategies. *Source: Part a, Vico Software Inc. Part b, Jill Fehrenbacher.*

in 3D in a format that can be shared with the entire project team. All drawings, specifications, and construction details are integral to the model, which incorporates building geometry, spatial relationships, geographic information, and quantity properties of building components. This allows team members to identify design issues/construction conflicts and resolve them in a virtual environment long before construction actually commences.

As BIM increases in popularity, it is rapidly becoming the foundation of building design, visualization studies, contract documents, cost analysis, 3D simulation, and facilities management. Autodesk Revit® is making considerable headway in its market penetration into architectural firms, and it is anticipated that within the next few years the majority of major projects designed in the United States will be modeled with BIM technology.

5.2 GREEN DESIGN STRATEGIES

Building construction and operations can have significant direct and indirect impacts on the environment, society, and economy. Clear design objectives are therefore critical to any project's success, yet to achieve maximum benefit, the project goals must be identified early on and held in proper balance during the design process. In addition to including sustainable design concepts in new construction, sustainable design advocates very often encourage retrofitting existing buildings when possible rather than building anew. But whether retrofitting or building anew, an integrated design approach to design and construction must be employed to achieve a high-performance building. By working together, for example, the civil engineer, landscape architect, and design architect can maneuver and direct the ground plane, building shape, section, and planting scheme to provide increased thermal protection and reduce wind loads and heat loss/gain. This allows them to reduce heating and cooling loads and, with the participation of the mechanical engineer, reduce the size of mechanical equipment necessary to achieve comfort. The architect, lighting, and mechanical engineers can work together to design, for example, an effective interior/exterior light-shelf that can not only serve as an architectural feature but also provide sunscreening, reducing summer cooling load while letting daylight penetrate deep into the interior. The resultant outcome is almost always more efficient environmental performance at the equivalent first cost, followed by a stream of ongoing operational savings. Should first cost exceed initial estimates, it should be easily offset by decreased life-cycle dollar and environmental costs.

Another feature of green design is "Smart Growth," which is an issue that concerns many communities around the country. It relates to the ability to control sprawl, reusing existing infrastructure, and creating walkable neighborhoods. The locating of places to live and work near public transportation is an obvious plus toward reducing energy. Likewise, it is more resource-efficient and logical to reuse existing roads and utilities than build new ones at considerable distances from cities in rural areas. Smart growth is needed to preserve open spaces, farmlands, and undeveloped land, and strengthens the evolution of existing communities and their quality of life.

However, before dwelling too deeply into green design and the integrated design process (IDP) and to understand what it is all about, it may be prudent to first describe the more conventional design process. The traditional approach to design, simply put, usually begins with the architect and the client agreeing on a budget and design concept, consisting of a general massing scheme, typical floor plan, schematic elevations and, usually, the general exterior appearance as determined by these characteristics as well as by basic materials. The mechanical and electrical engineers are then asked to implement the design and to suggest appropriate systems.

This conventional design approach, though greatly oversimplified, in fact continues to be employed by many if not the majority of general-purpose design consultant firms, which in turn unfortunately tends to limit the achievable performance to conventional levels. The traditional design process consists mainly of a linear structure because of the sequential contributions of the members of the design team. The opportunity for optimization is limited in the traditional process, while optimization in the later stages of the process is often difficult if at all viable. Thus, although this process may seem adequate and even appropriate, the actual results often tell a different story, with high operating costs and a substandard interior environment. These factors can have a significant impact on a property's ability to attract quality tenants or achieve desirable long-term rentals in addition to a reduced asset value of the property. Moreover, because conventional design methods do not typically use computer simulations of predicted energy performance, the consequential dismal performance and steep operating costs will often come as an unpleasant surprise to the owners, operators, or users.

5.2.1 The Integrated Design Process

Practitioners of an integrated process are required to develop new skills that may not have been required as part of their previous professional work

Figure 5.2 *Diagram Showing the Various Elements That Impact the Design of High-Performance Buildings Using the Integrated Design Approach.*

because the integrated process requires a different way of thinking and working. The IDP approach differs from the conventional design process in a number of ways. For example, the owner takes on a more direct role in the process and the architect takes on the role of a team leader rather than the sole decision maker. Other consultants like the structural, electrical, mechanical, lighting, and other relevant players become part of the team from the outset and participate in the project's decision-making process, and not after completion of the initial design (Figure 5.2). Moreover, the IDP requires an increase of time and collaboration during the early conceptual and design phases than for conventional practices. Tiffany Coyle of the USGBC says, "An integrated process is highly collaborative. This approach requires the whole project team to think of the entire building and all of its systems together, emphasizing connections and improving communication among professionals and stakeholders throughout the life of a project. It breaks down disciplinary boundaries and rejects linear planning and design processes that can lead to inefficient solutions. Although the term integrated design is most often applied to new construction or renovations, an integrated process is applicable to any phase in the life cycle of a building." Coyle also says that "time must be spent building the team, setting goals, and doing analysis before any decisions are made or implemented. This upfront investment of time, however, reduces the time it takes to produce construction documents. Because the goals have been thoroughly explored and woven throughout the process, projects can be executed more thoughtfully, take advantage of building system synergies, and better meet the needs of their occupants or communities, and ultimately save money, too."

To ensure success of any building project, the following strategies should be carefully considered:

- Establish a vision that embraces sustainable principles in order to apply an integrated design approach to a project.
- Articulate a clear statement of the project's requirements, objectives, design criteria, and priorities.
- Develop a budget for the project that encompasses green building measures. Allocate contingencies for possible incorporation of additional options.
- Establish performance goals for siting, energy, water, materials, and indoor environmental quality along with other sustainable design goals, and ensure the incorporation of these goals throughout the design and life cycle of the building.
- Incorporate BIM concepts to enable members of the project team to create a virtual model of the project and its various systems in 3D in a format that can be shared with the entire project team.
- Form a design and construction team that is committed to the project vision. During the selection process, ensure that contractors are appropriately qualified and capable of implementing an integrated system of green building measures.
- Include in the contract documents a project schedule that allows for systems testing and commissioning.
- Develop contract plans and specifications that ensure that the building design is at an appropriate level of building performance.
- Augment a building management plan that ensures that operating decisions and tenant education are implemented in a satisfactory manner with regard to integrated, sustainable building operations and maintenance.
- Consider all stages of the building's life cycle, including deconstruction.

Depending on a project's size, type, and complexity, the decision to appoint a leader for the project usually rests with the user/client by identifying the need for a person that is proficient and able to lead a team to design and build the project on the basis of quantifiable requirements for space and budgetary capacity to undertake the activity. A project brief should accompany this planning activity that describes existing space use; develop realistic estimates of both spatial and technical requirements, and arrive at a space program around which design activity can develop. In the case of larger projects, engaging a construction manager or a general contractor at this point may be appropriate.

Once the *Predesign* activities are completed, the architect, designer of record (DOR), other prime consultants, in collaboration with the other team members or subconsultants, may produce initial graphic suggestions for the project or portions of it via a 3D modeling program (e.g., BIM) or manually. Such suggestions are meant more to stimulate thought and discussion than to describe the final outcome, although normally the fewer the changes made before bidding the project, the more cost effective will the project be. Involvement of subconsultants is a critical part of the process at this stage, and their individual insights can prevent costly changes further along in the process. A final design gradually emerges that embodies the interests and requirements of all participants, including the owner, while also meeting the overall area requirements and project budget that was established during *predesign* stage.

The resulting *schematic design* produced at this stage should show site location and organization, a 3D model of the project, space allocation, and an outline specification including an initial list of components and systems to be designed and/or specified for the final result. A preliminary cost estimate should also be made, and depending on the size of the project, it may be performed by a professional cost estimator or computer program at this point. For smaller projects, the DOD or one or more possible builders may perform this service as part of a preliminary bidding arrangement. On larger projects, a cost estimate can be linked to the selection process for a builder, assuming other prerequisites like bonding capacity, experience, and satisfactory references are met.

The *design development* phase entails going into greater detail for all aspects of the building, including systems, materials, etc. The integrated collaborative process that is in place continues with the architect working hand in hand with the owner and the various contributors. The conclusion of this phase is a detailed design on which a consensus exists of all players and who may be asked to sign off. When the design is developed in a spirit of collaboration among the main contributors, the end result is usually a design that is highly efficient with minimal, if not zero, incremental capital costs, and reduced long-term operating and maintenance costs.

This is followed by the development and production of *contract documents*, which involves converting the design development data into formats that can be used for pricing and bidding, permitting, and construction. Although no set of construction documents can ever be perfect, an efficient set can be achieved by careful scrutiny and accountability to the initial program needs as outlined by the design team and the client, in addition to careful coordination and collaboration with the technical consultants on the design team. Design, budgetary, and other decisions continue to be made with the appropriate contributions of the various players. Any changes in scope during this phase can significantly impact the project, and once pricing has begun can invite confusion, errors, and added costs. Cost estimates may be made at this point, prior to or simultaneous with bidding, in order to ensure compliance with the budget and to check the bids. Accepted bids at this point may be used as a basis for selecting a contractor.

On selection of the general contractor and during the *construction phase*, the DOD and other members of the project team must remain fully involved. Previous decisions may require clarification, supplier samples and information must be reviewed for compliance with the contract documents, and proposed substitutions need evaluation. Whenever proposed changes affect the operation of the building, it is imperative to inform and involve the user/client. Any changes in user requirements may necessitate changes in the building; such changes will require broad consultation among the consultants and subconsultants, pricing, as well as incorporation into the contract documents and the building.

The design team is ultimately responsible for ensuring that upon the building's completion it meets the requirements of the contract documents. The building's success or otherwise of meeting the program performance requirements can be assessed by an independent third party in a process known as commissioning where the full range of functions in the building are evaluated and the design and construction team can be called upon to make any required changes and adjustments. This is discussed in greater detail in Section 5.5.

Sometimes a *postoccupancy evaluation* is conducted after the building is fully operational, basically to assess how well the building meets the original and emerging requirements for its use. Such information is particularly useful if the owner/developer anticipates further construction of the same type by avoiding mistakes and repeating successes.

Because buildings consist of a number of complex systems of interacting elements, careful combinations of design strategies are necessary to produce a successful and sustainable project. Intelligent green design takes into account the effects of the various elements on each other, and on the building as a whole. For example, the need for mechanical and electrical systems is greatly influenced by a building's form, orientation, and the design of its envelope. Allied strategies such as daylighting, natural cooling, solar load control, and natural ventilation can all work in unity to reduce lighting, heating, and cooling loads as well as the cost of necessary equipment to meet them.

It has been noted that when buildings are designed to adapt to changing uses over long periods of time, the life-cycle resource consumption is reduced. Adaptable structural elements that provide generous service space and accommodate open plans with nonload-bearing partitions can last centuries and easily outlast buildings that cannot adapt to changing times and functions. Durable well-designed envelope assemblies reduce life-cycle maintenance and energy costs while improving indoor air quality and comfort. Mechanical and electrical systems that are readily accessible and which allow easy modifications save materials and money whenever tenant improvements or renovations are needed. This is why durability and use remain valuable criteria in evaluating aspects of sustainability and green design. Also preventing problems from the outset instead of fixing them after the fact always makes practical and economic sense. Thus, the use of low-toxicity building materials and installation practices is much more effective than diluting indoor air pollution from toxic sources by the use of increased air ventilation. Likewise, using efficient designs that minimize heating, cooling, and lighting loads is far more cost-effective than installing bulky or larger HVAC and electrical equipment to address the problem.

Climate-responsive design has come of age as it rediscovers the powerful relationship of buildings to place. Buildings that respond to local topography, microclimate, vegetation, and water resources are typically better suited to meet today's challenges; they are more comfortable and efficient than conventional designs that ignore their surroundings by relying on technological fixes to address the various environmental and other issues that continuously crop up. Many regions have excellent solar and wind resources for passive solar heating, natural cooling, ventilation, and daylighting. Local water supplies on the other hand are rapidly being depleted or polluted. Taking advantage of available natural resources, and conserving commodities that are being rapidly depleted are two of the best ways to reduce costs and connect occupants to their surroundings.

5.2.2 The Green Building Project Delivery Process

For conventional (nongreen) buildings, the various specialties associated with project delivery, from design and construction through building occupancy, are responsive in nature, and utilizing restricted approaches to address particular problems. Each of these specialties typically has wide-ranging knowledge in their specific fields, and they seek solutions to any problems that arise by solely using their knowledge of their specific fields. For example, an air-conditioning specialist when asked to address a problem of an unduly warm room will immediately suggest increasing the cooling capacity of the HVAC system servicing that room, rather than investigate the source of the problem of why this room is unduly warm. The excessive heat gain could, for example, be due to incorporating external louvers or operable windows. The end result is often a functional but highly inefficient building comprising different materials and systems with little integration between them.

From a design perspective, the key process difference between green building design and conventional design is the concept of integration. Green buildings typically use an IDP, which uses a multidisciplinary team of building professionals who typically collectively work together from the predesign phase through the postoccupancy phase. The delivery team would therefore consist of those professionals that were involved in the programming, planning, design, construction, and subcontractor roles for the project. It goes without saying that it is crucial to assemble the correct team in order to produce a superior, cost-effective, environmentally friendly, high-performance building while minimizing risk and loss. It is also important to understand that the essence of integrated design is that buildings consist of interconnected or interdependent systems, all of which impact each other to some degree. Also, if you are planning a project that is intended to target LEED certification, selecting the right delivery method can have a significant impact on your ability to achieve your LEED objectives as well as on the long-term operating costs for your building.

Properly engineered systems will help ensure the comfort and safety of all building occupants. They also enable designers to create environments that are healthy, efficient, and cost-effective. Integrated design is a critical and consistent component in the design and construction of green buildings. The brief description highlights the benefits of integrated design and highlights the main differences between conventional and integrated design.

5.2.3 The Integrated, Multidisciplinary Project Team

Some of the professions required to join the Project Team in this type of process may include the following:

The *owner's representative* (OR) is the person that represents the owner and speaks on his/her behalf. The OR must be prepared to devote the time needed to be able to advocate, defend, clarify, and develop the owner's interests. The OR may come from within the organization commissioning the project or may be hired by the owner as a consultant.

The *construction manager* (CM) is normally hired on a fee basis to represent the logistics and costs of the construction process and can be an architect, a general contractor, or specifically a consulting construction manager. It is important for this person to be involved from the beginning of the project.

In most building projects, it is the architect who typically leads the design team and who coordinates with subconsultants, other experts, etc., and ensures compliance with the project brief and budget. In some cases, the architect has the authority to hire some or all of the subconsultants; in larger projects, the owner may decide to contract directly with some or all of them. The architect usually manages the production of the contract documents and oversees the construction phase of the project, ensuring compliance with the contract documents by conducting appropriate inspections, and managing submissions approvals, and evaluations by the subconsultants. The architect also oversees the evaluation of requests for payment by the builder and other professionals.

Early involvement of the *civil engineer* is essential for understanding the land, soil, and regulatory aspects of a construction project; the civil engineer may be hired directly by the owner or the architect. The role of the civil engineer usually includes preparation of the civil engineering section of the contract documents in addition to assessing compliance of the work with the contract documents.

Consulting structural, mechanical, and electrical engineers may be engaged by the architect as part of his work or sometimes on larger or more complex projects, may be engaged directly by the owner. They are responsible for designing the structural, heating, ventilation, and air-conditioning and the power, and illumination aspects of the project. Each produces his or her own portions of the contract documents and all participate in assessing their part of the work for compliance with those documents.

The *landscape architect* depending on the type and size of the project is often hired as an independent consultant. He/she should be involved early in the design process to assess existing natural systems, how they will be impacted by the project, and the best ways to accommodate the project to those systems. The landscape architect should have extensive experience in sustainable landscaping, including erosion control, green roofs, indigenous plant species, managing stormwater runoff, etc.

Specialized consultants are adopted into the design team as needed by the special requirements of the project. These may include lighting consultants, specifications writers, materials and component specialists, sustainability consultants, and technical specialists such as audiovisual, materials handling, and parking. The size, complexity, and specialization of the project will suggest the kinds of any additional experts that may be needed. As with all contributors to the IDP, they should be involved early in the design process to include their suggestions and requirements in the design so as to ensure that their contributions are taken into account and do not become remedial.

5.3 DESIGN PROCESS FOR HIGH-PERFORMANCE BUILDINGS

The world is rapidly changing, and building construction practices and advances in architectural modeling technologies have reached a unique crossroad in history with changing needs and expectations. Professional interest and investment in high-performance green/sustainable buildings have grown substantially over the last two decades. And with many successful new building projects taking shape throughout the country today, it calls into question the performance level of our more typical construction endeavors and makes us wonder just how far our conventional buildings are falling short of the mark and what needs to be done to meet the new challenges. With this increased attention, these buildings now account for a significant percentage of the markets across the United States and Canada.

Many clients/investors fail to appreciate the extent to which the traditional project delivery process affects their capital and operating costs. They also fail to realize that the process of green building design and construction differs fundamentally from current standard practice. The features that are pivotal to successful green building design require a commitment to stringent health, commitment to the environment, and resource use performance targets by developers, designers, and builders. Measurable targets challenge the design and construction team and allow progress to be tracked and managed throughout development and beyond. Postoccupancy evaluations are often used as a market tool as well as to demonstrate performance achievement of ambitious targets.

It is no secret that the more integrated and better the project delivery process is, the greater the product both for client and for the bottom line. The integrative design process (IDP) is essentially better business, independent of green building. But high performance outcomes necessitate a far more

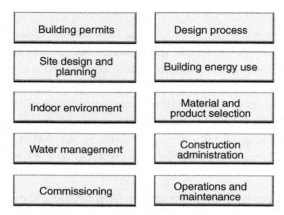

Figure 5.3 *Elements of High-Performance Building Design.*

integrated team approach to the design process, marking a departure from traditional practices, where emerging designs are handed sequentially from architect to engineer to subconsultant. Close collaboration by multidisciplinary teams, from the commencement of the conceptual design process, throughout design and construction, is critical for success. Typically, the design team is expanded to include additional members, such as energy analysts, materials consultants, environmentalists, lighting designers, and cost consultants (Figure 5.3).

An IDP requires the inclusion of contractors, operating staff, and prospective occupants to increase the prospects of success. The enlarged design team offers fresh perspectives and approaches and feedback on performance and cost. This results in a typically more unified, more team-driven design and construction process that encompasses different experts early in the design setting process. This in turn increases the likelihood of creating high-performance buildings that achieve significantly higher targets for energy efficiency and environmental performance. The design process becomes one continuous, sustained team effort from conceptual design through commissioning and final occupancy. Moreover, a new version of LEED® v4 green building rating system now includes a credit toward certification for integrative project delivery.

A team-driven approach is considered to be a "front-loading" of expertise. The process should typically begin with the consultant and owner leading a green design charrette with all stakeholders (design professionals, operators, and contractors) in a brainstorming session or "partnering" approach that encourages collaboration in achieving high-performance green goals for the new building, while breaking down traditional adversarial roles. During the design development phase, continuous input from users and operators can advance progress, instigate commitment to decision making, minimize errors, and identify opportunities for collaborating.

Superior results can be achieved in building design and construction through the application of best practices and an integrated team-driven approach. It is by applying integrated design methods that elevates mere energy and resource efficiency practices into the realm of high performance. This approach differs from the typical planning and design process of relying on the expertise of various specialists who work in their respective specialties somewhat isolated from each other. The IDP encourages designers from all the disciplines – architectural, civil, electrical, mechanical, landscape, hydrology, interior design, etc. – to be collectively involved in the design decision-making process. Coming together at the beginning of the project's preliminary design phase, and at key stages in the project development process, allows professionals to work together to achieve exceptional and creative design solutions that yield multiple benefits at no extra cost while achieving very beneficial synergies.

The Integrated Design Team Process: In the absence of an interactive approach to the design process, it would be extremely difficult to achieve a successful high-performance building. The successful design of buildings requires the integration of many kinds of information into a synthetic whole. The process draws its strength from the knowledge pool of all the stakeholders (including the owner) across the life cycle of the project and their collaborative involvement from defining the need for a building, through planning, design, construction, operation, and maintenance of the facility and building occupancy. The best buildings result from active, consistent, organized collaboration among all players, which is why the stakeholders need to fully understand the issues and concerns of all the parties and be able to interact closely throughout the various phases of the project.

A design charrette basically consists of a focused and collaborative brainstorming session held at the beginning of a project to encourage an exchange of ideas and information and to allow integrated design solutions to take shape. Team members (the stakeholders) are expected to discuss and address problems beyond their field of expertise. The charrette has been found to be particularly useful in complex situations where many people represent the interests of the client and conflicting needs and constituencies

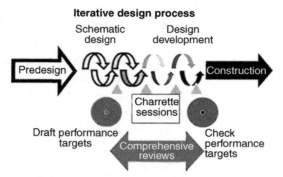

Figure 5.4 *Diagram Illustrating How a Charrette Fits Into the Integrated Design Process.* A charrette, whether used for community planning or building design, is an interdisciplinary session, involving project stakeholders, design experts, regulatory authorities, etc. In a charrette, there are cooperative tasks set with time limits for achieving a synthesis of design ideas and "road map" in sketch form. *Source: David Rousseau, Archemy Consulting Ltd.*

(Figure 5.4). Participants are educated about the issues, and resolution enables them to "buy into" the schematic solutions. A final solution isn't necessarily produced, but important, often interdependent, issues are explored and clarified.

Designing the project in a holistic manner alone is inadequate because the effectiveness and outcome of the integrated design solution also has to be determined. Conducting a facility performance evaluation to ensure that the high-performance goals have been met and will continue to be met over the life cycle of the project should be considered. Likewise, retrocommissioning to ensure that the building will continue to optimally perform through continual adjustments and modifications should be considered.

Computer energy simulations should be conducted as early as possible in the design process and continue until the design is complete, to enable proper assessment of energy conservation measures. The collaboration of the expanded design team early in conceptual design allows the design team to generate several alternative concepts for the building's form, envelope, and landscaping and concentrating on reduction of peak energy loads, demand, and consumption. Computer energy simulation is an excellent tool to assess the project's effectiveness in energy conservation, as well as its construction costs. By employing sustainable approaches that reduce heating and cooling loads, such as more efficient glazing, insulation, lighting systems, daylighting, and similar measures, the mechanical engineer can follow suit by designing a more appropriate, more efficient, and less expensive HVAC equipment and systems that in turn would result in minimal if any increase in construction cost compared to conventional designs.

Simulations are most often used to see how a design can be improved so that energy conservation and capital cost goals are met and to ensure that a design complies with regulatory requirements. Design alternatives are typically evaluated either on capital cost or on the basis of reduced life-cycle cost. In either case, the principal aim of producing alternative designs is to simultaneously minimize both a building's life-cycle cost and its construction cost. But a more accurate estimation of these costs requires a comprehensive approach that includes costs and environmental impacts of everything from resource extraction, materials and assembly manufacture, construction, operation, and maintenance in use to final reuse, recycling, or disposal. Computerized energy simulation is but one of several tools employed to include operational costs into the analysis; other computer tools are also obtainable to facilitate performing life-cycle cost analysis.

The IDP reflects a deeper analysis of the project than would be typical of a conventional design practice and requires greater effort from design consultants. Design fees for this additional work typically reflects the increased work being implemented, but such an investment is insignificant when compared to the environmental and cost impacts over the life of a typical building.

High-performance, sustainable/green buildings have now entered the mainstream and are emerging as an important market sector in the United States and globally. At the same time, this increased demand for high-performance buildings is encouraging facility owners and design professionals to rethink the design process. This re-evaluation of emerging patterns and key process on successful high-performance building projects is having a substantial impact on both the private and government sectors.

For example, a Federal Leadership in High Performance and Sustainable Buildings Memorandum of Understanding (MOU) was signed in January 2006, for which the signatory agencies commit to federal leadership in the design, construction, and operation of high-performance and sustainable buildings. An important element of this strategy is the implementation of common strategies for planning, acquiring, siting, designing, building, operating, and maintaining high performance and sustainable buildings. Included in the MOU are a number of guiding principles to be adopted for Federal Leadership in High Performance and Sustainable Buildings. These included more detailed guidance on the principles for optimizing energy performance, conserving water, improved IEQ, integrated design, and reduction of the impact of materials. Since the signing of the MOU, many federal facilities have followed these guidelines and have succeeded in creating high-performance buildings that save energy and reduce the environmental impact on our lives. This is of particular importance because according to the *Whole Building Design Guide* (WBDG), "The Federal Government is the nation's single largest landlord and energy consumer, operating more than 500,000 facilities comprising more than 3 billion square feet. Historically, approximately $30 billion is spent annually on acquiring or substantially renovating Federal facilities, and about $7 billion is spent on energy for Federal facilities."

Several executive orders and legislative mandates were introduced to direct Federal Agencies to achieve specific HPSB goals, and on December 5, 2008, a new guidance on high-performance federal buildings was issued. It includes revised Guiding Principles for new construction as well as new Guiding Principles for existing buildings. It also includes clarification of reporting guidelines for entering information on the sustainability data element (#25) in the Federal Real Property Profile and provides clarification on the method to calculate the percentage of buildings and square footage that are compliant with the Guiding Principles for agencies' scorecard input. Thus, Executive Orders 13514 and 13423 were introduced, which require all Federal agencies to comply with the Guiding Principles for New Construction and Major Renovation. This section includes technical guidance needed to meet each of these Guiding Principles. Likewise, Executive Order 13514 requires at least 15% of each agency's existing facilities and building leases (above 5000 ft.^{2} gross) to meet the Guiding Principles by 2015. To meet this goal, most agencies must upgrade their existing building stock, which means compliance with a separate set of Guiding Principles for Sustainable Existing Buildings. And on March 19, 2015, President Obama signed Executive Order 13693: Planning for Federal Sustainability in the Next Decade. This Executive Order supersedes Executive Order 13423 and Executive Order 13514 in authority.

5.4 SUSTAINABLE SITE SELECTION

This section should be read in conjunction with Section 3.2.2.

5.4.1 General

Site considerations are critical to creating an environmentally friendly and aesthetically pleasing project. Siting of green buildings should take into account the natural environmental features of the site and minimize any potential adverse impacts on these features. Attentive design and construction of a green building within an undeveloped natural setting such as a "greenfield" site can mitigate the damage inflicted on natural systems, wildlife habitat, and biodiversity. Exploiting previously developed "brownfield" sites that have been disturbed and contaminated by past uses, both usually less expensive to purchase and also benefits the environment because it averts impacts on natural resources, provides for site remediation, and takes advantage of existing infrastructure.

Site design and building placement should also encourage sustainable transportation alternatives such as walking and bicycling, thereby minimizing automobile use. The implementation of sustainable landscaping practices that use hardy, indigenous plant species that do not require potable water for irrigation, or the use of pesticides is another strategy to be encouraged.

Creation of sustainable buildings begins typically with the selection of a suitable site. A building's location and orientation on a site impacts a wide range of environmental factors in addition to other important factors like security and accessibility. Likewise, a building's location can impact the energy likely to be consumed by occupants for commuting and their choice of transportation methods and can have a major impact on local ecosystems and the degree to which existing structures and infrastructures are employed. This is why it is preferable to locate buildings in areas of existing development and, whenever possible, to consider reuse or retrofitting existing buildings and rehabilitating historic properties. Whether the project consists of a single building or group of buildings, a campus or military base, it is important to incorporate smart growth principles in the project's development process.

Physical security has recently taken center stage to become a critical issue in optimizing site design and should not be an afterthought. Site security requires that the determination and location of access roads, parking, vehicle barriers, and addressing perimeter lighting all be integrated into the design at the earliest phase of the design, along with sustainable site considerations.

The Sustainable Sites credits in LEED v4 are generally similar to those in version 2009 (v3). Some credits, however, including "Bicycle Facilities," "Access to Quality Transit," and "Green Vehicles," have been moved to a new "Location and Transportation" credit category in LEED v4. The Sustainable Sites credit category has retained credits for construction activity pollution prevention, light pollution reduction, heat island reduction, protect/

restore habitat, and open space. Claire Moloney, a LEED Consultant with the Poplar Network™, says, "One notable change is that the stormwater management credits are now referred to as "Rainwater Management." The credit is quite different from the previous stormwater credits, in that it allows two options for compliance: (1) percentile of rainfall events and (2) natural land cover conditions. For percentile of rainfall events, the project must manage the runoff on the site for a certain "percentile of regional or local rainfall events." For the natural land cover conditions option, the project must "manage on site the annual increase in runoff volume from the natural land cover condition to the post developed condition." Another new credit for new projects is called "Site Assessment." It awards one point for projects that assess the site's condition before design for features such as topography, hydrology, climate, vegetation, soils, human use, and human health effects. The project team will be able to use the assessment to make better, more sustainable decisions about the building."

To achieve LEED credits, sustainable site planning should typically employ an integrated system approach that seeks to:

- Minimize the development of open space through the selection of disturbed land, brownfields, or building retrofits.
- Consider energy implications in site selection and building orientation to take advantage of natural ventilation and maximum daylighting. Buildings should be sited to allow integration of passive and active solar strategies. Investigate the potential impact of possible future developments surrounding the site, including daylighting, ventilation, solar energy, etc.
- Control/prevent erosion through improved landscaping practices and stabilization techniques (e.g., grading, seeding and mulching, installing pervious paving) and/or retaining sediment after erosion has occurred (e.g., earth dikes and sediment basins). This mitigates the negative impacts on water and air quality and reduces potential damage to a building's foundation and structural system that may be caused by natural hazards such as floods, mudslides, and torrential rainstorms. Installing of retention ponds and berms should also be considered to control erosion, manage stormwater runoff on site, and reduce heat islands while also serving as physical barriers to control access to a building. Use of vegetated swales and depressions will facilitate reducing runoff.
- Reduce heat islands using landscaping techniques such as the use of existing trees and other vegetation to shade walkways, parking lots, and other open areas. However, the site work and landscaping should be

integrated with security and design safety. Landforms and landscaping should be integrated into the site planning process to enhance resource protection. Employing light-colored materials to cover the facility's roof helps reduce energy loads and helps to extend the life of the roof. In hot, dry climates, consider creating shading by the covering of walkways, parking lots, and other open areas that are paved or made with low-reflectivity materials. For security reasons, any shading devices in place must not block critical ground-level sight lines.

- Consider incorporating green roofs into the project in warm climates, in addition to using roofing products that meet or exceed Energy Star standards. The main site benefits of employing green/vegetated roofs include helping to control stormwater runoff, improving water quality, and mitigating urban heat island effects.
- Minimize habitat disturbance by limiting site disturbance to a minimal area around the building perimeter, including locating buildings adjacent to existing infrastructure, and retaining prime vegetation features to the extent possible. Habitat disturbance can also be positively impacted by reducing building and paving footprints, and planning construction staging areas with the environment in mind.
- Restore the health of degraded areas by increasing and improving the presence of healthy habitat for indigenous species through native plants and closed-loop water systems, in addition to conserving water use through xeriscaping with native plants and native or climate-tolerant trees to improve the quality of the site as well as provide visual protection by obscuring assets and people.
- Consider the installation of retention ponds and berms to control erosion, manage stormwater, and reduce heat islands, and which can also serve as physical barriers to control access to a building.
- Incorporate transportation solutions along with site plans that acknowledge the need for bicycle parking, carpool staging (Figure 5.5), and proximity to mass transit. Alternatives to traditional commuting can be encouraged by siting the proposed building, taking into account availability and access of public transport and limiting on-site parking. Adequate provisions for bicycling, walking, carpool parking, and telecommuting should be provided as well as provisions for refueling/recharging facilities for alternative fuel/electric vehicles (Figure 5.6).
- Address site security concurrently with other key sustainable site issues like location of access roads, parking, vehicle barriers, and perimeter lighting, among others.

Figure 5.5 *Drawing Illustrating the Placement of Carpool and Vanpool Parking Spaces.* They should be closer to the building entrance than other automobile parking. Prominent signage may be required to draw attention to the location of car- and vanpool parking and pickup areas. Attractive and comfortable waiting areas should be provided to encourage car- and vanpool commuters. *Source: City of Santa Monica Green Building Design & Construction Guidelines.*

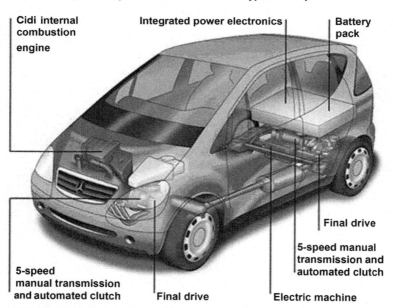

Figure 5.6 *Rendering of the Mercedes Hyper Hybrid Car Concept Cutaway Showing Major Components.* An example of an energy-efficient car designed to run on biodiesel fuels. EPA studies suggest that using biodiesel reduces many of the toxic substances present in diesel exhaust – including many cancer-causing compounds. *Source: PhysOrg.com.*

5.4.2 Site Selection

The USGBC says that "the Sustainable Sites (SS) category rewards project teams for designing the site to minimize adverse effects." Note that the new Location and Transportation (LT) category is essentially an outgrowth of the Sustainable Sites category, which formerly covered location–related topics. The Location and Transportation category is designed to reward project teams for choosing a preferable site location. The LEED™ Sustainable Site section consists of various credits depending on which category is being considered (e.g., New Construction, Schools, Commercial Interiors, etc.). Discussed next are some of the major issues to be considered.

The general intent of selecting a sustainable site is essentially to reduce environmental impacts by avoiding development of inappropriate sites and from the location of a building on a site. The intent for LEED™ Commercial Interior differs somewhat from say, LEED™ New Construction and LEED™ Schools in that the purpose here is mainly to encourage tenants to select buildings with best practices systems and employed green strategies. The USGBC says,

How a building is incorporated into the site can benefit or harm local and regional ecosystems and reduce or increase demand for water, chemicals, and pesticides for site management. Good decisions, made early in the design process, can result in attractive, easy-to-maintain landscaping that protects native plant and animal species and contributes to the health of local and regional habitats.

Rain that falls on a site can cause soil erosion and runoff of chemicals and pesticides – or it can offset potable water demand and recharge underground aquifers. Plant growth can be a burden, requiring regular upkeep, watering, and chemicals – or it can enhance property values while improving occupants' comfort, absorbing carbon, enriching the soil, and providing shade, aesthetic value, and habitat for native species.

Site design should take into consideration not only the aesthetic and functional preferences of the occupants but also long-term management needs, preservation principles, and potential effects on local and regional ecosystems.

Main LEED™ Requirements: Developing building projects, hardscape areas, roads, or parking areas on segments of sites that meet any one of the following criteria is to be avoided:

- Prime farmland (as defined by the United States Department of Agriculture).
- Previously undeveloped land with an elevation lower than 5 ft. above FEMA's 100-year flood elevation as defined by FEMA (Federal Emergency Management Agency).
- Land that is specifically identified as habitat for any endangered or threatened species (that are on Federal or State threatened or endangered lists).

- Within 100 ft. of any wetlands, and isolated wetlands or areas of special concern identified by state or local rule, *or* within setback distances from wetlands prescribed in state or local regulations, as defined by local or state rule or law, whichever is more stringent.
- Previously undeveloped land that is within 50 ft. of a water body, defined as seas, lakes, rivers, streams, and tributaries which support or could support fish, recreation, or industrial use, consistent with the terminology of the Clean Water Act.
- Land which prior to acquisition for the project was public parkland, unless the public landowner accepts in trade parkland that is of equal or greater value (Park Authority projects are exempt).
- Consider reuse of existing building space whenever possible.

Potential Technologies and Strategies: During the site selection process, give preference to those sites that do not include sensitive site elements and restrictive land types and formulate during the design phase an erosion and sedimentation control plan for the project's development. Select a suitable building location and design the building with the minimal footprint to minimize site disruption of those environmentally sensitive areas identified earlier. Once a site is selected, consider employing programs that encourage temporary and permanent seeding, mulching, earthen dikes, sediment basins, etc.

During the site selection process, preference should be given to urban and urbanized sites. Direct development of urban areas with existing infrastructure should be encouraged. Greenfields should be protected and habitat and natural resources should be preserved. Buildings should be sited near mass transit whenever possible to minimize transportation impacts, and walking and/or the use of bikes should be encouraged. The building should be programed to be used by the community by integrating building function with a variety of uses.

During the site selection process, give preference to brownfields redevelopment sites and also to the rehabilitation of previously utilized and/ or damaged sites with suspected or confirmed environmental contamination, thereby reducing pressure on undeveloped land. Develop and implement a site remediation plan.

5.4.3 Development Density and Community Connectivity

The *LEED™ Intent* here is to preserve natural resources by developing urban areas with existing infrastructure and protecting greenfields.

LEED Requirements: Option 1. Development Density – Construct/ renovate on previously developed land in a community with a development density of at least 60,000 ft.^{2} per acre net. The calculation is to be based on 2-story downtown development. This option is now a Location and Transportation credit. *Option 2.* Community Connectivity — Construct/renovate on previously developed land that (1) is within a half mile of a residential area or neighborhood with an average density of 10 units per acre, (2) is within a half mile of at least 10 basic services, and (3) has pedestrian access between the building and the services.

For Commercial Interiors, the LEED Requirement is to find a space in a building that is located in an existing, walkable community with a development density of at least 60,000 ft.^{2} per acre net.

Basic Services as per LEED requirements include, but are not limited to:

1. Bank
2. Place of worship
3. Convenience grocery
4. Day care
5. Cleaners
6. Fire station
7. Beauty
8. Hardware
9. Laundry
10. Library
11. Medical/dental
12. Senior care facility
13. Park
14. Pharmacy
15. Post office
16. Restaurant
17. School
18. Supermarket
19. Theater
20. Community center
21. Fitness center
22. Museum

Potential Technologies and Strategies: Site Selection — Preference should be given to urban sites with pedestrian access to a variety of services.

5.4.4 Brownfield Redevelopment

The *LEED Intent* is to protect greenfields by encouraging the cleanup of contaminated lands and developing sites that have been identified as environmentally contaminated or damaged.

LEED Requirements: Option 1. Develop on a contaminated site as determined by ASTM E1903-97 Phase II Environmental Site Assessment or a local voluntary cleanup program. *Option 2.* Develop on a brownfield site determined by a local, state, or federal government agency.

Potential Technologies and Strategies: Site Selection — give preference to brownfield sites and identify all tax incentives and property cost savings. Also coordinate site development plans with remediation activity, as required.

5.4.5 Alternative Transportation

LEED v4 has introduced several changes to the scope of work for the CxA. These scope changes affect the breadth, duration, and types of commissioning agents required for LEED v4 projects (check with the USGBC for latest updates).

1. *Public transport access*

   The *LEED intent* is to reduce the impacts of greenhouse gas emissions, air pollution, and other environmental and public health harms associated with motor vehicle use and land development on the environment by reducing motor vehicle use.

   LEED requirements: (1) Develop within a half mile (800-m) walking distance of existing, planned, or funded commuter rail, light rail, subway station or ferry terminals, and/or (2) develop within a quarter mile (400-m) walking distance of at least one stop for at least two public or private/campus bus lines, rideshare stops, or streetcar stops that can be used by building occupants.

   Exemplary performance: Institute a comprehensive transportation management plan that quadruples subway, commuter rails, light rails, or bus service and doubles ridership.

   Potential technologies and strategies: Identify transportation needs of future building occupants by performing a transportation survey. Site proposed building near mass transit.

2. *Bicycle storage and changing rooms*

   The *LEED intent* is to reduce the impacts of pollution and land development impacts on the environment by reducing the automobile use.

   LEED requirements: For Commercial or Institutional Projects: Provide secure bicycle racks and/or storage within 200 yards of a building entrance for at least 5% or more of full-time equivalent (FTE) occupants and transient occupants during peak building hours. Also, provide shower and changing facilities in the building, or within 200 yards (180 m) of a building entrance, for 0.5% of full-time equivalent (FTE) occupants. And for Residential Buildings: Provide covered bicycle storage, for at least 15% of the building occupants. This is now a Transportation and Location credit (1 point).

Walking and bicycling distance: According to the USGBC (LEED v4),

Walking and bicycling distances are measurements of how far a pedestrian and bicyclist would travel from a point of origin to a destination, such as the nearest bus stop. This distance, also known as shortest path analysis, replaces the simple straight-line radius used in LEED 2009 and better reflects pedestrians' and bicyclists' access to amenities, taking into account safety, convenience, and obstructions to movement. This in turn better predicts the use of these amenities.

Walking distances must be measured along infrastructure that is safe and comfortable for pedestrian: sidewalks, all-weather-surface footpaths, crosswalks, or equivalent pedestrian facilities.

Bicycling distances must be measured along infrastructure that is safe and comfortable for bicyclists: on-street bicycle lanes, off-street bicycle paths or trails, and streets with low target vehicle speed. Project teams may use bicycling distance instead of walking distance to measure the proximity of bicycle storage to a bicycle network in LT Credit Bicycle Facilities.

When calculating the walking or bicycling distance, sum the continuous segments of the walking or bicycling route to determine the distance from origin to destination. A straight-line radius from the origin that does not follow pedestrian and bicyclist infrastructure will not be accepted.

Refer to specific credits to select the appropriate origin and destination points. In all cases, the origin must be accessible to all building users, and the walking or bicycling distance must not exceed the distance specified in the credit requirements.

3. *Low-emitting and fuel-efficient vehicles*

According to Poplar Network, "With transportation producing almost 30% of all US global warming emissions, low emitting and fuel-efficient vehicles, aka 'green vehicles,' have become more and more important to the American economy and to mitigating global climate change." It goes on to say that

Low emitting vehicles are those that produce little emission of atmospheric pollutants while fuel-efficient vehicles are those that produce power at a rate considered optimal with regard to the amount of fuel consumed. Since cars and light trucks contribute 60% of the US transportation emissions, and 1 gallon of gasoline is about 24 pounds of greenhouse gases in the atmosphere, the Leadership in Energy and Environmental Design (LEED) green building certification program has incorporated low emitting and fuel efficient vehicles into its best-in-class building strategies and practices by providing preferred or reduced cost parking for these vehicles. The relevant credit is called Sustainable Sites: Alternative Transportation – Low Emitting and Fuel Efficient Vehicles.

This credit is focused mainly on limiting environmental impacts from automobile use. And while it targets commuting specifically, it also addresses company vehicle fleets, maintenance vehicles, and buses. The use of low emitting and fuel-efficient vehicles is becoming increasingly necessary.

The *LEED intent* is to reduce pollution and land development impacts from automobile use.

LEED™ v4 requirements:

1. NC
   - *Provide preferred parking for low emitting and fuel efficient vehicles for 5% of total parking. OR*
   - *Install alternative fuel fuelling station for 3% of total parking. OR*
   - *Provide low emitting and fuel efficient vehicles for 3% FTE. OR*
   - *Provide low emitting and fuel efficient vehicle sharing program.*
2. CS
   - *Provide preferred parking for low emitting and fuel-efficient vehicles for 5% of total parking. OR*
   - *Install alternative fuel fuelling station for 3% of total parking.*
3. Schools
   - *Provide preferred parking for low emitting and fuel-efficient vehicles for 5% of total parking, AND at least one designated carpool drop off for low emitting and efficient vehicles. OR*
   - *Develop a plan for buses and maintenance vehicles to use 20% (by vehicles, fuel or both) natural gas, propane, biodiesel or be a low emitting and fuel-efficient vehicle.*

Source: USGBC – LEED.

For credit purposes, LEED defines low–emitting and fuel-efficient vehicles as "vehicles that are either classified as Zero Emission Vehicles (ZEV) by the California Air Resources Board or have achieved a minimum green score of 40 on the American Council for an Energy Efficient Economy (ACEEE) annual vehicle rating guide."

Exemplary performance: Institute a comprehensive transportation management plan demonstrating a quantifiable reduction in personal automobile use through the use of alternative options.

Potential technologies and strategies: Provide transportation amenities such as alternative fuel refueling stations. Consider sharing the costs and benefits of refueling stations with neighbors.

Green vehicles (LEED v4 BD+C: New Construction): The LEED v4 intent is to reduce pollution by promoting alternatives to conventionally fueled automobiles.

Requirements: As per the USGBC (LEED v4),

Designate 5% of all parking spaces used by the project as preferred parking for green vehicles. Clearly identify and enforce for sole use by green vehicles. Distribute preferred parking spaces proportionally among various parking sections (e.g., between short-term and long-term spaces).

Green vehicles must achieve a minimum green score of 45 on the American Council for an Energy Efficient Economy (ACEEE) annual vehicle rating guide (or local equivalent for projects outside the U.S.).

A discounted parking rate of at least 20% for green vehicles is an acceptable substitute for preferred parking spaces. The discounted rate must be publicly posted at the entrance of the parking area and permanently available to every qualifying vehicle.

In addition to preferred parking for green vehicles, meet one of the following two options for alternative-fuel fueling stations:

Option 1. Electric vehicle charging

Install electrical vehicle supply equipment (EVSE) in 2% of all parking spaces used by the project. Clearly identify and reserve these spaces for the sole use by plug-in electric vehicles. Parking spaces that include EVSE must be provided separate from and in addition to preferred parking spaces for green vehicles.

Option 2. Liquid, gas, or battery facilities

Install liquid or gas alternative fuel fueling facilities or a battery switching station capable of refueling a number of vehicles per day equal to at least 2% of all parking spaces.

4. *Parking capacity*

The *LEED intent* is to minimize pollution and land development impacts associated with parking facilities, including automobile dependence, land consumption, and rainwater runoff.

LEED requirements:

Option 1. Nonresidential – Parking capacity to meet, but not exceed, minimum local zoning requirements, and reserve 5% of the preferred parking spaces for carpools or vanpools. *Option 2.* Nonresidential – For projects providing parking for less than 5% of FTE building occupants, reserve 5% of the preferred parking spaces for carpools or vanpools. Providing a discounted parking rate is an acceptable substitute for preferred parking for carpool or vanpool vehicles. Providing a discounted parking rate is an acceptable alternative for preferred parking for low-emitting/fuel-efficient vehicles providing the discounted parking rate is at least 20% in order to establish a meaningful incentive. *Option 3.* Residential – Parking capacity not to exceed minimum local zoning requirements, and infrastructure and support programs are to be provided to facilitate shared vehicle usage, designated parking for vanpools, car-share services, ride boards, and shuttle services to mass transit. *Option 4.* All – No new parking to be provided. *Option 5.* Mixed Use (Commercial and Residential) – For mixed use buildings in which the commercial area is less than 10%, the entire building should then be considered residential and adhere to the residential requirements of Option 3. For mixed use buildings in which the commercial area is more than 10%, the commercial space is to adhere to Nonresidential requirements, while the residential component is to adhere to residential requirements. This option only applies to mixed use projects that have residential and commercial/retail components.

LEED v4: The credit calculations must include all existing and new off-street parking spaces that are leased or owned by the project, including parking that is outside the project boundary but is used by the project. On-street parking in public rights-of-way is excluded from these calculations. For projects that use pooled parking, calculate compliance using the project's share of the pooled parking.

Provide preferred parking for carpools for 5% of the total parking spaces after reductions are made from the base ratios. Preferred parking is not required if no off-street parking is provided.

Mixed use projects should determine the percentage reduction by first aggregating the parking amount of each use (as specified by the base ratios) and then determining the percentage reduction from the aggregated parking amount.

Do not count parking spaces for fleet and inventory vehicles unless these vehicles are regularly used by employees for commuting as well as business purposes.

Potential technologies and strategies: Minimize size of parking lot/garage and consider possibility of sharing parking facilities with adjacent buildings as well as alternatives that will limit the use of single-occupancy vehicles.

5.4.6 Site Development

1. *Protect or restore habitat*

   The *LEED intent* is to promote biodiversity and provide habitat by conserving existing natural areas and restore damaged areas.

   LEED requirements: Preserve and protect from all development and construction activity 40% of the greenfield area on the site (where such areas exist).

   Option 1. On-site restoration (2 points except health care, 1 point health care)

   Using native or adapted vegetation, restore 30% (including the building footprint) of all portions of the site identified as previously disturbed.

Projects that achieve a density of 1.5 floor–area ratio may include vegetated roof surfaces in this calculation if the plants are native or adapted, provide habitat, and promote biodiversity.

Restore all disturbed or compacted soils that will be revegetated within the project's development footprint to meet the following requirements[2]:

- Soils (imported and *in situ*) must be reused for functions comparable to their original function.
- Imported topsoils or soil blends designed to serve as topsoil may not include the following:
  - Soils defined regionally by the Natural Resources Conservation Service web soil survey (or local equivalent for projects outside the United States) as prime farmland, unique farmland, or farmland of statewide or local importance; or
  - Soils from other greenfield sites, unless those soils are a byproduct of a construction process.
- Restored soil must meet the criteria of reference soils in categories 1–3 and meet the criteria of either category 4 or 5:
  1. organic matter;
  2. compaction;
  3. infiltration rates;
  4. soil biological function; and
  5. soil chemical characteristics.

Project teams may exclude vegetated landscape areas that are constructed to accommodate rainwater infiltration from the vegetation and soils requirements, provided all such rainwater infiltration areas are treated consistently with SS Credit Rainwater Management.

Option 2. Financial support (1 point)

Provide financial support equivalent to at least $0.40 ft.$^{-2}$ (US\$4 m^{-2}) for the total site area (including the building footprint).

Financial support must be provided to a nationally or locally recognized land trust or conservation organization within the same EPA Level III ecoregion or the project's state (or within 100 miles of the project (160 km) for projects outside the United States). For US projects, the land trust must be accredited by the Land Trust Alliance. (Source: LEED™ – USGBC)

Potential technologies and strategies: For greenfield sites, perform a site survey to identify site elements and adopt a master plan for development of the project site. Minimize any disruption to existing ecosystems by carefully siting the building and designing it to minimize its building footprint. Establish clearly marked construction boundaries and restore previously degraded areas to their natural state.

2. *Maximize open space*

The *LEED Intent* is to provide sites with a high ratio of vegetated open space to development footprint to promote biodiversity and recreation.

LEED requirements: Option 1. To reduce the development footprint (which is defined as the total area of the building footprint, hardscape, access roads, and parking) and/or provide vegetated open space that exceeds local zoning requirements for the site by 25%. *Option 2.* For sites where there are no local zoning ordinance in place (e.g., military bases), provide vegetated open space area that is equal to the building footprint. *Option 3.* Where a zoning ordinance exists, but there is no requirement for open space (zero), a vegetated open space equal to 20% of the project's site area is to be provided.

Projects earning SSc2: Development Density and Community Connectivity can count green roof space and pedestrian-oriented hardscape toward credit compliance. At least 25% of the open space counted must be vegetated.

Wetlands and naturally designed ponds can also count toward open space if the side slope is vegetated with an incline of 1:4 (vertical: horizontal) or less.

One EP point is available for demonstrating double the required open space. (Source: LEEDuser – BuildingGreen, Inc.)

Potential technologies and strategies: A site survey is required to identify site elements and adopt a master plan for development of the project site. To minimize site disruption, select a suitable building location and design the building with a minimal footprint.

5.4.7 Stormwater Design

1. *Quantity control*

The *LEED intent* is to support natural water hydrology by increasing porosity, increasing on-site infiltration, and reducing or eliminating pollution from stormwater runoff, and contaminants.

LEED requirements:

Case 1: Existing cover is less than or equal to 50% imperviousness. – *Option 1.* Implement a stormwater management plan in which the postdevelopment peak discharge rate and quantity must not exceed the predevelopment peak discharge rate and quantity for 1- and 2-year, 24-h design storms. *Option 2.* Put in place a stormwater management plan that protects receiving stream channels from excessive erosion by implementing a stream channel protection strategy and quantity control strategies.

Case 2: Existing cover is greater than 50% imperviousness. – Implement a stormwater management plan that results in a 25% decrease in the volume of stormwater runoff from the 2-year, 24-h design storm.

Potential technologies and strategies: During the project site design, maintain natural stormwater flows by promoting infiltration and specify vegetated roofs, pervious paving, and other measures to minimize impervious surfaces. Encourage reuse stormwater volumes generated for nonpotable uses, including landscape irrigation, toilet and urinal flushing, and custodial uses.

2. *Quality control*

The *LEED intent* is to reduce water pollution by increasing pervious area and on-site infiltration, eliminating contaminants and removing pollutants from stormwater runoff.

LEED requirements: Implement a stormwater management plan that reduces impervious cover, promotes infiltration, and captures and treats 90% of the average annual rainfall using best management practices (BMPs).

BMPs used to treat runoff must be capable of removing 80% of the average annual postdevelopment total suspended solids (TSS) load. BMPs are considered to be acceptable if (1) they are designed in accordance with standards and specifications from a state or local program that has adopted these performance standards, or (2) there exists in-field performance-monitoring data that demonstrate compliance with the criteria. Data must conform to accepted protocol for BMP monitoring.

Potential technologies and strategies: Use alternative sustainable surfaces and nonstructural techniques such as rain gardens, vegetated swales, and rainwater recycling to reduce imperviousness and promote infiltration so as to reduce pollutant loadings.

Use sustainable design strategies to design integrated natural and mechanical treatment systems such as constructed wetlands, vegetated filters, and open channels to treat stormwater runoff.

5.4.8 Heat Island Reduction

The "Heat Island Reduction" credit now encompasses both roof and nonroof measures but was previously divided into two credits, "Heat Island Effect – Roof" and "Heat Island Effect – Nonroof." The intent is to minimize effects on microclimates and human and wildlife habitats by reducing heat islands.

1. *Nonroof*

   The *LEED intent* is to minimize impact on microclimate and human and wildlife habitat by reducing heat islands (thermal gradient differences between developed and undeveloped areas).

   LEED requirements: *Option 1*. Provide for 50% of the site hardscape, including roads, sidewalks, courtyards, and parking lots using any combination of the strategies listed as follows:

   a. Use existing trees or plant material or install plants that provide shade over paving areas (including playgrounds) on the site within 10 years of planting. Install vegetated planters. Plants must be in place at the time of occupancy permit and cannot include artificial turf.

   b. Provide shade from structures covered by energy generation systems such as solar thermal collectors, photovoltaics, and wind turbines that produce energy used to offset some nonrenewable resource use.

   c. Provide shade with architectural devices or structures that have a 3–year aged solar reflectance index (SRI) value of at least 0.28. If 3–year aged value information is not available, use materials with an initial SRI of at least 0.33 at installation.

   d. Provide shade with vegetated structures.

   e. Employ hardscape materials with a 3–year aged solar reflectance (SR) value of at least 0.28. If 3–year aged value information is not available, use materials with an initial SRI of at least 0.33 at installation.

   f. Employ an open-grid pavement system that is at least 50% pervious; or

   Option 2. Cover a minimum of 50% of the parking spaces under (defined as underground, under deck, under roof, or under a building). Roofs used to shade or cover parking must have a minimum SRI of 28, be a vegetated green roof, or be covered by solar panels that produce energy used to offset some nonrenewable resource use (Figure 5.7).

   Potential technologies and strategies: Employ strategies, materials, and landscaping techniques that reduce heat absorption of exterior materials. Shade is to be provided from use of native or adapted trees and large shrubs, vegetated trellises, or other exterior structures supporting vegetation. Photovoltaic cells should be positioned to shade impervious surfaces.

2. *Roof*

   The *LEED intent* is to minimize impact on microclimate and human and wildlife habitat by reducing heat islands (thermal gradient differences between developed and undeveloped areas).

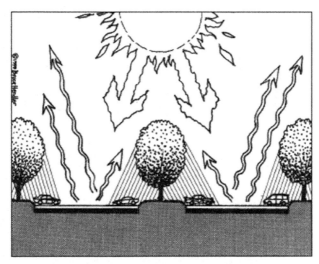

Figure 5.7 *Sketch Showing the Mitigation of Urban Heat Island Effect. Illustration: Bruce Hendler.*

Typically used on roofs and sometimes at grade, the USGBC specifies green roofs as a means of mitigating the urban heat island effect.

LEED requirements: Option 1. (1) Low slope (less than 2:12) – use roofing materials having a solar reflectance index (SRI) equal to or greater than 78 for a minimum of 75% of the roof surface. (2) Steep slope (greater or equal to 2:12) – roofing materials having a lower SRI value of 78 or greater and for low slopes greater than 29 for steep slopes may be used if the weighted rooftop SRI average meets the following criteria: (area SRI roof/total roof area) (SRI of installed roof/required SRI) ≥ 75%. *Option 2.* Install a vegetated roof for at least 50% of the roof area. *Option 3.* Install high albedo and vegetated roof surfaces that, in combination, meet the following criteria: (area of SRI roof/0.75) + (area of vegetated roof/0.5) ≥ total roof area, where the SRI for low-sloped roofs is 78 or greater, and for steep-sloped roofs, the SRI is greater than 29 (Figure 5.8).

Potential technologies and strategies: Installation of high–albedo and vegetated roofs to reduce heat absorption should be considered. SRI is calculated according to ASTM E 1980. Reflectance is measured according to ASTM E 903, ASTM E 1918, or ASTM C 1549. Emittance is measured according to ASTM E 408 or ASTM C 1371. For default values, consult the LEED v4 and v3 (2009) *Reference Guide* (www.usgbc.org). Product information is available from the Cool Roof Rating Council website (www.coolroofs.org) and the Energy Star website (www.energystar.gov).

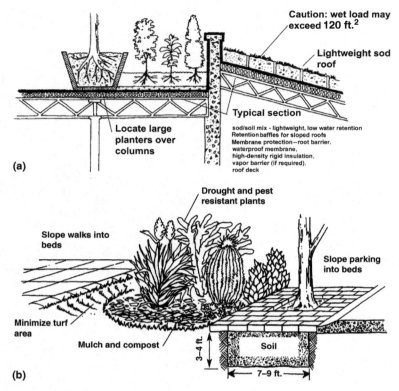

Figure 5.8 (a) Sketch showing use of planting on balconies, terraces, and roofs, which can significantly enhance the environment and provide habitat for wildlife and opportunities for food and decorative gardens. Moderately sloped and flat roofs can be planted with drought-tolerant perennial grasses and groundcovers that require minimal maintenance. (b) Applying environmental landscape principles. These involve selecting slow-growing, drought-tolerant plants that require less water and maintenance, significantly reducing water consumption. *Source: City of Santa Monica Green Building Design & Construction Guidelines.*

5.4.9 Light Pollution Reduction

Light pollution is often caused by the way light is emitted from lighting equipment. Choosing proper equipment and carefully mounting and aiming it can make a significant difference.

Referenced code/standards: (1) ANSI/ASHRAE/IESNA Standard 90.1-2007, Energy Standard for Buildings Except Low Rise Residential Lighting, Section 9.

The *LEED intent* is to minimize light trespass from the building and site, reduce sky-glow to increase night sky access, improve nighttime visibility through glare reduction, and reduce development impact from lighting on nocturnal environments.

Objective:

- To minimize light trespass from building and site.
- Reduce sky glow.
- Improve nighttime visibility through glare reduction.
- Reduce development impact from lighting on nocturnal environments.

LEED requirements:

1. *For interior lighting*

   Option 1. Fixture shielding. Shield all exterior fixtures (where the sum of the mean lamp lumens for that fixture exceeds 2500) such that the installed fixtures do not directly emit any light at a vertical angle more than 90° from straight down.

   All nonemergency interior luminaires, with a direct line of sight to any openings in the envelope (translucent or transparent), shall have its input power automatically reduced by at least 50% between the hours of 11 p.m. and 5 a.m. Afterhours override may be provided by a manual or occupant-sensing device provided that the override does not last in excess of 30 min; or

   Option 2. Perimeter measurements. Measure the night illumination levels at regularly spaced points on the project boundary, taking the measurements with the building's exterior and site lights both on and off. At least eight measurements are required, at a maximum spacing of 100 ft. (30 m) apart. The illumination level measured with the lights on must not be more than 20% above the level measured with the lights off. All openings in the envelope with a direct line of sight to any nonemergency luminaires shall have shielding that will be controlled/closed by automatic device between the hours of 11 p.m. and 5 a.m.

2. *For exterior lighting*

   Only light areas as required for safety and comfort will be used. Lighting Power Densities shall not exceed ASHRAE/IESNA Standard 90.1-2007 for the classified zone. Meet exterior lighting control requirements from ASHRAE/IESNA Standard 90.1-2007, Exterior Lighting section, without amendments.

   For zone classification as defined in IESNA RP-33, see the USGBC LEED™ reference guide for further details and the relevant category.

   Potential technologies and strategies: To maintain safe light levels while avoiding off-site lighting and night sky pollution, site lighting criteria should be adopted and the possible use of a computer model of the site lighting.

Technologies to reduce light pollution include full cutoff luminaires, low-reflectance surfaces and low-angle spotlights.

5.5 COMMISSIONING PROCESS

5.5.1 Overview

Commissioning is a systematic process that ensures that all building systems perform interactively according to the design intent and the Owner's operational needs. Ideally it starts at a project's inception, that is, at the beginning of the design process, and culminates in the inspection, testing, balancing, and actual verification that new building systems are operating properly and efficiently with the agreed warranty period. Commissioning is designed to make sure that the heating, ventilating, air-conditioning, and refrigeration (HVAC&R) systems (mechanical and passive) and associated controls, lighting controls, including daylighting, domestic hot water systems, renewable energy systems (photovoltaic, wind, solar, etc.), and other energy-using building systems meet the owner's performance requirements, function the way they were intended, and operate at maximum efficiency. Commissioning results can identify changes that will dramatically reduce operating and maintenance costs, provide better occupant conditions, and facilitate upgrades, as well as fulfill LEED requirements.

Many buildings today contain highly sophisticated conservation and environmental control technologies, which, in order to function correctly, require careful supervision of installations, testing and calibration, and instruction of building operators. Some buildings possess unusual electrical or air-conditioning systems, or employ certain sustainable features that may require particular attention to ensure they operate as designed. Commissioning can reduce operating and maintenance costs, improve the comfort of a building's occupants, and extend the useful life of equipment. In addition, commissioning new construction projects will help reduce construction delays, ensure the correct equipment is installed, reduce future maintenance costs, and reduce employee absenteeism. It is important that complete as-built information and operating and maintenance information are passed on to owners and operating staff upon completion of the project. Returns for these commissioning services often pay for themselves in energy savings within a year of completion.

Commissioning is a fairly recent concept that includes what was formerly referred to as "testing, adjusting and balancing" but it goes much further. Commissioning is particularly important when complex mechanical and electrical systems are involved, to ensure that these systems operate as intended, and to realize energy savings and a quality-building environment and thereby justify the incorporation and installation of more complex systems. When special building features are installed to generate renewable energy, recycle waste, or reduce other environmental impacts, commissioning is often necessary to ensure optimum performance. More importantly, commissioning practices should be tailored to the size and complexity of the building and its systems and components in order to verify their performance to ensure that all requirements are met as per the construction documents and specifications. In addition to verification of the installation and performance of systems, commissioning will include producing a commissioning report.

Few experiences can be more frustrating than finding out that essential systems in a new building do not operate according to specifications. To prevent this sort of outcome, total building commissioning should be included as part of the design, construction, and operation process. The degree of commissioning required should be appropriate to the complexity of the project and its systems, the owner's need for assurances, and the budget and time available. HVAC commissioning costs vary but are usually from 1% to 4% of the value of the mechanical contract.

Commissioning of building systems will vary from one project to another but will generally encompass and coordinate equipment startup, HVAC systems, electrical, plumbing, communications, security and fire management systems, and their controls and calibration. Other systems and components may be included, particularly in large or complex projects. Commissioning begins by checking documentation and design intent for future reference. This is followed by performance testing of components when they first arrive on the jobsite, and again after installation is complete. Balancing of air and water distribution systems to deliver services as designed, followed by checking and adjusting controls systems to ensure energy savings and environmental conditions meet design intent. Providing maintenance training and manuals is typically the final step of commissioning.

A final complete commissioning report must be prepared and submitted to the owners along with drawings and equipment manuals. The commissioning report should contain all documentation pertaining to the commissioning procedures and testing results, in addition to deficiency notices and records of satisfactory corrections of deficiencies. System commissioning is typically conducted by a mechanical consultant with appropriate experience and training, ideally hired by and responsible directly to the project owner and independent of the mechanical consultant firm and general contractor. In cases where commissioning is required for very complex projects, a special commissioning coordinator may be designated to be responsible for conducting such commissioning. The architect or designer of record (DOR) typically has the responsibility of overseeing the completion of commissioning.

Buildings requiring commissioning: Formal commissioning is strongly recommended for buildings incorporating complex and digitally controlled HVAC systems, or natural ventilation systems integrated with HVAC systems, those with renewable energy, on-site water treatment systems, occupancy sensor lighting controls, or incorporating other high-technology systems. Commissioning is rarely required for projects with minimal mechanical or electrical complexity, such as typical residential projects. In addition, commissioning also serves an important construction quality-control function for all building types, and helps consultants track the progress of contracts.

For USGBC LEED certification, commissioning is an integral and prerequisite component. For the New Construction, Commercial Interiors, Schools and Core and Shell categories, LEED has two commissioning components:

1. Fundamental Commissioning of Building Systems, which is a prerequisite (i.e., obligatory), and
2. Enhanced Commissioning, which receives a credit but which is not a prerequisite.

5.5.2 Fundamental Commissioning

The *LEED intent* of fundamental commissioning is to support the design, construction, and subsequent operation of a project that meets the owner's project requirements, including energy, water, indoor environmental quality, and durability. The intention should also ensure that installation, calibration, and performance of energy systems meet project requirements, basis of design, and construction documents.

LEED requirements: The following commissioning process activities shall be implemented by the commissioning team (by the end of the design development phase):

1. The owner or project team must designate an individual as the Commissioning Authority (CxA) to lead, review, and oversee the commissioning process activities until completion. This individual should be independent of the project's design or construction management unless the project is less than 20,000 ft.^{2}. This has been reduced from previous versions of LEED, which cited a cut-off point of 50,000 ft.^{2}.
2. The CxA may be a qualified employee of the owner, an independent consultant, or an employee of the design or construction firm who is

not part of the project's design or construction team, or a disinterested subcontractor of the design or construction team. For projects smaller than 20,000 ft.^{2} (1860 m^2), the CxA responsible for fundamental commissioning may be a qualified member of the design or construction team. In all cases, the CxA must report his or her findings directly to the owner.

3. The designated CxA should have documented commissioning authority experience in at least two building projects with a similar scope of work. The experience must extend from early design phase through at least 10 months of occupancy. Additionally, the designated CxA should ideally meet the minimum qualifications of having an appropriate level of experience in energy systems design, installation and operation, as well as commissioning planning and process management. LEED also recommends that a designated CxA have hands-on field experience with energy systems performance, start-up, balancing, testing, troubleshooting, operation, and maintenance procedures and energy systems automation control knowledge.

4. CxA should clearly document and review owner's project requirements and the basis of design for the building's energy-related systems (usually by A/E). Updates to these documents shall be made during design and construction by the design team. The commissioning process does not absolve or diminish the responsibility of the contractor to meet the contract documents requirements.

Commissioning during the design phase is intended to achieve the following specific objectives:

- The design and operational intent are clearly documented and fully understood.
- Ensuring that the recommendations are communicated to the design team during the design process to facilitate the development of commissioning and avoid later contract modifications.
- That the commissioning process for the construction phase is appropriately reflected in the construction documents.

Commissioning during the construction phase is intended to achieve the following objectives in line with the contract documents:

- The commissioning requirements should be incorporated into the construction documents.
- Develop and put in place a commissioning plan.
- Verify and document that the installation and performance of energy-consuming equipment and systems meet the owner's project requirements and basis of design.

- Verify that applicable equipment and systems are installed according to the manufacturer's recommendations and to industry-accepted minimum standards.
- A complete summary commissioning report should be executed.
- Verify that operations and maintenance (O&M) documentation left on site is complete.
- Verify that training of the Owner's operating personnel is adequate.

Commissioning is a LEED™ prerequisite and should be conducted for both new construction and major retrofits, as well as for medium or large energy management control systems that incorporate more than 50 control points. Commissioning is also necessary when large or very complex mechanical or electrical systems are in place and where on-site renewable energy generation systems, such as solar hot water heaters or photovoltaic arrays are in place. Likewise, commissioning is useful where innovative water-conservation strategies, such as graywater irrigation systems or composting toilets are in place. Most building systems over the years become less efficient, because of changing occupant needs, building renovations, wear and tear, and obsolete systems, which can cause significant occupant discomfort and complaints. These issues drive up a facility's operating costs and make it less attractive to new and existing tenants. But these problems can be mitigated, by investing in a commissioning process that typically pays for itself in less than a year.

The following checked systems should be commissioned:

HVAC equipment and system
- Variable speed drives
- Hydronic piping systems
- HVAC pumps
- Boilers
- Chemical treatment system
- Air-cooled condensing units
- Makeup air systems
- Air handling units
- Underfloor air distribution
- Centrifugal fans
- Ductwork
- Fire/smoke dampers
- Automatic temperature controls
- Laboratory fume hoods
- Testing, adjusting, and balancing

- Building/space pressurization
- Ceiling radiant heating
- Underfloor radiant heating

Electrical equipment and system
- Power distribution system
- Lighting control systems
- Lighting control programs
- Engine generators
- Transfer switches
- Switchboard
- Panelboards
- Grounding
- Fire alarm and interface items with HVAC
- Security system

Plumbing system
- Domestic water heater
- Air compressor and dryer
- Storm water oil/grit separators

Building envelope
- Building insulation installation
- Building roof installation methods
- Doors and windows installation methods
- Water infiltration/shell drainage plain

Note: Commissioning needs may differ by project; however, commissioning the building envelope systems, domestic water heating, power distribution, ductwork, and any hydronic piping systems is strongly recommended for any project (Source: BuildingGreen, Inc.)

Retro-commissioning (RCx): While building commissioning is an important aspect of new construction projects as a means of ensuring that all installed systems perform as intended, we find that most buildings have never encountered the commissioning or quality assurance process and not surprisingly are therefore performing well below their potential. Retro-commissioning (RCx) is simply the commissioning of existing buildings and continues to witness increasing prominence as a cost-effective strategy for improving energy performance. Retro-commissioning is an independent systematic process designed to be used on buildings that have never been previously commissioned, to improve the current conditions and operations of an existing building, and to check the functionality of equipment and systems as well as to optimize how they operate and function together in order to reduce energy waste and improve building operation and comfort.

The RCx process, therefore, basically reviews the functionality of installed equipment and systems and optimizes how they function together to facilitate the reduction of energy waste, increase comfort, and improve building operation. RCx may also be required to address issues such as modifications to system components; function/space changes from the original design intent; failure to operate according to designed benchmarks; and complaints regarding indoor air quality (IAQ), temperature, BRI, SBS, and the like. This inefficiency is confirmed by an LBNL study of 60 different types of buildings, which showed that

- >50% had control problems,
- 40% had HVAC equipment problems,
- 15% had missing equipment, and
- 25% had BAS with economizers, VFDs, and advanced applications that were simply not operating correctly.

Source: Association of State Energy Research Technology Internships and US Department of Energy.

Other advantages of retro–commissioning, in addition to energy savings, is the ability to address issues like modifications made to system components, space changes from original design intent, building systems not operating efficiently against benchmarks, complaints regarding IAQ, temperature, etc.

Retro-commissioning phases: This phase requires an assessment of BAS capabilities, potential building opportunities, and Investigation approach. A determination is needed on how systems are intended to operate; measure and monitor how they operate, and prepare a prioritized list of the operating opportunities. Implement opportunities and verify proper operation. Operating improvements made should be recorded and the building operator trained on how to sustain efficient operation and implement capital improvements.

Recommissioning (ReCx) applies mainly to buildings that have previously been commissioned or retro-commissioned. The original commissioning process documented that the building systems performed as intended at one point in time. The intent of recommissioning is to help ensure that the benefits of the initial commissioning or retro-Cx process endure. The need for recommissioning depends on several factors, among them changes in function and use of the facility, quality and schedule of preventive maintenance activities, and frequency of operational problems. In some cases, ongoing-commissioning (ongoing Cx) may be necessary as an "ongoing process" to resolve operating problems, improve comfort, optimize energy use, and identify energy and operational retrofits for existing buildings.

5.5.3 Enhanced Commissioning

For a LEED credit, enhanced commissioning is required in addition to the Fundamental Commissioning prerequisite. The *intent* of enhanced commissioning *according to LEED* is to start the commissioning process early during the design process and execute additional activities upon completion of the systems performance verification. For the Commercial Interiors category, the intent is to verify and ensure that the tenant space is designed, constructed, and calibrated to operate as intended.

*LEED requirement:*This requires the implementation, or to have a contract in place requiring implementation, further commissioning process activities in addition to the Fundamental Commissioning prerequisite requirements, and in accordance with the relevant LEED *Reference Guide.* Duties of the Commissioning Authority (CxA) include:

1. Prior to the end of design development and commencement of the construction documents phase, a commissioning authority (CxA) independent of the firms represented on the design and construction team must be designated to lead, review, and oversee the completion of all commissioning process activities. This person can be an employee or consultant of the owner, although this requirement has no deviation for project size. Furthermore, this person must
   a. have documented commissioning authority experience in at least 2 building projects;
   b. be independent of the project's design and construction management;
   c. not be an employee of the design firm, though the individual may be contracted through them; and
   d. not be an employee of, or contracted through, a contractor or construction manager of the construction project.
2. The CxA must report all results, findings, and recommendations directly to the owner.
3. The CxA must conduct a minimum of one commissioning design review of the owner's project requirements, basis of design, and design documents prior to the midconstruction documents phase, and back-check the review comments in the subsequent design submission.
4. The CxA must review contractor submittals and confirm that they comply with the owner's project requirements and basis of design for

systems being commissioned. This review must be conducted in parallel with the review of the architect or engineer of record and submitted to the design team and the owner.

5. The CxA or other members of the project team are required to develop a systems manual that provides future operating staff with the necessary information to understand and optimally operate the commissioned systems. For Commercial Interiors, the manual must contain the information required for recommissioning the tenant space energy-related systems.

6. The task of verifying that the requirements for training operating personnel and building occupants have been completed, may be performed by either the CxA or other members of the project team.

7. The CxA must be involved in reviewing the operation of the building with O&M staff and occupants and having a plan in place for resolving outstanding commissioning-related issues within 10 months after substantial completion. For Commercial Interiors, there must also be a contract in place to review tenant space operation for O&M staff and occupants.

5.6 GREEN PROJECT COST MANAGEMENT

For the best results, we need to align project management with caring for the environment and integrate the project's activities, and manage costs collaboratively, with the integrated project team engaged wherever possible at the earliest phases of design – using target costing, value management, and risk management. The temptation to put in place a guaranteed maximum price on the project before the design stage is complete should be resisted, to ensure quality and functionality for the building owner or stakeholder; if the price has to be fixed at an earlier stage, it may be prudent to agree on an incentive scheme for the sharing of benefits. A clear understanding of actual construction costs should be aimed for in terms of labor, plant, and materials. All underlying costs should be identified and separated from risk allowances, and a distinction should be made between profit and overhead margins. Also all parties in the supply chain, including material and component suppliers, should have at their disposal reliable data on the operational costs of their products, including running and maintenance costs.

Cost management overview: Management of the running and overall cost of the project is generally the responsibility of the project manager, who in turn reports to the owner (or lender depending on the Contract Documents). Limits of authority for these two roles should be agreed upon before commencement of the project, so that everyone knows exactly what they are empowered to do in managing project costs.

According to the UK Office of Government Commerce, the main ingredients for successful project cost management are the following:

- To manage the base estimate and risk allowance.
- To operate change control procedures.
- To produce cost reports, estimates, and forecasts. The project manager is directly responsible for understanding and reporting the cost consequences of any decisions and for initiating corrective actions if necessary.
- To maintain an up-to-date estimated outturn cost and cash flow.
- To manage expenditure of the risk allowance.
- To initiate action to avoid overspend.
- To issue a monthly financial status report.

The cost management objectives during the construction phase include:

- Delivering the project at the appropriate capital cost using the value criteria established at the commencement of the project.
- Making sure that, throughout the project, full and proper accounts are monitored of all transactions, payments and changes.

The key areas that cost management teams should consider during the design and execution of the project are:

- Identifying elements and components to be included in the project and constricting expenditure accordingly.
- Defining the project program from inception to completion.
- Making sure that designs meet the scope and budget of the project and delivering quality is appropriate and conforms to the brief.
- Checking that orders are properly authorized.
- Certifying that the contracts provide full and proper control and that all incurred costs are as authorized. All materials are to be appropriately specified to meet the project's scope and design criteria and that materials can be procured effectively.
- Monitoring all expenditure relating to risks to ensure that it is appropriately allocated from the risk allowance and properly authorized. Also monitor use of risk allowance to assess impact on overall outturn cost.
- Maintaining strict planning and control of both commitments and expenditure within budgets to help prevent any unexpected cost over/under-runs. All transactions are to be properly recorded and authorized and, where appropriate, decisions are justified.

Risk allowance management: It is no secret that the property development business contains many potential risks, which can be caused by unexpected and uncontrollable issues (e.g., adding "green" features). This is why, it is necessary to incorporate risk allowances. The project owner or someone representing the owner should normally manage the risk allowance since they would be the party most capable of managing these risks, with support and advice from the project manager. Management of the risk allowance consists essentially of a procedure to transfer costs out of the risk allowance into the base estimate for the project work as risks materialize or actions are taken to manage the risks. Formal procedures are required to be put in place for controlling quality, cost, time, and changes. Risk allowances should not be disbursed unless the identified risks to which they relate actually occur. When risks occur that have not previously been identified, they should be treated as changes (variation orders) to the project. Likewise, risks that materialize but have insufficient risk allowance allocated for them will also need to be treated as changes.

Risks and risk allowances should be reviewed on a regular basis, particularly when formal estimates are prepared, but also throughout the design, construction, and equipping stages. A risk–allocated cost contingency is normally set up and included in the total project cost estimate to assist in mitigating any potential risks. With the increase in firm commitments being entered into and the work being carried out, the risks in future commitments and work are naturally reduced. Any risk allowance estimate should reflect this.

Every effort should be made to avoid introducing changes after the briefing and outline design stages are complete. Changes can be minimized by ensuring that from the very beginning the project brief and contract documents are as complete and comprehensive as possible and that the stakeholders have approved it. This may involve early discussions with planning authorities to anticipate their requirements and to ensure that designs are adequately developed and coordinated before construction begins. In the case of existing buildings to be renovated, site investigations, or condition surveys may be required.

Cost planning: In an elemental cost plan the estimate is broken down into a number of components that can then be compared with later estimates, or with actual costs in line with the project's progress. In using this approach, each element is treated as a separate cost center, although money can be transferred between elements as long as a reasonable balance between elements is maintained and the overall target budget is not exceeded. Often the initial cost plan is based on estimates, which provide a fair basis for determining the validity of future assessments. Control by the project manager is achieved by an ongoing review of estimates for each cost center against its target budget. As the design develops and is priced, any differential in cost from the cost plan is identified. Decisions are then taken on whether that element can be authorized to increase in cost, which would require a corresponding reduction elsewhere, or whether the element needs to be redesigned in order to keep within the budget. It is important to note that green buildings need comprehensive planning to achieve optimal results, but the extra effort is usually well worth it when you consider the operating costs over the life of the project.

Continuous and stage estimates: One of the responsibilities of the project manager is to maintain ongoing reviews of designs as they develop and provide advice on costs to the integrated project team. This continuous cost oversight is of particular benefit in assessing individual decisions and is especially useful on large and complex schemes. It may also prove useful to schedule in periodic formal assessments of the whole scheme, as budgetary estimates, at each phase of the project.

Cost control during the design development phase: The owner's designated representative normally has overall responsibility for the project, including the estimated cost, and as such has to be satisfied that appropriate systems are in place and operating for controlling the project's cost. Where substantial monies are earmarked to a design, these costs must be aptly assessed against the budget decision and appropriately authorized. The owner's representative will often delegate a level of financial authority for design development decisions to the integrated project team, appropriate to the project. For complex projects, the owner may prefer to have different delegated levels for each cost center. The payment process is typically managed in a similar manner to the design/construction process including the final payment. Payments for variations, provisional sums, etc., should be discharged after formal approval is given and as the work is carried out. All payments should be made as per the contract documents and on time.

Cost management during construction: During the construction process, instructions issued to the integrated project team, whether for change via a formal change order procedure or for clarification of detail, will have an immediate impact on the project's cost, and possibly other impacts such time delay. For this reason, the project team needs to establish specific procedures and protocols for issuance of instructions and information that ensure that instructions are issued within delegated authority, that before being issued, their cost is estimated and their impact fully evaluated. Any issued instruction should be justified in terms of value for money and overall impact on the project. Furthermore, the cost of all issued instructions is to be constantly monitored, and where costs are estimated to be outside the delegated authority, specific approval is required.

Payment: The normal procedure is for the client, who is the contracting party and therefore legally obligated for paying the integrated project team the interim and final payments as per the contract. These payments usually take the form of stage payments (e.g., monthly intervals) after inspection and assessment of the work in place that are assessed during the course of the work. The client organization's finance division (e.g., bank) should be kept aware of future payment requirements by means of updated reports and cash flow forecasts.

The terms of the contract may include clauses that allow the integrated project team to claim additional payments in certain circumstances as defined in the contract conditions. The justification for these additional payment claims may be the result of or caused by occurring risks that are essentially client risks under the contract, requesting additional/varied work, or failure by the client to comply with its contract obligations – often expressed as disruption to the integrated project team's work program due to modifications or other reasons.

CHAPTER 6

Green Building Materials and Products

6.1 OVERVIEW

Building material is essentially any material that is used for the purpose of construction. The US Green Building Council (USGBC) states that building material choices are important in sustainable design because of the extensive network of extraction, processing, and transportation steps required to process them; further, the activities involved in producing building materials may pollute the air and water, destroy natural habitats, and deplete natural resources. Also, incorporating green products into a project does not imply a sacrifice in performance, or aesthetics, and does not necessarily entail higher costs as exemplified by recent research.

Tiffany Coyle (USGBC) outlines the main areas of focus around materials and resources choices:

- *Conservation of material.* A building generates a large amount of waste throughout its life cycle. Meaningful waste reduction begins with eliminating the need for materials during the planning and design phases.
- *Environmentally preferable materials.* Materials that are locally harvested, sustainably grown, made from rapidly renewable materials, biodegradable, and free of toxins – all these designations demonstrate awareness for sustainability.
- *Waste management and reduction.* The goal is to reduce the waste that is hauled to and disposed of in landfills or incineration facilities. During construction or renovation, materials should be recycled or reused whenever possible. During the building's daily operations, recycling, reuse, and reduction programs can curb the amount of material destined for local landfills.

Many naturally occurring substances, such as clay, sand, wood, and stone, have been used to construct buildings. Apart from naturally occurring materials, many man-made products are also in use, some more and some less synthetic. The manufacture of building materials has long been an established industry in many countries around the world, and the materials'

Copyright © 2016 Elsevier Inc.
All rights reserved.

221

use is typically segmented into specific specialty trades, such as carpentry, plumbing, roofing, and flooring.

Repairing and reuse of a building instead as opposed to tearing it down and building a new structure is one of the most compelling strategies for minimizing environmental impacts. And rehabilitation of existing buildings' components saves natural resources, including the raw materials, energy, and water resources required to build new; prevents pollution that might take place as a by-product of extraction, manufacturing, and transportation of virgin materials; and it also minimizes the creation of solid waste that could end up in landfills.

Some states, such as North Carolina, have initiated grant programs like the "Building Reuse and Restoration Grants Program" to renovate vacant buildings in rural counties or in economically distressed urban areas. The 40 most *distressed counties* are designated as Tier 1, the next 40 as Tier 2, and the final 20 least distressed are classified as Tier 3. This tier system is assimilated into various state programs to inspire economic activity in the less prosperous areas of the state. The USGBC's LEED™ Rating System likewise recognizes the importance of building reuse. Reusing a building can contribute to earning points under LEED™ Materials Resource Credits on Building Reuse.

6.1.1 What is a Green Building Material?

The concept of green building materials is somewhat elusive and defies easy definition. In general, building materials are called green because they are good for the environment. The ideal building material would have no negative environmental impacts, and might even have positive environmental impacts, including air, land, and water purification. There is no perfect green material but in practice, there are a growing number of green materials that reduce or eliminate negative impacts on people and the environment.

Green materials may be made from recycled materials, which puts waste to good use and reduces the energy required to make the materials. Good examples include counter tiles made from recycled automobile windshields and carpeting made from recycled soda bottles. Building materials are also classified as green because they can be recycled once their useful life span is over, for example, aluminum roofing shingles. If that product itself is made from recycled materials, the benefit of recyclability is multiplied.

Building materials are generally considered green when they are made from renewable resources that are sustainably harvested. Flooring made from sustainably grown and harvested lumber or bamboo is a good example.

Some building materials are even considered green because they are durable. A durable form of siding, for instance, outlasts less-durable products, resulting in a significant savings in energy and materials over the lifetime of a property. If a durable product is made from environmentally friendly materials, such as recycled waste, it offers even greater benefits. Siding made from cement and recycled wood fiber (wood waste) is a good example of a green building material that offers at least two significant advantages over less environmentally friendly materials.

6.1.2 Natural Versus Synthetic

Building materials can be normally categorized into two categories, natural and synthetic. Natural building materials are those that are unprocessed or minimally processed by industry, such as timber or glass. Synthetic materials on the other hand are made in industrial settings after considerable human manipulations, such as plastics and petroleum-based paints. Both have their uses as their advantages and disadvantages – pros and cons.

Mud, stone, and fibrous plants are among the most basic natural building materials. These materials are being used together by people all over the world to create shelters and other structures to suit their local weather conditions. In general, stone and/or brush are used as basic structural components in these buildings, while mud is generally used to fill in the space between, acting as a form of concrete and insulation.

Synthetic materials: Plastic is an excellent example of a typical synthetic material. The term plastic covers a range of synthetic or semisynthetic organic condensation or polymerization products that can be molded or extruded into objects, films, or fibers. The name is derived from malleable nature when in a semiliquid state. Plastics vary immensely in heat tolerance, hardness, and resiliency. Combined with this adaptability, their general uniformity of composition and lightness of plastics facilitates their use in almost all industrial applications.

The evaluation of a material's greenness is generally based on certain criteria, such as whether the material is renewable and resource efficient in its manufacture, installation, use, and disposal.

Likewise, whether the material supports the health and well being of occupants, construction personnel, the public, and the environment and whether the material is appropriate for the application and what the environmental and economic trade-offs among alternatives are needs to be evaluated.

Much research remains to be conducted to satisfactorily evaluate alternatives and select the best material for a project. Material selection ideally considers the impacts of a product throughout its life cycle (from raw material extraction, to use, and then to reuse, recycling, or disposal). The areas of impact to consider at each stage in the life cycle of a material include the following:

- Natural resource depletion, air and water pollution, hazardous and solid waste disposal, durability
- Energy required for extraction, manufacturing, and transport of materials
- Energy performance in useful life of material, durability
- Effect on indoor air quality; exposure of occupant, manufacturer, or installer to harmful substances; moisture and mold resistance; cleaning and maintenance methods.

Among the properties that typical green building materials and techniques may have are the following:

- Durable
- Readily recyclable or reusable when no longer needed
- Sustainably harvested from renewable resources
- Can be salvaged for reuse, refurbished, remanufactured, or recycled
- Manufactured from a waste material such as straw or fly ash, or a waste-reducing process
- Minimally packaged or wrapped with recyclable packaging
- Locally extracted and processed
- Energy-efficient in use
- Less energy used in extraction, processing, and transport to the job site
- Generate renewable energy
- Water-efficient
- Manufactured with a water-efficient process

6.1.3 Storage and Collection of Recyclables

A Materials and Resources prerequisite in most of the LEED Rating System categories is the "Storage and Collection of Recyclables." The intent of this prerequisite is essentially to reduce waste sent to landfills by encouraging storage and collection of recyclables. The main differences between LEED v4 and LEED v2009 are the following:

- Batteries, mercury-containing lamps, and e-waste all require dedicated storage; project teams must choose any two of the three.
- For retail projects, the required number of waste streams with dedicated storage has increased from three to four.

The USGBC has put forward general guidelines (Table 6.1) and says that the area should be easily accessible, serve the entire building(s), and

Table 6.1 Recycling storage area guidelines based on overall building square-footage

Commercial Building Square-footage (ft.^{2})	Minimum Recycling Area (ft.^{2})
0 to 5,000	82
5,001 to 15,000	125
15,001 to 50,000	175
50,001 to 100,000	225
100,001 to 200,000	275
200,001 or greater	500

(*Source:* USGBC).

be dedicated to the storage and collection of nonhazardous materials for recycling. Items that are considered recyclable material include mixed paper, corrugated cardboard, glass, plastics, and metals.

6.2 LOW-EMITTING MATERIALS

LEED addresses low-emitting materials within the Indoor Environmental Quality section and as outlined in Chapters 3 and 4 of this handbook. LEED says the intent of low-emitting materials "is to reduce concentrations of chemical contaminants that can damage air quality, human health, productivity, and the environment."

6.2.1 Adhesives, Finishes, and Sealants

For LEED™ credits, all adhesives and sealants with volatile organic compound (VOC) content must not exceed the VOC content limits of South Coast Air Quality Management District (SCAQMD) Rule #1168 limits scheduled for 2007 as indicated in Table 6.2. Aerosol adhesives not covered by Rule 1168 must meet Green Seal Standard GS–36 requirements. Carpet adhesive must meet the requirements of IEQ Credit 4.1: Adhesives and Sealants, which includes a VOC limit of 50 g/L. All indoor air contaminants that are odorous, potentially irritating, and/or harmful to the comfort and well-being of installers and occupants should be minimized.

A characteristic of sealants is that they increase the resistance of materials to water or other chemical exposure, while caulks and other adhesives can help control vibration and strengthen assemblies by spreading loads beyond the immediate vicinity of fasteners. Both properties enhance durability of surfaces and structures, although they do so at a cost as they are often hazardous in

Table 6.2 Adhesives, sealants and sealant primers: South Coast Air Quality Management District (SCAQMD) Rule #1168

Architectural Applications (SCAQMD 1168)	VOC Limit (g/L less water)	Specialty Applications (SCAQMD 1168)	VOC Limit (g/L less water)
Ceramic tile	65	Welding: ABS (avoid)	325
Contact	80	Welding: CPVC (avoid)	490
Drywall and panel	50	Welding: plastic cement	250
Metal to metal	30	Welding: PVC (avoid)	510
Multipurpose construction	70	Plastic primer (avoid)	650
Rubber floor	60	Special-purpose contact	250
Wood: structural member	140		
Wood: flooring	100	**Sealant primers (SCAQMD**	
Wood: all other	30	**1168)**	
All other adhesives	50	Architectural porous primers	775
Carpet pad	50	(avoid)	
Structural glazing	100	Sealants and nonporous primers	250
		Other primers (avoid)	750
Substrate-specific applications			**VOC limit (%)**
Fiberglass	80	**Aerosol adhesives (GS-36)**	
Metal to metal	30	General purpose mist spray	65
		General-purpose web spray	55
		Special-purpose aerosol adhesives	70

Note: VOC limits are listed in the table and correspond to an effective date of July 1, 2005, and rule amendment date of January 7, 2005. In 2015, the Environmental Protection Agency proposed to approve revisions to the South Coast Air Quality Management District (SCAQMD) and Sacramento Metropolitan Air Quality Management District (SMAQMD) portions of the California State Implementation Plan (SIP). These revisions concern particulate matter (PM) emissions from particulate matter air pollution control devices and residential wood burning.
Note: VOC weight limit is based on grams/liter of VOC minus water. Percentage is by total weight
Source: Based on USGBC.

manufacture and application. Many construction adhesives contain more than 30% volatile petroleum-derived solvents, such as hexane, to maintain liquidity until application. Because of this, workers become exposed to toxic solvents, and as the materials continue to outgas during curing, the occupants can also be potentially exposed to emissions for extended periods, which will have a negative on occupants' health.

Water-based adhesives are available from a number of different manufacturers. Industry tests indicate that these products work as well as or better than solvent-based adhesives, can pass all relevant American Society for Testing and Materials (ASTM) and the American Plywood Association – The Engineered Wood Association (APA) performance tests, and are available at comparable costs to common solvent-based adhesives. When adhesives are purchased in bulk, larger containers can often be returned to vendors for refill.

Stains and sealants also normally emit potentially toxic VOCs into indoor air. The most efficient method to manage this problem is to employ materials that do not require additional sealing, such as stone, ceramic and glass tile, and clay plasters. The toxicity, and the air and water pollution generated through the manufacture of chlorinated hydrocarbons such as methylene chloride, emphasizes strongly for using responsible, effective alternatives, such as plant-based, nontoxic, or low-toxicity sealant formulations.

According to LEED™ requirements, all adhesives and sealants used on the interior of the building (defined as inside of the weatherproofing system and applied on-site) should comply with the reference standards shown in Table 6.2.

Maintenance and cleaning products: Environmentally preferable cleaning methods and products can reduce indoor air pollution and solid/liquid waste generation. Safe cleansers are readily available and are competitively priced and environmentally friendly. The improper use and disposal of some common cleaning and maintenance products can contribute to indoor air contamination, water pollution, and toxic waste.

Biodegradability is a key factor for surfactants, which are the active ingredients in cleaners. Even low surfactant concentrations in runoff can increase the ability to pose greater risks to the environment. Generally, petroleum-derived surfactants break down more slowly than vegetable oil–derived fatty acids; some materials are even resistant to municipal sewage treatment.

Important considerations to keep in mind to minimize the harmful effects of toxins should include the following:

- Select materials that have a durable finish and that do not require frequent stripping, waxing, or oiling (e.g., colored concrete, linoleum, or cork).
- Hazardous materials should be stored outside the building envelope.
- Select whenever possible biodegradable, nontoxic cleansers.
- Avoid selecting cleansers, waxes, and oils that are labeled as toxic or highly toxic, poisonous, harmful or fatal if swallowed, corrosive, flammable, explosive, volatile, causing cancer or reproductive harm, or which require "adequate ventilation" or safety equipment.
- Select products that have approved third-party or government agency certification:
  - Green Seal
  - Scientific Certification Systems (SCS)

- US EPA Environmentally Preferable Purchasing Program
- General Services Agency
- CIWMB Recycled Content Product Directory
- Minimize stripper use by placing mats at all building entrances and clean regularly; dust mop and/or vacuum frequently, and wet mop with a liquid cleaner; refinish only areas where the finish surface is wearing.

6.2.2 Paints and Coatings

The intent here is to reduce the quantity of indoor air contaminants that are odorous, irritating, and/or harmful to the comfort and well being of installers and occupants. Paints basically consist of a mixture of solid pigment suspended in a liquid vehicle and applied as a thin, usually opaque coating to a surface for protection and/or decoration. Primers are basecoats applied to a surface to increase the adhesion of subsequent coats of paint or varnish. Sealers are also basecoats but are applied to a surface to help reduce the absorption of subsequent coats of paint or varnish, or to prevent bleeding through the finish coat. Green Seal requires that all paints and coatings, including primers and undercoats, shall meet certain performance requirements. All tests shall be performed on products produced by the manufacturer and do not include additives at the point of sale.

The USGBC requires that paints and coatings applied on site and used on the interior of the building (defined as inside of the weatherproofing system) shall comply with the following referenced standards:

1. Architectural paints, coatings, and primers applied to interior walls and ceilings: Do not exceed the VOC content limits established in Green Seal Standard GS-11, Paints, third edition, August 17, 2011.
   a. Flats: 50 g/L
   b. Nonflats: 100 g/L
2. Anticorrosive and antirust paints applied to interior ferrous metal substrates: Do not exceed the VOC content limit of 250 g/L established in Green Seal Standard GC-03, Anti-Corrosive Paints, second edition, January 7, 1997.
3. Clear wood finishes, floor coatings, stains, and shellacs applied to interior elements: Do not exceed the VOC content limits established in South Coast Air Quality Management District (SCAQMD) Rule 1113, Architectural Coatings, rules in effect on January 1, 2004.
   a. Clear wood finishes: varnish 350 g/L; lacquer 550 g/L
   b. Floor coatings: 100 g/L

c. Sealers: waterproofing sealers 250 g/L; sanding sealers 275 g/L; all other sealers 200 g/L

d. Shellacs: Clear 730 g/L; pigmented 550 g/L

e. Stains: 250 g/L

Table 6.3 shows the allowable VOC levels stipulated by SCAQMD.

According to Santa Cruz County, paint has significant environmental and health implications in its manufacture, application, and disposal. Most paint, even water-based "latex," is derived from petroleum. Its manufacture requires substantial energy and water and creates air pollution and solid/liquid waste. VOCs are typically the pollutants of greatest concern in paints. VOCs from the solvents found in most paints (including latex paints) are released into the atmosphere during manufacture, application, and for weeks or months after application. VOCs emitted from paint and other building materials are associated with eye, lung, and skin irritation, headaches, nausea, respiratory problems, and liver and kidney damage.

Exposure to solvents emitted by finished products can be significant. However, renewable alternatives, such as milk paint, address many of these concerns but often at a premium price, and some products are only suitable for indoor applications. Reformulated low- and zero-VOC latex paints with excellent performance in both indoor and outdoor applications are now available at the same or lower price than older high-VOC products. Paints that meet GS-11 standards are low in VOCs and aromatic solvents, do not contain heavy metals, formaldehyde, or chlorinated solvents, and meet stringent performance criteria.

Silicate paints provide an alternative type of paint that is solvent-free and that may be used on concrete, stone, and stucco. Silicate paint has many advantages, being odorless, nontoxic, vapor-permeable, naturally resistant to fungi and algae, noncombustible, colorfast, light-reflective, and even resists acid rain. These paints cannot spall or flake off, and will only crack if the substrate cracks. Though silicate paints are expensive, their extraordinary durability can be a significant compensation.

Table 6.3 Allowable VOC levels (grams/liter less water and exempt compounds) according to the South Coast Air Quality Management District (SCAQMD)

Paint Type	VOC Limit (g/L)
Flat	50
Nonflat	50
Primers, sealers, and undercoats	100
Quick-dry enamels	50

6.2.3 Flooring Systems

1. *Carpet*

Carpet manufacture, use, and disposal have significant environmental and health implications.

Most carpet products are synthetic, usually derivatives of nonrenewable petroleum products; its manufacture requires substantial energy and water, and creates harmful air and solid/liquid waste. However many carpets are now becoming available with recycled content, and a growing number of carpet manufacturers are refurbishing and recycling used carpets into new carpet. At the end of its useful life, most carpet tends to end up in landfills; in California, for example, roughly 840,000 tons of carpet − roughly 2% of its waste stream − ends up in landfills every year. And it is estimated that more than four billion pounds of carpet enter the solid waste stream annually in the United States, accounting for more than 1% by weight and about 2% by volume of all municipal solid waste.

Leasing arrangements in which the manufacturer will recycle worn or stained carpets also help to reduce waste. Resources are required for recycling and, unless it can be recycled indefinitely, the carpet will still end up in a landfill after a finite number of uses. Carpet tiles limit waste because only worn or stained tiles need to be replaced and are available for both commercial and residential applications.

A variety of VOCs can be emitted from carpet materials, although VOC emissions from new carpet typically fall to very low levels within 4872 h after installation when accompanied by good ventilation. Synthetic carpets, backings, and adhesives typically emit VOCs, all of which will pollute indoor and outdoor air. Redesigned carpets, new adhesives, and natural fibers are available that emit low or zero amounts of VOCs. For improved air quality, selected carpets and adhesives should meet a third-party standard, such as the Carpet and Rug Institute (CRI) Green Label Plus or the State of California's Indoor Air Emission Standard 1350. VOCs often have an odor and are often characterized as the "new carpet smell."

Natural fibers are an environmentally preferable carpeting option because they are renewable and biodegradable. Options include jute, sisal, coir, and wool floor coverings. Biodegradable carpets made from plant extracts and plant-derived chemicals are also available. One of the disadvantages of carpets is that they tend to harbor more dust, allergens, and contaminants than many other materials (Figure 6.1a). Durable

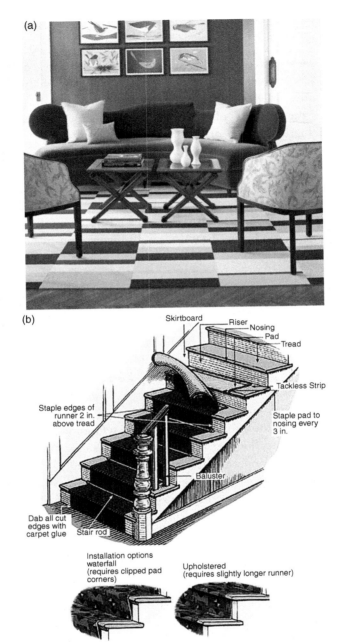

(a)

(b)

Skirtboard
Riser
Nosing
Pad
Tread

Tackless Strip

Staple edges of
runner 2 in.
above tread

Staple pad to
nosing every
3 in.

Baluster

Dab all cut
edges with
carpet glue

Stair rod

Installation options
waterfall
(requires clipped pad
corners)

Upholstered
(requires slightly longer runner)

Figure 6.1 (a) Photo of FLOR carpet squares laid in a flexible and practical "tile" format. The tiles are made from renewable and recyclable materials and are available in a range of colors, textures, and patterns. More than 2 million tons of carpets are landfilled in the United States each year. (b) There are several reasons to include a stair runner to your staircase. For example, in addition to being more attractive, Stair runners are a good way to protect the stair treads from wear. Additionally, Stair runners are safer particularly when coming down a staircase. The runner provides a better gripping surface than a polyurethane finished stair tread. *Courtesy: Part a, FLOR Inc.; part b, Architectural Turnings.*

flooring, such as a concrete finish floor, linoleum, cork, or reclaimed hardwoods, is generally preferable in helping to improve indoor air quality.

The intent of low-emitting carpet systems according to LEED is to reduce the quantity of indoor air contaminants that are odorous, potentially irritating and or harmful to the comfort and well-being of installers and occupants. In addition, for LEED credits, all carpet installed within a building's interior must meet or exceed testing and product requirements of the Carpet and Rug Institute (CRI) Green Label Plus program for VOC emission limits. Carpet pads installed within a building's interior must meet or exceed CRI's Green Label program for VOC emission limits. Adhesives and sealants used in carpet system installation must comply with South Coast Air Quality Management District (SCAQMD) Rule #1168.

It should be noted that in LEED™ 2009 NC, C&S and CI, the title of EQ 4.3 has been changed from "Carpet" to "Flooring Systems," and this credit has been substantially expanded in LEED. From LEED 2009 and moving forward, all of the hard surface flooring must be certified as compliant with the FloorScore standard by an independent third party. FloorScore is a program developed by the Resilient Floor Covering Institute and SCS, which tests and certifies hard surface flooring and flooring adhesive products for compliance with indoor air quality emission targets. Flooring products covered by FloorScore include vinyl, linoleum, laminate flooring, wood flooring, ceramic flooring, rubber flooring, wall base, and associated sundries.

2. *Vinyl/PVC*

Polyvinyl chloride (PVC, often referred to as "vinyl") deserves special attention because it accounts for almost 50% of the total plastic used in construction and because it is increasingly recognized as problematic. Vinyl is ubiquitous – some 14 billion pounds are produced every year in North America – and artificially cheap, and not all of its alternatives have yet worked out all their negative issues. Moreover, as the USGBC suggested in its long-awaited report on PVC, all materials have potential pitfalls, from indoor air quality to disposal. In addition to flooring, PVC is common in pipes, siding, wire insulation, conduit, window frames, wallcovering, and roofing – as well as a host of other products, including fabrics.

Vinyl has been extensively used because of its benefits: good strength relative to its weight, durability, water resistance, and adaptability. Vinyl tends to be inexpensive, in part because vinyl production typically requires roughly half the energy required to produce other plastics.

Products made from vinyl can be resistant to biodegradation and weather, and are effective insulators. The physical properties of vinyl can be tailored for a wide variety of applications. Many firms are increasingly concerned about the dramatic environmental liabilities of vinyl but are struggling to find appropriate substitutes. It should also be noted that the vinyl industry has been vocally opposed to the new LEED v4 MR credits although LEED-NC v4 does not actually ban any building materials.

Vinyl environmental concerns: Vinyl's life cycle begins and ends with hazards, most stemming from chlorine, its primary component. Chlorine makes PVC more fire resistant than other plastics. Vinyl chloride, the building block of PVC, causes cancer. Lead, cadmium, and other heavy metals are sometimes added to vinyl as stabilizers, and phthalate plasticizers, which give PVC its flexibility, pose potential reproductive risks. Also, over time, phthalates can leach out or offgas, exposing building occupants to materials linked to reproductive system damage, and cancer in laboratory animals. Manufacturing vinyl or burning it in incinerators produces dioxins, which are among the most toxic chemicals known to man. Research has shown that the health effects of dioxin, even in minute quantities, include cancer and birth defects. But in addition to PVC, "vinyls" in building materials also include

a. ethylene vinyl acetate (EVA), which is used in films, wire coating and adhesives,
b. polyethylene vinyl acetate (PEVA) which is a copolymer of polyethylene and EVA and is used in shower curtains, body bags, etc.,
c. polyvinyl acetate (PVA), is used in paints and adhesives,
d. polyvinyl butyryl (PVB), which is used in safety glass films.

PVC is produced from vinyl chloride monomer (VCM) and ethylene dichloride (EDC), which are carcinogenic and acutely toxic. During the production of PVC, VCM and EDC are released into the environment, and for which there is no safe vinyl chloride exposure level. Clean air regulation and liability concerns have been effective in reducing total VCM releases since 1980, whereas PVC use has roughly tripled. Although PVC resin is inert in normal use, older PVC products are frequently contaminated with traces of VCM, which can leach into the surrounding environment and contaminate drinking water.

Environmentally preferable alternatives to vinyl: Substitutes may cost more or require different maintenance, but many will also outlast plastics with proper care. Moreover, for many applications, particularly indoors where occupants can be directly exposed to off-gassing plasticizers, substitution of vinyl would clearly be prudent for the sake of health and well-being.

Examples of possible alternatives include:

a. Flooring made from natural linoleum, cork, tile, finished concrete, or earth.

b. Windows framed with fiberglass, Forest Stewardship Council (FSC)-certified wood, or possibly wood-based composites utilizing formaldehyde-free binders.

c. Stucco, lime plaster, reclaimed wood, fiber-cement, and FSC-certified wood siding.

d. Glass shower doors instead of vinyl curtains.

e. Natural wallcoverings instead of vinyl wallpaper.

Also, we are now witnessing the beginning of some of the nonchlorinated vinyls (EVA, PEVA, PVA, and PVB) being used as direct substitutes for PVC.

3. *Tile*

Tile is made primarily from fired clay (porcelain and other ceramics), glass, or stone, and provides a useful option for flooring, countertops, and wall applications whose principle environmental benefit is durability. Tile can last indefinitely even in high-traffic areas, eliminating the waste and expense of replacements. Tile production however is energy intensive, although tile from recycled glass requires less energy than tile from virgin materials. Among tiles' attributes is that they do not burn, will not retain liquids, and do not absorb fumes, odors, or smoke and, when installed with low- or zero-VOC mortar, can contribute to good indoor air quality.

Tile can offer such performance only if it has the appropriate surface hardness for the location. Tile hardness is measured on the PEI (Porcelain Enamel Institute) scale of 0–5, with 0 being the least hard, indicating a tile should not be used as flooring and 5 signifying a surface designed for very heavy foot traffic and abrasion. If kept relatively free of sand and grit, floor tile can easily last as long as the building it is in.

The impacts of mining, producing, and delivering a unit of tile are important considerations. Approximately 75% of the tiles used in the United States are imported. The remaining amount represents about 650 million ft.^{2} of ceramic tile produced by US factories each year, together with the billions of square feet manufactured throughout the world. This requires mining millions of tons of clay and other minerals, and substantial energy to fire material into hardened tile. Once the raw materials are processed, a number of steps are needed to obtain the finished product. These steps include batching, mixing and grinding, spray-drying, forming, drying, glazing, and firing. Many of these steps are now accomplished using automated equipment. Stone requires relatively little energy to process but significant energy to quarry and ship. Selecting tile made locally and regionally may dramatically reduce the energy use and pollution of transport.

Glazed versus unglazed tiles: Nearly 99% of the tile industry's product share consists of glazed tile, unglazed floor tile, wall tile (including quarry tile), and ceramic mosaic tile (Figure 6.2). Because of the industry being so focused on decorative tiles, it has become completely dependent on the economic health of the construction and remodeling industries.

Figure 6.2 *Interior Showing Installation of Crystal Micro Double Loading Polished Tile.*
Courtesy: Foshan YeSheng Yuan Ceramics Co.

Ceramic glazed tile production resembles the production of ordinary ceramic tile, except that it includes a step known as glazing. Glazing ceramic tile requires a liquid made from colored dyes and a glass derivative known as flirt that is applied on the tile, either using a high-pressure spray or by directly pouring on the tile. This in turn gives a glazing look to the ceramic tile. Tile in low-traffic areas, particularly roofing, may use lower-impact water-based glazes. Though they can be more slippery, glazed tiles are practically stainproof. In addition to their slip resistance, the integral color and generally greater thickness of unglazed tiles tend to make them more durable than glazed tiles. However, unglazed tiles may require a sealant (factory-sealed tiles can help minimize or eliminate a source of indoor VOC emissions). Glass floor tile can also offer a nonskid surface appropriate for ADA compliance. When installing stone tile, especially for countertop applications, use a nontoxic sealer for the grout and tile surface.

6.2.4 Earthen Building Materials

The use of earthen building materials goes back to Sumerian times, before the invention of writing, and includes adobe bricks – made from clay, sand, and straw; rammed earth – compressed with fibers for stabilization; and cob – clay, sand, and straw that is stacked and shaped while wet. Provided they are obtained locally, earthen building materials can reduce or eliminate many of the environmental problems posed by conventional building materials since they are plentiful, nontoxic, reusable, and biodegradable.

Well-built earthen buildings are known to be durable and long lasting and require little maintenance. For thousands of years, people throughout the world have crafted cozy homes and communities with earthen materials that provide excellent shelter after centuries of use. A key element of American architectural vernacular, the Great Plains, are home to sod and straw bale construction, and in the Southwest, adobe construction provides protection from extremes in summer and winter. Though the domestic popularity of earthen materials waned during the twentieth century, a revival can be witnessed since the 1970s. By contrast, modern "stick-frame" construction, which requires specialized skills and tools, became standard practice in the United States only since the end of World War II and today remains uncommon in many parts of the world.

Considerations in earthen construction:
- Earthen walls are thick and may comprise a higher percentage of floor area on a small site.

- Construction is generally labor-intensive, although minimal skill is required.
- Multistory and cob structures require post-and-beam designs.
- May be more difficult to obtain necessary permits although the necessary code recognition, structural testing, etc., is available
- If labor is done primarily by building professionals, the square-foot cost of earthen construction is comparable to conventional building methods.

Benefits of earthen materials include the following:

- Environmental impact is minimal, provided materials come from local sources.
- Durable and maintenance required is low.
- Thermal mass helps keep indoor temperatures stable, particularly in the mild to warm climates.
- Biodegradable or reusable.
- Can be easy to build, requiring few special skills or tools.
- If well designed, provides pleasing aesthetics.
- Highly resistant to fire and insects.
- Require no toxic treatments, and does not offgas hazardous fumes; good for chemically sensitive individuals.

Earthen flooring: Earthen flooring, also called adobe floor, is a durable, inexpensive, environmentally friendly, and uniquely aesthetic complement to a home or office. Because "dirt" is plentiful and locally available, earthen flooring virtually eliminates the waste, pollution, and energy necessary to manufacture a floor, and can save money. However, in the United States earth floors are still most often used in outbuildings and sheds, but if properly installed, can also be used in interior spaces (Figure 6.3). For interior use, earth floors must be properly insulated and moisture sealed and must be protected from capillary action of water by sealing with a water-tight membrane underlayment.

Construction preparation includes removal of any vegetation under the floor area followed by ramming of the area. The ground must be dry before installation of the floor. After the surface is moistureproofed, a foundation of stone, gravel or sand is installed, 20–25 cm deep. Then, an insulating layer is installed, such as a straw clay mixture. Key to a good earthen floor is the proper mixture of dirt, clay, and straw. (Stabilizers such as starch paste, casein, glues, or Portland cement may be added for a harder floor.) Earthen floors are first troweled to a smooth finish and then usually sealed with an oxidizing oil such as linseed or hemp oil, which hardens it and allows it to be swept and mopped.

Figure 6.3 *Illustration Showing Method for Installing an Earth Floor. Source: Dancing Rabbit.*

Earthen floor considerations include:
- Elimination of construction waste; any excess earth can be reincorporated into the landscape.
- Materials are generally inexpensive when found locally.
- Minimal to zero pollution – earthen materials require only simple processing and little or no transport. Even when produced by a machine, a finished earthen slab is estimated to have 90% lower embodied energy than finished concrete.
- Durable with proper care, and repairable.
- Low maintenance and when properly sealed can be swept or moist-mopped; stabilized earthen flooring is not dusty.
- Labor intensive to install
- High traffic areas such as entries or work spaces may require flagstones or other protective materials.
- More vulnerable to scratching and gouging than hard tile or cement – but earthen flooring is more durable than vinyl because it is repairable.
- Few local contractors are experienced with earthen flooring.

6.2.5 Windows

Windows are an essential element in construction because they provide ventilation, light, views, and a connection to the outside world. Drafty,

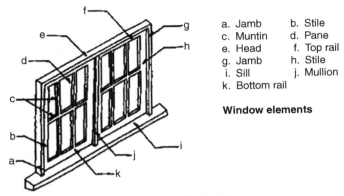

a. Jamb	b. Stile
c. Muntin	d. Pane
e. Head	f. Top rail
g. Jamb	h. Stile
i. Sill	j. Mullion
k. Bottom rail	

Window elements

Figure 6.4 *Drawing of a Window Showing Individual Elements.*

inefficient, poorly insulated, or simply poorly chosen windows can compromise the energy efficiency of a building's envelope. The fabrication of windows whether made of wood, aluminum, plastic or steel, as with any other manufactured product will require energy and will likely generate air pollution. Energy efficiency is one of the main considerations in reducing the environmental impacts of a window, followed by waste generated in manufacturing and general durability. Figure 6.4 shows the various components of a window.

Windows are available in a variety of glazing options. Each option offers a different thermal resistance or *R*-value. *R*-values are approximate and vary with temperature, type of coating, type of glass, and distance between glazings. From lowest resistance to greatest, they are as follows:

1. Single glazing and acrylic single glazing are similar. $R = 1.0$
2. Single glazing with a storm window and double glazing are similar. $R = 2.0$
3. Double glazing with a low-emissivity (low-e) coating and triple glazing are similar. $R = 3.0$
4. Triple glazing with a low-e coating. $R = 4.0$
   Note that for a conventional, insulated stud wall, $R = 14.0$.

Older, single-pane windows are very unlikely to perform as well as new windows and should preferably be reused only in unheated structures such as greenhouses. Residential window frames are typically made from wood, vinyl, aluminum, or fiberglass or from combinations of wood and aluminum or vinyl (i.e., clad). Each has different cost, insulating ability, and durability as shown later:

- Wood requires continuous maintenance for durability. The wood source should be certified by accredited organization such as by the FSC.

- Fiberglass is energy intensive to manufacture, but is strong, durable, and has excellent insulating value.
- Aluminum and steel are poor insulators and very energy intensive to manufacture. When using metal-framed windows, look for recycled content and seek frames with "thermal breaks" to limit the loss of heat to outdoors.
- Vinyl offers good insulation, but is highly toxic in its manufacture and if burned. High-efficiency windows typically utilize dual or triple panes with low-e coatings and gas fill (typically argon) between panes to help control heat gain and loss. Factory-applied low-e coatings on internal glass surfaces are more durable and effective than films. High-quality, efficient windows are widely available from local retailers. To make an informed choice, consider only windows that have National Fenestration Rating Council (NFRC) ratings. The EPA ENERGY STAR® label for windows can be a useful summary of these factors.

NFRC (National Fenestration Rating Council) ratings include the following:

- The U-factor summarizes a window's ability to keep heat inside or outside a building. The lower the U-factor, the better its insulating value; look for values of 0.4 or lower.
- Solar heat gain coefficient (SHGC) summarizes a window's capability to block heat from sunlight. Seek out SHGC values of less than 0.4.
- Visible light transmittance (VLT) is a measure of how much light gets through a window. Desired VLT varies with taste and application.
- Low values for air leakage are best.
- The higher the condensation resistance, the better; values range from 0 to 100. Condensation can contribute to mold growth, although new, high-quality windows (with low U-factor) are generally better equipped to resist condensation than older windows.

6.2.6 Miscellaneous Building Elements

1. *Gypsum wallboard*
   According to the Gypsum Association,

   Gypsum board is the generic name for a family of panel products that consist of a noncombustible core, composed primarily of gypsum, and a paper surfacing on the face, back and long edges. Gypsum board is one of several building materials covered by the umbrella term "gypsum panel products." All gypsum panel products contain gypsum cores; however, they can be faced with a variety of different materials, including paper and fiberglass mats.

Gypsum wallboard, also known as plasterboard, or Drywall, is a plaster-based wall finish that is available in a variety of sizes. The standard size gypsum boards are 48 in. wide and 8, 10, 12, or 14 ft. long. The 48-in. width is compatible with standard framing methods in which studs or joists are spaced 16 in. and 24 in. o.c. Thicknesses vary in 1/8 in. increments from 1/4 in. to 3/4 in. It differs from other panel-type building products, such as plywood, hardboard, and fiberboard because it contains a noncombustible core and paper facers. When used in interiors, joints and fastener heads are covered with a joint compound system, thereby creating a continuous surface suitable for most types of interior decoration. A typical board application is shown in Figure 6.5a.

In the United States and Canada, gypsum board is manufactured to comply with ASTM Specification C 1396. This standard must be met whether the core is made of natural ore or synthetic gypsum.

The vast majority of the synthetic gypsum used by the industry is a by-product of the process used to remove pollutants from the exhaust created by the burning of fossil fuels for power generation. Nearly 100% of the fiber used in the production of gypsum board face and back paper comes from newsprint and postconsumer waste materials.

Advantages of gypsum board include low cost, ease of installation and finishing, fire resistance, sound control, and availability. Disadvantages include difficulty in curved surface application and low durability when subject to damage from impact or abrasion.

Because of its ease of installation, familiarity, fire resistance, nontoxicity, and sound attenuation, gypsum wallboard, known by its proprietary names Drywall® and Sheetrock®, is ubiquitous in construction. Gypsum wallboard is a benign substance (basically paper-covered calcium sulfate), but it has significant environmental impacts because it is used on a vast scale; domestic construction uses an estimated 30 billion ft.^{2}/year.

The primary environmental impacts of gypsum are habitat disruption from mining, energy use, and associated emissions in processing and shipment, and solid waste from disposal. Using "synthetic" or recycled gypsum board can significantly reduce several of these impacts. Synthetic gypsum accounts for approximately 20% of US raw gypsum use and is made from the by-product of manufacturing and energy-generating processes, primarily from desulfurization of coal power plant exhaust gases. In excess of 80% of coal fly ash sold in the United States is used in gypsum board manufacturing.

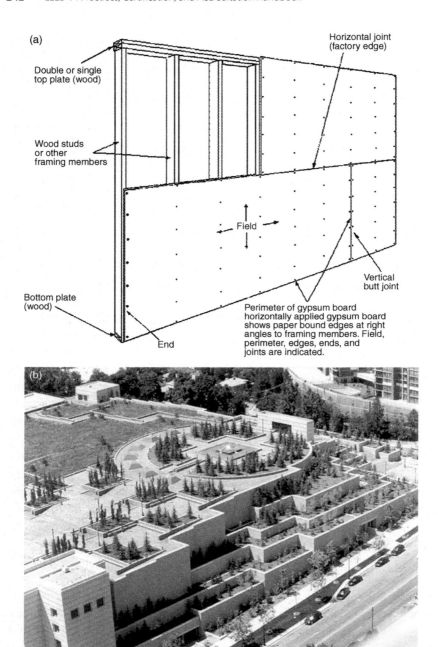

Figure 6.5 (a) Horizontally applied gypsum wallboard showing joints and framing. (b) Example of 70,000 ft.^{2} of vegetated roof on the LDS Assembly Hall building, Salt Lake City, UT. *Source: Part a, Gypsum Association; part b, American Hydrotech, Inc.*

Synthetic gypsum, which is now used in about 30% of drywall, is a by-product of coal-fired power plants and comprises about 95%, by weight, of American Gypsum wallboard. American Gypsum say it is processed almost identically as you would natural gypsum rock. Though synthetic gypsum board use is growing in popularity, diverting drywall from the waste stream is proving more challenging. Reclaimed gypsum board can easily be recycled into new gypsum panels that conform to the same quality standards as natural and synthetic gypsum, but doing this may not be practical because gypsum is an inexpensive material but can require significant labor to separate and prepare it for recycling. Gypsum board face paper is commonly 100% recycled, from newsprint, cardboard, and other postconsumer waste streams, but most recycled gypsum in wallboard products is postindustrial from gypsum board manufacture. Gypsum board should be purchased in sizes that minimize the need for trimming (saving time and waste). Working crushed gypsum off cuts (that have not been painted, glued, or otherwise contaminated) into soil helps reduce waste while improving the workability and calcium availability of many soils.

Normally, synthetic gypsum has lower trace metal content then what is typically found in residential soil standards. Synthetic gypsum wallboard products by American Gypsum, for example, are just as safe as natural gypsum and are certified as low-VOC products by the GREENGUARD Environmental Institute.

2. *Siding*

Siding is an external protection element that provides protection for wall systems from moisture and the heat and ultraviolet radiation of the sun. Selecting siding that is reclaimed, recyclable, and biodegradable in a landfill or incorporates recycled material are key considerations that will reduce waste and pollution. Maintenance, too, is an important consideration. High-maintenance materials requiring regular upkeep, such as repainting, and use additional resources and energy over their life cycle are less sustainable. There are many types of siding, and the environmental impacts of siding products vary considerably.

Considerations: Earth or lime plasters last a long time and require relatively little maintenance. Cement or lime is commonly added for improved hardening and durability, but the comparatively small (or zero) overall cement content of natural plasters means the material causes relatively small amounts of pollution and energy use to prepare and install. Deep eaves or overhangs that protect the siding from extended moisture exposure are critical to the longevity of natural plasters.

Fiber-cement siding has proven to be very durable, and many products are backed by 50-year or lifetime warranties. It is fire and pest resistant and emits no pollutants in use. However, it possesses a high embodied energy because of its cement content and because it is manufactured with wood fiber from overseas.

Cement stucco is another extremely durable material that helps minimize long-term waste, but cement is also energy intensive to manufacture. Cement substitutes such as fly ash or rice hull ash can mitigate the environmental cost of stuccos. In coastal regions, salt spray can accelerate corrosion of reinforcing meshes.

Metal siding is very durable, recyclable, and typically contains significant postconsumer recycled content. It is energy intensive to manufacture, but recycled steel and aluminum require far less energy than virgin ore. Some types of metal siding are prone to being easily damaged.

Composite siding (hardboard) is made of newspaper or wood fiber mixed with recycled plastic or binding agents. It is highly durable, resists moisture and decay, often has significant recycled content, and is not prone to warping or cracking like wood. Composites require less frequent repainting and some need not be painted at all, saving waste and resources.

Wood siding requires more maintenance than many of the other siding options, but it is renewable and requires relatively little energy to harvest and process. If it is not well maintained, wood can easily be the least durable option, generating significant waste. The most durable solid wood siding comes unfortunately from old growth and tropical forests.

Siding selection considerations:

a. The most durable siding product that is appropriate should be selected. Siding failures that allow water into the wall cavity can lead to expensive repairs, the waste of damaged components, and the environmental costs of replacement materials. Fire resistance is a feature that helps reduce the financial and environmental impact of rebuilding, particularly in high-risk areas.

b. For existing buildings, consider refinishing existing siding to minimize waste, pollution, and energy use.

c. Select materials that are biodegradable, have recycled content, and/or are recyclable.

d. Reclaimed or remilled wood siding should be used to minimize demand for virgin wood and reduce waste (painted wood should be tested for lead contamination prior to use).

e. New wood siding should display an FSC-certified products label.

f. Vinyl is somewhat durable, but it is not a green building material. Attributed disadvantages include pollution generated in manufacturing, air emissions, human health hazards of manufacturing and installation, the release of dioxin and other toxic persistent organic pollutants in the event of fire, and the difficulty recycling.

6.2.7 Roofing

For buildings and homes especially, a roof performs multiple functions, all of which are tied into providing protection from the elements. A roof's main role is protecting the structural members and interior materials from deterioration. Moisture, dependability, and durability are the most critical characteristics of roofing materials.

The extraction, manufacture, transport, and disposal of roofing materials pollutes air and water, depletes resources, and damages natural habitats. Roofing materials comprise an estimated 12–15% of construction and demolition (C&D) waste. An environmentally sustainable roof must first be durable and long-lasting, but it may also contain recycled or low-impact materials. Roofs that are environmentally friendly can provide aesthetically pleasing design options, reduced life cycle costs, and environmental benefits such as reduced landfill waste, energy use, and impacts from harvest or mining of virgin materials. It takes roughly the same materials, energy, and labor to manufacture and install a 50-year warranted roof as a 30-year roof, yet disposal and replacement is delayed. A well-installed 50+-year-rated roof can reduce roofing waste by 80–90% over its lifetime, relative to a roof with a 20-year warranty.

Mild climates are well suited for passive temperature controls that reduce winter heating and reduce or eliminate the need for mechanical cooling. Air conditioning is used less in mild climates because operable windows and skylights are often employed that can easily provide ventilation and cooling, particularly in smaller buildings. Larger, commercial buildings can also be effectively cooled in mild climates without air conditioning, but more careful design is required, and roofing that minimizes heat gain is a key consideration.

Low-slope commercial roofing: Ideally, roofing should renew our natural resources. For example, the potential habitat for birds and native plants on a green or living roof can be an island of safety in the urban environment, and can help provide pathways for migration through fragmented ecosystems. Similarly, electricity from solar photovoltaic panels displaces demand for fossil fuels. Cool roofing does not renew resources, but is often a highly cost-effective way to conserve them.

Roofing systems should be chosen that are durable, reduce waste, liability, and frustration. Some of the key considerations that will impact the type of roof chosen should include the following:

- Its capacity to reflect sunlight and re-emit surface heat. Cool roofs can reduce cooling loads and urban heat-island effects while providing longer roof life.
- Ability to resist the flow of heat from the roof into the interior, whether through insulation, radiant barriers, or both.
- Ability to reduce ambient roof air temperatures through evaporation and shading, as in the case of vegetated green roofs.
- Recyclability and/or capability of being reusable, to reduce waste, pollution, and resource use. Options with high postconsumer recycled content of 30% are preferable.
- Nonhalogenated roofing membranes (i.e., materials that do not contain bromine or chlorine) are preferable. In the event of fire burning PVC and thermoplastic polyolefin (TPO) produce strong acids and toxic persistent organic pollution, including dioxin.

Roofs may require protective ballast such as concrete tile to comply with local fire codes. Existing PVC and TPO roofing membranes, as well as underlying polystyrene insulation, can sometimes be recycled, and this practice is expected to become more prevalent as Federal construction specification requirements generate increased demand.

Residential and commercial roofing options:

- *Clay or cement tiles* are very durable and made from abundant materials, but heavy and expensive.
- *Recycled plastic, rubber, or wood composite shingles* are durable, lightweight, and sometimes recyclable, but not biodegradable.
- *Composite shingles* can get 50-year warranties, which are better than 20- to 40-year products. Can be recycled, but typically landfilled.
- *Fiber cement* is durable and fire/insect-proof, but heavy and not renewable or biodegradable. May be ground up and used as inert fill at demolition.
- *Metal* is a durable and fire/insect-proof, recyclable material. It typically contains recycled content, but manufacture is energy intensive and causes pollution and habitat destruction.
- *Built-up roofing* – its durability depends largely on the structure, installation, flashing, and membrane chosen. Most membranes are not made from renewable resources, but some may contain recycled content. High-VOC products emit air pollution during installation.

- *Vegetated green roofs* are most commonly installed on roofs with slopes less than 30°.
- *Wood shakes* are a biodegradable material, but flammable, and not very durable. It is not typically considered to be a "green" option for fire-prone areas.

Green roofing: Emerging new technologies are helping to promote Green Building, and there is increasing efforts to make useable space of existing rooftops and/or new roofs to allow additional living space. As extensive green roofs became subjects in scientific research, additional benefits began to gain importance. Ecological benefits such as *reduced energy consumption* for heating and cooling, *stormwater retention*, and *heat island mitigation* are some of the reasons for the increased popularity of green roofs. Other benefits include *habitat restoration*, filtration of acid rain and air pollutants, *noise pollution reduction*, and the therapeutic effects found from being in the presence of nature. Green roof systems are a natural way of providing additional clean air through the transference of CO_2 and oxygen between the plants and vegetation with the atmosphere. Traditional drainage systems using pipe and stone are not plausible.

Green or "living" roofing is the use of vegetation as the weathering surface for a roof. It has proved very successful because it reduces extremes in rooftop temperature, saves energy, and extends the useful life of the roof. High temperatures shorten the life of a roof (leading to increased C&D waste) and increase summer cooling costs.

Green roofing has proven to be effective for three principle reasons: the large surface area of soil and plants helps to reradiate heat; it provides shade and insulation for the waterproof roof membrane and the plants' transpiration provides cooling. The net result is a 25–80° F decrease in peak roof temperature and up to 75% reduction in cooling energy demand.

Green roofs provide additional environmental and aesthetic benefits. The soil and vegetation in many common extensive designs can detain up to 75% of a 1-in. rainfall event and will filter the remainder. This on-site stormwater management helps reduce demand on stormwater infrastructure, saving resources and money for the entire community. Green roofs provide urban wildlife microhabitat. Although not a replacement for wildlands, a vegetated roof can accommodate birds, beneficial insects, and native plants far better than tar and gravel. Cooler roof temperatures also reduce the urban heat island effect, helping to reduce the cooling load for surrounding buildings. It can also benefit local property values.

Contemporary green roof designs generally contain a mixture of hard and soft landscaping such as the green roof of the LDS Assembly Hall building in Salt Lake City, Utah (Figure 6.5b). It is very important that the selected drainage/retention layer is capable of supporting any type of landscape – from roadways and paths, to soil and trees, so as to permit excess water to drain unobstructed underneath. (Figure 6.6a,b).

Green roof components. Green roof systems generally contain certain essential components, the location of these components, the materials they are made from, and whether they are present or not vary widely from one installation to another. AWD, a leader in roof garden and prefabricated drainage systems notes that the main components, materials, and locations to take into consideration if contemplating the installation of a green or vegetated roof are as follows: *Structural support.* Roof and roof garden systems are required to have an underlying structural system in place to

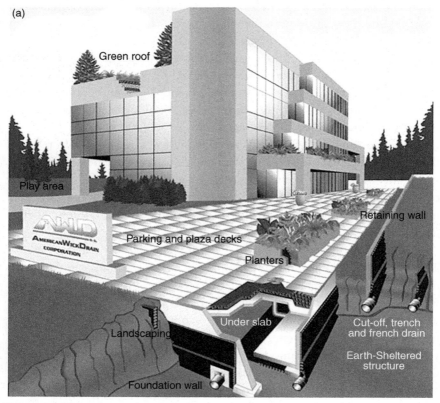

Figure 6.6 (a) Rendering showing commercial application of green elements.

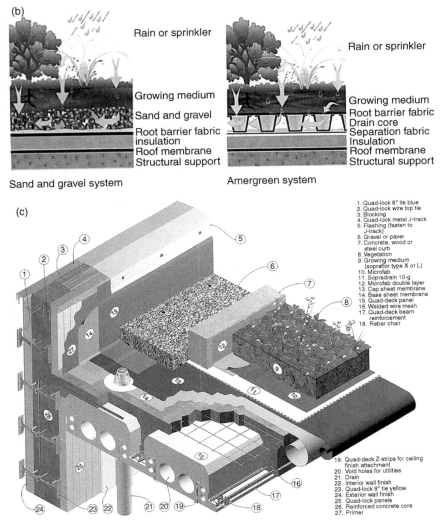

(b)

Rain or sprinkler

Growing medium
Sand and gravel
Root barrier fabric
insulation
Roof membrane
Structural support

Sand and gravel system

Rain or sprinkler

Growing medium
Root barrier fabric
Drain core
Separation fabric
Insulation
Roof membrane
Structural support

Amergreen system

(c)

1. Quad-lock 6″ tie blue
2. Quad-lock wire top tie
3. Blocking
4. Quad-lock metal J-track
5. Flashing (fasten to J-track)
6. Gravel or paver
7. Concrete, wood or steel curb
8. Vegetation
9. Growing medium (sopraflor type X or L)
10. Microfab
11. Sopradrain 10-g
12. Microfab double layer
13. Cap sheet membrane
14. Base sheet membrane
15. Quad-deck panel
16. Welded wire mesh
17. Quad-deck beam reinforcement
18. Rebar chair
19. Quad-deck Z-strips for ceiling finish attachment
20. Void holes for utilities
21. Drain
22. Interior wall finish
23. Quad-lock 8″ tie yellow
24. Exterior wall finish
25. Quad-lock panels
26. Reinforced concrete core
27. Primer

Figure 6.6 (cont.) (b) Drawing showing two types of roofing systems: (1) Sand and gravel system and (2) the Amergreen roofing system. (c) Drawing showing 27 typical roofing system components. *Source: American Wick Drain Corporation.*

support additional weights resulting from use of normal building materials such as concrete, wood, etc. The structural engineer must consider the load of the roof garden system in the initial design. *Roofing membrane.* Various roofing membrane options are available to the design engineer. The final membrane choice may be affected by loads imposed by the rooftop garden, by available membrane protection elements in the rooftop garden system, by root penetration properties of the membrane, and by membrane drainage and aeration requirements. *Membrane protection.* The roofing membrane may require protection from installation damage, long-term water exposure, UV exposure, drainage medium loads, or growing medium loads or chemical properties. *Root barrier.* A root barrier may be required between the growing medium and lower components to eliminate or mitigate potential root penetration into the roofing membrane, drainage medium, or water storage medium. The optimum location is above the membrane, drainage medium, and water storage medium so that all three components are protected. Alternate locations may require multiple roof guard elements to avoid long-term root penetration. *Insulation.* Insulation may be installed above the structural support, depending on the thermal design of the structure. It may be installed above or below the roofing membrane.

Drainage medium. Some sort of drainage medium is necessary to ensure the proper range of water content in the growing medium. Too much water will have an adverse effect and cause root rot. Too little water will result in poor vegetation growth. Drainage medium options range from gravel to various materials designed specifically for this purpose. Plastic materials usually combine the drainage function with water storage and aeration, plus also provide some root protection to the roofing membrane and protect the membrane from potentially damaging materials in the growing medium. *Aeration medium.* Aeration is an essential element and required to promote optimal vegetation growth. The aeration medium usually serves as both drainage and aeration medium. Open channels in the prefabricated drain core are designed to provide air to the plant roots. *Water storage medium.* Although the growing medium will store a certain amount of water, additional water storage may be required to provide more successful growth of vegetation. Most prefabricated plastic roof garden products have water storage capability. The plastic cones of the prefabricated core provide positive water storage reservoirs. Sand and gravel may also hold some water or it may mostly drain out. *Growing medium.* Various natural and manufactured materials may be used as a growing medium. Soil is often mixed with other material to reduce weight, to provide better structure for roots, and to provide essential nutrients, water, and oxygen. *Vegetation.* The upper and most visible layer of the roof garden system is the vegetation. A wide variety of typical landscaping and garden plants may be employed for a rooftop garden. An experienced landscape architect should be involved in plant selection to ensure vegetation is appropriate both for the geographic region and for a rooftop garden environment. The type of vegetation affects the selection of the other components of the roof garden system. Such issues as root protection, water and aeration needs, and drainage requirements are affected by plant selection and vice versa.

6.2.7.1 Classes of Green Roofs: Extensive and Intensive

Extensive green roofs typically contain a layer of soil media that is relatively thin (2–6 in.), and lightweight (10–50 lb/ft.^{2} for the entire system (when saturated with water).

It is lightweight, relatively easy to install, durable, and cost-effective.

Intensive green roofs are designed to accommodate trees and gardens. Soil can be as deep as needed to accommodate the desired tree or plant species, but deeper, denser soil dramatically increases dead load, requiring a stronger and more expensive structure, greater maintenance, and either terracing or a relatively flat roof. But green roof options are available for virtually any building type or location.

Some of the benefits attributed to green roofs include the following:

- Enhanced insulation and more moderate rooftop temperatures, which reduce cooling and heating requirements, saving energy and money.
- Facilitates filtration and detention of stormwater, reduces pollution, and the cost of new and expanded infrastructure as paved areas increase.
- Absorption of dust and airborne pollutants.
- Reduces ambient air temperatures, reduces urban heat island effects, and helps to enhance the microclimate of surrounding areas.
- Extended life of roof membranes, which are protected from ultraviolet radiation, extreme temperatures, and mechanical damage. (Plant species, soil depth, and root-resistant layers are carefully matched to ensure the roof membrane is not damaged by the roots themselves.)
- Lightweight extensive systems can be designed with dead loads comparable to standard low-slope roofing ballast. Structural reinforcement may not be necessary, and cost can be comparable to conventional high-quality roofing options.

6.2.7.2 How Green Roofs can Contribute to LEED Certification

The wide variety of benefits associated with green roofs is captured to varying degrees in the USGBC's LEED rating system. Every green roof project is unique, and the extent to which a green roof on any building can help earn credits and LEED certifications varies. Overall, Green Roof systems installed on 50% or more of the roof surface virtually guarantees 2 LEED points and can contribute an additional 7+ points toward LEED certification, almost 20% of the total needed for a project to be LEED certified.

6.2.8 Wood

Pressure-treated lumber: The popularity over the decades for pressure treated lumber has been partially due to its resistance to rot and insects. The familiar copper chromated arsenate (CCA) was largely phased out in a cooperative effort between manufacturers and the US Environmental Protection Agency. Arsenic is acutely toxic and carcinogenic and was shown to be leaching into surrounding soils, which prompted the ceasing of its production for residential use at the end of 2003. It is unfortunate that these problems were not identified before it was put into widespread use.

Existing CCA-treated lumber poses a challenge. It remains resistant to rot and insects and its reuse would help conserve forest resources and keep a potentially useful resource out of landfills. However, permitting its reuse would allow it to continue to leach arsenic into soils. CCA-treated wood should not be composted or disposed of in green waste or wood waste bins. Burning CCA-treated wood is highly toxic. Disposal of CCA-treated wood is now mandated to be in a lined landfill or as class I hazardous waste. The newer, much less toxic wood treatments (copper azole (C-A) and alkaline copper quaternary (ACQ)) are more corrosive than CCA. Manufacturer-recommended fasteners should be employed to minimize rust and prevent staining.

Points to consider:

- Repair and/or refinish existing decks, railing, or fencing.
- Reuse wood in good condition.
- Build with durable materials such as plastic lumber.
- For structural elements that will be in contact with soil and water, consider:
  - Heartwood from decay-resistant species such as redwood or cedar that has been FSC certified as harvested from a responsibly managed forest.
  - If pressure-treated lumber is required, the two water-resistant preservatives currently used are C-A and ACQ, which are significantly less toxic than CCA.
  - Avoid the few remaining stocks of CCA.

- For fencing, consider friendly alternatives such as a living fence of bushes, shrubs, and live bamboo, in urban settings, or fencing made of cut bamboo (a rapidly renewable material).

Lumber and engineered wood: Wood is a renewable material, and requires less energy than most materials to process into finished products. However, logging, manufacture, transport, and disposal of wood products have substantial environmental impacts. Standard logging practices cause erosion, pollute streams and waterways with sediments, damage sensitive ecosystems, reduce biodiversity, and lead to loss of soil carbon. The key to reducing these impacts is the minimization of wood use by substituting suitable alternatives such as preferable materials or building systems, reusing salvaged wood, selecting wood from responsibly managed forests, controlling waste, and minimizing redundant components.

Certified forest products: Where salvaged or reclaimed wood is not available or applicable (i.e., structural applications), specify products that are certified by an approved and accredited organization such as the FSC. LEED v4 continues to promote responsible forest management and incentivizing forest conservation. For the FSC community, the key credit in LEED v4 is known as "Building product disclosure and optimization – sourcing of raw materials" (MRc3). This credit contains two options, with the second focused on leadership extraction practices. To receive this credit, a project must use at least 25%, by cost, of the overall value of permanently installed building products. For wood products, the credit requires products certified by the FSC- or USGBC-approved equivalent. Reused or recycled materials are another means to achieve this credit. Products sourced within 100 miles are valued at 200% of their cost.

Engineered wood products: Engineered lumber also known as *composite wood, man-made wood* consists of a range of derivative wood products that are manufactured by pressing or laminating together the strands, particles, fibers, or veneers of wood, with a binding agent, to form composite materials. The superior strength and durability of engineered lumber allows it to displace the use of large (and increasingly unavailable), mature timber. They are also less susceptible to humidity-induced warping than equivalent solid woods, although the majority of particle and fiber-based boards require treatment with sealant or paint to prevent water penetration.

These products are engineered to meet precise application-specific design specifications, which are tested to meet national or international standards. Using engineered lumber instead of large dimension rafters, joists, trusses, and posts can save money and reduce total wood use by as much as 35%.

The wider spacing of members possible with engineered lumber also has the advantage of increasing the insulated portion of walls. Other advantages include their ability to form large panels from fibers taken from small-diameter trees, and the small pieces of wood, and wood that has defects, can be used in many engineered wood products, especially particle and fiber-based boards. Engineered wood products are used in a number of ways, usually in applications similar to solid wood products, but many builders prefer engineered products because they are economical, typically longer, stronger, straighter, more durable, and lighter than comparable solid lumber.

Engineered wood products also have some disadvantages; for example, they require more primary energy for their manufacture than solid lumber. Furthermore, the adhesives used may be toxic. A concern with some resins is the release of formaldehyde in the finished product, often seen with urea-formaldehyde-bonded products. Cutting and otherwise working with engineered wood products can therefore expose workers to toxic constituents that could potentially cause them harm.

The primary types of adhesives used in engineered wood include urea-formaldehyde resins which is not waterproof but is popular because it is inexpensive; melamine-formaldehyde resin, which is white in color and is heat and water resistant, and whose use is preferred for exposed surfaces in more costly designs; phenol-formaldehyde resins, which has a yellow/brown color, and commonly used for exterior exposure products; and methylene diphenyl diisocyanate or polyurethane resins, which are expensive, generally waterproof, and do not contain formaldehyde.

Structural sheathing: Sheathing is the structural covering of plywood or oriented strand board (OSB) that is applied to studs and roof/floor joists to provide shear strength and serves as a base for finish flooring or a building's weatherproof exterior. OSB relies on smaller (aspen and poplar) trees, which are a more rapidly renewable resource than the mature timber required for plywood. Nonetheless, sheathing is considered to be the second most wood-intensive element of wood-frame construction.

Wood is a renewable product and requires less energy than most materials to process into finished products. However, the logging, transport, and disposal of wood have substantial environmental impacts. Standard logging techniques lead to erosion and fouling of streams and waterways with sediments and lead to loss of carbon stored in soil, damage sensitive ecosystems, and reduce biodiversity. Minimizing wood use is the key to reducing these impacts, through a combination of substitute materials, selection of wood from responsibly managed forests, and design and construction practices that control waste and minimize redundant components. Salvaged, remanufactured, or certified wood products (by accredited organizations such as the FSC) should be considered.

Engineered wood sheathing materials do have some environmental trade-offs because the wood fibers are typically bound with formaldehyde-based resins. Interior-grade plywood typically contains urea formaldehyde, which is less chemically stable than the phenol formaldehyde found in water-resistant exterior-grade plywood and OSB. This advantage makes exterior-grade plywood preferable for indoor applications, as it emits less toxic and suspected carcinogenic compounds.

There are other alternatives to these wood-intensive conventional and engineered materials.

Fiberboard products rated for structural applications (such as Homasote™'s 100% recycled nailable structural board) are alternatives to plywood and OSB. Structural-grade fiber cement composite siding combines sheathing and cladding, providing shear strength and protection from the elements in a single component while reducing labor costs for installation. Water-resistant exterior-grade gypsum sheathing can be one of the options employed under brick and stucco exterior finishes to reduce wood requirements.

Designs that combine bracing with nonstructural sheathing can provide necessary strength while enhancing insulation and reduce wood requirements. Structural insulated panel (SIP) construction provides interior and exterior sheathing as well as insulation, in precut, factory-made panels. And by designing for disassembly, sheathing materials can be readily reused or recycled. For example, while adhesives distribute loads over larger areas than fasteners alone, materials attached with removable fasteners (in appropriate designs) are more easily deconstructed than materials installed with adhesives.

Medium-density fiberboard (MDF): MDF is a composite panel product typically consisting of low-value wood by-products such as sawdust combined with a synthetic resin such as urea formaldehyde resins or other suitable bonding systems and joined together under heat and pressure. Additives may be introduced during manufacturing to impart additional characteristics. MDF is primarily used for internal use applications, mainly because of its poor moisture resistance. MDF is widely used in the manufacturing of furniture, kitchen cabinets, door parts, moldings, millwork, and laminate flooring. MDF panels are manufactured with a variety of physical properties and dimensions, providing the opportunity to design the end product with the specific MDF characteristics and density needed. MDF is available in raw form with a fine-sanded surface or with decorative overlay such as wood veneer, melamine paper, or vinyl.

Although methylene diphenyl diisocyanate is highly toxic to manufacture, it does not emit VOCs in use. MDF will accept a wide range of sealers, primers, and coatings to produce a hard, durable tool surface, but it is not suitable for high-temperature applications. It is also entirely possible to make MDF-type panels using waste wood fiber from demolition wood and waste paper.

The properties of MDF can vary based on the country or region of the country where it is produced. It also comes in a variety of densities depending on the application it is to be used for.

The surface of MDF is flat, smooth, uniform, dense, and free of knots and grain patterns. The homogeneous density profile of MDF allows intricate and precise machining and finishing techniques for superior finished products. Trim waste is significantly reduced when using MDF compared to other substrates. Stability and strength are important assets of MDF, which can be machined into complex patterns that require precise tolerances.

Agricultural residue panels (Ag): One of the more recent developments in North American composites has been the introduction of boards made from agricultural residues. Increasing constraints on residue burning have been a prime motivator for its introduction. Their manufacture entails compressing agricultural residue materials with nonformaldehyde glues; the panels provide an excellent alternative to plywood sheets 3/8 in. and thicker and can be used in much the same way as MDF. They can also replace OSB and MDF for interior walls and partitions work. Agricultural residue (ag-res) boards are made from waste wheat straw, rice straw, or even sunflower seed husks. Ag boards are both aesthetically pleasing and often stronger than MDF, and just as functional. Under heat and pressure, microscopic "hooks" on the straws link together, reducing or eliminating the need for binders. The use of new soy adhesives promises both improved performance and economics to the Ag composites industries. They are also expected to be safer to handle and to reduce VOC emissions.

Homasote is a panel product made of 100% postconsumer recycled newspaper fiber, and has actually been in production longer than plywood and OSB. It has many potential applications including walls, structural roof decking, paintable interior panels, and concrete forms. At the time of writing, agricultural straw panels are difficult to find due to a combination of intense demand, lack of supply, and the time required for bringing a manufacturing facility online. Some of Homasote's physical properties include being a weather-resistant, structural, insulating, extremely durable board with two to three times the strength of typical light-density wood fiberboards.

Advanced framing techniques: Advanced framing, also known as optimum value engineering (OVE), refers to a variety of framing techniques designed to reduce the amount of lumber used and waste generated in the construction of a wood-framed house, thereby reducing material cost and use of natural resources, while at the same time increasing energy efficiency through increased space for insulation. The extraction, manufacture, transport, and disposal of lumber deplete resources, damage natural habitats, and pollute air and water. Dimensional lumber supplies depend upon larger trees that require decades to mature.

OVE advanced framing techniques include studs spaced at 24 in. o.c.; 2-ft. modular design that reduces cut-off waste from standard-sized building materials (Figure 6.7); in-line framing that reduces the need for double top plates; building corners with two studs; and using insulated headers over exterior building openings (or using no headers for nonload-bearing walls). The advantage of spacing studs at 24 in. o.c. rather than 16 in. o.c. reduces the amount of framing lumber required to construct a home and replaces framing members with insulation. This allows the wall to achieve a higher overall insulating value and costs less to construct than a conventionally framed wall while still meeting the structural requirements of the home. However, to be successful, advanced framing techniques must be considered early in the design process.

Benefits from advanced framing: Both builders and homeowners can benefit from advanced framing. Advanced framing techniques create a structurally sound home that has lower material and labor costs than a conventionally framed house. Additional construction cost savings result from the generation of less waste that needs to be disposed of, which also helps the environment.

Advanced framing improves energy efficiency by replacing lumber with insulation material. The whole-wall *R*-value is improved and insulation values are enhanced (whole-wall *R*-value) because fewer studs means the insulated wall area is maximized, the deeper wall cavity allows for thicker insulation, which reduces thermal bridging (conduction of heat through framing).

In many cases, conventional framing is structurally redundant, using wood unnecessarily for convenience. The US Department of Energy's Office of Building Technology notes that through advanced framing

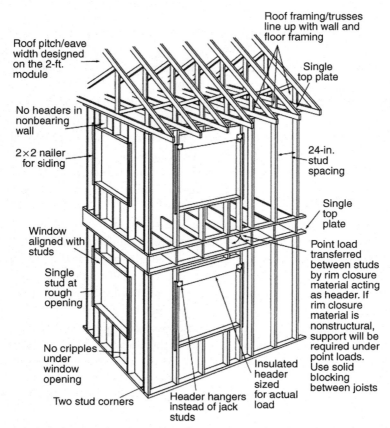

Figure 6.7 *Isometric Drawing Illustration Advanced Framing Techniques Used in Residential Construction. Source: Adapted from Building Science Corporation.*

techniques, $500 in material costs per 1200 ft.² of house can be saved, in addition to 3%–5% off of labor costs, and an annual heating and cooling costs saving by 5%. While advanced framing is more wood-efficient than conventional framing, keep in mind that alternative structural technologies such as insulated structural systems, straw bale or earthen construction, and high-recycled content steel framing with thermal breaks place fewer demands on our forest resources than stick framing.

Although advanced framing techniques have been researched extensively and proven effective, some techniques may not be allowed under certain circumstances (i.e., prone to high winds or seismic potential) or in some jurisdictions. Local building officials should be consulted early in the design phase to verify or obtain acceptance of these techniques.

Structural insulated panels: SIPs are factory-built walls/roof panels that consist of two sheets of rigid structural facing board, for example, OSB or plywood that is applied to both sides of a rigid expanded polystyrene (EPS) insulating foam core that is 4 or more in. thick. The result of this simple sandwich forms a strong structural panel for building walls, roofs, and floors, and which is significantly more energy-efficient, yielding increased *R*-values as compared to traditional framing. In addition to SIPs' excellent insulation properties, it offers airtight assembly, noise attenuation, and superior structural strength. Though SIP panels cost more per square foot of wall, floor, or ceiling than conventional construction, total construction costs are often minimized as a result of reduced labor and faster completion.

The reduction in construction waste and superior insulation from SIP use is especially significant when compared to conventional stick or steel stud systems. They can be delivered precut to the precise dimensions required, and each panel contains the structure, insulation, and moisture barrier of the wall system. OSB is the most common sheathing and facing material in SIPs, reducing wood use by as much as 35% and reducing pressure on mature forests by allowing the use of smaller farm-grown trees for structural applications. The exterior-grade plywood used in some products requires more mature timber.

SIP wall assemblies tend to be well sealed, enhancing energy efficiency. As with any tightly sealed structure, moisture control and well-designed ventilation are critical. SIP construction can contribute to very good indoor air quality; the plastic insulating foams (EPS or polyurethane/polyisocyanurate) are very chemically stable, and OSB is a low-emitting material.

SIPs are engineered and custom manufactured to give the designer greater control over the project, including materials and costs. SIPs are usually chosen for their versatility, strength, cost-effectiveness, and energy efficiency. They are suitable for a wide range of residential and commercial projects, and SIP panels are custom made according to specifications and drawings. Another advantage of insulated structural systems is that they integrate a building's structure and insulation into a single component.

SIP core materials: There are several materials used for SIP cores. EPS is the most common SIP core material. It requires less energy to manufacture than other options and is more recyclable than polyurethane or polyisocyanurate. Many products offer a 1-h fire rating when installed with 5/8-in. or thicker gypsum sheathing. EPS foam is expanded with pentane, which does not contribute to ozone depletion or global warming, and is often recaptured at the factory for reuse.

Polyurethane or polyisocyanurate possesses greater insulation properties per inch of foam than EPS, and offers greater resistance to thermal breakdown. However, polyurethane and polyisocyanurate are unlikely to be recycled. Polyurethane and polyisocyanurate use HCFC blowing agents, which contribute to global warming and ozone depletion (though to a lesser degree than CFCs). Research is currently underway on another alternative – new resins (polyurethane/polyisocyanurate) derived from soy and which may soon be available for use in SIPs.

Straw core SIPs are made from waste agricultural straw. They are renewable and recyclable, and the pressed straw core does not require a binding agent. However, straw-core SIPs offer less insulation per inch of thickness and are considerably heavier than other options; energy use in shipping is a significant consideration.

Some of the factors to consider when building with SIPs:

- Design to minimize waste by ordering SIP panels precut to meet project requirements, including window and door openings. Likewise, design should be to standardized panel dimensions.
- Inquire whether the SIP supplier or manufacturer will take back any off-cuts for recycling.
- When sizing your heating system, consider thermal performance of SIPs to save money upfront and energy over time. Oversized heating and cooling systems are inefficient.
- SIP roofs do not necessarily require ventilation, making them appropriate for low-slope roofs. If local jurisdiction mandates ventilated roofs, consider SIPs with integrated air channels, or upgrading from composition roofing.

Insulated concrete forms (ICFs): At their most basic level, ICFs serve as a forming system for poured concrete walls and consist of stay-in-place formwork for energy-efficient, cast-in-place, reinforced concrete walls. The forms are interlocking modular units that are dry-stacked (without mortar) and filled with concrete. The forms lock together somewhat like Lego bricks and serve to create a form for the structural walls of a building. Concrete is pumped into the cavity to form the structural element of the walls. ICFs use an insulating material as permanent formwork that becomes a part of the finished wall. All ICFs can be considered "green" materials; they are durable, produce little or no waste during construction, and dramatically improve the thermal performance of concrete walls.

ICFs usually employ reinforcing steel (rebar) before concrete placement to give the resulting walls flexural strength, similar to bridges and high-rise buildings made of concrete. The forms are filled with concrete every several feet in order to reduce the risk of blowouts. After the concrete has cured, or firmed up, the forms are left in place permanently to increase thermal and acoustic insulation, increased fire protection and provide backing for gypsum boards on the interior and stucco, brick, or other siding on the exterior. It can also accommodate electrical and plumbing installation (Figure 6.8).

There are several ICF systems on the market, which vary in the shape of the resulting concrete within the wall. Some examples are as follows:

- The "flat" system forms an even thickness of concrete throughout the walls, like a conventionally poured wall.
- The "waffle grid" systems create a waffle pattern where the concrete is thicker at some points than others.
- The "post-and-beam" system forms detached horizontal and vertical columns of concrete.

One of the attributes of standard concrete is that it is a dense material with a high heat capacity that can be used as thermal mass, reducing the energy required to maintain comfortable interior temperatures. However, concrete is not a good insulator, and standard formwork is waste intensive,

Figure 6.8 *Photo Depicting Application of Insulating Concrete Formwork Quadlock System. Source: Quad-Lock Building Systems Ltd.*

and toxic materials are frequently needed to separate formwork from the hardened product. ICF address these weaknesses by reducing solid waste, air and water pollution, and potentially reducing construction cost. ICF wall systems are thermally superior, which enhances their usefulness for passive heating and cooling; comfort is also enhanced, and energy costs are reduced while first costs can be minimized through the resulting reduced heating/cooling system size.

ICF systems can be manufactured from a variety of materials, such as lightweight foamed concrete panels, rigid foams like EPS, polyurethane, and composites that combine concrete with mineral wool, wood waste, paper pulp, or EPS beads. There are also ICF systems, such as BaleBlock and Faswall, that substitute straw bales or fiber-cement for polystyrene. Rigid foams used in ICFs generally do not have significant recycled content and are less likely to be recyclable at the end of life, but may be reused in fill or other composite concrete products.

ICFs offer the structural and fire-resistance benefits of reinforced concrete; structural failure due to fire is rare to nonexistent. Because of the addition of flame-retardant additives, polystyrene ICFs tend to melt rather than burn, and interior ICF walls tend to contain fires much better than wood frame walls, improving overall fire safety. Some ICFs like "Rastra" are made from recycled postconsumer polystyrene (aka foam) waste products.

As is the case in most heated structures, moisture control is a key design consideration for ICF walls. Solid concrete walls sandwiched in polystyrene blocks tend to be very well sealed to enhance energy efficiency, but they consequently also tend to seal water vapor within the structure. Potential mold growth and impaired indoor air quality are serious health concerns, which require addressing. The easiest method to resolve this is by incorporating mechanical ventilation. Certain systems such as straw bale and Rastra tend to be more vapor permeable, thereby reducing this concern.

6.2.9 Concrete

Concrete is a composite building material made from a combination of aggregates such as sand and crushed stone and a binder or paste such as cement. The most common form of concrete consists of mineral aggregate such as stones, gravel and sand, cement and water. The cement hydrates after mixing and hardens into a stone-like material. Concrete has a low tensile strength and is generally strengthened by the addition of steel reinforcing bars: this is commonly referred to as reinforced concrete.

Concrete is a strong, durable, and inexpensive material that is the most widely used structural building material in the United States. Because of the vast scale of concrete demand, the impacts of its manufacture, use, and demolition are widespread. Habitats are disturbed from materials extraction; significant energy is used to extract, produce, and ship cement; and toxic air and water emissions result from cement manufacturing. Cement manufacture in particular is energy intensive.

Estimates indicate that approximately one ton of carbon dioxide is released for each ton of cement produced, resulting in 7%–8% of man-made CO_2 emissions. And although concrete is typically only 9%–13% cement, yet it accounts for 92% of concrete's embodied energy. Cement dust contains free silicon dioxide crystals, the trace element chromium, and lime, all of which can adversely impact worker health. Mixing concrete requires a great deal of water, and generates alkaline wastewater and run-off that can contaminate waterways and vegetation.

Minimizing environmental effects: Incorporating local and/or recycled aggregate (such as ground concrete from demolition) is an excellent way to reduce the impacts of solid waste, transit emissions, and habitat disturbance. Environmental impacts can also be reduced substantially by substituting alternative pozzolan ash (industrial by-product such as fly ash, silica fume, rice husk ash, furnace slag, and volcanic tuff) for Portland cement. Fly ash, a residue from coal combustion, is quite popular as a cement substitute that generally decreases porosity, increases durability, and improves workability and compressive strength, although the curing time is increased. Fly ash generally constitutes 10%–15% of standard mixes, but many applications will allow substituting of up to 35%–60% of cement, and with certain types of fly ash (e.g., Class C), cement can be completely replaced for some projects.

In nonstructural applications, concrete use may be reduced by trapping air in the finished product or through the use of low-density aggregates. Trapped air displaces concrete while enhancing insulation value and reducing weight and material costs, without compromising the durability and fire resistance of standard concrete. Low-density aggregates such as pumice, vermiculite, perlite, shale, polystyrene beads, or mineral fiber provide similar insulation and weight-reduction benefits.

Cast-in-place or precast concrete, and concrete masonry unit (CMU) considerations:
- Design for the reuse of portions of existing structures, such as slabs or walls that are in satisfactory condition.
- Recycle demolished concrete on site for use as aggregate or fill material for new projects, or recycle at local landfills.

- Incorporate the maximum amount of fly ash, blast furnace slag, silica fume, and/or rice husk slag appropriate to the project, thereby reducing cement use by 15–100%.
- Use of precast systems will minimize waste of forming material and reduce the impact of wash water on soils.
- Consider alternative or possible material substitutes for concrete such, as ICFs, that reduce waste, enhance thermal performance, and may reduce construction schedules. Likewise, cellular, foamed, autoclaved aerated, and other lightweight concretes add insulation value while reducing the weight and concrete required. The use of earthen and rapidly renewable materials, such as rammed earth, cob, or straw bale reduce the need for insulation and finish materials in both residential and commercial projects.
- Use nontoxic form-release agents.
- Waste can be minimized by carefully planning concrete material quantities.
- For footings, consider fabric-based form systems for fast installation and wood savings.
- Reduce wood waste and material costs by employing steel or aluminum concrete forms, which can be reused many more times than wood forms.

Pervious/porous concrete: Up to 75% of urban surface area is covered by impermeable pavement, which inhibits groundwater recharge, contributes to erosion and flooding, conveys pollution to local waters, and increases the complexity and expense of stormwater treatment. One of the main characteristics of pervious paving is that it contains voids that allow water to percolate through to the base materials below. It also reduces peak storm water flow and water pollution and promotes groundwater recharge. Pervious paving may incorporate recycled aggregate and fly ash, which helps reduce waste and embodied energy. Pervious paving is suitable for use in parking and access areas having a compressive strength of up to 4000 psi. It also mitigates problems with tree roots, and percolation area encourages roots to grow deeper. Enhanced heat exchange with the underlying soil can decrease summer ambient air temperature by 2–4°F.

Concrete formwork: Poured-in-place applications require on-site formwork to give shape to walls and slabs and other project elements as it cures (Figure 6.9). Plywood and milled lumber are the most common form materials, contributing to construction waste, and the impacts of timber harvest and processing. Wooden formwork can be made from salvaged wood and typically be disassembled and reused several times. Disassembling

Figure 6.9 *Photo of Carpenters Setting Concrete Formwork for the High Level Waste Facility Pit Walls. Source: Bechtel Corporation.*

construction-grade lumber and exterior-grade plywood forms should be considered for reuse within the project.

Form-releasers or parting agents are materials that facilitate the separation of forms from hardened concrete. Such materials prevent concrete from bonding to the form, which can mar the surface when forms are disassembled. Traditional form releasers such as diesel fuel, motor oil, and home heating oil are carcinogenic, which limits the potential for reuse of wood formwork because it exposes construction personnel to VOCs (and potentially occupants as well). They are now prohibited by a variety of state and federal regulations, including the Clean Air Act. Low- and zero-VOC water-based form release compounds that incorporate soy or other biologically derived oils dramatically reduce health risks to construction staff and occupants, and often make it easier to apply finishes or sealants, when necessary. Many soy-based options are less expensive than their petroleum-based counterparts.

When designing concrete formwork, one should also consider factors that will adversely affect and impact concrete formwork pressure. These factors include the rate of placement, concrete mix, and temperature. The rate of placement should generally be lower in the winter than in the summer. Basically, it does not matter how many cubic yards are actually placed per hour or how large the project is. What does matter is the rate of placement per height and time (height of wall poured per hour).

6.3 BUILDING AND MATERIAL REUSE

6.3.1 Building Reuse

1. *Maintain existing walls, floors, and roof:*

   The *LEED™ intent* is to protect virgin resources by reusing building materials and products, thereby reducing impacts associated with the extraction and processing of virgin resources. It further aims to extend the life cycle of existing building stock, conserve resources, retain cultural resources, reduce waste, and lessen the environmental impacts of new buildings as they relate to materials manufacturing and transport. The Credit for "Maintain Existing Walls, Floors and Roof" requirements has been moved to "Building Life Cycle Impact Reduction" credit in LEED v4.

   LEED requirements: For New Construction, maintain a minimum of 50, 75, or 95% (for up to 3 points) of the existing building structure to include structural floor and roof decking as well in addition to the envelope (exterior skin and framing, but excluding window assemblies and nonstructural roofing material). It is possible to achieve a credit by maintaining a minimum of 50% (by area) of interior nonstructural elements such as interior walls, doors, floor coverings, and ceiling systems. Hazardous materials that are remediated as a part of the project scope are to be excluded from the calculation of the percentage maintained. The credit will not apply if the project includes an addition to an existing building, where the square-footage of the addition is more than 2 times the square footage of the existing building.

The table describes the minimum percentage of building structure reuse requirements for achieving credits achievement for New Construction:

Percentage building reuse	Points
55	1
75	2
95	3

However, for Core and Shell, you are required to maintain a minimum of 25, 33, 42, 50, or 75% of Existing Walls, Floors and Roof for up to 5 credits. For Schools must maintain 55% and 75% of Existing Walls, Floors and Roof for up to 2 credits. Because credit requirements are continuously changing, it is strongly advised to check with LEED for the updated requirements of a particular category.

Potential technologies and strategies: LEED v4 has a number of key changes that impact the role of technology in green buildings. For example, LEED v4 emphasizes performance by using building technologies to help fulfill prerequisites and credit requirements. Performance standards for new construction means an emphasis on installing technologies or technology infrastructure that prepare buildings to continuously monitor and improve performance.

Consider the use of salvaged, refurbished, or reused materials from previously occupied buildings, including structure, envelope, and elements. Remove elements that pose contamination risk to building occupants and upgrade components that would improve energy and water efficiency such as windows, mechanical systems, and plumbing fixtures. However, mechanical, electrical, plumbing, or specialty items and components should not be included for this credit. Furniture may be included only if it is included in the other MR credits 3–7.

2. *Maintain 50% of interior nonstructural elements for new construction and schools and 40% and 60% for commercial interiors*
The intent here according to LEED is to extend the life cycle of existing building stock, conserve resources, retain cultural resources, reduce waste, and lessen the environmental impacts of new buildings as they relate to materials manufacturing and transport.

LEED requirements: Maintain at least 50% (by area) of existing interior nonshell, nonstructural elements (interior walls, doors, floor coverings, and ceiling systems) of the completed building (including additions). If the project includes an addition to an existing building, this credit is not applicable if the square-footage of the addition is more than 2 times the square-footage of the existing building.

In terms of *Potential technologies and strategies*, LEED requires that consideration be given to the reuse of existing buildings, including structure, envelope, and interior nonstructural elements. Also elements that pose contamination risk to building occupants are to be removed and components that would improve energy and water efficiency, such as mechanical systems and plumbing fixtures, should be upgraded. For the LEED credit, the extent of building reuse needs to be quantified.

The owner/developer shall provide a report prepared by a suitably qualified person outlining the extent to which major building elements from a previous building were incorporated into the existing building. The report shall include preconstruction and postconstruction details highlighting and quantifying the reused elements, be it foundations, structural elements, or facades. Windows, doors, and similar assemblies may be excluded.

The rehabilitation of old buildings highlights the many successful commercial redevelopments being executed in many cities around the world. There is potential to lower building costs and provide a mix of desirable building characteristics. However, the reuse of existing structural elements depends on many factors, not least of which includes fire safety, energy efficiency, and regulatory requirements; all of these should have been taken into account for reuse in the existing building.

6.3.2 Material Reuse 5% and 10% for New Construction, Schools, Commercial Interiors (Also 30% for Furniture and Furnishing), and 5% for Core and Shell

The intent is to reuse building materials and products in order to protect and reduce demand for virgin material resources and to reduce waste, thereby reducing impacts associated with the extraction and processing of virgin resources.

LEED requirements: Use salvaged, refurbished, or reused materials such that the sum of these materials constitutes at least 5, 10, or 30% (only for Commercial Interiors – Furniture and Furnishings) based on the cost of the total value of materials on the project. Mechanical, electrical, and plumbing components and specialty items such as elevators and equipment shall not be included in this calculation. Include only materials permanently installed in the project. Furniture may be included, providing it is included consistently in MR Credits 3–7. Most credits in the Material and Resource category are calculated using a percentage of total building materials.

LEED™ potential technologies and strategies include the identification of opportunities to incorporate salvaged materials into the building design and research potential material suppliers. Salvaged materials such as beams and posts, flooring, paneling, doors and frames, cabinetry and furniture, brick, and decorative items should be considered.

Reusable, recyclable, and biodegradable materials: Sometimes there is confusion when identifying the difference between "reused" and "recycled." The terms are frequently (incorrectly) used interchangeably – but there is a distinct difference, especially for the purposes of LEED. Reuse is essentially the salvage and reinstallation of materials in their original form. Recycling is the collection and remanufacture of materials into a new material or product, typically different from the original. Biodegradable material is that which breaks down organically and may be returned to the earth with none of the damage associated with the generation of typical waste materials.

Reusing materials slated for the landfill has become one of the most environmentally sound ways to build because the extraction, manufacture/transport, and disposal of virgin building materials pollutes air and water, depletes resources, and damages natural habitats. Construction and demolition are estimated to be responsible for about 30% of the US solid waste stream. Real-world case studies by the Alameda County Waste Management Authority, for example, have concluded that more than 85% of that material, from flooring to roofing to packaging, is reusable or recyclable.

Salvage and reuse: By salvaging materials from renovation projects and specifying salvaged materials, material costs can be reduced while adding character to projects and maximizing environmental benefits, such as reduced landfill waste, reduced embodied energy, and reduced impacts from harvest/mining of virgin materials (e.g., logging old-growth or tropical hardwood trees, mining metals, etc.).

It should be noted that some materials require remediation or should not be reused at all. For example, materials contaminated by hazardous substances such as asbestos, arsenic and lead paint must be treated and/or disposed of properly. Avoiding materials that will cause future problems is critical to long-term waste reduction, as well as the health of communities and the planet (Figure 6.10a).

(a)

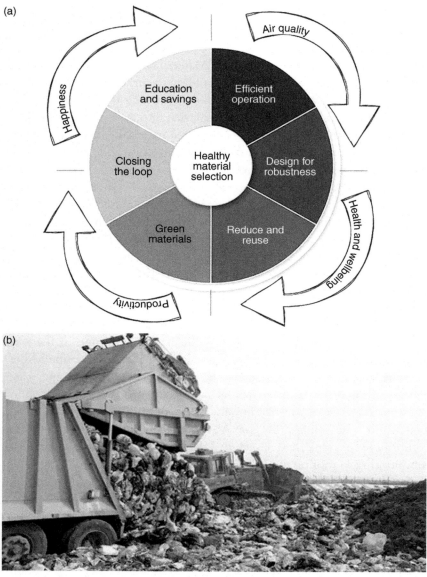

(b)

Figure 6.10 (a) Life on the planet continues to grow in complexity and this is true in the architectural and construction realms as well. Areas of concern, into which decisions are expanding, include occupant health, the overall environment and resource diminution. The materials and techniques used in creating a "green building" should ideally have no negative impact on the environment (b) Photo of typical landfill showing debris. *Courtesy: Part a, Deborah Hart Yemm.*

Recycling, salvage, and disposal calculations: A system should be developed for tracking the volume or weight of all materials salvaged, recycled, and disposed of during the project. At the end of the project, or more frequently if stipulated in the specifications, submit a calculation documenting the project achieved a 50 or 75% diversion rate. This rate will depend on the targeted goals set for the project.

Considerations for selecting reusable building materials (Check LEED website for latest updates):

- Reusing existing building shells, when appropriate, can yield the greatest overall reduction in project impacts. Additionally, for remodels/renovations, materials may be reused on site.
- For remodels and redevelopment, adequate time should be allowed in the construction schedule for deconstruction and recycling.
- Replace inefficient fixtures, components, and appliances (e.g., toilets using more than 1.6 gallons per flush, single-pane windows, and refrigerators or other appliances more than 5 years old).
- Products containing hazardous materials such as asbestos, lead, or arsenic should either be disposed of properly or remediated prior to reuse.
- Note that salvaged materials can vary in availability, quality, and uniformity. Ensure that materials are readily available to meet project needs before specifying them.
- Building materials composed of one substance (e.g., steel, concrete, wood, etc.) or that are readily disassembled are generally easiest to reuse or recycle.
- Materials composed of many ingredients, such as vinyl siding, OSB, or particleboard are generally not recyclable or biodegradable.
- Materials should be carefully evaluated to ensure that they offer best choice for the application. They need to be durable and can preferably be readily disassembled for reuse, recycling, or biodegrading at the end of the useful life of the building.

6.4 CONSTRUCTION WASTE MANAGEMENT

The broad intent of the Construction Waste Management credit is to avoid materials going to landfills during construction by diverting the construction waste, demolition, and land clearing debris from landfill disposal; redirect recyclable recovered resources back to the manufacturing process; and redirect reusable materials to appropriate sites.

LEED requirement: Recycle and/or salvage at least 50–75% of nonhazardous C&D and packaging debris. Develop and implement a construction waste management plan that, at a minimum, identifies and quantifies the materials generated during construction and that is to be recycled or that is to be diverted from disposal and whether such materials will be sorted on-site or comingled. Typical items would include brick debris, concrete, steel, ductwork, clean dimensional wood, paperboard, and plastic used in packing, etc. (Figure 6.10b). Excavated soil and land-clearing debris do not contribute to this credit. Calculations can be done by weight or by volume, but must be consistent throughout.

Each credit a project attempts to achieve using the LEED system requires documentation to prove the activity was completed. LEED™ Letter Templates are to be used to certify that LEED requirements are met for each prerequisite and credit. Additional documentation may also be required. As the contractor is responsible for construction waste management, the contractor will generally be responsible for completing the required LEED Letter Template documentation for these two credits. The documentation required includes the following:

1. Providing the LEED Letter Template, signed by the architect, owner, or other responsible party, tabulating the total waste material, quantities diverted and the means by which diverted, and declaring that the credit requirements have been met. A portion of the credits in each application may be audited, and the contractor should be prepared with backup documentation for credits they are involved in implementing.

2. A Waste Management Plan: LEED™ projects will generally require a plan and regular submittals tracking progress. The plan should show how the required recycling rate is to be achieved, including materials to be recycled or salvaged, cost estimates comparing recycling to disposal fees, materials handling requirements, and how the plan will be communicated to the crew and subcontractors. All subcontractors are required to adhere to the plan in their contracts.

Design considerations relating to construction waste reduction include the following:

- Smaller projects use less material, reducing both solid waste and operating costs.
- Assemblies should be designed to match the standard dimensions of the materials to be used.
- Disassembly design should be considered so that materials can be readily reused or recycled.

- Clips and stops should be employed to support drywall or wood paneling at top plates, end walls, and corners. Clips can provide the potential for two-stud corners, reducing wood use, easing electrical and plumbing rough-in, and improving thermal performance.
- Materials attached with removable fasteners are generally quicker, cheaper, and more feasible to deconstruct than materials installed with adhesives. However, adhesives distribute loads over larger areas than when fasteners are used alone.
- Remodel to make use of existing foundations and structures in good condition, reducing waste, material requirements, and possible labor costs.
- Design for flexibility and changing use of spaces.
- Specify materials such as SIPs, panelized wood framing, and precast concrete that can be delivered precut for rapid, almost waste-free installation.
- For wood construction, consider 24 in. on center framing with insulated headers, trusses for roofs and floors, finger-jointed studs, and engineered wood framing and sheathing materials.
- Whenever practical, specify materials with high recycled content.

It is estimated that commercial construction typically generates between 2 and 2.5 pounds of solid waste per square foot, the majority of which can be recycled. Salvaging and recycling C&D waste reduces demand for virgin resources and the associated environmental impacts. Effective construction waste management, including appropriate handling of nonrecyclables, can reduce contamination from and extend the life of existing landfills. Whenever feasible, reducing initial waste generation is environmentally preferable to reuse or recycling.

The Construction Waste Management Plan should from the outset recognize project waste as an integral part of overall materials management. The premise that waste management is a part of materials management, and the recognition that one project's wastes are materials available for another project, facilitates efficient and effective waste management.

Waste management requirements should be taken into account early in the design process and be the topic of discussion at both preconstruction and ongoing regular job meetings, to ensure that contractors and appropriate subcontractors are fully informed of the implications of these requirements on their work prior to and throughout construction.

Waste management should also be coordinated with or be part of the standard quality assurance program, and waste management requirements should be addressed regularly at each phase of the project. Any topical application of processed clean wood waste and ground gypsum board as a soil amendment must be done in accordance with local and state regulations. If possible, adherence to the plan would be facilitated by tying completion of recycling documentation to one of the payments for each trade contractor.

6.5 RECYCLED MATERIALS

Recycling is the practice of recovering used materials from the waste stream and then incorporating those same materials into the manufacturing process. Recycled content refers to that portion of material used in a product that has been diverted from the solid waste stream, and is the most widely cited attribute of green building products. The LEED intent is to protect virgin resources by increasing demand for building products and recycled content. If the materials are diverted during the manufacturing process, they are referred to as *preconsumer* recycled content (sometimes also known as postindustrial). If they are diverted after consumer use, they are termed *postconsumer*. Postconsumer content is generally viewed as offering greater environmental benefit than preconsumer content. While preconsumer waste is much greater, it is also more likely to be diverted from the waste stream. Postconsumer waste is more likely to fill limited space in municipal landfills and is typically mixed, making recovery more difficult.

LEED requirements: Use materials with recycled content such that the total material costs of postconsumer recycled content plus one-half of the preconsumer content represents at least 10%–20% (based on cost) of the total value of the materials used in the project. The recycled content value of the material used shall be determined by weight. The recycled content value is arrived at by multiplying the recycled fraction of the assembly by the cost of the assembly. Mechanical, electrical, and plumbing components and specialty items such as elevators shall not be included in this calculation. Furniture may be included, providing it is included consistently in MR Credits 3–7.

For a company to claim that it is using preconsumer recycled content, it must be able to substantiate that the material it is using would have become garbage had it not purchased it from another company's waste stream. However, if a manufacturer routinely collects scraps and feeds them back into its own process, that material does not qualify as recycled.

Numerous federal, state, and local government agencies have "buy recycled" programs aimed at increasing markets for recycled materials. These programs usually have a specific goal of supporting recycling programs to reduce solid-waste disposal, and many communities in the United States now offer curbside collection or drop-off sites for certain recyclable materials. Collecting materials however is only the first step toward making the recycling process work. Successful recycling also depends on manufacturers producing products from recovered materials and, in turn, consumers purchasing products made of recycled materials. The LEED Rating System applies credits for using recycled materials.

Recyclability is the ability of a product to be recycled and can only describe products that can be collected and recycled through an established process. This definition has often stretched beyond credibility by numerous manufacturers who make that claim based on a laboratory process. Sometimes they attempt to justify this lax approach with new products by noting that they are not expected to enter the waste stream in large quantities for years. In fact, many national and international companies seek an environmental marketing edge by advertising the recycled content of their products, which is often undocumented or certified.

Such claims come under the jurisdiction of the Federal Trade Commission (FTC), which first published definitions for common environmental terms in its Green Guides in 1992. The LEED Rating System offers credit for recycled-content materials, referencing definitions from ISO 14021. These definitions can still leave a lot of gray areas, which many manufacturers tend to interpret in their own favor. Third-party certification of recycled content is useful in maintaining a high standard and offering the ability to verify any claims that are made.

In some industries, recycling pathways have become very reliable. Products like engineered lumber and fuel pellets have created a reliable demand for waste from the forest products industry, including sawdust and woodchips. Although these materials may at one time have been part of the waste stream, the active market for them means that they are nearly always diverted. That does not change their status as preconsumer recycled content, however.

Postconsumer recycled content does not always refer to the average individual consumer; the consumer may be a large business or manufacturer. For example, Timbron International, Inc., an interior molding company, offers interior decorative wood-alternative moldings (crowns, bases, quarters, and rounds) that are 90% recycled EPS. Timbron's recycled content has been certified by SCS as 75% postconsumer and 15% preconsumer. Most of that comes from large industrial facilities that receive shipments packed in polystyrene. Since those facilities are the end-users of that packaging, it can qualify as postconsumer recycled content.

Waste has a cost, and we all bear it. Moreover, the extraction, manufacture/transport, and disposal of building materials clogs our landfills, pollutes our air and water, depletes resources, and damages natural habitats. The California Integrated Waste Management Board (CIWMB) notes that C&D wastes are responsible for about 28% of California's solid waste stream. More than 85% of that material, from flooring to roofing, is either reusable or recyclable. In addition to C&D waste, the material in our recycling bins, our used bottles, paper, cans, and cardboard, are also considered to be suitable raw materials for recycled-content products.

Keeping a substance out of the landfill is only the first step to putting "waste" back into productive use. The material must be processed into a new, high-quality item, and that product must be sold to a builder or homeowner who recognizes its benefits. By mining our "waste" as the raw material for new products, we increase demand for recycling, and convince manufacturers to use more recycled material, continuously strengthening this cycle.

Benefits of recycled content materials include reduced solid waste, reduced energy and water use, reduced pollution, reduced greenhouse gas emissions, and a healthier economy. The following materials are readily recyclable and generally cost less to recycle than to dispose of as garbage:

- Cardboard
- Clean wood (includes engineered products)
- Land clearing debris
- Metals
- Window glass
- Plastic film (sheeting, shrink wrap, packaging)
- Acoustical ceiling tiles
- Asphalt roofing
- Plastic and wood-plastic composite lumber from plastic and wood chips, ideal for outdoor decking and railings.
- Fluorescent lights and ballasts
- Insulation, such as cotton made from denim, newspaper processed into cellulose, or fiberglass with some recycled glass content.
- Carpet and carpet pad made of plastic bottles (or sometimes from used carpet). Up to half of all polyester carpet made in the United States contains recycled plastic.
- Tile containing recycled glass.

- Concrete containing ground up concrete as aggregate, or fly ash – a cementitious waste product from coal-burning power plants. Also asphalt, brick, and other cementitious materials
- Countertops made with everything from recycled glass to sunflower seed shells
- Drywall made with recycled gypsum, and Homasote wallboard made from recycled paper.
  For maximum benefit, when selecting a recycled-content building material:
- Choose material with the highest recycled content available. For example, some recycled products may only be 5% recycled and 95% virgin material.
- Seek high postconsumer recycled content. Some "recycled" content is waste from manufacturing processes. Reducing manufacturing waste is the first step, but recycling postconsumer material is necessary to close the loop.
- Check that the materials are appropriate for the application in hand.
- Salvaging (reusing) whole materials is preferable to recycling. Reuse all but eliminates waste, energy, water use, and pollution.
- Seek out materials that are not only recycled but recyclable or biodegradable at the end of their useful life. Ideally, a material may be recycled again and again back into the same product.

Reclaimed wood – an alternative to conventional hardwood: Reclaimed wood has many applications including for flooring, trim, siding, furniture, and in some cases, as structural members. Consider reusing wood from an existing building on site, or look to salvage yards and on-site deconstruction sales for a portion of your material needs.

Reclaimed wood flooring is manufactured from timbers salvaged from old buildings, bridges, or other timber structures or it may be manufactured from logs retrieved from river bottoms, or from trees being removed in urban and suburban areas. The character and aesthetic of reclaimed flooring can be particularly attractive. But equally important, the salvaging or reusing wood can reduce solid waste, save forest resources, and save money. Moreover, reclaimed wood is often available in dimensions, species, and with old-growth quality that is no longer obtainable from virgin forests at any price. However, planning and research are necessary, as available species, dimensions, and lumber quality can vary considerably from location to location.

Examples of reusable (RU), recyclable (RC), and biodegradable (B) building materials include the following:

- Asphalt (RC)
- Bricks (RU, RC)

- Concrete – ground and used as aggregate (RC)
- Steel, aluminum, iron, copper (RU, RC)
- Gypsum wallboard1 (RU, B)
- Earthen materials (RU, B)
- Wood and dimensional lumber, including beams, studs, plywood, and trusses (RU, RC, B)
- Straw bales (B)
- Wool carpet (B)
- Linoleum flooring (B)
- Doors and windows (RU)
- Plumbing and lighting fixtures (RU)
- Unique and antique products that may no longer be available (RU)

Deconstruction consists of the dismantling of a building to preserve the useful value of its component materials. Deconstruction is preferable to demolishing; the combination of tax breaks, new tools, and increasing local expertise are making it easier to keep materials out of the landfill, and money in owner's wallets. Although deconstruction takes longer and may initially cost more than demolition, it is nevertheless likely to reduce the overall project cost.

Best construction practices:

- Basic but true: Measure twice and cut once. Protect materials from the elements.
- Consider deconstructing and salvaging existing materials.
- Develop a waste reduction plan that includes waste prevention. This should be followed by assigning responsibility for its implementation to a motivated individual on the construction team. Post the plan and set up on-site locations for recycling, with color-coding for separation. Be sure to include time in the schedule for salvage and recycling. Require participation of all team members, including subcontractors.
- Delineate and limit the construction footprint and coordinate construction with a landscape professional to minimize grading and retain native soils and vegetation.
- Drywall clips should be used to fasten drywall. Recycled-content polyethylene clips are available as an alternative to metal.
- Surplus materials can be donated to organizations like Habitat for Humanity.

Avoiding waste: From an Environmental and financial point of view, avoiding generating waste in the first place is a far better practice than recycling. Waste reduction has the benefits of minimizing energy use, conserving resources, and easing pressure on landfill capacity.

6.6 REGIONAL MATERIALS

The main *LEED intent* here is to reduce material transport by increasing demand for building products made with the region. This is achieved by increasing the demand for building materials and products that are extracted and manufactured within the designated region. It will also support the regional economy and reduce the environmental impacts emanating from transportation.

With that in mind, Chrissy Macken, USGBC's assistant program manager for LEED v4, says,

> *The approach to LEED v4 was shaped by several overarching factors:*
> - *Global focus: In recognition of the worldwide spread of LEED, the new version provides more flexibility for local and regional equivalencies. This includes eliminating double documentation for designers in areas with more stringent requirements for certain categories, flexibility in some jurisdictions to use local codes, and a wider variety of reference standards that come from outside the U.S.*
> - *More markets: LEED v4 recognizes more market sectors via 21 space type adaptations, including data centers, existing schools, hospitality, and mid-rise residential buildings.*
> - *All about performance: The changes made were intended to increase the technical rigor of the rating system so that project teams achieve higher performance along a variety of different paths, such as reducing contribution to climate change, protecting water, and human health. Credits are more closely linked to improved environmental outcomes. For example, instead of just getting points for bike racks and changing rooms, the credit also requires locating projects within a bikeable area; water and energy meters are now prerequisites; and material credits are focused on holistic rather than single attributes.*

LEED requirement: Use a minimum of 10% or 20% (based on cost) of total building materials and products that are manufactured* (extracted, harvested, recovered or manufactured) regionally within a radius of 500 miles of the site (Figure 6.11). Either the default 45% rule or actual materials cost may be used. All mechanical, electrical, plumbing, and specialty items such as elevator equipment should be excluded. If only a fraction of the product/material is *extracted, harvested, recovered* or *manufactured* within 500 miles of the site, then only that percentage (based on weight) will contribute to the regional value. Furniture may be included only if it is included throughout MR Credits 3–7.

*Manufacturing alludes to the final assembly of components into the building product that is furnished and installed by the contractor. For example, if the hardware comes from Los Angeles, California, the lumber from Vancouver, British Columbia, and the joist is assembled in Fairfax, Virginia then the location of the final assembly is Fairfax, Virginia.

Figure 6.11 *A Map Program That is Capable of Drawing any Required Radius for any Chosen Location. Source: Free Map Tool.*

It should be noted that LEED v4 approaches materials content credits in a fundamentally different manner than previous editions of LEED. What were 8 MR credits for materials reuse, recycled content, regional materials, rapidly renewable materials, and certified wood are now 6 credits (8 for Healthcare).

The 20% credit threshold can often be attained with no cost impact, simply by tracking the materials that are normally produced and supplied within 500 miles of a project site. In some cases, however, the 20% threshold can only be achieved by targeting certain materials (e.g., specific types of stone or brick) or limiting the number of manufacturers whose products will be considered in the project bids. In these cases, cost premiums may be incurred.

In order to verify that the extraction/harvest/recovery site is located within a 500-mile radius of the project site, project teams are required to indicate the actual mileage between the project site and the manufacturer and, similarly, the distance between the project site and the extraction site for each raw material in the submittal template. Alternatively, a statement on the manufacturer's letterhead indicating that the point of manufacture is within 500 miles of the LEED™ project site will also be accepted as part of the LEED™ documentation and credit submittals.

6.7 RAPIDLY RENEWABLE MATERIALS

There have been many changes in Rapidly Renewable Materials in LEED v4. In LEED 2009, MR credits were usually heavily weighted on individual attributes of materials, such as recycled content, or rapid renewability. This does not apply with LEED v4, where the USGBC has taken more of a lifecycle cost analysis approach to all the products and materials that comprise a building. LEED 2009 states that the intent of using rapidly renewable materials is to "reduce the use and depletion of finite raw materials and long-cycle renewable materials by replacing them with rapidly renewable materials." Materials and Resources Credit 6, Rapidly Renewable Materials, awards 1 point to projects that uses rapidly renewable materials for at least 2.5% of all building materials used on the project, based on cost. LEED considers these materials to have harvest cycles of 10 years or less.

Such materials include:

- Bamboo
- Cork
- Insulation
- Natural linoleum (marmoleum)
- Straw (straw–bale construction)
- Wheatboard
- Agrifiber
- Wool
- Strawboard

6.7.1 Bamboo

Bamboo is emerging as an alternative resource to other types of wood commonly used in the United States and abroad. It was previously intuitively used as a basic material for making different household objects and small structures. But ongoing research and engineering efforts are enabling bamboo's true value to be realized as a renewable, versatile, and readily available economic resource.

Bamboo is a giant grass and comes in 1500 varieties that produces hard, strong, dimensionally stable wood. It is commonly found in the tropical regions of Asia, Africa, and South America. It has been used as both a building material and for furniture construction for thousands of years. Bamboo is considered to be the fastest-growing woody plant on earth and is also one of the most versatile and sustainable building materials available. It grows remarkably fast and can reach maturity in months in a wide range of climates and is exceedingly strong for its weight. It can be used both structurally and as a finish material. Bamboo is characterized as a totally renewable resource; you can cut it, and it grows right back.

The swift growth, strength, and durability of bamboo make it an environmentally superior alternative to conventional hardwood flooring. Some hardwood trees may require decades to reach maturity, whereas bamboo is typically harvested on a 4- or 5-year cycle and the mature forest will continue to send up new shoots for decades. Pine forests have the most rapid growth among tree species, but bamboo grass species used in flooring can grow more than 3 ft. per day and produce almost twice as much harvestable fiber per year. Though bamboo grows rapidly, it can nonetheless yield a product that is 13% harder than rock maple, with durability comparable to red oak. Bamboo canes are beautiful when exposed and may also be used for paneling, furnishings, and cabinetry (Figure 6.12a,b). Vertically laminated flooring and plywood products consist of layers of bamboo compressed with a binder, creating a durable, resilient finish material.

There is a long vernacular tradition to the use of bamboo in structures in many parts of the world, especially in more tropical climates, where it grows into larger-diameter canes. When using bamboo, care should be given to joinery details since its strength comes from its integral structure, which means it cannot be joined using many of the traditional methods used with wood. In this respect, the old ways of building with bamboo can be especially informative.

Bamboo is an extremely strong fiber that has twice the compressive strength of concrete and also roughly the same strength-to-weight ratio of steel in tension. The strongest bamboo fibers also have greater shear strength than structural woods, and they take much longer to come to ultimate failure. However, this ability of bamboo to bend without breaking makes it unsuitable for building floor structures due to our low tolerance for deflection, and unwillingness to accept a floor that feels "alive." However, bamboo as a ¾-in.-thick finish floor is an appropriate substitute for the standard oak because it installs the same way, is harder, and expands less. Like most interior-grade hardwood plywood, bamboo flooring is typically made with a urea formaldehyde binder, which can emit tiny amounts of formaldehyde. Choosing high-quality products, particularly from manufacturers that provide independent air-quality testing data, can help to minimize this source of indoor air pollution. And when well maintained, bamboo floors can last decades.

Figure 6.12 (a) Kitchen cabinets made from bamboo. (b) Photo showing use of bamboo as exterior siding. *Source: Bamboo Technologies.*

6.7.2 Cork

Cork is a natural, sustainable product harvested from the outer bark of the evergreen cork oak, *Quercus suber*, which grows primarily in Portugal, Spain, and surrounding areas. Cork is completely sustainable and is an ideal flooring choice for the health of the environment and the home or building. Cork oak can be first harvested when it is 25 years old, when the virgin bark is carefully cut from the tree. From this point on, the tree can be "stripped" of its cork every 9 years for nearly 200 years without harm coming to the tree, and which helps to encourage long-term management of this renewable resource. An 80-year-old cork tree can produce up to about 500 pounds of cork. Cork harvesting remains one of the best examples of a sustainable agroforestry system.

Cork is becoming increasingly popular because of its extraordinary combination of beauty, durability, insulation, and renewability. Modern cork floors are durable, fire resistant, provide thermal and acoustic insulation, and are soft on the feet. They are typically covered with an acrylic finish, but may be covered with polyurethane for bathroom or kitchen applications. Cork floors can last for decades and the material is biodegradable at the end of its useful life (Figure 6.13a,b).

By contrast, the extraction, manufacture and transport, and disposal of synthetic flooring materials pollute air and water, deplete resources, damage natural habitats, and can have negative health impacts. Hardwood flooring requires logging slow-maturing trees that require decades to centuries to mature.

(a)   (b)

Figure 6.13 (a) Residential interior using cork flooring. (b) Cork pattern detail. *Source: Globus Cork.*

Although cork is mainly considered to be a resilient flooring option, cork is also a natural alternative to carpet. Moreover, although carpet can attract and hold indoor pollutants in its fibers, cork is easier to thoroughly clean, and is inherently resistant to mold and mildew, sheds no dust or fibers, and is naturally antistatic. These hypoallergenic properties, combined with thermal and acoustic insulation, allow cork floors to provide the majority of the benefits of carpet, without its liabilities. Furthermore, the benefits of cork extend beyond human health; they include reduced landfill waste, low embodied energy and local availability for many products, excellent aesthetics, and reduced impacts from the harvest or mining of virgin materials.

6.7.3 Insulation

Well-insulated building envelopes are primary considerations in comfort and sustainability. Insulation helps to protect a building's occupants from heat, cold, and noise; in addition, it reduces pollution while conserving the energy needed to heat and cool a building. Environmentally preferable insulation options offer additional benefits, such as reduced waste and pollution in manufacture and installation, more efficient resource use, recyclability, enhanced R-value, and reduced or eliminated health risks for installers and occupants. The comfort and energy efficiency of a home and to a lesser degree, an office, depends on the R-value of the entire wall, roof, or floor (i.e., Whole-Wall R-value), not just the R-value of the insulation.

Techniques such as Advanced Framing increase the wall area covered by insulation, thereby increasing whole wall's effectiveness. Framing conducts far more heat than insulation, in the same manner that most window frames conduct more heat than double-paned glass. An additional layer of rigid insulation between framing and exterior sheathing can help improve the whole-wall R-value by insulating the entire wall, and not just the clear space. In non-"breathable" wall designs, closed-cell rigid foam with taped seams can provide an effective vapor barrier.

Fiberglass: the conventional choice: For economic reasons, fiberglass is often the insulation of choice. Any fiberglass insulation used should be a formaldehyde-free product with a minimum 50% total recycled content (minimum 25% postconsumer). Some products are manufactured with heavier, intertwined glass fibers to reduce airborne fibers and mitigate the fraction of fibers that can enter the lungs. Like all glass products, fiberglass insulation is made primarily from silica heated to high temperatures, requiring significant energy and releasing formaldehyde. Short-term effects include irritation to eyes, nose, throat, lungs, and skin during installation or other contact. Longer-term effects are controversial, but OSHA requires fiberglass insulation to carry a cancer-warning label. Binders in most fiberglass batts contain toxic formaldehyde that is slowly emitted for months or years after installation, potentially contaminating indoor air.

Some of the environmentally preferable insulation options include the use of recycled cotton insulation, which insulates as well as fiberglass and offers superior noise reduction. Cotton insulation poses no health risk and is not irritating during installation. Likewise, cellulose (recycled newspaper) insulation is an acceptable alternative. It is sprayed in wet or dry, poses no health risk, and offers superior *R*-value per inch. Both cotton and cellulose are treated with borate, which is not toxic to humans, and makes both materials more resistant to fire and insects than fiberglass. Sprayed polyurethane foams expand to fill cracks, and also makes a good alternative, providing insulation, vapor barrier, and additional shear strength. Sprayed cementitious foams such as Air-Krete have similar properties. Finally, structural insulating systems integrate a building's structure and insulation into a single component. They produce little or no waste during construction and provide excellent thermal performance.

6.7.4 Linoleum

Linoleum is a highly durable resilient material used mainly for flooring. It is made from natural materials, a mixture of linseed oil, wood flour, powdered cork, and pine resin, which is pressed onto a jute fiber backing.

Flexible vinyl flooring that displaced linoleum from the marketplace in the 1960s is often incorrectly referred to as "linoleum." The two materials are quite different. First, costs of linoleum are higher, but linoleum offers performance that is in many ways superior to vinyl: linoleum lasts longer, is inherently antistatic, and is antibacterial. All natural, linoleum requires less energy and creates less waste in its manufacture, and can be chipped and composted at the end of its useful life. Maintenance of linoleum is also less labor intensive and less costly because it does not require sealing, waxing, or polishing as often as vinyl.

By contrast, flexible vinyl flooring is a more prolific generator of solid waste because it is manufactured from toxic materials, and typically lasts less than 10 years; also, it is neither biodegradable nor recyclable. Linoleum emits far fewer VOCs when installed with a low-VOC adhesive than flexible vinyl, and does not exude the phthalate plasticizers that are an increasing concern for human health.

The durability of hard vinyl composition tile is comparable to linoleum, but recycling it is impractical, which is why vinyl composition tile will ultimately end up in a landfill. Moreover, vinyl products can be harmful because they involve toxic manufacturing chemistry that generates hazardous wastes and air pollution, and vinyl manufacturing also consumes petroleum.

Linoleum attributes:

- Very durable and can often last for decades; this helps reduce waste associated with the frequent replacement of flexible vinyl flooring.
- It is quiet and comfortable.
- Is made from natural, nontoxic components – does not contain formaldehyde, asbestos, or plasticizers.
- Biodegradable at the end of its useful life.
- Easy to clean and maintain, using minimal water and gentle detergent.
- Is resistant to temporary water exposure, making it appropriate for use in kitchens. However, its sensitivity to standing water can be a concern in bathrooms.
- Being naturally antistatic, helps control dust.
- It has very low VOC emissions when installed with appropriate adhesives, although the scent of curing linseed oil may not agree with chemically sensitive persons.
- Square-foot cost can be comparable to high-quality flexible vinyl flooring. However, flexible vinyl is commonly replaced within 10 years and is toxic to manufacture. Also, vinyl is neither biodegradable nor recyclable.
- As linoleum is the same color all the way through, gouges and scratches can be buffed out – reducing long-term costs and waste.

6.7.5 Straw Bale Construction

Straw bale construction consists of using compressed blocks (bales) of straw, either as fill for a wall cavity (nonload-bearing) or as a structural component of a wall (load-bearing). A post-and-beam framework that supports the basic structure of the building, with the bales of straw used as infill, is the most common nonload-bearing approach. This is also the primary method that is permitted in many jurisdictions, although many localities have specific codes for straw bale construction, and some banks are even willing to lend on this technique. Straw is a renewable resource that acts as excellent insulation and is fairly easy to build with.

The interior and exterior sides of the bale wall are typically covered by stucco, plaster, clay, or other treatment. This type of construction can offer structural properties superior to the sum of its parts. Both load- and nonload-bearing straw bale design divert agricultural waste from the landfill for use as a building material with many exceptional qualities.

Building with bales of straw has become popular in many parts of the country and there are now thousands of straw bale homes in the United States. However, with load-bearing straw bale buildings, care must be taken to consider the possible settling of the straw bales as the weight of the roof, etc. compresses them. Care should also be taken to ensure that the straw is kept dry, or it will eventually rot. For this reason, it is generally best to allow a straw bale wall to remain breathable; any moisture barrier will invite condensation to collect and undermine the structure. Other possible concerns with straw bale walls are infestation of rodents or insects, so the skin on the straw should be made to resist these.

Straw bale houses typically save about 15% of the wood used in a conventionally framed house. Also, because of the specialized work that goes into plastering both sides of straw bale walls, the cost of finishing a straw bale house sometimes exceeds that of standard construction. Nevertheless, the result is usually worth it because of the superior insulation and wall depth that is achieved. In many cases, straw bales provide an excellent "alternative" building material that helps to reduce or eliminate many environmental problems because they use nontoxic, reusable, and biodegradable elements that build durable, comfortable, and healthy places to live and work (Figure 6.14).

Figure 6.14 *An Interior of a Residence Built of Straw Bale. Source: StrawBale Innovations, LLC.*

Issues to consider when using straw bale construction include the following:

- Straw bale walls are thick and may constitute a high percentage of the floor area on a small site.
- Straw bales may be plastered on both sides to provide thermal mass and, similar to standard construction, the walls must be protected from moisture.
- When labor is executed primarily by building professionals, the square-foot cost of straw bale construction may be equivalent to conventional building methods.
- Straw bale can be more resistant to termites and vermin than stick construction, but as is the case with other types of construction, elimination of cracks and holes is critical.
- Straw bales do not hold nails as well as wood and thus nailing surfaces need to be provided.

Benefits of using straw bale construction methods include the following:

- Very good thermal and acoustic insulation, thereby enhancing occupant comfort throughout the year. The interior plaster of straw bale houses increase the "thermal mass" of the home, which helps to stabilize interior temperature fluctuations.
- A traditional "stick frame" home of 2×6 construction usually has an insulating value of R14, whereas with a properly insulated roof, straw bale enables a person to build a warm winter home with an R factor of R35–R50.
- Does not require toxic treatment, which helps chemically sensitive individuals.
- Can be economical. Straw bales are inexpensive (or free), and owner/builders and volunteers can contribute significantly to labor.
- Produces reduced construction waste: the main building material is a "waste," and excess straw can be used on-site in compost or as soil-protecting ground cover.
- Material is biodegradable or reusable at the end of its useful life.
- Potential for major reduction in wood and cement use, particularly in load-bearing straw bale designs.
- Conventional foundations and roofs can be used with straw bale buildings.
- Highly resistant to vermin (including termites).
- Aesthetic flexibility – from conventional linearity to organic undulation.
- Can be used in new construction, additions, and remodels.

6.7.6 Wheatboard

Environmentally friendly furniture comes from a variety of resources. One example is wheatboard, which is a by-product made from wheat straw and is one of the more popular renewable materials. This material has no formaldehyde and can be used to create, among other things, quality furniture and cabinets. Wheatboard is a durable material and is produced in sheets. These sheets can be filled, sealed, painted, stained, or varnished and can be shaped in a wide variety of designs.

Typically, wheatboard is an MDF substitute that is made from rapidly renewable material with at least 90% preconsumer recycled content. Wood MDF is usually made with urea formaldehyde, which offgases and negatively affects indoor air quality, but not all wheatboard contains the chemical. Wheatboard is versatile, and can be used among other things for architectural millwork, wall coverings, retail displays, furniture, cabinetry, and flooring.

Wheatboard has traditionally been burned or added to landfills. A number of manufacturers produce wheatboard – both in the United States and Canada. It is a viable substitute for wood and benefits the environment by reducing deforestation and also lessens both air pollution and landfill use. Along with its versatility, using wheatboard can help a building project earn crucial LEED credits for rapidly renewable materials.

6.7.7 Agrifiber Board

Agrifiber boards are composite panels made using agricultural materials. The intent of Agrifiber board is to reduce the quantity of indoor air contaminants that are odorous, irritating, and/or harmful to the comfort and well being of installers and occupants. Normally, the board is melded under high heat and pressure, which makes the board strong without the use of adhesives. This means that it does not offgas VOCs or contain formaldehyde. It is also nearly fireproof, nontoxic, carbon negative, and mold and pest resistant. It is energy-efficient, with R values ranging from 6 to 25, and sound absorbent, with STC ranging from 33 to 66.

Agrifiber board is versatile and can be used to replace various building materials, including MDF, gypsum board, particle board, soundproofing panels, and fiberglass insulation. This means that it can be used in a variety of applications, from SIPs, to furniture, kit houses, door cores, and soundproofing. It is available in various thicknesses and widths, making it suitable for these diverse projects. Also, otherwise unused farm waste such as rice straw, wheat stalks, sunflower husks, sugarcane bagasse, and walnut shells can find new life in agrifiber boards.

6.7.8 Wool

Wool is the textile fiber obtained from sheep and certain other animals, including cashmere and mohair from goats. It is a rapidly renewable material since a sheep can be shorn about every 2 years with no harm to the animal. Sheep shearing is the process by which the woollen fleece of a sheep is cut off. Wool has several qualities that distinguish it from hair or fur: it is crimped, it is elastic, and it grows in clusters. Wool is usually used for furniture or carpeting in buildings.

6.7.9 Strawboard

Strawboard is manufactured from straw such as wheat or rice by a patented process of heat and pressure that fuses the straw using its internal resins, thus making it a totally natural product. It is normally produced by continuous extrusion and has a tough, bonded external surface of recycled paper. The high density and low oxygen content of the resulting panels makes them resistant to combustion and, since they contain no added resins, alcohol, or other chemicals, they do not release flammable vapors. A standard 2¼-in. panel has a 1-h fire rating, an R-value of 3, and noise level reductions from one side to the other of 32 dB. A standard strawboard panel measures 4 ft. by 8 ft. by 2¼ in. thick and weighs 120 lb. Strawboard panels make an excellent alternative to timber for frame structures, interior partition walls, floors, load and nonbearing ceilings, roofs, and prefabricated buildings.

6.8 USE AND SELECTION OF GREEN OFFICE EQUIPMENT

The intent is to reduce the environmental and economic harms of excessive energy use by achieving a minimum level of energy efficiency for the building and its systems. People continue to search for ways to reduce their environmental footprint and act greener, both at home, and at the office. With buildings contributing nearly 40% of the US CO_2 emissions, even small changes at the office can potentially lead to a dramatic impact. In fact, the impact of energy costs directly affects the bottom-line of both building owners and tenants alike. Energy represents 30% of operating expenses in a typical office building, which makes it the single largest and most manageable operating expense in the provision of office space. The cost of energy to power an appliance over time is typically many times the original price of the equipment. By choosing the most energy-efficient models available, a positive impact on the environment is achieved while at the same time saving money.

The energy and water that appliances and office equipment consume directly translates into more fuel being burned at power plants, which in turn contributes to air pollution, global climate change, and waste of our limited natural resources. The good news is that efficient new appliances and office equipment can use only one-half to one-third as much energy as those purchased a decade ago. The Lawrence Berkeley Laboratory reports that replacing older refrigerators, clothes washers, dishwashers, thermostats, heating equipment, and incandescent lighting with ENERGY STAR equipment can save enough energy and water to provide an average after-tax return on investment of more than 16%, which is substantially better than the stock market. When appliances are well maintained and in good condition, one should weigh the waste, energy, and pollution required to make a new piece of equipment against gains in efficiency. There are no tangible savings or environmental benefits if older appliances remain in use, as is common with refrigerators.

Retailers typically carry efficient, durable appliances and office equipment. Appliances and equipment with the ENERGY STAR label are preferred although these are not necessarily the most efficient of all available models. Nevertheless, ENERGY STAR® products do usually perform significantly better than federal minimum efficiency standards. According to a 2002 EPA report, ENERGY STAR® labeled office buildings generate utility bills 40% less than the average office building. There are often rebates and/or incentives for the purchase of energy- and water-saving appliances.

Things to consider when selecting equipment:

- Appliances and equipment that use the least energy and/or water.
- Appliances and equipment that have the ENERGY STAR label.
- Durable appliances and equipment that meet long-term needs.
- Natural gas appliances for space and water heating; gas is often more cost-effective and can reduce overall energy use. But natural gas, like other fossil fuels, is not a renewable resource.
- Sealed combustion and direct vent furnaces and water heaters increase indoor air quality.
- Consider use of occupancy sensors in offices to minimize unnecessary lighting, as well as "smart" power-strips that combine an occupancy sensor with a surge protector. Smart power-strips will shut down devices (such as monitors, task lights, space heaters, and printers) that can be safely turned off when space is unoccupied.

ENERGY STAR Green Building guidelines: ENERGY STAR, a national program from the US EPA and US DOE, offers a comprehensive approach for new and existing homes and commercial buildings that reduce energy costs, while creating a more comfortable, healthier building. With relation to residential construction ENERGY STAR offers homebuyers the features they want in a new home, plus energy-efficient improvements that deliver better performance, greater comfort, and lower utility bills. To earn the ENERGY STAR, a home must meet strict guidelines for energy efficiency set by the US Environmental Protection Agency. Such homes are typically 20–30% more efficient than standard homes.

Commercial buildings qualify as ENERGY STAR after demonstrating that they are as energy-efficient as the top 25% of comparable facilities. A residential or commercial building design can earn the Designed to Earn the ENERGY STAR qualification by making informed decisions about energy efficiency during the design process. There is also the potential to achieve LEED™ credit points for using "green" office equipment. For example, in the LEED for existing buildings, the requirements for electronics can be found in the Materials and Resources (MR 2.1) section under "Sustainable Purchasing: Durable Goods, Electric," which states:

One point is awarded to projects that achieve sustainable purchases of at least 40% of total purchases of electric-powered equipment (by cost) over the performance period. Examples of electric-powered equipment include, but are not limited to, office equipment (computers, monitors, copiers, printers, scanners, fax machines), appliances (refrigerators, dishwashers, water coolers), external power adapters, and televisions and other audiovisual equipment. Sustainable purchases are those that meet one of the following criteria:

- The equipment is ENERGY STAR labeled (for product categories with developed specifications). This will reduce the carbon footprint of the equipment as well as the office's operating costs.
- The equipment (either battery or corded) replaces conventional gas-powered equipment.

When applying for project certification, you should constantly consult the USGBC website (http://www.usgbc.org).

6.9 CERTIFIED WOOD

Forest *certification* is a means for independent organizations to develop standards of good forest management, and independent auditors to issue certificates to forest operations that comply with those standards. This certification verifies that forests are well managed, as defined by a particular standard, and ensures that certain wood and paper products come from responsibly managed forests.

The Intent of the LEED™ Certified Wood credit is to encourage such environmentally responsible forest management programs.

After several years of studying the issue, the USGBC has proposed a major change for certified wood in its LEED Rating System. Previously, LEED awarded credit to projects that used wood certified to the standards of the FSC for at least half of their wood-based materials. The USGBC has now broadened the credit to recognize any forest-certification program that meets its criteria. The change is partly in response to criticism that LEED favors one forest-certification program, FSC, over others – particularly the Sustainable Forestry Initiative (SFI), a rival to the FSC that is portrayed by some environmentalists as less rigorous. Nevertheless, the revision brings the credit into line with a trend in LEED toward using transparent criteria to decide which third-party certification programs to reference.

Under the newly proposed credit language, wood certification systems would be evaluated for eligibility to earn points toward LEED certification against this measurable benchmark that includes:

- Governance
- Technical/standards substance
- Accreditation and auditing
- Chain of custody and labeling

LEED says that it will recognize wood certification programs that are, after thorough objective analysis, deemed compliant with the benchmark. On the other hand, wood certification programs that are not found to be in alignment with the benchmark would have a clear and transparent understanding of what modifications are necessary to receive recognition under LEED. The FSC appears likely to meet the criteria, but a number of other programs, including SFI and the Canadian Standards Association (CSA), may face some difficulty with parts of the benchmark system, such as the ban on genetically modified organisms, the emphasis on integrated pest management and bans on certain chemicals. All of the benchmark criteria must be met for recognition in LEED.

The implementation requirements to achieve the LEED credit include the following:

1. Use of a minimum of 50% of wood materials and products certified in accordance with an approved certifier (e.g., FSC). Principles and criteria for wood building components:
   a. Framing
   b. Flooring and subflooring
   c. Wood doors
   d. Other

2. Based on cost of certified wood products compared to total wood material cost.
3. Exclude MEP and elevator equipment
4. Furniture that is okay may be included in this calculation
5. Contractor does not need the certification number, but supplier does.

Meaning of forest certification: The concept of forest certification was launched over a decade ago to help protect forests from destructive logging practices. Forest certification was intended as a seal of approval and a means of notifying consumers that a wood or paper product comes from forests managed in accordance with strict environmental and social standards. For example, a person shopping for wall panel or furniture would seek a certified forest product to be sure that the wood was harvested in a sustainable manner from a healthy forest, and not procured from a tropical rainforest or the ancestral homelands of forest-dependent indigenous people.

Forest certification involves the green labeling of companies and wood products that meet standards of "sustainable" or "responsible" forestry. The primary purpose of forest certification is to provide market recognition for forest producers who meet a set of agreed upon environmental and social standards.

Forest conservation program: The SCS has been providing global leadership in third-party environmental and sustainability certification, auditing, testing, and standards development for three decades. The SCS first developed its Forest Conservation Program in 1991 and has since developed and grown to emerge as a global leader in certifying forest management operations and wood product manufacturers. The FSC in 1996 accredited SCS as a certification body, enabling it to evaluate forests according to the FSC Principles and Criteria for Forest Stewardship. SCS is accredited to provide services under a wide range of nationally and internationally recognized certification programs, and through a well-developed network of regional representatives and contractors, SCS offers timely and cost-effective certification services worldwide.

6.10 LIFE-CYCLE ASSESSMENT OF BUILDING MATERIALS AND PRODUCTS

This subject Life-Cycle Cost Analysis (LCCA) is discussed in greater detail in Chapter 10 (Economics of Green Design).

6.10.1 Life-Cycle Cost Analysis

The National Institute of Standards and Technology (NIST) defines Life-Cycle Cost (LCC) as "the total discounted dollar cost of owning, operating, maintaining, and disposing of a building or a building system" over a period of time. Life-cycle costing is therefore an economic evaluation technique that determines the total cost of owning and operating a facility over a period of time and takes into consideration relevant costs of alternative building designs, systems, components, materials, or practices, in addition to the multiple impacts on the environment (both positive and negative) that building materials and certain products have.

This method takes into account the impacts of every stage in the life of a material, from first costs, including the cost of planning, design, extraction/harvest, and installation as well as future costs, including costs of fuel, operation, maintenance, repair, and replacement, recycling, or ultimate disposal. It also takes into account any resale or salvage value recovered during or at the end of the time period examined. This type of analysis allows for comprehensive and multidimensional product comparisons. Potential health and safety issues that emerge during the construction, occupation, maintenance, alteration, and disposal of the facility should be taken into consideration. Initial failure to address the ease with which the built environment can be safely maintained can lead to unnecessary costs and risks to health and safety at a later date.

Life-cycle cost analysis (LCCA) can be employed on both large and small buildings as well as on isolated building systems. LCCA is especially useful when applied to project alternatives that fulfill the same performance requirements but differ with respect to initial costs and operating costs, and have to be compared in order to select the one that maximizes net savings. For example, LCCA will help determine whether the incorporation of a high-performance HVAC or glazing system, which may increase initial cost but result in dramatically reduced operating and maintenance costs, is cost-effective or not. LCCA is not useful for budget allocation.

Many building owners apply the principles of LCCA in decisions they make regarding construction or improvements to a facility. From the building owner who opts for aluminum windows in lieu of wood windows to the strip mall developer who chooses paving blocks over asphalt, both owners are taking into consideration the future maintenance and replacement costs in their selections. While initial cost has always been a factor in the decision making process, it is not the only factor.

Related to LCCA, life-cycle costing is the systematic evaluation of financial ramifications of a material, a design decision, or a whole building. LCC tools can help calculate payback, cash flow, present value, internal rate of return, and other financial measures. Such criteria can go a long way toward comprehending how a modest up-front cost for environmentally preferable materials or design features can provide a very sound investment over the life of a building.

Lowest LCC is considered the most straightforward and easy-to-interpret system of economic evaluation. There are other commonly used measures available such as Net Savings (or Net Benefits), Savings-to-Investment Ratio (or Savings Benefit-to-Cost Ratio), Internal Rate of Return, and Payback Period. All of these systems are consistent with the Lowest LCC measure of evaluation if they use the same parameters and length of study period. Building economists, architects, cost engineers, quantity surveyors, and others might use any or several of these techniques to evaluate a project. The approach to making cost-effective choices for building-related projects can be quite similar whether it is called cost estimating, value engineering, or economic analysis. Open book accounting, when shared across the whole project team, helps everyone to see the actual costs of the project.

LCCA methodology: It is worth reiterating that the purpose of an LCCA is to estimate the overall costs of project alternatives and to select the design that ensures the facility will provide the lowest overall cost of ownership consistent with its quality and function. The LCCA should be performed early in the design process while there remains an opportunity to refine the design in a manner that results in a reduction in LCC.

The first and most challenging task of an LCCA, or any economic evaluation method for that matter, is to determine the economic effects of alternative designs of buildings and building systems and to quantify these effects and express them in dollar amounts. When viewed over a 30-year period, initial building costs have been shown to generally account for approximately just 2% of the total, whereas operations and maintenance costs equal 6%, and personnel costs reflect the lion's share of 92% (Source: *Sustainable Building Technical Manual*).

General costs: There are various costs associated with acquiring, operating, maintaining, and disposing of a building or building system. Building-related costs usually fall into one of the following categories (see Chapter 10):

- First costs – purchase, acquisition, construction costs
- Operation, maintenance, and repair costs
- Fuel costs
- Replacement costs
- Residual values – resale or salvage values or disposal costs
- Finance charges – loan interest payments
- Nonmonetary benefits or costs

Only costs within each category that are relevant to the decision and significant in amount are needed to make a valid investment decision. To be relevant, costs must differ from one alternative compared with another; costs achieve significance when they are large enough to make a credible difference in the LCC of a project alternative. All costs are entered as base-year amounts in today's dollars; the LCCA method accelerates all amounts to their future year of occurrence and discounts them back to the base date to convert them to present values.

Construction costs: Detailed estimates of construction costs are not required for preliminary economic analyses of alternative building designs or systems, although a preliminary budget is advisable to help avoid or minimize cost overruns. More realistic cost estimates are usually available when the design is relatively advanced and the opportunities for cost-reducing design modifications have been missed. LCCA can be repeated throughout the design process whenever more detailed cost information becomes available. Initially, construction costs are estimated by reference to historical data from similar facilities. They can also be determined from government or private sector cost-estimating guides and databases.

Detailed cost estimates are prepared at the submittal stages of design (typically at 30%, 60%, and 90%) based on quantity take-off calculations. These estimates rely mainly on cost databases such as the *R.S. Means Building Construction Cost Database.* Testing organizations such as ASTM International and trade organizations also have reference data for materials and products they test or represent.

The most astute and logical way for the owner/developer to avoid cost overruns is to have the following:

- Objectives that are reasonable and that are not subject to modification during the course of the project.
- A complete design that meets planning and statutory requirements and will not require later modification.
- Clear leadership and appropriate management controls in place.
- Project estimates that are realistic and not unduly optimistic.
- A project brief that is comprehensive, clear, and consistent.
- A coordinated design that takes into account maintenance, health and safety, and sustainability.
- An appropriate risk allocation and sufficient contingency that is unambiguous and clear.
- A payment mechanism that is simple and that incentivize the parties to achieve a common and agreed goal.

6.10.2 Third-Party Certification

The LEED™ third-party certification program is recognized as an internationally accepted benchmark for the design, construction, and operation of high-performance green buildings. According to Alice Soulek, VP of LEED™ development, "Third-party certification is the hallmark of the LEED™ program," and "Moving the administration of LEED™ certification under GBCI will continue to support market transformation by delivering auditable third-party certification. Importantly, it also allows UGSBC to stick to the knitting of advancing the technical and scientific basis of LEED™."

Moving administration of the LEED certification process to the Green Building Certification Institute (GBCI), a nonprofit organization established in 2007 with the support of USGBC will likely have far-reaching ramifications for the USGBC and its influential LEED rating systems. Working together with the selected certification bodies, GBCI will be in a position to deliver a substantially improved, ISO-compliant certification process that will be able to grow with the green building movement.

USGBC has essentially outsourced LEED certification to independent, accredited certifiers overseen by GBCI. In that respect, the USGBC's Leadership in Energy and Environmental Design (LEED™) v3, has identified the certification bodies for the updated LEED™ Green Building Rating System. The companies are well known and respected for their role in certifying organizations, processes, and products to ISO and other standards. These members include the following:

- ABS Quality Evaluations, Inc., Houston, TX
- BSI Management Systems America, Inc., Reston, VA
- Bureau Veritas North America, Inc., San Diego, CA
- DNV Certification, Houston, TX
- Intertek, Houston, TX
- KEMA-Registered Quality, Inc., Chalfont, PA
- Lloyd's Register Quality Assurance Inc., Houston, TX
- NSF-International Strategic Registrations, Ann Arbor, MI
- SRI Quality System Registrar, Inc., Pittsburgh, PA
- Underwriters Laboratories-DQS Inc., Northbrook, IL

In the spirit of the many successful LEED projects, this development in the certification process has been undertaken as an integrated part of a major update to the technical rating system, which was put in place as LEED 2009. However, the latest version (LEED v4) is bolder, more specialized, and designed for an improved user experience. Moreover, according to the Green Building Council, LEED v4 is better because it:

- *Includes a focus on materials that goes beyond how much is used to get a better understanding of what's in the materials we spec for our buildings and the effect those components have on human health and the environment*
- *Takes a more performance-based approach to indoor environmental quality to ensure improved occupant comfort*
- *Brings the benefits of smart grid thinking to the forefront with a credit that rewards projects for participating in demand response programs*
- *Provides a clearer picture of water efficiency by evaluating total building water use.*

The update will also include a comprehensive technology upgrade to LEED™ Online aimed at improving the user experience and expanding its portfolio management capabilities.

Third-party testing and certification is required for LEED certification to provide an independent analysis of manufacturers' environmental performance claims, based on established standards. It provides building owners and operators with the tools necessary to have an immediate and measurable impact on their buildings' performance. Sustainable building strategies should be considered early in the development cycle. An integrated project team will include the major stakeholders of the project such as the developer/owner, architect, engineer, landscape architect, contractor, and asset and property management staff. Making choices based on third-party analysis is simpler than LCA, but care should be taken to evaluate the independence, credibility, and testing protocols of the third-party certifiers.

Michelle Moore, senior vice president of policy and public affairs at USGBC, says that, "We believe in third-party certification," and the "USGBC provides independent third-party verification to ensure a building meets these high performance standards. As part of this process, USGBC requires technically rigorous documentation that includes information such as project drawings and renderings, product manufacturer specifications, energy calculations, and actual utility bills. This process is facilitated through a comprehensive online system that guides project teams through the certification process. All certification submittals are audited by third-party reviewers." Moore also believes that separating the LEED from the certification process will bring LEED into alignment with norms established by the International Organization for Standardization for certification programs.

To be healthy, green building materials and techniques should typically have zero or low emissions of toxic or irritating chemicals and be moisture and mold resistant. They are typically manufactured with a low-pollution process and from nontoxic components and have low maintenance requirements and do not require the use of toxic cleansers. Most green materials do not emit VOCs, particularly indoors, and are free of toxic materials such as chlorine, lead, mercury, and arsenic. It should be noted that individual products do not carry LEED points; instead they can contribute toward LEED points. Green building techniques include the monitoring of indoor pollutants and poor ventilation with the use of radon and carbon monoxide detectors. The use of ozone-depleting gases is avoided (e.g., free of HCFCs and halons).

CHAPTER 7

Indoor Environmental Quality (IEQ)

7.1 GENERAL OVERVIEW

Over the past few decades, architects, building developers, building owners, and the general public have become increasingly aware of the health hazards related to contaminated air. They now realize that healthier and more comfortable building occupants are happier and more productive. As a result, many of today's sustainable building designs take the issue of indoor environmental quality (IEQ) that includes comfort into consideration. To achieve optimum IEQ, designers need to focus on comfort from four different perspectives:

- Thermal comfort
- Visual comfort
- Acoustical comfort, and
- Indoor air quality

There are a variety of factors that can contribute to poor indoor air quality in buildings, the primary factor being indoor pollution sources that release gases or particles into the air. Other major sources include outdoor pollutants near the building; pollution carried by faulty or inadequate ventilation systems; a variety of combustion sources such as tobacco smoke, gas, oil, kerosene, coal, and wood; and emissions from building materials, furnishings, and various types of equipment. The relative importance of a particular source depends mostly on the amount of a given pollutant it emits and how hazardous those emissions are. The people who are affected most by poor indoor air quality are those that are exposed to it for the longest periods of time. These groups typically consist of the young, the elderly, and the chronically ill, especially those suffering from respiratory or cardiovascular disease.

The majority of buildings today suffer from inadequate performance, including excessive energy consumption, thermal comfort issues, and insufficient daylighting. These deficiencies are often caused by an inability of the design team to consider a variety of design options for all these criteria in an integrated and holistic way. This in turn is producing poor

LEED v4 Practices, Certification, and Accreditation Handbook
http://dx.doi.org/10.1016/B978-0-12-803830-7.00007-4
Copyright © 2016 Elsevier Inc.
All rights reserved.

indoor environmental quality, which has become a major concern in the home, schools, and the workplace; it can lead to poor health, learning difficulties, and productivity problems. Because the majority of us spend up to 90% of our time indoors (especially in the United States), it is not surprising that we should expect our indoor environment to be healthy and free from the plethora of hazardous pollutants. Yet studies by the American College of Allergies show that roughly 50% of all illness is aggravated or caused by polluted indoor air. Moreover, cases of building-related illness (BRI) and sick building syndrome (SBS) appear to be on the increase. The main reason is that the indoor environment we live in is often contaminated by various toxic or hazardous substances as well as pollutants of biological origin (Figure 7.1). In fact, recent studies point to the presence of more than 900 possible contaminants, from thousands of different sources, in a given indoor environment.

Indoor air pollution is now generally acknowledged as having a greater potential negative impact on public health than most types of outdoor air pollution, causing numerous health problems from respiratory distress to

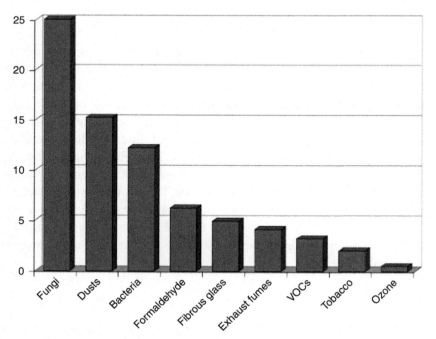

Figure 7.1 *Percentage of Buildings With Inappropriate Concentration of Contaminants.*
Source: HBI Database.

cancer. Furthermore, a building interior's air quality is one of the most pivotal factors in maintaining building occupants' safety, productivity, and well-being.

This heightened public awareness has led to a sudden increase of building occupants demanding compensation for their illnesses. Tenants are not only suing building owners but also architects, engineers, and others involved in the building's construction. To shift the blame, building owners have made claims against the consultant, the contractor, and others involved in the construction of the facility.

Although architects and engineers to date have not been a major target of publicity or litigation arising out of IAQ issues, the potential scope and cost of some of the incidents have led to everyone associated with a project being blamed when the indoor air of a building appears to be the cause of its occupants becoming sick. This is causing great concern among design professionals because it can result in a loss of reputation, as well as cost time and money.

Because of the intense competition to maintain high occupancy rates, forward-thinking property developers, owners, and managers of offices and public buildings find themselves under increasing pressure to meet or exceed the demands of the marketplace in attracting and retaining tenants. Moreover, in today's increasingly litigious society, they add another factor to be addressed – protecting their investment from liability because of air quality issues.

These negative factors can be mitigated by taking the following into consideration when evaluating the conditions for optimal thermal comfort:

- Air temperature
- Relative humidity
- Surface temperatures that influence radiation
- Occupant's personal metabolic rates
- Amount of clothing worn by occupants
- Air speed across body surfaces

Technological breakthroughs are bringing down the cost of facility monitoring systems and making them more affordable for a wider range of building types. By reducing the cost of facility monitoring, many financial and maintenance obstacles are removed, making permanent monitoring systems an appropriate consideration for a broader range of facilities managers. Schools, healthcare facilities, and general office buildings can benefit from measuring many of the environmental conditions and use that information to respond to occupant complaints, optimize facility performance, and keep energy costs in check. In addition, feedback from the indoor environment can be used to establish baselines for building performance and document improvements to indoor air quality.

Facility monitoring systems can be valuable instruments for improving indoor air quality, identifying energy savings opportunities, and validating facility performance. Automating the process of recording and analyzing relevant data and providing facilities managers adequate access to this information can improve their ability to meet the challenge of maintaining healthy, productive environments.

Buildings are dynamic environments – but sometimes the original design calculations, which may have been sound, need modifying after they are operational. Experience tells us that a building's function and occupancy rates may change. These changes can have a significant impact on the heating, ventilation, and air-conditioning (HVAC) system's ability (as originally designed) to maintain a balance between occupant comfort, health, productivity, and operating costs. Indoor pollution is found to exist under many diverse conditions, from dust and bacterial buildup in ductwork to secondhand smoke and the off-gassing of paint solvents, all of which are potential health hazards.

Causes of indoor pollution: The primary cause of poor indoor air quality is the result of sources that release gases or particles into the air. Inadequate ventilation is considered the single most common cause of pollutant buildup (Figure 7.2a) because it can increase indoor pollution levels by not bringing enough outdoor air in to dilute emissions from indoor sources and by not removing indoor air pollutants to the outside. High temperature and humidity levels can also increase concentrations of some pollutants. The second most common cause of pollutant buildup is inefficient filtration

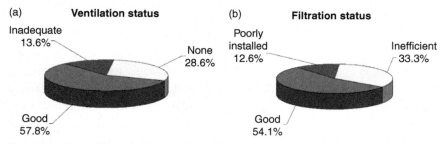

Figure 7.2 (a) Inadequate ventilation is the single most common cause of pollutant buildup. (b) Inefficient filtration is the second most important factor in the cause of indoor pollution. *Source: HBI Database.*

(Figure 7.2b). But despite fundamental improvements in air filter technology, far too many buildings persist in relying on inefficient filters or continue to be negligent in the maintenance of acceptable filters.

An investigation into indoor air quality contamination can be triggered by several factors, including presence of biological growth (mold), unusual odors, adverse health concerns of occupants, as well as a variety of other symptoms or observations, such as respiratory problems, headaches, nausea, irritation of eyes, nose, or throat, fatigue, etc.

Indoor environmental quality (IEQ) and energy efficiency may be classified into three basic categories: (1) Comfort and ventilation, (2) air cleanliness, and (3) building pollutants. Within these basic categories, facility-wide monitoring systems are available that can provide independent measurement of a range of parameters, such as temperature, humidity, total volatile organic compounds (TVOCs), carbon dioxide (CO_2), carbon monoxide (CO), and airborne particulates. Any information that is extracted from continuous monitoring can help minimize the total investigative time and expense needed to respond to occupant complaints; the information can also be used proactively in the optimization of building performance.

Unfortunately, to this day there has been insignificant federal legislation controlling indoor air quality. This has led several engineering societies such as the American Society of Heating, Refrigerating, and Air-Conditioning Engineers (ASHRAE) to establish guidelines that are now generally accepted by designers as minimum air design requirements for commercial buildings. ASHRAE has established two procedures for determining minimum acceptable ventilation rates: (1) the ventilation rate procedure. This stipulates a minimum ventilation rate based on space functions within a specified building type and is based on respiration rates resulting from occupants' activities. (2) The indoor air quality procedure. This procedure requires the monitoring of certain indoor air contaminants below specified values.

Air sampling techniques require the use of a devise to impinge organisms from a specific volume of air and place it onto a sterile agar growth medium. The sample is then incubated for a specified period of time (say, 7 days). The colonies are then counted and the results recorded. When testing the air of a potentially contaminated area, it is best practice to have comparative samples of air from both the contaminated area and the air outside of the potentially contaminated building.

The United States Environmental Protection Agency says that the

EPA standards based on technology performance have been successful in achieving large reductions in national emissions of air toxics. As directed by Congress, EPA

has completed emissions standards for all 174 major source categories, and 68 categories of small area sources representing 90% of emissions of 30 priority pollutants for urban areas. In addition, EPA has reduced the benzene content in gasoline, and has established stringent emission standards for on-road and nonroad diesel and gasoline engine emissions that significantly reduce emissions of mobile source air toxics. As required by the Act, EPA has completed 23 residual risk assessments and technology reviews covering over 40 regulated source categories to assess whether more protective air toxics standards are warranted. EPA has updated standards as appropriate. Additional residual risk assessments and technology reviews are currently underway.

[The EPA] also encourages and supports area-wide air toxics strategies of state, tribal and local agencies through national, regional and community-based initiatives. Among these initiatives are the National Clean Diesel Campaign, which through partnerships and grants reduces diesel emissions for existing engines that EPA does not regulate; Clean School Bus USA, a national partnership to minimize pollution from school buses; the Smartway Transport Partnership to promote efficient goods movement; wood smoke reduction initiatives; a collision repair campaign involving autobody shops; community-scale air toxics ambient monitoring grants; and other programs including Community Action for a Renewed Environment (CARE). The CARE program helps communities develop broad-based local partnerships (that include business and local government) and conduct community-driven problem solving as they build capacity to understand and take effective actions on addressing environmental problems.

Sick building syndrome: Sometimes, building occupants complain of symptoms that do not appear to fit the pattern of any specific illness and are difficult to trace to any specific source. This has been labeled SBS and is a fairly recent phenomenon. It is a term used to describe situations in which building occupants experience acute health and discomfort effects that appear to be attributed to time spent in a building, but often no specific illness or cause can be identified. Complaints may be localized in a particular room or zone, or may be widespread throughout the building. Factors that may impact SBS include, noise, poor lighting, thermal discomfort, and psychological stress. The US Environmental Protection Agency's (EPA's) Indoor Air Quality Web site contains pertinent information regarding strategies for identifying the causes of SBS as well as finding possible solutions to the problem.

SBS indicators may include:
• Acute discomfort
• Headaches and dizziness
• Nausea
• Erythema

- Aches and pains
- Fatigue or lethargy
- Poor concentration and forgetfulness
- Shortness of breath or chest tightness
- Eye and throat irritation
- Irritated, blocked, or runny nose
- Dry or skin irritation

The symptoms of SBS can occur either on their own or in combination with each other. Likewise, they may be at variance from day to day. Different individuals in the same building may experience dissimilar symptoms. They typically improve or disappear altogether shortly after leaving the building and often return upon re-entering the building.

In contrast "building-related illness" or (BRI) is the general term used to describe symptoms for a medically diagnosable illness that is caused by or related to occupancy of a building and which can be attributed directly to airborne building contaminants. BRI is usually linked to the environment of modern airtight buildings that are characterized by sealed windows and dependence on HVAC systems for circulation of air. The causes of BRIs can be determined and are typically related to allergic reactions and infections.

BRI indicators may include:

- Building occupants complain of symptoms such as cough, fever, chills, chest tightness, and muscle aches.
- Symptoms can be clinically defined and have clearly identifiable causes.
- Complainants may require extended recovery times after leaving the building.

It should be noted that there is no particular manner in which these health problems appear. Sometimes, they begin to appear as workers enter their offices and diminish or disappear as workers leave; other times, symptoms persist until the illness is treated. On occasions, there are outbreaks of illness among many workers in a single building; in other cases, they show up only in individual workers.

Complaints may also result from other causes. Since SBS symptoms are quite diverse and nonspecific, it seems they could all arise from numerous other ailments completely unrelated to indoor air quality. In light of these types of complaints, technical and medical scholars have searched for a satisfactory definition of Sick Building, especially since it is the most common indoor air quality problem in commercial buildings today. To this end, some World Health Organization (WHO) experts estimated that 30% of all new and renovated commercial buildings in the United States suffer from SBS, and furthermore, estimate the expenses in terms of medical costs and lost production to be in the tens of billions of dollars per year in the United States alone.

According to industry IAQ standards, SBS is diagnosed if significantly more than 20% of a building's occupants complain of adverse health effects such as headaches, eye irritation, fatigue, and dizziness, etc., over a period of 2 weeks or more, but without a clinically diagnosable disease being identified, and the SBS symptoms disappear or are diminished when the complainant leaves the building.

We are surrounded by microorganisms both indoors and outdoors. Some microorganisms are good for our health whereas others can harm us, but most microorganisms are benign. It is important to prevent buildings from becoming breeding grounds for harmful microbes, and to do that one needs to deprive them of the essentials they need to live. Water is one of microbes' main survival ingredients. It can come from leaky roofs and burst water pipes to the presence of warm moist air, when water gets absorbed by ceiling tiles, drywall, and carpet padding among other things, we have the main necessary ingredient for microbes to flourish.

The greatest amount of our time is spent indoors, and more than half of today's work force spends much of it sitting in front of a computer screen. It is not surprising that we should witness a steady increase in chronic work-related illness, including repetitive stress injuries, asthma, and cardiovascular disease – all suggesting that much of our artificial environment is hazardous to workplace productivity.

It has been known for some time that indoor environments strongly affect human health. The EPA, for example, has estimated that pollutant concentration levels (such as volatile organic compounds (VOCs)) inside a building may be two to five times higher than outside levels. Fisk and Rosenfeld (1997) estimated that the cost to the nation's workforce of upper respiratory diseases in 1995 reached $35 billion in lost work, which doesn't include an estimated additional $29 billion in health care costs. The report suggests that these costs could be reduced by 10–30% by just having healthier and more efficient indoor environments.

Certain attributes indoor environments can also have a strong impact on occupant well-being and functioning; these include personal control over environmental conditions, the amount and quality of light and color, the sense of privacy, access to window views, connection to nature, and sensory variety. Proper design that takes into consideration occupants' psychological well-being will have a positive impact on worker productivity and effectiveness as well as on other high-value issues, such as stress reduction, job satisfaction, and organizational loyalty.

In order to fully profit from the fiscal, physical, and psychological benefits of healthy buildings, projects need to incorporate a comprehensive, integrated design and development process that seeks to

- ensure adequate ventilation,
- provide maximum access to natural daylight and views to the outdoors,
- eliminate or control sources of indoor air contamination,
- prevent water leaks and unwanted moisture accumulation, and
- improve the psychological and social aspects of space.

There are currently several environmental rating methods for buildings, but it is not always clear whether these methods assess the most relevant environmental aspects or whether other considerations lie behind the specific methods chosen. The Swedish rating method included basing the selection on a number of factors such as the severity and extent of problems, on mandatory rules, on official objectives, and on current practice. Upon identifying the presence of building-related health problems, possible indicators for monitoring these problems can be tested with respect to the theoretical and practical criteria applied in order to better understand the strengths and limitations of different indicators.

Concern for occupant health continues to increase with increased awareness, and this has translated into public demand for more exacting performance requirements for materials selection and installation, improved ventilation practices, and better commissioning and monitoring protocols.

Many of the building products used today contain chemicals that evaporate or "offgas" for significant periods of time after installation. When substantial quantities of these products are utilized inside a building, or products are used that have particularly strong emissions, they pollute the indoor air and can be hazardous. Some products readily trap dust and odors and release them over time. Building materials particularly when damp can also support growth of mold and bacteria, which can cause allergic reactions, respiratory problems, and persistent odors, that is, SBS symptoms.

The Insurance Information Institute (III) reports a dramatic increase in IAQ-related lawsuits within the United States and that there are currently more than 10,000 IAQ-related cases pending. This follows several lawsuits with large damage awards that have been won in recent years by building occupants suffering from health problems linked to chemicals off-gassed from building materials that is setting legal precedents across the country.

In recent years, IAQ problems and related lawsuits has prompted insurance companies to re-examine their policies and their clients' design and building methods. Personal injury claims, which makes use of remedies under insurance disability and workers' compensation laws, the Occupational Safety and Hazard Act, which includes property damage, constructive eviction, and design and construction defect claims, may relate to IAQ problems. An effective way to reduce health risks and thus minimize potential liability is to follow a rigorous selection procedure for construction materials aimed at minimizing harmful effects on occupants.

The WHO estimates that nearly 30% of all new and remodeled buildings worldwide may be afflicted with indoor air quality problems. This and studies performed by the EPA rank IAQ concerns among the top environmental risks to public health. Not surprisingly, commentators expect IAQ litigation to proliferate. Perhaps 8% of commercial buildings in the United States fall short of compliance with engineering standards for acceptable indoor air according to the National Energy Management Institute (Source: O'Neal-Coble, Leslie).

7.2 INDOOR ENVIRONMENTAL QUALITY AND FACTORS AFFECTING THE INDOOR ENVIRONMENT

A report on Indoor Environmental Quality points out that there is an increasing number of people suffering a range of debilitating physical reactions from exposures to everyday materials and chemicals found in building products, floor coverings, cleaning products, and fragrances, among others. In addition, there are those who have developed an acute sensitivity to various types of chemicals, a condition known as multiple chemical sensitivity (MCS). The range and severity of these reactions vary as do the potential triggering agents.

7.2.1 Indoor Air Quality

The health and productivity of employees and tenants are greatly influenced by the quality of the indoor environment, and studies consistently reinforce the correlation between improved indoor environmental quality and occupants' health and well-being. The adverse effects to building occupants caused by poor air quality and lighting levels, the growth of molds and bacteria, and off-gassing of chemicals from building materials can be significant. One of the chief characteristics of sustainable design is to support the well being of building occupants by reducing indoor air pollution. This can best be achieved through the selection of materials with low off-gassing potential, appropriate ventilation strategies, providing adequate access to daylight and views, and providing for optimum comfort through control of lighting, humidity, and temperature levels.

7.2.1.1 Inorganic Contaminants – Asbestos, Radon, and Lead

Inorganic substances such as asbestos, radon, and lead are among the leading indoor contaminants whose exposure can create significant health risks.

7.2.1.1.1 Asbestos

Asbestos is a generic term given to a variety of naturally occurring, hydrated fibrous silicate minerals that possess unique physical and chemical properties that distinguishes them from other silicate minerals. Such properties include thermal, electric, and acoustic insulation, chemical and thermal stability, and high tensile strength, all of which have contributed to their wide use by the construction industry (Figure 7.3).

High concentrations of airborne asbestos can occur during demolition and after asbestos–containing materials are disturbed by cutting, sanding,

Sample asbestos-containing material list

- Acoustical ceiling texture
- Asphalt flooring
- Base flashing
- Blown-in insulation
- Boiler/tank insulation
- Breaching insulation
- Brick mortar
- Built-up roofing
- Caulking/putties
- Ceiling tiles/panels/mastic
- Cement board
- Cement pipes
- Cement roofing shingles
- Chalkboards
- Construction mastics
- Duct tape/paper
- Ductwork flexible connections
- Electrical cloth
- Electrical panel partitions
- Electrical wiring insulation
- Elevator brake shoes
- Fire blankets
- Fire curtains/hose
- Fire doors
- Fireproofing
- Furnace insulation

- Gray roofing paint
- High temperature gaskets
- HVAC duct insulation
- Incandescent light fixture backing
- Joint compound/wallboard
- Laboratory hoods/table tops
- Laboratory fume hood
- Mudded pipe elbow insulation
- Nicolet (white) roofing paper
- Packing materials
- Paper fire box in walls
- Pipe insulation/fittings
- Plaster/wall joints
- Poured flooring
- Rolled roofing
- Roofing shingles
- Sink insulation
- Spray-applied insulation
- Stucco
- Subflooring slip sheet
- Textured paints/coatings
- Vapor barrier
- Vermiculite
- Vinyl floor tile/mastic
- Vinyl sheet flooring/mastic
- Vinyl wall coverings

Note: This is a sample list of products that may contain asbestos. It is intended as a general guide to show which types of materials may contain asbestos and is not all inclusive.

Figure 7.3 *Partial List of Materials that may Contain Asbestos.*

and other activities. Inadequate attempts to remove these materials could release asbestos fibers into the air, increasing asbestos levels and thereby endangering people living or working in these spaces.

Asbestos-containing material has become a high-profile public concern after federal legislation known as AHERA (Asbestos Hazard Emergency Response Act) was enacted in 1987. As an example, for schools, The Asbestos Hazard Emergency Response Act (AHERA) and its regulations require public school districts and nonprofit schools including charter schools and schools affiliated with religious institutions to inspect their buildings for asbestos-containing material. Schools are also required to prepare management plans and to take action to prevent or reduce asbestos hazards.

The EPA (Environmental Protection Agency) and CPSC (U.S. Consumer Product Safety Commission) have also banned several asbestos products. Manufacturers have too voluntarily limited uses of asbestos. Today, asbestos can still be found in older homes, in pipe and furnace insulation materials, asbestos shingles, millboard, textured paints and other coating materials, and floor tiles. Asbestos is considered the most widely recognized environmentally regulated material (ERM) during building evaluations.

Asbestos-containing materials are also found in concealed areas such as wall cavities, below ground level, and other hidden spaces. In many older establishments, asbestos-based insulation was used on heating pipes and on the boiler. An adequate Asbestos Survey requires the Inspector to perform destructive testing (i.e., opening walls etc.) to inspect areas likely to contain suspect materials.

Health hazards: The most dangerous asbestos fibers are too small to be visible. The health danger of asbestos fibers depends mainly on the quantity of fibers in the atmosphere and the length of exposure. Asbestos is made up of microscopic bundles of fibers that may become airborne when asbestos-containing materials are damaged or disturbed. Impaction and abrasion are typically the chief causes of increased airborne fiber levels. The type, quantity, and physical condition of the asbestos-based material have a significant bearing on the degree of risk.

The risk of airborne asbestos fibers is generally low when the material is in good condition. However, when the material becomes damaged or if it is located in a high activity area (family room, work shop, laundry, etc.) the risk increases. Increased levels of exposure to airborne asbestos fibers will cause disease. When these fibers get into the air they may be inhaled and remain and accumulate in the lung tissue, where they can cause substantial health problems including lung cancer, mesothelioma (a cancer of the chest and abdominal linings), and asbestosis (irreversible lung scarring that can be fatal). Symptoms of these diseases do not show up until many years after exposure began. Studies show that people with asbestos-related diseases were usually exposed to elevated concentrations on the job although some developed disease from exposure to clothing and equipment brought home from job sites. While the process is slow, and years may pass before health problems are evidenced, the result and, thus, the risk are well established.

7.2.1.1.2 Radon

Radon is a natural odorless, tasteless, radioactive gas that is emitted from the soil as a carcinogenic by-product of the radioactive decay of radium-226, which is found in uranium ores (although radon itself does not react with other substances). The by-product can however cling to dust particles, which when inhaled, settle in bronchial airways. Generally, radon is drawn into a building environment by the presence of air pressure differentials. The ground beneath a building is typically under higher pressure than the basement or foundation. Air and gas move from high-pressure areas to low-pressure areas. The gas can enter the building through cracks in walls and floors, as well as penetrations associated with plumbing, electrical openings, sump wells, etc., in building spaces coming in close contact with uranium-rich soil. Vent fans and exhaust fans also facilitate to put a room under negative pressure and increase the draw of soil gas, which can increase the level of radon within a building. Radon exposure becomes a concern when it becomes trapped in buildings and indoor levels of concentrations build up.

Adequate ventilation is necessary to prevent radon from accumulating in buildings to dangerous levels as this can pose a serious health hazard. Where radon is suspected, a survey should be conducted to measure the concentrations of radon in the air and determine whether any actions will be required to reduce the contamination. Radon levels will vary from region to region, season to season, and one building to another. Radon levels are typically at their highest during the coolest part of the day when pressure differentials are at their greatest. It should be noted that radon is found throughout the United States and can vary substantially from one home to another. Homes with high levels have been found in all states. Radon testing for homes is simple and inexpensive. Many state radon programs are in place that offer free radon test kits, and additionally there are inexpensive ways to fix and prevent high radon levels in homes.

According to the United States Environmental Protection Agency, Radon is the second leading cause of lung cancer in the United States and is associated with 15,000–22,000 lung cancer deaths annually. The primary factors affecting radon concentration levels in the air in addition to ventilation is the radon source. The most common radon source is the presence of radium-226 in the soil and rock surrounding or adjoining basement walls and cellar floor slabs. Although you cannot see radon, it is not hard to find out if you have a radon problem in your home or office. There are many kinds of inexpensive, state-certified do-it-yourself radon test kits that one can get through the mail or in hardware stores and other retail outlets. When radon decays and is inhaled into the lungs, it releases energy that can damage the DNA in sensitive lung tissue and cause cancer.

When high concentrations of radon in air are detected, it is often an indication of possible radon contamination of the water supply (if a private water supply is present). In this case, a water test for radon is the prudent first step. Should high concentrations of radon be found in the water, then an evaluation of ventilation rates in the structure as well as air quality tests for radon are highly recommended. Generally speaking, high radon concentrations are more likely to exist where there are large rock masses, such as in mountainous regions. The United States Environmental Protection Agency (EPA) recommends that buildings should be tested every few years to assess the safety of radon levels.

Radon mitigation: Everything being equal, elevated radon levels should not necessarily deter investors from purchasing a property, as the problem can usually be easily resolved – even in existing buildings, without having to incur great expense. However, lowering high radon levels requires technical knowledge and special skills, which means a trained radon reduction contractor who understands how to fix radon problems should be used.

With new construction, some builders have started to incorporate radon prevention techniques in their designs. Some municipalities also have local building codes requiring prevention systems. The EPA has published several brochures and instructional aides regarding radon-resistant construction. This is perhaps the most cost-effective way to handle a radon problem, as it is easier to build the system into the building rather than retrofit it later. If your building has a radon system built in, the EPA recommends periodic testing to ensure that the system is working properly and that the radon level in your building has not changed. The development of foundation cracks is one example of how the radon level could change in your building.

EPA studies suggest that elevated radon levels are more likely to exist in energy-efficient buildings than otherwise. The reason being that radon can become hazardous indoors is the limited exposure to outside air to dilute the indoor concentrations. Although energy-efficient construction may save energy bills, it may also increase occupants' exposure to radon and other indoor air pollutants. It is also recommended that testing for radon should also be conducted when any major renovations are made to the building.

Health risks: The principal health hazard associated with exposure to elevated levels of radon is lung cancer. Research suggests that while swallowing water with high radon levels may also pose risks, these are believed to be much lower than those from breathing air containing radon.

During the decaying process, radon emits alpha, beta, and gamma radiation. However, the real threat is not so much from the gas itself but from the products that it produces such as lead, bismuth, and polonium when it decays. These products are microscopic particles that readily attach themselves to dust, pollen, smoke, and other airborne particles in a building and are inhaled. Once in the lungs, the particles become trapped and as they begin to decay, they expose the sensitive lung tissue to the harmful radiation they emit during the decaying process. The effect is cumulative, and elevated levels of radon exposure causes lung cancer in both smokers and nonsmokers alike. The risk would be greatest for people with diminished lung capacity, asthma sufferers, and smokers, etc.

Radon testing methods: For *short-term testing*, consultants typically use electret ionization chambers that generally last about a week. The chambers method work by incorporating a small, charged Teflon plate screwed into the bottom section of a small plastic chamber. When the radon gas enters the chamber, it begins to decay and creates charged ions that deplete the charge on the Teflon plate. By registering the voltage prior to deployment and then reading the voltage on recovery, a mathematical formula is used to calculate the radon concentration levels within the building.

For *long-term testing*, consultants prefer to use either long-term electret chambers (like those used for the short-term measurements) or alpha tracks. The alpha track when deployed in the building records the number of alpha particles that scratch the plastic surface inside the detector. The laboratory counts these microscopic indentations and then mathematically calculates the radon levels within the building.

7.2.1.1.3 Lead

For many years now, lead has been recognized as a harmful environmental pollutant to the extent that in late 1991, the Secretary of the Department of Health and Human Services described lead as the "*number one environmental threat to the health of children in the United States."* There are many ways in which humans may get exposed to lead: as mineral particles in ambient air, drinking water, food, contaminated soil, deteriorating paint, and dust. Lead is a heavy metal and does not break down in the environment and continues to be used in many materials and products to this day. Lead is a natural element and most of the lead in use today is inorganic lead, which enters the body when an individual breathes (inhales) or swallows (ingestion) lead particles or dust once it has settled. Lead dust or particles cannot penetrate the skin unless the skin is broken. Organic lead, however, such as the type used in gasoline, can penetrate the skin.

Lead levels in the indoor environment: Because of its widespread use and the nature of individual uses, lead has for some time been known to be a common contaminant of interior environments. For centuries, lead compounds like white lead and lead chromate have been used as white pigments in commercial paints. In addition to their pigment properties, these lead compounds were valued because of their durability and weather resistance, which made their use more viable, particularly in exterior white paints. In addition to lead's durability properties, it was also added to paint to improve its drying characteristics.

Old lead–based paint is considered the most significant source of lead exposure in the U.S. today. In fact, the majority of homes and buildings built before 1960 contained heavily leaded paint. Even as recently as 1978, there were homes and buildings that used lead paint. This paint may have been used on window frames, walls, the building's exterior, or other surfaces. The improper removal of lead-based paint from surfaces by dry scraping, sanding, or open-flame burning can create harmful exposures to lead. High concentrations of airborne lead particles in a space can also result from lead dust from outdoor sources, including contaminated soil tracked inside. Harmful effects can also result from the use of lead in certain indoor activities such as soldering and stained-glass making.

Because of potentially serious health hazards and negative publicity, lead content was gradually reduced until it was eliminated altogether in 1978 (in the United States). In commercial buildings, lead was used primarily as a paint preservative.

Lead piping has sometimes been used in older buildings, and although not legally required to be replaced, it can create a health hazard because frequently the piping is found to be deteriorating and leaching into the building's drinking water. In some buildings, lead solder has also been used in copper pipe installation, but in most jurisdictions this procedure has now been banned because of water contamination resulting from the deterioration of the solder. In any case, any evaluation for lead contamination requires that the water content be analyzed by a laboratory for lead concentrations, and action then taken to mitigate the hazards to reduce contamination if lead content is found to be in excess of regulated limits. The potential for water contamination can often be removed by chemical treatment of the water. Where this cannot be accomplished, the piping may have to be replaced.

Unfortunately, lead is still allowed in paint for bridge construction and machinery and thus it remains a significant source of lead exposure. Its continued use is mainly due to its ability to resist corrosion and its capability to expand and contract with the metal surface of a structure without cracking. But even if its use were banned today, there would still be potential exposure to workers and surrounding communities for many years to come because of the many metal structures, such as bridges, that have been coated with it.

Health risks: Lead is a highly toxic substance that affects a variety of target organs and systems within the body, including the brain and the central nervous system, renal, reproductive, and cardiovascular systems. High levels of lead exposure can cause convulsions, coma, and even death. However, the nervous system appears to be the main target organ system for lead exposure.

One of the best methods to assess human exposure lead is to measure the blood lead levels because it is a sensitive indicator of exposures that has been correlated with a number of health endpoints. It also gives an immediate estimate of the level of a person's recent exposure to lead. The negative aspect of measuring blood lead levels is that while it will tell you how much lead is in the bloodstream, it will not tell you what is stored in soft tissues or bones. Additionally, the test will not spell out your body burden of lead or any damage, if any of that has occurred.

Effects of lead poisoning depend largely on dose exposure. Contact with lead-contaminated dust is the primary method in which most children are exposed to harmful levels of lead. It enters a child's body mainly through ingestion. Lead-contaminated dust is often hard to see but they can get into the body and create substantial health risks. Pregnant women, infants, and children are more vulnerable to lead exposure than adults because lead is more easily absorbed into growing bodies, and the tissues of small children are more sensitive to the damaging effects of lead. Ingestion of lead has been proven to cause delays in children's physical and mental development, as well as negatively impacting their developing nervous systems, causing lower IQ levels, increased behavioral problems, and learning disabilities.

In adults, high lead levels has many adverse effects, including causing kidney damage, digestive problems, high blood pressure, headaches, diminishing memory and concentration, mood changes, nerve disorders, sleep disturbances, and muscle or joint pain. A single, very high exposure to lead can also result in lead poisoning. Adults' bones and teeth contain about 95% of the body burden (total amount of lead that is stored in the body). Likewise, lead can seriously impact the ability of both men and women to bear healthy children.

Testing for lead paint: There are various methods and procedures that inspectors employ to assist them in identifying if lead paint is present. In the field, the most widely applied method is the use of an X-ray fluorescent (XRF) lead-in-paint analyzer. The XRF analyzer is normally held up to the surface being tested for several seconds. The analyzer then emits radiation, which is absorbed and then fluoresces (is emitted) back to the analyzer. The XRF unit breaks down the signals to determine if lead is present and, if so, in what concentration. An XRF analyzer can normally read through up to 20 layers of paint, but these analyzers are expensive and should only be used by trained professionals.

7.2.1.2 Contaminates Generated by Combustion

Some examples of combustion by-products include fine particulate matter, carbon monoxide (CO), and nitrogen oxides. Tobacco smoke is another source.

Combustion (burning) by-products are essentially gases and tiny particles that are created by the incomplete burning of fuels. These fuels (such as natural gas, propane, kerosene, fuel oil, coal, coke, charcoal, wood, and gasoline, and materials such as tobacco, candles, and incense) when burned produce a wide variety of air contaminants. If fuels and materials used in the combustion process were free of contaminates and combustion were complete, emissions would have been limited to carbon dioxide (CO_2), water vapor (H_2O), and high-temperature reaction products formed from atmospheric nitrogen (NOx) and oxygen (O_2). Sources of combustion-generated pollutants in indoor environments are many and include wood heaters and woodstoves, furnaces, gas ranges, fireplaces, and car exhaust (in an attached garage).

There are several combustion-generated contaminates including CO_2, H_2O, carbon monoxide (CO), nitrogen oxides (NOx) such as nitric oxide (NO) and nitrogen dioxide (NO_2), respirable particles (RSP), aldehydes like formaldehyde (HCHO) and acetaldehyde, as well as a number of

VOCs; fuels and materials containing sulfur will produce sulfur dioxide (SO_2). Particulate-phase emissions may include tar and nicotine from tobacco, creosote from wood, inorganic carbon, and polycyclic aromatic hydrocarbons (PAHs).

Carbon dioxide (CO_2) is a colorless, odorless, heavy, incombustible gas that is found in the atmosphere and formed during respiration. It is typically obtained from the burning of gasoline, oil, kerosene, natural gas, wood, coal, and coke. It is also obtained from carbohydrates by fermentation, by reaction of acid with limestone or other carbonates, and naturally from springs. CO_2 is absorbed from the air by plants in a process called photosynthesis. Although carbon dioxide is not normally a safety problem, a high CO_2 level can indicate poor ventilation, which in turn can lead to a buildup of particles and more harmful gases such as carbon monoxide that can negatively impact people's health and safety. CO_2 is used extensively in industry as dry ice, or carbon dioxide snow, in carbonated beverages, fire extinguishers, etc.

Carbon monoxide (CO) is an odorless, colorless, lighter-than-air, nonirritating gas that interferes with the delivery of oxygen throughout the body. CO is the leading cause of poisoning deaths in the United States, and occurs when there is incomplete combustion of carbon-containing material such as coal, wood, natural gas, kerosene, gasoline, charcoal, fuel oil, fabrics, and plastics.

At low concentrations levels, healthy people may experience fatigue and shortness of breath during exertion. Flushed skin, tightness across the forehead, and slightly impaired motor skills may also occur. People with heart disease may encounter chest pain. The first and most obvious symptom is usually a headache with throbbing temples. Infants, children, pregnant women, the elderly, and people with heart or respiratory problems are most susceptible to carbon monoxide poisoning. The fact that CO cannot be seen, smelled, or tasted makes it especially dangerous because you are not aware when you are being poisoned. Moreover, CO poisoning is frequently misdiagnosed by both victims and doctors.

Symptoms of mild to moderate CO poisoning may cause flulike symptoms or gastroenteritis, particularly in children, and include nausea, lethargy, and malaise. As the CO level or exposure time increases, symptoms become more severe and additional ones appear: irritability, chest pain, fatigue, confusion, dizziness, and impaired vision and coordination. Higher levels cause fainting upon exertion, marked confusion, and collapse. At very high concentrations, we may witness coma and convulsion as well as permanent damage to the brain, central nervous system, or heart, and finally death, by reducing the amount of oxygen red blood cells carry.

The four primary sources of CO in the environment are as follows:

- Automobile exhaust from attached garages combined with inadequate ventilation is responsible for two-thirds of all accidental CO deaths. Lethal levels of the gas can accumulate in as little as 10 min in a closed garage. Certain occupations are exposed regularly to elevated levels of the CO. These include toll-booth attendants, professional drivers, highway workers, traffic officers, and tunnel workers. Likewise, certain indoor events, such as tractor pulls, car and truck exhibitions, if not adequately ventilated can expose spectators and participants to elevated CO levels.

- Faulty heating equipment accounts for almost one-third of accidental CO fatalities. These fatalities can be caused by home or office heating systems (e.g., leaking chimneys and furnaces; back-drafting from furnaces, gas water heaters, woodstoves, and fireplaces; gas stoves), but also improperly vented or unvented kerosene and gas appliances, or propane space heaters, charcoal grills or hibachis, and Sterno-type fuels. Dangerous amounts of carbon monoxide can be released when there is inadequate fresh air or a flame is not sufficiently hot to completely burn a fuel.

- Fires have been found to raise CO levels in the blood of unprotected persons to 150 times normal in a single minute; CO poisoning is the most frequent cause of immediate death associated with fire. Environmental tobacco smoke can also cause elevated CO levels in both smokers and nonsmokers who are exposed to the smoke.

- Methylene chloride is a solvent used in some paints and varnish strippers that is readily absorbed by the body and changes to CO. Using products that contain methylene chloride for more than a few hours can raise CO levels in the blood 7 to 25 times normal. It is particularly dangerous for persons with pre-existing cardiac conditions who use these products in unventilated spaces as they risk heart attack and death.

Testing for CO: An electronic device by the name of a carbon monoxide alarm is the only reliable method currently used to test for the presence of carbon monoxide. Most fire departments, gas companies, and some specialized contractors have sophisticated equipment that can measure and record carbon monoxide levels. In the home, detectors should be placed in areas where the family spends most of its time such as the family room, bedroom, or kitchen. Detectors should be placed far enough away from obvious and predictable sources of CO, such as a gas stove, to avoid false alarms.

Steps to reduce exposure to CO poisoning: CO poisoning can be prevented by following a number of simple steps:

- Keep gas appliances properly adjusted and have them checked periodically for proper operation and venting.
- Open flues when fireplaces are being used and ensure flues, chimneys, and vents are clear of debris and working properly.
- Have CO monitors installed at home and the workplace and ensure that they are properly maintained.
- Refrain from using unvented space heaters, gas stoves, charcoal grills, or Sterno-type fuels as sources of heat. Woodstoves when used should be properly sized and certified to meet EPA emission standards.
- When working around CO sources such as propane-powered forklifts and space heaters, ensure that adequate ventilation is in place. Whenever exposure is unavoidable, CO-monitoring badges should be worn.
- The car's exhaust system should be checked regularly and properly maintained at all times. Cars or other gasoline-powered engines should not run inside a garage, even with the doors open.
- Do not use paint strippers that contain methylene chloride. If the use of solvents containing this substance is unavoidable, make sure the area is properly ventilated.
- A trained professional should be brought in at least once a year to inspect, clean, and tune up the central heating system (furnaces, flues, and chimneys) and any leaks promptly repaired.

Nitrogen dioxide (NO_2) is a colorless, odorless gas that irritates the mucous membranes in the eye, nose, and throat and causes shortness of breath when exposed to high concentrations. Nitrogen dioxide is also a major concern as an air pollutant because it contributes to the formation of photochemical smog, which can have significant impacts on human health. Documented evidence indicates that high concentrations or continued exposure to low levels of nitrogen dioxide increases the likelihood of respiratory problems. Because nitrogen dioxide is relatively nonsoluble in tissue fluids, it enters the lungs, where it may expose lower airways and alveolar tissue. Nitrogen dioxide inflames the lining of the lungs and can reduce the immunity to lung infections. There is also documented evidence from animal studies that repeated exposures to elevated nitrogen dioxide levels may lead, or contribute, to the development of lung disease such as emphysema. People at particular risk from exposure to nitrogen dioxide include children with asthma and older people with heart disease or other respiratory diseases.

This can cause problems such as wheezing, coughing, colds, flu, and bronchitis. Increased levels of nitrogen dioxide can also have significant impacts on people with asthma because it can cause more frequent and more intense attacks.

Irritants: There are numerous combustion-generated contaminants, including several mucous membrane and upper respiratory system irritants. Aldehydes like HCHO are the most common, although in some cases acrolein, RSP, and SO_2 may also be included. Aldehydes cause irritation to the eyes, nose, throat, and sinuses and are discussed in greater detail in the following section. Respirable particles vary in composition, and their primary effect is irritating the upper respiratory passages and bronchi. Because of its solubility in tissue fluids, SO_2 can cause bronchial irritation.

7.2.1.3 Organic Contaminants – Aldehydes, VOCs/SVOCs, Pesticides

In today's environment, natural and synthetic organic chemicals comprise many different types and can be found virtually everywhere, including soil, groundwater, surface water, plants, and our bodies. Modern industrialized societies have developed such a massive array of organic pollutants that it is becoming increasingly difficult to generalize in a meaningful way as to the sources, uses, or impacts. These contaminants find their way into the natural environment through accidental leakage or spills, such as leaking underground storage tanks, or through planned spraying of pesticides to agricultural land, etc.

The main organic compounds include VOCs, the very volatile organic compounds (VVOCs), semivolatile organic compounds (SVOCs), and solid organic compounds (POMs). Solid organic compounds may comprise components of airborne or surface dusts. Organic compounds often pose serious indoor contamination problems and include the aldehydes; VOCs/SVOCs, which include a large number of volatile as well as less volatile compounds and pesticides and biocides, which are largely SVOCs.

Organic compounds that are known to be contaminates of indoor environments include a large variety of aliphatic hydrocarbons, aromatic hydrocarbons, oxygenated hydrocarbons (such as aldehydes, ketones, alcohols, ethers, esters, and acids), and halogenated hydrocarbons (primarily chlorine and fluorine containing).

VOC concentration levels are generally higher in indoor environments than outdoor air. Studies suggest that indoor air may contain several hundred different VOCs. Moreover, VOCs can be released from products while being used and to a lesser degree while being in storage. Fortunately, the amounts of VOCs emitted tend to decrease as the product ages and dries out.

There has been a steady increase in the number of identified VOCs in recent years. They are characterized by a wide range of physical and chemical attributes – the most important of which is their water solubility and whether they are neutral, basic, or acidic. VOCs are released into the indoor environment by extensive sources, including building materials, furnishings, paints, solvents, air fresheners, aerosol sprays, adhesives, fabrics, consumer products, building cleaning and maintenance materials, pest control and disinfection products, humans, office equipment, tobacco smoking, plastics, lubricants, refrigerants, fuels, solvents, pesticides, and many others.

Among the health hazards many VOCs pose is that they are potent narcotics that cause a depression in the central nervous system, and others can cause eye, nose, and throat irritation; headaches; loss of coordination; nausea; and damage to the liver, kidneys, and central nervous system (Figure 7.4). A number of these chemicals are suspected or known to cause cancer in humans.

Formaldehyde (HCHO) is a colorless, pungent-smelling gas, and one of the more common VOCs found indoors, which is an important chemical used widely by industry to manufacture building materials and numerous household products. It is also a by-product of combustion and certain other natural processes and, thus, may be present in substantial concentrations both indoors and outdoors.

Formaldehyde is the most common of the aldehydes and is considered by many as possibly the single most important indoor pollutant because of

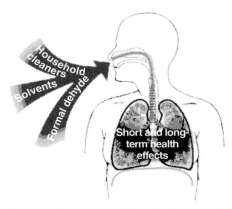

Figure 7.4 *Diagram Showing Inhalation of Volatile Organic Compounds. Source: Air Advice Inc.*

its common occurrence and its strong toxicity. Formaldehyde is a colorless, gaseous substance with an unpleasant smell and is omnipresent in both ambient and indoor environments. On condensing, it forms a liquid with a high vapor pressure. Because of its high reactivity, it rapidly polymerizes with itself to form paraformaldehyde. Because of this, liquid HCHO needs to be kept at low temperature or mixed with a stabilizer (such as methanol) to prevent or minimize polymerization.

Formaldehyde, by itself or in combination with other chemicals, serves a number of purposes in manufactured products. For example, it is used to add permanent-press qualities to clothing and draperies, as a component of glues and adhesives, and as a preservative in some paints and coating products. It is also used in a variety of deodorizing commercial products, such as lavatory and carpet preparations.

Formaldehyde is commercially available as both paraformaldehyde and as formalin, an aqueous solution that typically contains 37–38% HCHO by weight and 6–15% methanol. HCHO is also used in many different chemical processes, such as the production of urea and phenol-formaldehyde resin.

Urea-formaldehyde copolymeric resins are present in many building materials such as wood adhesives which are used in the manufacture of pressed wood products, including particle board, medium-density fiber board (MDF), plywood, finish coatings (acid-cured), textile treatments, as well as in the production of urea-formaldehyde foam insulation (UFFI).

People are often unaware that formaldehyde is given off by materials other than UFFI. Certain types of pressed wood products (composition board, e.g., MDF, paneling, etc.), carpeting, and other material can be formaldehyde sources. Many of these products use a urea-formaldehyde-based resin as an adhesive. Some of these materials will continue to give off formaldehyde much longer than UFFI. Like the majority of VOCs, formaldehyde levels will decrease substantially with time and/or with increased ventilation rates.

Health risks: For some people, formaldehyde can be a respiratory irritant, and continuous exposure to it can be dangerous. More specifically, chronic, low-level, continuous, and even intermittent exposure to formaldehyde can cause chemical hypersensitivity and provides an accelerating factor in the development of chronic bronchitis and pulmonary emphysema. HCHO also has the ability to cause irritation and inflammatory-type symptoms and symptoms of the central nervous system such as headache, sleeplessness, and fatigue. Elevated HCHO exposures (above 0.1 ppm) can trigger asthma attacks, nausea, diarrhea, unnatural thirst, and menstrual irregularities.

Acetaldehyde is a two-carbon aliphatic aldehyde with a pungent, fruity odor. It is used in a number of industrial processes and is a major constituent of automobile exhaust fumes. It is also a predominant aldehyde found in tobacco smoke. Compared to HCHO, it is a relatively mild irritant of the eye and upper respiratory system.

Acrolein is a three-carbon aldehyde with one double bond. It is highly volatile gas with an unpleasant choking odor. It is primarily used in the production of a variety of compounds and products and is released into the environment as a combustion oxidation product from oils and fats containing glycerol, wood, tobacco, and automobile/diesel fuels. At relatively low exposure concentrations, acrolein is a potent eye irritant causing lacrimation (tearing).

Glutaraldehyde is a five-carbon dialdehyde. It is found in liquid form and has a sharp, fruity odor. Its main application is as an active ingredient in disinfectant formulations widely used by the medical and dental professions. The main health effects associated with glutaraldehyde exposures include irritant symptoms of the nose and throat, nausea and headache, as well as pulmonary symptoms such as chest tightening and asthma.

Polychlorinated biphenyls (*PCBs*) are oils used primarily as a coolant in electrical transformers. Although production and sale of PCB was banned by the Environmental Protection Agency (EPA) in 1979, a large number of PCB-filled transformers remain in use. It has also been estimated that some 2,000,000 mineral oil transformers still contain some percentage of PCB. PCBs may also be found in light ballasts and elevator hydraulic fluids. PCBs are a suspected carcinogen, but properly sealed or contained PCBs do not pose a hazard. However, PCBs do become a hazard when they catch fire, creating carcinogenic by-products that can contaminate the air, water, finishes, and contents of a building. Leaking PCB can also contaminate building materials and soil. PCB evaluations typically focus on identifying the presence of, or potential for, PCB leakage, measuring of PCB concentrations levels, and determining the presence of combustible materials adjacent to PCB-containing equipment.

Hydrocarbons are a class of organic chemical compounds consisting only of the elements hydrogen (H) and carbon (C). They are cardinal to our modern way of life and its quality and being one of the Earth's most important energy resources. Hydrocarbons are the principal constituents of petroleum, natural gas, and are also derived from coal. The bulk of the world's hydrocarbons are used for fuels and lubricants, as well as for electrical power generation, and heating. The chemical, petrochemical, plastics, and rubber industries are also dependent on hydrocarbons as raw materials for their products. It is a colorless, flammable, toxic liquid.

Symptoms associated with exposure to aliphatic hydrocarbons may include watery eyes, nausea vomiting, dizziness, weakness, central nervous system effects such as depression, convulsions, and in extreme cases, coma. Other symptoms may include pulmonary and gastrointestinal irritation, pulmonary edema, bronchial pneumonia, anorexia, anemia, nervousness, pain in the limbs, and numbness. Noticeable odors similar to gasoline and oil all suggest that some hydrocarbon contamination may be present. Likewise, leaking subsurface tanks at fuel stations and other facilities have created significant health and safety problems by contaminating the soil around the buildings. Benzene is found in most hydrocarbons and is considered to be one of the more serious contaminants and is known to cause leukemia. Air quality tests may be necessary as well as tests for contaminants in the soil around the foundation.

Pesticides are chemical poisons, designed to control, destroy, or repel plants and animals such as insects (insecticides), weeds (herbicides), rodents (rodenticides), and mold or fungus (fungicides). They include active ingredients (those intended to kill the target) and inert ingredients, which are often not "inert" at all. Pesticides are generally toxic and can be absorbed through the skin, swallowed, or inhaled and as such are unique contaminants of indoor environments.

Studies show that approximately 16 million Americans are sensitive to pesticides because their immune systems have been damaged as a result of prior pesticide exposure. In addition, pesticides have been linked to a wide range of serious and often fatal conditions: cancer, leukemia, miscarriages, genetic damage, decreased fertility, liver damage, thyroid disorders, diabetes, neuropathy, still births, decreased sperm counts, asthma, and other autoimmune disorders (e.g., lupus).

The Federal Government, in cooperation with the States, carefully regulates pesticides to ensure that they do not pose unreasonable risks to human health or the environment. There are currently more than 1055 active ingredients registered as pesticides, which are formulated into thousands of different pesticide products that are available in the marketplace, including some of the most widely used over the past 60 years that are persistent and have become globally distributed. These include aldrin, chlordane, DDT, dieldrin, endrin, heptachlor, mirex, toxaphene, and lindane (hexachlorocyclohexane (HCH)). Many pesticides (most notably Chlordane, used for termite treatment) are serious hazards. It is hoped that ecological methods of pest control will in the future replace the overdependence on chemicals that now threatens our ecosystem.

7.2.1.4 Biological Contaminants – Mold and Mildew, Viruses, Bacteria, and Exposures to Mite, Insect, and Animal Allergens

Biological pollutants arise from various sources such as microbiological contamination, for example, fungi, bacteria, viruses, mites, pollens, and the remains and dropping of pests such as cockroaches and other pests. Such pollutants of biological origin can significantly impact indoor air quality and cause infectious disease through airborne transmission. Of particular concern are those biological contaminants that cause immunological sensitization manifested as chronic allergic rhinitis, asthma, and hypersensitivity pneumonitis.

Moisture in buildings is one of the major contributors to poor indoor air quality, unhealthy buildings, and mold growth but by controlling the relative humidity level, the growth of some sources of biologicals can be minimized. Standing water, water-damaged materials, rainwater leaks, or wet surfaces also serve as a breeding ground for molds, mildews, bacteria, and insects. Likewise, damp or wet areas such as cooling coils, humidifiers, condensate pans, or unvented bathrooms can serve as suitable breeding grounds as well as contaminated central HVAC systems, which can then distribute these contaminants through the building.

Proactive preventive and remedial measures include rainwater tight detail design; preventing uncontrolled air movement, reducing indoor air moisture content, reducing water vapor diffusion into walls and roofs, selecting building materials that have appropriate water transmission characteristics, and maintaining proper field workmanship quality control.

A proven method for deterring rainwater intrusion into walls is the rain screen approach, which incorporates cladding, air cavity, drainage plane, and airtight support wall to offer multiple moisture-shedding pathways. The concept of the rain screen principle is simple; it is to separate the plane in a wall where the rainwater is shed and where the air infiltration is halted. In terms of construction, this means that there is an outer plane that sheds rainwater but allows air to freely circulate and an inner plane that is relatively airtight.

Mold and mildew are types of musty smelling fungi that thrive in moist environments. Their function in nature is primarily to break down and decompose organic materials such as leaves, wood, and plants. They grow, penetrate, and infect the air we breathe. There is no normal way to eliminate all mold and mold spores in the indoor environment, and thus the key to controlling indoor mold growth is to control moisture content. Exposure to fungus in indoor air settings has emerged as a significant health problem of great concern in both residential environments as well as in workplace settings. There are three major health hazards associated with fungi: infection, allergies, and toxins.

Fungi are primitive plants that lack chlorophyll and therefore feed on organic matter that they digest externally and absorb or must live as parasites. True fungi include yeast, mold, mildew, rust, smut, and mushrooms. Given sufficient moisture, they can grow on almost any material even inorganic ones when sufficient dirt or materials congregate on them. When mold spores land on a damp spot indoors, they can grow and start digesting whatever they are growing on. There are molds that can grow on wood, paper, carpet, fabric, and foods, but they usually grow best in dark moist habitats, especially if organic matter is available.

There are thousands of species of molds, which include pathogens, saprotrophs, aquatic species, and thermophiles. Molds are part of the natural environment growing on dead organic matter and are present everywhere in nature; their presence is only visible to the unaided eye when mold colonies grow (Figure 7.5).

Molds produce small spores to reproduce, which float through the indoor and outdoor air continually. These spores may contain a single nucleus or be multinucleate. Mold spores can be asexual or sexual; many species can produce both types. Some species can remain airborne indefinitely, and many are able to survive extremes of temperature (some molds can begin growing at temperatures as low as 2°C) as well as extremes of pressure. When conditions do not foster or enable growth, most molds have the ability to remain alive in a dormant state, within a wide range of temperatures before they die. Different mold species will vary enormously in their tolerance to temperature and humidity extremes.

But for mold to grow or establish itself, it needs four vital elements; they are viable spores, a nutrient source (organic matter like wood products, carpet, and drywall), moisture, and warmth. The mere presence of humid air, however, is not necessarily conducive to fostering mold growth, except where air has a relative humidity (RH) level at or above 80% and is in contact with a surface. Mold spores are carried by air currents and can reach all surfaces and cavities of buildings. When the surfaces and/or cavities are warm, and contain the right nutrients and amounts of moisture, the mold spores will grow into colonies and gradually destroy the things they grow on.

Typical molds found in damp buildings

Fungal species	Substrate	Possible metabolites	Potential health effects
Alternaria alternata	Moist window-sills, walls	Allergens	Asthma, allergy
Aspergillus versicolor	Damp wood, wallpaper glue	Mycotoxins, VOCs	Unknown
Aspergillus fumigatus	House dust, potting soil	Allergens	Asthma, rhinitis, hypersensitivity pneumonitis
		Many mycotoxins	Toxic pneumonitis infection
Cladosporium herbarum	Moist window-sills, wood	Allergens	Asthma, allergy
Penicillium chrysogenum	Damp wallpaper, behind paint	Mycotoxins	Unknown
		VOCs	Unknown
Penicillium expansum	Damp wallpaper	Mycotoxins	Nephrotoxicity?
Stachybotrys chartarum (*atra*)	Heavily wetted carpet, gypsum board	Mycotoxins	Dermatitis, mucosal irritation, immunosuppression

Figure 7.5 *Table Showing Partial List of Typical Types of Molds Found in Damp Buildings.* Source: *California Department of Health Services, Environmental Health Investigations Branch.*

The key to controlling mold growth is the control of indoor moisture and the temperatures of all surfaces, including interstitial surfaces within walls. Mold generally needs a temperature range between 40°F and 100°F to grow, and maintaining relative humidity levels between 30% and 60% will help control mold and many of these known biological contaminants. Humidity control prevents the indoor growth of mold, mildew, viruses, and dust mites. Winter humidification and summer dehumidification controls/modules can supplement central HVAC systems when climate excesses require additional conditioning measures. Likewise, by removing any of the four essential growth elements, the growth process will be inhibited or nonexistent.

The first step in a mold remediation project includes determining the root cause of the mold growth. The next step is to evaluate the order of magnitude of the mold growth; this is usually done through visual examination. Since old mold growth may not always be visible, investigators may need to use instruments such as moisture meters, thermal imaging equipment, or borescope cameras to identify moisture in building materials or "hidden" mold growth within wall cavities, HVAC ducts, etc. Mold assessments and inspections should always include HVAC systems and their air-handler units, drain pans, coils, and ductwork in their surveys. In addition, depending on the age of the building, the inspector should take samples of the building materials, such as ceiling tiles, drywall joint compound, and sheet floor for the presence of asbestos. All these organisms may contribute to poor indoor air quality and can cause serious health problems.

Fungi in indoor environments comprise of microscopic yeasts and molds, called micro fungi, whereas plaster and wood–rotting fungi are referred to as macro fungi because they produce spores that are visible to the naked eye. Some molds produce toxic liquid or gaseous compounds, known as mycotoxins, in addition to infectious airborne mold spores, which often cause serious health problems to residents and workers. Toxicity can arise from inhalation or skin contact with toxigenic molds. Some of these toxic molds can only produce the dangerous mycotoxins under specific growing conditions. Mycotoxins are harmful or lethal to humans and animals when exposed to high concentrations. Toxic molds and fungi are a significant source of airborne VOCs that create IAQ problems as can be seen in Figure 7.6a,b.

Bacteria and viruses: Tens of millions of people around the world suffer daily from viral infections of varying degrees of severity at immense cost to the economy including the costs of medical treatment of infected people, costs of lost income due to inability to work, and costs of decreased productivity of those who are infected. In fact, viruses have been identified as the most common cause of infectious diseases acquired within indoor environments, particularly those causing respiratory and gastrointestinal infection. The most common viruses causing respiratory infections include influenza viruses, rhinoviruses, corona viruses, respiratory syncytial viruses and parainfluenza viruses, whereas viruses responsible for gastrointestinal infections include rotavirus, astrovirus, and Norwalk-like viruses. Some of these infections like the common cold are very widely spread but are not severe, while infections like influenza are relatively more severe.

Bacteria and viruses are minute in size and readily become airborne, remaining suspended in air for hours. Airborne bacteria and viruses in

Figure 7.6 (a) Example of mold on a ceiling that is growing out of control, which can be found in damp buildings. (b) Example of an extreme case of toxic mold growth in the process of being treated. *Source: Part a, mold-kill.com; part b, Applied Forensic Engineering, LLC.*

interior spaces are a cause of considerable concern because of their ability to transmit infectious diseases. Bacteria, viruses, and other bioaerosols that are common in both the home and the workplace may increasingly contribute to SBS if humidity levels are either too low or too high, depending on how their growth and our respiratory system are affected.

There are many pathways to infection spread, and among the most significant, from an epidemiological point of view, is airborne transport. Microorganisms can become airborne when droplets are given off during speech, coughing, sneezing, vomiting, or atomization of feces during sewage removal. Q fever is another emerging infectious disease among US soldiers serving in Iraq. Fever, pneumonia, and/or hepatitis are the most common signs of acute infection with Q fever.

Liquid and solid airborne particles (aerosols) in indoor air originate from many indoor and outdoor sources. These particles may differ in size, shape, and chemical and biological composition. Particle size signifies the most important characteristic affecting particle fate during transport and it is also significant in affecting their biological properties. Bacterial aerosols have also been found to be a means to transmit several major diseases as shown in Table 7.1.

Professor Lidia Morawska of Queensland University of Technology, Australia, says that the degree of hazard created by biological contaminants including viruses in indoor environments is controlled by a number of factors, for example:
- The type of virus and potential health effects it causes
- Mode of exit from the body
- Concentration levels
- Size distribution of aerosol containing the virus

Table 7.1 Major infectious diseases associated with bacterial aerosols

Disease	Causal organism
Tuberculosis	*Mycobacterium tuberculosis*
Pneumonia	*Mycoplasma pneumoniae*
Diphtheria	*Corynebacterium diphtheriae*
Anthrax	*Bacillus anthracis*
Legionnaires' disease	*Legionella pneumophila*
Meningococcal meningitis	*Neisseria meningitides*
Respiratory infections	*Pseudomonas aeruginosa*
Wound infections	*Staphylococcus aureus*

Table 7.2 Some of the major infectious diseases associated with viral aerosols

Disease	Virus/bacteria type
Influenza	*Orthomyxovirus*
Cold	*Coronavirus*
Measles	*Paramyxovirus*
Rubella	*Togavirus*
Chicken pox	*Herpes virus*
Respiratory infection	*Adenovirus*

- Physical characteristics of the environment (temperature, humidity, oxygenation, ultraviolet light, suspension medium, etc.)
- Air circulation pattern
- Operation of the HVAC system

The physical characteristics of the indoor environment as well as the design and operation of building ventilation systems are of paramount importance. Ducts, coils, and recesses of building ventilation systems often provide fertile breeding grounds for viruses and bacteria, which have been proven to cause a wide range of ailments from influenza to tuberculosis. Likewise, a number of viral diseases may be transmitted in aerosols derived from infected individuals. A number of infectious viral diseases and associated causal viruses transmitted through air are shown in Table 7.2.

Mites are microscope bugs that thrive on the constant supply of shed human skin cells (commonly called dander) that accumulate on carpeting, drapes, furniture coverings, and bedding (Figure 7.7). The proteins in that combination of feces and skin shedding are what cause allergic reactions in humans. Dust mites are perhaps the most common cause of perennial allergic rhinitis. Dust mites are the source of one of the most powerful

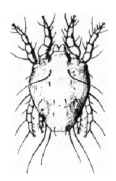

Figure 7.7 *Drawing of a Dust Mite. Source: EHSO.*

biological allergens, and flourish in damp, warm environments. Depending on the person and level of exposure, estimates given suggest that dust mites may be a factor in 50–80% of asthmatics, as well as in numerous cases of eczema, hay fever, and other allergic ailments.

It is estimated that up to 15% of people are allergic to dust mites, which, due to their very small size (250–300 μm in length) and translucent bodies, are not visible to the naked eye. To be able to give an accurate identification, one needs at least a 10× magnification. Dust mites have eight hairy legs, no eyes or antennae, and a mouthpart group in front of the body (resembles head) and a tough, translucent shell. Dust mites, like their insect cousins, have multiple developmental stages. These commence with an egg, and develop into larva, several nymph stages, and finally the adult.

Biological contaminants such as dust mites prefer warm, moist surroundings like the inside of a mattress, particularly when someone is lying on it, but they may also accumulate in draperies, carpet, and other areas where dust collects. The favorite food of mites appears to be dander (both human and animal skin flakes). Humans generally shed about 1/5 ounce of dander (dead skin) a week.

Dust mite populations are usually highest in humid regions and lowest in areas of high altitude and/or dry climates. Control measures are needed to reduce concentrations of dust-borne allergens in the living/working environment by controlling both allergen production and the dust that serves to transport it.

Rodent, insect, and animal allergens: According to the Illinois Department of Public Health, a typical large city in the United States annually receives more than 10,000 complaints about rodent problems and performs tens of thousands of rodent control inspections and baiting services. Effective measures should therefore be taken to prevent entry by rodents, insects, and pests from entering the home or office. Cockroaches, rats, termites, and other pests have plagued commercial facilities for far longer than computer viruses. Increasingly, research has confirmed and pinpointed pest infestation as the trigger or cause of a host of diseases, and according to the National Pest Management Association, pests can cause serious threats to human health, including such diseases as rabies, salmonellosis, dysentery, and staph. But in addition to pests presenting a serious health concern to a building's occupants, they also detract from a facility's appearance and value.

Today, communities of rats exist within and beneath cities, traveling unnoticed from building to building, along sewers and utility lines. Each rat colony has its own territory, which can span an entire city block and harbor

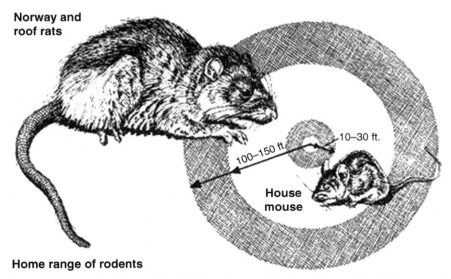

Norway and roof rats

100–150 ft.

10–30 ft.

House mouse

Home range of rodents

Figure 7.8 *Drawing Showing Typical Home Range of Rodents. Source: Illinois Department of Public Health.*

more than 100 rats. As they explore their territories, rats and mice discover new food sources and escape routes. A rat's territory or "home range" is generally within a 50–150-ft. radius of the nest, while mice usually have a much smaller range, living within a 10–30-ft. radius of the nest. In places where all their needs (food, water, shelter) are met, rodents have smaller territories (Figure 7.8).

Rodents are known to carry disease and fleas and leave waste. Wild and domestic rodents reportedly harbor and spread as many as 200 human pathogens. Diseases include the deadly hantavirus and arena virus. Hantavirus is usually contracted by inhaling airborne particles from rodent droppings, urine, or saliva left by infected rodents or through direct contact with infected rodents.

Insects: Today, more than 900,000 species of insects have been identified, and additional species are being identified every day. Some of these insects are known sources of inhalant allergens that may cause chronic allergic rhinitis and/or asthma. They include cockroaches, crickets, beetles, moths, locusts, midges, termites, and flies. Insect body parts are especially potent allergens for some people. Cockroach allergens are also potent allergens and are commonly implicated as contributors to SBS in urban housing and facilities with poor sanitation. Most of the allergens from cockroaches come from the insect's discarded skins. As the skins disintegrate over time, they become airborne and are inhaled. This type of allergen can be resolved by eliminating the cockroach population. The following are some of the insects commonly encountered in today's living and commercial environments:

Cockroaches are known carriers of serious diseases, such as salmonella, dysentery, gastroenteritis, and other stomach complaint organisms. They adulterate food and spread pathogenic organisms with their feces and defensive secretions. Cockroaches have been reported to spread at least 33 kinds of bacteria, six kinds of parasitic worms, and at least seven other kinds of human pathogens. They can pick up germs on their bodies as they crawl through decaying matter and then carry these onto food surfaces (Figure 7.9a). Cockroaches molt regularly throughout their life cycle. The discarded skin becomes airborne and can cause severe asthmatic reactions, particularly to children, the elderly, and people with bronchial ailments.

Ants: There are in excess of 20 varieties of ants invading homes and offices throughout the United States, particularly during the warm months of the year (Figure 7.9b). Worldwide, there are more than 12,000 species, but of these, only a limited number actually cause problems. The one trait all ants share is that they are unsightly and contaminate food. Ants range in color from red to black.

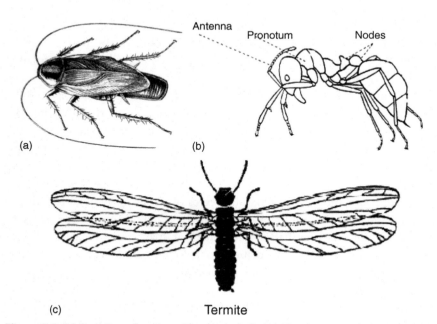

Figure 7.9 (a) Drawing of an American cockroach. (b) Illustration identifying primary features of an ant. (c) Illustration of a termite.

Destructive ant species include fire and carpenter ants. Others ant types include the Pharaoh, the honey, house, Argentine, and the thief ant. Fire ants are vicious, unrelenting predators and have a powerful, painful sting. More than 32 deaths in the United States have been attributed to severe allergic reactions to fire ant stings.

Termites can pose a major threat to structures, which is why it is important to address any termite infestation as soon as possible. A qualified termite control company or inspector should look for the many tell-tale signs termites usually provide such as small holes in wood, straw shaped mud tubes, crumbling drywall, termite insect wings, and sagging doors or floors (Figure 7.9c).

Allergens are produced by many mammalian and avian species and can be inhaled by humans, and cause immunological sensitization as well as symptoms of chronic allergic rhinitis and asthma. These allergens are normally associated with dander, hair, saliva, and urine of dogs, cats, rodents, and birds. Likewise, pollens, ragweed, and a variety of other allergens find their way indoors from the outdoors. Ragweed is known to cause what is commonly referred to as "hay fever," or what allergists/immunologists refer to as allergic rhinitis. In the United States, seasonal allergic rhinitis (hay fever) caused by breathing in allergens such as pollen affects more than 35 million people. Sufferers of allergic rhinitis exposure often experience sneezing, runny noses, and swollen, itchy, watery eyes. These allergy symptoms can have a major impact on a person's quality of life, including his or her ability to function well at school or work.

7.2.1.5 Reducing Exposure Measures

Resolving air quality problems in buildings requires first to consider all of the various contaminants, causes, and concentrations to be able to better address the problems and find appropriate solutions. The location where a person is exposed may also play a pivotal role. For while the average worker in the workplace is exposed approximately 40 h/week, many individuals are in their home 24/7. Exposure times, therefore, as well as concentration levels become part of the equation.

General steps to reducing pollutant exposure: Pollutant source removal or modification is the best approach whenever sources are identified and control is possible. These may include the following:

- Routine maintenance of HVAC systems
- Applying smoking restrictions in the home and the office
- Venting contaminant source emissions to the outdoors

- Proper storage and use of paints, pesticides, and other pollutant sources in well-ventilated areas and their use during periods of nonoccupancy
- Allowing time for building materials in new or remodeled areas to off-gas pollutants before occupancy

Most mechanical ventilation systems in large buildings are designed and operated not only to heat and cool the air but also to draw in and circulate outdoor air, and one cost-effective method to reduce indoor pollutant levels is to increase ventilation rates and air distribution. At a minimum, HVAC systems should be designed to meet ventilation standards in local building codes. In practice, however, many systems are not operated or adequately maintained to ensure that these design ventilation rates are in place. Often, IAQ can be improved by operating the HVAC system to at least its design standard, and to ASHRAE Standard 62-2001 if possible. When confronted with strong pollutant sources, local exhaust ventilation may be required to exhaust contaminated air directly from the building. The use of local exhaust ventilation is particularly advised to remove pollutants that accumulate in specific areas such as rest rooms, copy rooms, and printing facilities. Air cleaners can also be a useful adjunct to source control and ventilation although they are somewhat limited in their application. Air cleaners are discussed in Section 7.3.6 of this chapter.

Because the average person spends so much of their life indoors (roughly 90%), the quality of the indoor air is of paramount importance. In fact, indoor air pollution is currently ranked among the top four environmental risks in America by the EPA. This may explain why for many forward-thinking real estate property managers it is becoming a standard of doing business to have their buildings routinely inspected as part of a Proactive IAQ Monitoring Program.

7.2.1.6 Some Advantages to Having a Proactive Program in Place

- Proactive programs build value into the property by having a professional record of indoor environmental conditions that are formally documented over time.
- Having a proactive IAQ program in place can be used as an effective marketing tool to attract quality tenants in a very competitive real estate market.
- Regular monitoring of ventilation rates, building airflows, and filtration efficiencies, can dramatically improve IAQ, thereby creating a more comfortable and productive work environment.

- A proactive program reduces the likelihood of labeling a "healthy" building as causing SBS. Problems or potential problems can be quickly identified and resolved at minimal expense.
- Proactive IAQ monitoring shields owners and facility managers against liability. It shows that the owner has applied due diligence to ensure that his building remains healthy. Additionally, owners with large building portfolios can benefit from these routine inspections as a means to demonstrate that a cohesive, organized, and coordinated IAQ policy is in place.
- A proactive IAQ program enhances management/tenant relations by demonstrating a genuine concern for the tenants and their employees. Less time is needed to investigate and resolve tenant complaints.
- The proactive IAQ report provides critical third-party documentation certifying effective maintenance standards and supporting engineering or operation budgets.

7.2.1.7 Asbestos: Steps to Reducing Exposure
- Leave undamaged asbestos material alone if it is not likely to be disturbed.
- Trained and qualified contractors should be used for control measures that may disturb asbestos and for cleanup.
- Proper procedures should be followed when replacing woodstove door gaskets that may contain asbestos.

7.2.1.8 Radon: Steps to Reducing Exposure
- Test home and office for radon.
- If radon level is found to be 4 picocuries per liter (pCi/L) or higher, it needs to be fixed.
- Radon levels less than 4 pCi/L still pose a risk, and in many cases may be reduced.
- For additional information on radon, contact your state radon office, or call 800-SOS-RADON.

7.2.1.9 Lead: Steps to Reducing Exposure
- Leave lead-based paint undisturbed if it is in good condition.
- Do not sand or burn off paint that may contain lead.
- Do not remove lead paint yourself.
- Avoid bringing lead dust into the home or workplace.

- If your work or hobby involves lead, change clothes and use doormats prior to entering your home or workplace.
- Keep areas where children play as dust-free and clean as possible.
- Eat a balanced diet, rich in calcium and iron.

7.2.1.10 Respirable Particles: Steps to Reducing Exposure
- Vent all furnaces to outdoors and keep doors to remaining space open when using unvented space heaters.
- Choose properly sized woodstoves, certified to meet EPA emission standards and ensure that doors on all woodstoves fit tightly.
- Use trained professional to inspect, clean, and tune up central heating system (furnace, flues, and chimneys) annually and promptly repair any leaks.
- Change filters on central heating and cooling systems and air cleaners according to manufacturer's directions.

7.2.1.11 Formaldehyde: Steps to Reducing Exposure
- "Exterior-grade" pressed wood products should be used because they are lower-emitting (because they contain phenol resins, not urea resins).
- Use air conditioning and dehumidifiers to maintain moderate temperature and reduce humidity levels.
- Increase ventilation, particularly after bringing new sources of formaldehyde into the home or workspace.

7.2.1.12 Pests and Pesticides: Steps to Reduce Exposure
Pesticide are products that contain hazardous chemicals designed to kill or repel household pests (insecticides, termiticides, and disinfectants). Products are also used on lawns and gardens that drift or are tracked inside the property. Evidence of pests is typically found in different forms such as droppings (especially from cockroaches and rodents) and frass (from wood borers), gnawing, tracks, and grease marks (from rodents), damage (such as powderpost beetle exit holes), and shed insect skins. The presence of feeding debris or frass is one indication of infestation. Pests are also to be found behind baseboards, under furniture, behind moldings, in floor cracks, behind radiators, and in air ducts. Many pesticides contain harmful VOCs that contribute to poor indoor air and environmental quality (IAQ/IEQ). To reduce exposure to pests and pesticides, the following steps should be followed:

- Pesticides should be applied in strict accordance with manufacturer's instructions.
- Pesticides should be used in recommended quantities only.
- Check around door jambs for evidence of cockroaches and spider webs.
- Window sills should be examined regularly as many pests fly or crawl toward light.
- Increase ventilation when using indoors. Take plants or pets outdoors when applying pesticides to them.
- Check for presence of moisture as this can lead to moisture-related pests such as carpenter ants, termites, or mold.
- Use nonchemical methods of pest control where possible.
- If you use a pest control company, select it carefully.
- Do not store unneeded pesticides inside home; dispose of unwanted containers safely.
- Store clothes with moth repellents in separately ventilated areas, if possible.
- Keep indoor spaces clean, dry, and well ventilated to avoid pest and odor problems.

In addition, evidence of damaged screens, doors, and walls, which could allow pest entry should be fixed. Sanitation problems where existing, should be recorded and rectified. Presence of heavy landscaping near the foundation and plants such as ivy growing on walls need to be controlled or avoided as it increases the risk of outdoor pests moving inside. Moisture problems around the foundation, gutters, or air conditioning units should be monitored and rectified as they can facilitate moisture-related pests to enter a building. Bright exterior lights attract insects to the building's exterior and which may then find their way indoors.

7.2.1.13 Biological Contaminants: Steps to Reduce Exposure

- Install and use exhaust fans in kitchens and bathrooms and have them vented to the outdoors, and clothes dryers should also be vented to the outdoors. One benefit to using kitchen and bathroom exhaust fans is that they can also reduce levels of organic pollutants that vaporize from hot water used in showers and dishwashers.
- Ventilate any attic or crawl spaces to prevent moisture buildup, and keep humidity levels in these areas below 50% to avoid water condensation on building materials.
- If using cool mist or ultrasonic humidifiers, clean appliances according to manufacturer's instructions and refill with fresh water daily. These

humidifiers are susceptible to becoming breeding grounds for biological contaminants, and can potentially cause diseases such as hypersensitivity pneumonitis and humidifier fever. Clean evaporation trays in air conditioners, dehumidifiers, and refrigerators regularly.

- Water-damaged carpets and building materials can harbor mold and bacteria and therefore should be thoroughly cleaned and dried (within 24 h if possible) or removal and replacement should be considered.
- Keep the interior clean. Dust mites, pollens, animal dander, and other allergy-causing agents can be reduced, although not eliminated, through regular cleaning. Allergic individuals should also leave the space while being vacuumed as vacuuming can actually increase airborne levels of mite allergens and other biological contaminants. Using central vacuum systems that are vented to the outdoors or vacuums with high-efficiency filters may also be of help.
- Biological pollutants in basements can be minimized by regularly cleaning and disinfecting the basement floor drain. Operate a dehumidifier in the basement if needed to keep relative humidity levels between 30% and 50%.

Below are a number of relevant standards, codes, and guidelines:

- ASHRAE Standard 62-1999: Ventilation for Acceptable Indoor Air Quality
- ASHRAE 129-1997: Measuring Air-Change Effectiveness
- South Coast Rule #1168, South Coast Air Quality Management District
- Regulation 8, Rule 51 of the Bay Area Air Quality Management District
- Canadian Environmental Choice/EcoLogo
- Best Sustainable Indoor Air Quality Practices in Commercial Buildings
- Guidelines for Reducing Occupant Exposure to Volatile Organic Compounds (VOCs) from Office Building Construction Materials
- Carpet and Rug Institute Green Label Indoor Air Quality Test Program

Investigation procedure into IAQ may be characterized as a cycle of information gathering, hypothesis formation, and hypothesis testing. It typically begins with a walkthrough inspection of the problem area to gather information relating to the four basic factors that influence indoor air quality namely:

- A building's occupants
- A building's HVAC system
- Possible pollutant pathways
- Possible sources of contamination

Typical IAQ investigations include documentation of readily obtainable information regarding the building's history and of complaints if any. It would also include identifying known HVAC zones and complaint areas, notifying occupants of the upcoming investigation and, identifying key individuals that may be needed for information and access. The walkthrough itself entails visual inspection of critical building areas and consultation with occupants and staff. However, if insufficient information is obtained from the walkthrough to formulate a hypothesis, or if initial tests fail to reveal the source of the problem, the investigator needs to move on and collect additional information to allow formulation of additional hypotheses. The process of formulating hypotheses and testing and evaluating them continues until the problem is resolved.

It is very likely that in the coming years we will witness national and state regulations stipulating architects, designers, facility managers, and property owners meet mandated indoor air quality standards. The EPA, Occupational Safety and Health Administration (OSHA), ASHRAE, ASTM, and other organizations are currently in ongoing discussions concerning national indoor air quality standards.

7.2.2 Thermal Comfort

Thermal comfort is a state of well-being that almost defies definition. It involves temperature, humidity, and air movement among other things. Perhaps the most often heard complaint facility managers hear from building occupants is that their office space is too cold. That would appear to be an easy enough problem to resolve, if it was not for the fact that the number two most often heard complaint is that it is too hot. Studies show that people from different cultures generally have different comfort zones; even people belonging to the same family may feel comfortable under different conditions, and keeping everyone comfortable at the same time is an elusive goal at best. Regarding levels of thermal satisfaction, the Center for the Built Environment states, "Current comfort standards specify a 'comfort zone,' representing the optimal range and combinations of thermal factors (air temperature, radiant temperature, air velocity, humidity) and personal factors (clothing and activity level) with which at least 80% of the building occupants are expected to express satisfaction." This is the goal outlined by the American Society of Heating, Refrigerating and Air-Conditioning Engineers, Inc. (ASHRAE®) in the industry's gold standard of comfort – *Standard 55, Thermal Environmental Conditions for Human Occupancy.* ASHRAE

Standard 55 also specifies what thermal conditions are deemed likely to be comfortable to occupants.

We have shown that employee health and productivity are greatly influenced by the quality of the indoor environment. Poor air quality and lighting levels, off-gassing of chemicals from building materials, and the growth of molds and bacteria can all adversely affect building occupants. Sustainable design supports the well being of building occupants and their desire to achieve optimum comfort by reducing indoor air pollution. This can be achieved by applying a number of strategies such as the selection of materials with low off-gassing potential, providing access to daylight and views, appropriate ventilation strategies, and controlling lighting, humidity, and temperature levels.

Based on the above, finding the right temperature to satisfy everyone in a space is probably impossible. However, when temperature extremes – too cold or too hot – become the norm indoors, all building occupants suffer. In spaces that are either very hot or very cold, individuals must expend physiological energy to cope with the surroundings – energy that could be better utilized to focus on work and learning, particularly since research has shown that people simply do not perform as well, and attendance declines, in very hot and very cold workplaces.

7.2.3 Noise Pollution

Environmental noise is a global problem. And the way the problem is dealt with varies considerably from country to country, depending on the culture, economy, and politics of the country in question. Yet the problem persists even in areas where extensive resources have been applied for regulating, assessing and damping noise sources or for creation of noise barriers. In many countries, enormous efforts have been applied to reduce traffic noise at its source. For example, today's cars are much quieter than those manufactured a decade ago, but the substantial increase in traffic volume has been such that the effect of this effort has been neutralized and the annoyance level has in fact increased. Manufacturing quieter cars might have eased the problem temporarily but has failed to remove it.

Acoustical control is an important method of enhancing occupant comfort that has started to receive considerable attention in the design community. Noise pollution is considered a form of energy pollution in which distracting, irritating, or damaging sounds are freely audible. Noise and vibration from sources such as HVAC systems, vacuums, pumps, and helicopters can often trigger severe symptoms, including seizures, in susceptible individuals.

In the United States, regulation of noise pollution was stripped from the federal Environmental Protection Agency and passed on to individual states in the early 1980s. Although two noise-control bills passed by the EPA remain in effect, the EPA can no longer form relevant legislation. Needless to say, a noisy workplace is not conducive to getting work done. What is not so apparent is that constant noise can lead to voice disorders for paraprofessionals in the office where many employees spend time on the telephone, or routinely use their voices at work. An increasing number of teachers and paraprofessionals, are seeking medical care because they are chronically hoarse.

The voice is one of the most important instruments for professionals. What's more, people living or working in noisy environments secrete increased stress hormones, and stress will cause significant distraction from the work and learning at hand. For example, studies have documented higher stress levels among children and staff whose schools are located on busy streets or near major airports. A number of studies have also shown that office workers consider noise pollution to be a major irritant.

Humans, whether tenants or building occupants, have a basic right to live in an environment relatively free from the intrusion of noise pollution. Unfortunately, this is not always possible in an industrialized/urbanized society that relies heavily on equipment that generates objectionable noise. Although good engineering design can mitigate noise pollution levels to some extent, it is often not to acceptable levels, particularly if a significant number of individual sources combine to create a cumulative impact.

Definition of noise: The City of Berkeley's Planning and Development Department states that to understand noise, one must first have a clear understanding of the nature of sound. It defines sound as pressure variations in air or water that can be perceived by human hearing, and the objectionable nature of sound could be caused by its pitch or its loudness. In addition to the concepts of pitch and loudness, there are several methods to measure noise. The most common is use of a unit of measurement called a decibel (dB). On the decibel scale, the zero represents the lowest sound level that a healthy, unimpaired human ear can detect. Sound levels in decibels are calculated on a logarithmic basis. Thus, an increase of 10 dB represents a 10-fold increase in acoustic energy, and a 20 dB increase is 100 times more intense (10×10), etc. The human ear likewise responds logarithmically, and each 10-dB increase in sound level is perceived as approximately a doubling of loudness.

The impact of noise: Sound is of great value; it warns us of potential danger and gives us the advantage of speech and the ability to express joy or sorrow.

But sometimes sound can also prove to be undesirable. Often sound may interfere with and disrupt useful activities. Sometimes too, sounds such as certain types of music (e.g., pop or opera), may become noise at certain times (e.g., after midnight), in certain places (e.g., a museum), or to certain people (e.g., the elderly). There is therefore the introduction of a value judgment among people as to when sound becomes unwanted noise, which is why it is difficult to offer a clear definition of "good" or "bad" noise levels in any attempt to generalize the potential impact of noise on people.

Health effects: According to Wikipedia, elevated noise in the workplace or home can "cause hearing impairment, hypertension, ischemic heart disease, annoyance, sleep disturbance, and decreased school performance. Changes in the immune system and birth defects have been attributed to noise exposure, but evidence is limited." Hearing Loss is potentially one of the disabilities that can occur from chronic exposure to excessive noise, but it may also occur in certain circumstances such as after an explosion. Natural hearing loss associated with aging may also be accelerated from chronic exposure to loud noise. In many developed nations, the cumulative impact of noise is capable of impairing the hearing of a large fraction of the population over the course of a lifetime. Noise exposure has also been known to induce dilated pupils, elevated blood pressure, tinnitus, hypertension, vasoconstriction, and other cardiovascular impacts.

The OSHA has a noise exposure standard that is set at just below the noise threshold, where hearing loss may occur from long-term exposure. The impact of noise on physical stress reactions can be readily observed when people are exposed to noise levels of 85 dB or higher. The safe maximum level is set at 90 dB averaged over 8 h. If the noise is above 90 dB, the safe exposure dose becomes correspondingly shorter. Adverse stress-type reaction to excessive noise can be broken down into two stages. The first stage is where noise is above 65 dB, making it difficult to have a normal conversation without raising one's voice. The second is the link between noise and socioeconomic conditions that may further lead to undesirable stress–related behavior, increase workplace accident rates, or in many cases, stimulate aggression and other antisocial behaviors when they are exposed to chronically excessive noise.

Economic impact of noise: Most people accept the premise that all other things being equal, it is preferable to live in a house that is quieter than in a noisy one. This implies that there is an economic penalty associated with noise exposure. However, noise is not the only factor that can influence this decision. People living along heavily traveled roads may experience greater problems with traffic safety, air pollution, exhaust odor, crime, or loss of privacy. Accumulatively, these factors can significantly depress property values. Commercial uses may be mixed in with residential use, which may further reduce the desirability of a property. Upon considering all of these factors together, it becomes difficult to isolate the level of economic impact directly attributable to noise alone. New purchasers and renters may be unaware of how intrusive the noise can be, so that the undesirability level of living in a noisy environment may increase over time. Noise levels therefore may not significantly impact property values, especially when you take into consideration all of the other variables, bearing in mind that there may be a significant negative reaction to the noise levels encountered in the future.

Major sources of noise: The prevailing sources of artificial noise pollution in today's urban communities that are outside the control of affected individuals include the following:

* Transportation – cars, trucks, buses, trains near railroad tracks and aircraft near airports
* Routine activities of daily life
* Construction activity
* Industrial plant equipment noise

The main difference between transportation (Figure 7.10a,b) and nontransportation noise sources is that a municipality can generally impose control over the level and duration of noise at the property line of any nontransportation source of noise. Cities can only adopt noise exposure standards for noise levels emanating from trucks, trains, or planes and then not permit land uses to be developed in areas with excessive noise for an intended use. Cities also play a role in enforcing state vehicle code requirements regarding muffler operation and may set speed limits or weight restrictions on streets that impact noise generation. However, a city's actions

Figure 7.10 (a) Traffic is the main source of noise pollution in cities. Photo of traffic jam in Sao Paulo, Brazil. (b) A Boeing 747-400 passes close to houses shortly before landing at London Heathrow Airport – aircraft noise can significantly impact human health. *Source: Wikipedia.*

are typically proactive with regard to nontransportation sources and reactive for sources outside the city's control.

Noise abatement and reduction of excessive noise exposure can be accomplished using three basic approaches:

- Reduce the noise level at the source,
- Increase the distance between the source and the receiver, and
- Place an appropriate obstruction between the noise source and the receiver.

A noise wall is sometimes the only practical solution since vehicular noise is exempt from local control and relocation of sensitive land uses away from freeways or major roads is not practical. Yet noise walls have both positive and negative aspects. On the positive side, they can reduce the noise exposure to affected persons or other sensitive uses by effectively blocking the line of sight between source and receiver. A properly sited wall can reduce noise levels by almost 10 dB, which for most people translates to being about one-half as loud as before. Unfortunately, the social, economic, and aesthetic costs of noise walls are high. While noise walls would screen the traffic from receivers, it may also block beautiful views of trees, parks and water, and may also give drivers a claustrophobic feeling of being surrounded by massive walls.

Noise walls are usually made of concrete and are typically found near public areas (such as parks) and residential homes. The walls range in height from 6 ft. to 20 ft., but are normally between 12 ft. and 15 ft. tall. The construction cost of a noise wall is not cheap, averaging between \$100 ft.$^{-1}$ and \$200 ft.$^{-1}$ (traditional wall). For a green noise barrier, the cost would be higher than a traditional concrete barrier (which is not unexpected). This essentially means that one mile of traditional wall would cost between 0.5 million dollar and 1 million dollars. More importantly, many people have expressed great disappointment after completion of a sound wall because while the noise problem was reduced, it did not disappear as had been their expectation. Caltrans, for example, has a number of noise abatement programs in place, which focus on employing walls or berms to reduce noise intrusion from State and/or Federal highways. Likewise, Caltrans will generally support design features that minimize local objections providing their design standards are met.

Those standards include the following:

- Walls must reduce noise levels by at least 5 dB.
- Walls must be capable of blocking truck exhaust stacks that are located 11.5 ft. above pavement levels.

- Walls constructed within 15 ft. of the outside of the nearest travel lane must be built upon safety-shaped concrete barriers.

Concrete or masonry are most preferred wall materials. The effectiveness of a material in stopping sound transmission is called the transmission loss (TL).

7.2.4 Daylighting and Daylight Factor

The sun has been our main source of light and heat for millions of years, and through evolution, man has come to totally depend on it for his health and survival. The world and in particular the sustainable design movement is now returning to nature because of its increasing concern with global warming, carbon emissions, and sustainable design, and has started to take positive steps to increase its use of managed admission of natural light in both residential and nonresidential buildings. Daylighting has come to play a pivotal role in programs such as LEED and now has increased recognition in California's Title 24 energy code.

Craig DiLouie of the Lighting Controls Association defines daylighting as "the use of daylight as a primary source of illumination to support human activity in a space." Daylight is simply visible radiation that is generated by the Sun and which can reach us in one of three different forms:

1. Direct sunlight,
2. Skylight (i.e., sunlight that has been scattered in the atmosphere),
3. Sunlight or skylight that has been reflected off the ground.

Of the three, direct sunlight is the most powerful source and has the greatest impact on our lives; it not only provides visible light but also provides ultraviolet and infrared (heat) radiation (Figure 7.11).

Evaluation of daylight quality in a room traditionally consisted of a manual average daylight factor calculation, or based on a computerized version of the manual method. But in light of recent technological advances, daylight design is rapidly moving forward and is now able to provide the kind of information that would accommodate all of the requirements of the daylight consultant, the architect, and the end user. The ideal package should integrate natural lighting and electrical lighting calculations and also take into account an evaluation of the thermal impact on window design. Figure 7.12 lists recommended illumination levels for various locations and functions.

The term "average daylight factor" is sometimes construed to be the average daylight factor on all surfaces, whereas the output from most computer-based calculations reflect average daylight factor calculations derived from a series of points on the working plane.

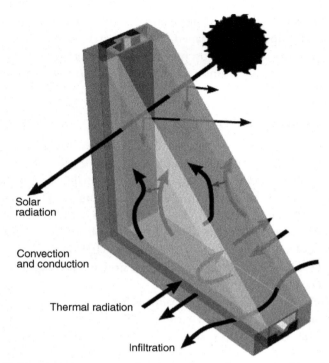

Solar
radiation

Convection
and conduction

Thermal radiation

Infiltration

Figure 7.11 Three major types of energy flows that occur through windows. (1) Nonsolar heat losses and gains in the form of conduction, convection, and radiation. (2) Solar heat gains in the form of radiation. (3) Airflow, both intentional as ventilation and unintentional as infiltration. *Source: US DOE.*

Area	Footcandles
Building surrounds	1
Parking area	5
Exterior entrance	5
Exterior shipping area	20
Exterior loading platforms	20
Office corridors and stairways	20
Elevators and escalators	20
Reception rooms	30
Reading or writing areas	70
General office work areas	100
Accounting/bookkeeping areas	150
Detailed drafting areas	200

Recommended illumination levels

Figure 7.12 *Table Showing Recommended Illumination Levels.*

The daylight factor (DF) is a very common and normally easy to understand measure for expressing the daylight availability in a room. Chris Croly, a building services engineering associate with BDP Dublin and Martin Lupton, a director with BDP Lighting, however, states that "the calculation of daylight factor using traditional methods becomes particularly difficult when trying to assess the effects of transfer glazing, external overhangs, or light shelves. Modern radiosity or ray tracing calculations are now readily available and are easy to use but still generally offer results in the form of daylight factor or lux levels corresponding to a particular static external condition." They define daylight factor as, "the ratio of the internal illuminance to that on a horizontal external surface located in an area with an unobstructed view of a hemisphere of the sky." The DF thus describes the ratio of outside luminance over inside luminance, expressed in percent. The higher the DF, the more natural light that is available in the room. The impact of direct sunlight to both illuminances must be considered separately and is not included. The DF can be expressed as follows: $DF = 100 \times E_{in}/E_{ext}$, where E_{in} represents the inside illuminance at a fixed point and E_{ext} represents the outside horizontal illuminance under an overcast (CIE sky) or uniform sky.

7.2.4.1 Daylighting Strategies

Documented research continues to demonstrate that effective daylighting saves energy while increasing the quality of the visual environment and reduces operating costs while improving occupant satisfaction. Thus, although daylight can reduce the amount of electric light needed to adequately illuminate a workspace and therefore reduce potential energy costs, allowing too much light or solar radiation into a space can have a negative effect, resulting in heat gain, offsetting any savings achieved by reducing lighting loads. Some architectural/design firms known for their sustainable design inclinations, like HOK and Gensler, design the majority of their buildings to be internally load-dominated, meaning the buildings need to be cooled for most of the year. It is important to fine-tune the glazing system to harvest as much daylight as possible without creating the negative effects of too much heat gain.

Strategies for improved daylighting include use of miniature optical light shelves, light-directing louvers, light-directing glazing, clerestories, roof monitors and skylights, light tubes, and heliostats. It is important to appreciate that whatever tools are applied in the daylighting design process, to be successful it will need the integration of several key disciplines including

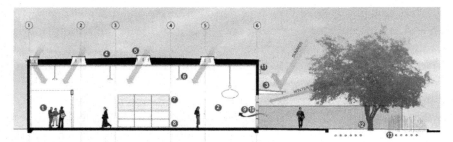

1. Private office   2. Open office   3. Sunshade with building integrated photovoltaics
4. Roof with building integrated photovoltaics   5. Skylight   6. Energy efficient and occupancy
sensor controlled light fixtures   7. Electrochromic glass   8. Radiant heat floor
9. Natural ventilation   10. High performance glass   11. Reduction of outdoor light pollution
12. Water-efficient landscaping   13. Ground source heat pump

Figure 7.13 *Schematic Drawing Illustrating an Integrated Approach to the Design of a Building's Various Systems in the New IDeAs Headquarters. Courtesy: Integrated Design Associates, Inc.*

architectural, mechanical, electrical, and lighting. As with sustainable design in general, these team members need to be brought into the design process early to ensure that daylighting concepts and strategies are satisfactorily implemented throughout the project (Figure 7.13).

The application of innovative, advanced daylighting strategies and systems can significantly improve the quality of light in an indoor environment as well as improve energy efficiency by minimizing lighting, heating, and cooling loads, thereby reducing a building's electricity consumption. By providing a direct link to the outdoors, daylighting helps create a visually stimulating and productive environment for building occupants, at the same time reducing as much as one-third of total building energy costs.

When light hits any surface, part of it is reflected back. This reflection is normally diffused (nondirectional) and is dependent on the object's reflection. The reflection of the outside ground is usually in the order of 0.2% or 20%. This signifies that in addition to the direct sunlight and skylight components, there also exists an indirect component that can make a significant contribution to the lighting inside a building, especially since the light reflected off the ground will hit the ceiling, which is usually very bright.

7.2.4.2 Daylighting and Visual Comfort

The challenge that designers often face with natural light, is the fluctuations in light levels, colors and direction of the light source. This led architects and engineers to make some unwise design decisions, which in the 1960s and 1970s culminated in hermetically sealed office blocks that were fully air conditioned and artificially lit. This in turn led to a sharp increase in complaints and symptoms attributed to BRIs and SBS. Gregg Ander, FAIA, says that "in large measure, the art and science of proper daylighting design is not so much how to provide enough daylight to an occupied space, but how to do so without any undesirable side effects."

The designer should adopt practical design strategies for sustainable daylighting design that will go a long way to achieving visual comfort by applying three primary approaches:

- Environmental – using the natural forces that impact design, resource, and energy conservation.
- Architectonic – what has made daylighting design so difficult until now is the lack of specific design tools. Today most large architectural practices have a diverse team of consultants and design tools that enable them to undertake complex daylighting analysis, whereas the typical school or small office do not have this capability or the budget for it.
- Human Factors – the impact on people and their experience. For visual comfort, designers need to achieve the best lighting levels possible while avoiding glare and high-contrast ratios. These can usually be avoided by not allowing direct sunlight to enter a workspace, for example, through the use of shading devices.

These basic lighting approaches reflect the strategies of sustainability and thus support the larger ecological goal.

A study conducted by the Heschong Mahone Group (HMG), a California architectural consulting firm, concluded that students who received their lessons in classrooms with more natural light scored as much as 25% higher on standardized tests than students in the same school district but whose classrooms had less natural light. This appears to confirm what many educationalists have suspected, that is, that children's capacity to learn is greater under natural illumination from skylights or windows than from artificial lighting. The logical explanation given is that "daylighting" enhances learning by boosting the eyesight, mood, and/or health of students and their teachers.

Another investigation by HMG looked at the relationship of natural light to retail sales. The study analyzed the sales of 108 stores that were part of a large retail chain. The stores were all one-story and virtually identical in layout, except that two-thirds of the stores had skylights while the others did not. The study specifically focused on skylighting as a means to isolate daylight as an illumination source, and avoid all of the other qualities associated with daylighting from windows. When they compared the sales figures for the various stores they discovered a statistically compelling connection between skylighting and retail sales performance, and found that stores with skylight systems had increased sales by 40% – even though the design and operation of all the store sites was remarkably uniform, except for the presence of skylights in some of the stores.

Skylights were found to have a major positive impact on the overall operation of the chain and were positively and significantly correlated to higher sales. The study showed that all other things being equal, an average nonskylit store in the chain would likely increase their sales by an average of about 40% just by adding skylights. HMG professes that this was determined with 99% statistical certainty.

Many architects are now specifying high-performance glass with spectrally selective coatings that only allows the visible light to pass through the glass while keeping out the infrared wavelength. This eliminates most of the infrared and ultraviolet radiation while allowing the majority of the visible light spectrum through the glass. But even with high-performance glass, much of the light can be converted to heat. Glass with high visible-light transmittance still allows light energy into a building, and when this light energy then hits a solid surface it is absorbed and reradiated into the space as heat.

The combination of daylighting and efficient electric lighting strategies can provide substantial energy savings. The building's planning module can often give indications on how best to organize the lighting. In any case, the lighting system must correlate to the various systems in place including structural, curtain wall system, ceiling system, and furniture system. Likewise, initial lighting costs may rise when designing for sustainability and implementing energy-efficient strategies. These energy-saving designs may require items like dimmable ballasts, photocells, and occupancy sensors, all of which are not typically covered in most initial project budgets (Figure 7.14). However, this could easily be compensated for by reducing building system and load costs, and even possibly reallocating some of this money to the lighting budget.

One of the main attributes of daylight is that it enhances the psychological value of space. Likewise, the introduction of daylight into a building reduces the need for electric lighting during the day while helping to provide a connection between indoor spaces and the outdoors for the building occupants. However, natural light is not without its negatives. These include glare, overheating, variability and privacy issues. The designer

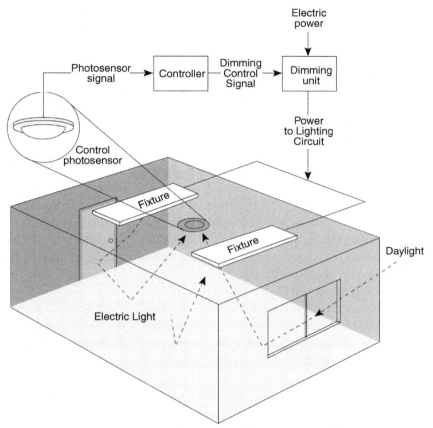

Figure 7.14 *Schematic Diagram of a Room Utilizing a Photoelectric Dimming System.* The ceiling-mounted photosensor reads both electric light and daylight within the space and adjusts the electric lighting as required to maintain the design level of total lighting. *Source: Ernest Orlando Lawrence Berkeley Laboratory.*

needs therefore to find ways to increase the positive aspects of using natural light in buildings, while at the same time minimizing the negative.

Addressing glare means keeping sunlight out of the field of view of building occupants while protecting them from disturbing reflections. Addressing overheating means adding appropriate exterior shading, filtering incoming solar radiation or even using passive control means such as thermal mass. Furthermore, addressing the variability and privacy issues requires creative ways to block or alter light patterns and compensate with other light sources.

It is only recently that daylighting strategies are starting to be considered at an early stage of a building design. This is because previously simple tools that can predict the performance of advanced daylighting strategies were not available to the designer. For best results, daylight design strategies should be in place at the outset of the building design process and the daylighting consultant should be involved very early in the design process. The data output from their daylighting studies could be extremely useful for fine-tuning and finalizing the building's orientation, massing, space planning, and interior finishes.

Innovative daylighting systems are designed to redirect sunlight or skylight to areas where it is most needed, with the avoidance of glare. These systems use optical devices that initiate reflection, refraction, and/or use the total internal reflection of sunlight and skylight. And with today's advancing technology, daylighting systems can be programmed to actively track the sun's movement or passively control the direction of sunlight and skylight.

The financial and competitive pressure of powerful market forces are driving some owners and design teams to seek architectural solutions such as the utilization of highly glazed, transparent façades. While these trends may offer clear potential benefits, such approaches also expose owners to real risks and costs associated with them as well. The general interest in potential benefits from these design solutions can be summarized as follows:

- Most building owners desire daylight, and find concepts and buildings that employ highly transparent façades preferable to dark tinted or reflective buildings of the 1970s and 1980s.
- Building owners are generally aware of the potential health and productivity benefits of daylight.
- The evident shift toward highly glazed façades is coupled with interior designs that reflect the desire of building owners to provide view and daylight to their employees. Open-plans, low-height partition furniture layouts, allow the daylit zones to be extended from a conventional 10–15 ft. (3.0–4.6 m) depth to a 20 ft. (6.1 m) or even 30 ft. (9.1 m) depth from the window wall (Figure 7.15).
- The increased use of low-reflectance, higher-brightness flat-screen LCD monitors, has allowed architects to employ design solutions that involve increasing daylight and luminance levels within buildings. But to take full advantage of natural daylight and avoid potential dark zones, it is critical that the lighting designer plans the lighting circuits and switching schemes in relation to the building's fenestration system.

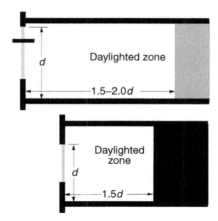

Figure 7.15 A rule of thumb for daylight penetration with typical depth and ceiling height is 1.5 times head height for standard windows and 1.5–2.0 times head height with light shelf, for south facing windows under direct sunlight. *Source: Ernest Orlando Lawrence Berkeley Laboratory.*

At the same time, there are potential risks associated with highly glazed façades. These include the following:

- Adequate tools not always available to reliably predict thermal and optical performance of components and systems, and to assess environmental quality.
- Increased sun penetration and excessive brightness levels that exceed good practice may heighten visual discomfort.
- Elevated cost of automated shading systems and purchasing lighting controls utilizing dimming ballasts and difficulty in commissioning systems after installation.
- High cost and technical difficulty of reliably integrating dimmable lighting and shading controls with each other and with building automation systems to ensure effective operation over time.
- Buildings utilizing transparent glazing generally use greater cooling loads and cooling energy, which has the potential for thermal discomfort (Figure 7.16).
- Uncertainty of occupant behavior with the use of automated, distributed controls in open landscaped office space and the potential for conflict between different needs and preferences.

To cash in on the potential benefits and minimize possible risks, there is a growing recognition that at least in workspaces (as distinct from corridors, lobbies, etc.), large glazed spaces require much better sun and glare control.

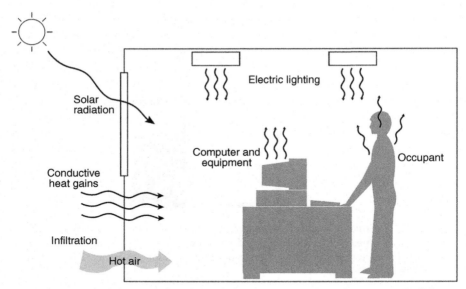

Figure 7.16 *Various Sources of Cooling Load. Source: Ernest Orlando Lawrence Berkeley Laboratory.*

Appropriate solutions must be delivered by systems that can rapidly respond to exterior climate and interior needs. One of the challenges facing manufacturers is how to provide such needed increased functionality at lower cost and risk to owners. Because of their various advantages and disadvantages, lighting consultants usually recommend the use of switching for spaces with nonstationary tasks such as corridors, and continuous dimming for spaces where users perform stationary tasks, such as offices. It has been shown that daylight harvesting using continuous dimming equipment automatically controlled by a photosensor, can generate 30–40% savings in lighting energy consumption, thereby significantly reducing operating costs for the owner.

7.2.4.3 Shades and Shade Controls

To achieve the greatest benefit of harvesting daylight, it is essential to implement a shading strategy tailored to the building. In hot climates, exterior shading devices have been found to work well to both reduce heat gain and diffuse natural light prior to entering the workspace (Figure 7.17). Examples of such devices include light shelves (Figure 7.18), overhangs, vertical louvers, horizontal louvers, and dynamic tracking or reflecting systems. Thus, for example, exterior shading of the glass can eliminate up to

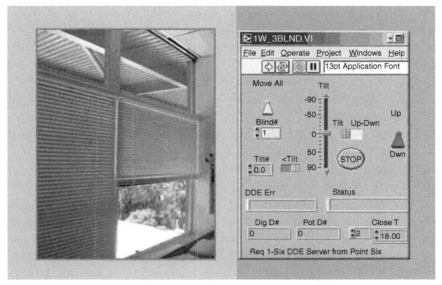

Figure 7.17 *A Venetian-Blind System at a Berkeley Lab Office Building is Equipped With a "Virtual Instrument" Panel for IBECS Control of Blinds Settings. Source: HPCBS.*

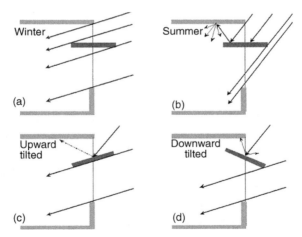

Figure 7.18 *Interior-Exterior Light Shelf-Source.* Light shelves are most effective for relatively clear climates, at midlatitudes. They are best used on south-facing façades, where they are most efficient. (1) Top Section (a,b) of an interior and exterior light shelf with specular surface, showing the path of sunlight rays in the winter and in the summer. (2) Bottom Section (c,d) shows how an upward- to downward-tilted reflective light shelf influences shading and daylight reflection. Note: In winter the light shelf alone does not adequately control glare.

80% of the solar heat gain. Shades and shade control strategy is based on the perception that occupants of commercial buildings typically prefer natural light to electric light, and the shade system goals would normally include the following:

- Maximizing use of natural light within a glare-free environment.
- Avoid direct solar radiation on occupants through interception of sunlight penetration.
- Facilitate occupant connectivity with the outdoors through increased glazing and external views.
- Provide manual override capability for occupants.

The overall determination is to ensure the shades are operational as much of the time as is possible without causing thermal or visual discomfort. Thermal comfort is maintained by solar tracking and the design of the external sunscreens. Visual comfort is affirmed by managing the luminance on the window wall. The specifications for the manual override system are based on postoccupancy surveys of office building occupants with automated shade systems. The prime recorded complaint in these studies was an occupant's inability to control the operation of a shade device when required.

7.2.5 Views

Recent studies show that windows providing daylight and ample views can dramatically affect building occupants' mental alertness, productivity, and psychological well-being. David Hobstetter, a principal in Kaplan-McLaughlin-Diaz, a San Francisco–based architectural practice, reaffirms this saying, "Dozens of research studies have confirmed the benefits of natural daylight and views of greenspace in improving a person's productivity, reducing absenteeism and improving health and well-being."

In many countries around the world, views, whether high-rise or otherwise, are normally considered mere perks, but recent research suggests that the view from a window may be even more important than the daylight it admits. The California Energy Commission's 2003 study of workers in the Sacramento Municipal Utility District's call center found that better views were consistently associated with better performance: "Workers with good views were found to process calls 7–12% faster than colleagues without views. Workers with better views also reported better health conditions and feelings of well being, while their counterparts reported higher fatigue." Another revealing study noted that computer programmers with views spent 15% more time on their primary task, while workers without views spent 15% more time talking on the phone or to one another.

Though some educators are of the opinion that views out of windows may be unnecessarily distracting to students, the CEC's 2003 study of the Fresno school district found that a varied view out of a window that included vegetation or human activity and objects in the far distance, supported better learning. Such findings confirm results of earlier research, such as a 1984 hospital study that concluded that postoperative patients with a view of vegetation took far fewer painkillers and experienced faster recovery times than patients looking at concrete walls.

Most building occupants relish contact with the outside world, even if only through a windowpane, and landscapes not surprisingly preferred to cityscapes. Researchers have concluded that views of nature improve attention spans after extended mental activity has drained a person's ability to concentrate. Among the main building types that can most benefit from the application of daylighting are educational buildings such as schools, administrative buildings such as offices, storage facilities such as warehouses, and maintenance facilities.

7.3 VENTILATION AND FILTRATION

Throughout ancient times, buildings, whether a Babylonian palace, an Egyptian temple, or a Roman castle, were ventilated naturally using wind shafts called "badgeer" (also known as "malqafs" – wind shafts/towers) or some other innovative method (Figure 7.19) since mechanical systems did not exist at the time. Andy Walker of the National Renewable Energy Laboratory says, "Wind towers, often topped with fabric sails that direct wind into the building, are a common feature in historic Arabic architecture, and are known as 'malqafs.' The incoming air is often routed past a fountain to achieve evaporative cooling as well as ventilation. At night, the process is reversed and the wind tower acts as a chimney to vent room air." It is not surprising that with today's increased awareness of the cost and environmental impacts of energy use, natural ventilation has once again come to the fore and become an increasingly attractive method for reducing energy use and cost and for providing acceptable indoor environmental quality.

Natural ventilation systems utilize the natural forces of wind and buoyancy, that is, pressure differences to move fresh air through buildings. These pressure differences can be a result of wind, temperature differences, or differences in humidity. The amount and type of ventilation achieved will depend to a large extent on the size and placement of openings in the building. Inadequate ventilation is one of the main culprits that cause

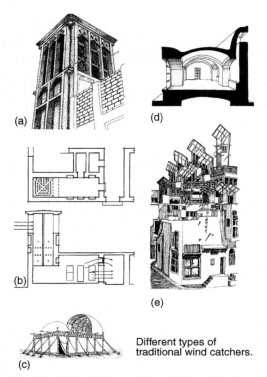

(a)

(d)

(b)

(e)

(c)

Different types of
traditional wind catchers.

Figure 7.19 *Drawings Depicting Various Types of Wind Catchers (Badgeer/Malqaf) Used in Traditional and Ancient Architecture.* (a) Multidirectional traditional Dubai wind catcher. (b) Plan and section of Dubai wind catcher. (c) Ancient Assyrian wind catcher. (d) Section through traditional wind scoop. (e) Traditional Pakistani wind catchers.

increased indoor pollutant levels by not bringing in enough outdoor air to dilute emissions from indoor sources and by not carrying indoor air pollutants out of the home or workplace.

7.3.1 Ventilation and Ductwork

Ventilation is vital for the health and comfort of building occupants. It is specifically needed to reduce and remove pollutants emitted from various internal and external sources. Good design combined with optimum air tightness is a prerequisite to ensuring healthy air quality, occupant comfort, and energy efficiency. Sufficient air supply and movement can be tested and analyzed to determine the efficiency of an HVAC system. Regular maintenance of ductwork is pivotal to achieving both better indoor environment and system stability. Ductwork can be evaluated, cleaned, and sealed to prevent airflow and quality issues. All ductwork should be analyzed by a professional trained and certified by the National Air Duct Cleaning Association.

7.3.2 Air Filtration

It is unfortunate that to date, no Federal standards have yet been adopted for air filter performance, although the Food and Drug Administration (FDA) has asked groups of experts to recommend national standards. So far, the FDA has concluded, there remains insufficient research data on the relationship between air filtration and actual health improvement to recommend national standards. Air cleaning filters are designed to remove pollutants from indoor air, so by properly filtering your facility's air of harmful particles, you can improve the indoor environment and breathe cleaner air. Proper filtration removes dirt, dust, and debris from the air you breathe. It also reduces pollen and other allergens, which can cause asthmatic attacks and allergic reactions. But while air cleaning devices may help to control the levels of airborne allergens, particles, and in some cases, gaseous pollutants in a facility, they may not decrease adverse health effects from indoor air pollutants.

There are several kinds of air cleaning devices to choose from − some are designed to be installed inside the ductwork of a facility's central HVAC system to clean the air in the whole facility. Other types include portable room air cleaners, which are designed to be used to clean the air in a single room or specific areas, and are not intended for complete facility filtration. There are various types of air filters currently on the market, such as mechanical filters, electronic filters, hybrid filters, gas phase, and ozone generators.

7.3.3 Air Purification

Some sources, such as building materials, furnishings, and household products like air fresheners, release pollutants more or less continuously. Other sources, related to activities carried out in the home or workplace, release pollutants intermittently. These include smoking, the use of unvented or malfunctioning stoves, furnaces, or space heaters, the use of solvents in cleaning and hobby activities, the use of paint strippers in redecorating activities, and the use of cleaning products and pesticides. High pollutant concentrations can remain in the air for long periods after some of these activities if action is not taken.

Although air filtration removes particulate matter, air purification is required to remove the things a filter does not, such as odors and gases.

Chemicals in paints, carpets, and other building materials (i.e., VOCs) are harmful to building occupants and should be removed through air purification. There is also an increasing concern regarding the presence of biological infectious agents, and air purification on a regular basis is one way to address these potential problems.

7.3.4 Amount of Ventilation

Outdoor air normally enters and leaves a building through various means, particularly infiltration, natural ventilation, and mechanical ventilation. Outdoor air can infiltrate a building through openings, joints, and cracks in walls, floors, and ceilings, and around windows and doors. Natural ventilation involves air moving through opened windows and doors. Air movement associated with infiltration and natural ventilation is a consequence of air temperature differences between the indoor and outdoor air and by wind. When insufficient outdoor air enters a home, pollutants can accumulate to elevated levels to a degree that they can pose health and comfort problems.

In the event that natural ventilation is insufficient to achieve good air quality, there are a number of mechanical ventilation devices, from outdoor-vented fans that will intermittently remove air from a room, such as bathrooms and kitchen, to air handling systems that utilize fans and duct systems to continuously remove indoor air and distribute filtered and conditioned outdoor air to strategic points throughout the building. The rate at which outdoor air replaces indoor air is known as the air exchange rate. Insufficient air infiltration, natural ventilation, or mechanical ventilation means that the air exchange rate is low and can experience rising pollutant levels.

Sometimes residents or occupants are in a position to take appropriate action to improve the indoor air quality of a space by removing the source, altering an activity, unblocking an air supply vent, or opening a window to temporarily increase the ventilation; in other cases, however, the building owner or manager are the only ones in a position to remedy the problem. Building management should be prevailed upon to follow guidance in EPA's IAQ Building Education and Assessment Model (I-BEAM). I-BEAM expands and updates EPA's existing Building Air Quality guidance and is considered to be a complete state-of-the-art guidance for managing IAQ in commercial buildings. Building management should also be encouraged to follow guidance in EPA and NIOSH's Building Air Quality: A Guide for Building Owners and Facility Managers. The BAQ guidance is available as PDF files that can be downloaded.

7.3.5 Ventilation Improvements

As previously stated, another approach to lowering the concentrations of indoor air pollutants is to increase the amount of outdoor air coming indoors. Many heating and cooling systems, including forced air heating systems, do not mechanically bring fresh air into the house. This can often be addressed by opening windows and doors, or running a window air conditioner with the vent control open, which increases the outdoor ventilation rate. In residences, local bathroom or kitchen fans that exhaust outdoors can be used to remove contaminants directly from the room where the fan is located while increasing the outdoor air ventilation rate.

Good ventilation is especially important when undertaking short-term activities that can generate high levels of pollutants, for example, painting, paint stripping, or heating with kerosene heaters. Such activities should preferably be executed outdoors whenever possible.

Advanced designs of new homes have recently come on the market that feature mechanical systems that bring outdoor air into the home as well as energy-efficient heat recovery ventilators (also known as air-to-air heat exchangers).

The following include a number of design recommendations to help achieve better ventilation in buildings:

- Naturally ventilated buildings should preferably be narrow, as wide buildings pose greater difficulty to distribute fresh air to all areas using natural ventilation.
- Occupants should be able to operate window openings.
- Use of mechanical cooling is advised in hot, humid climates.
- Determine whether an open- or closed-building ventilation approach potentially offers the best results. A closed-building approach will work better in hot, dry climates where there is a large diurnal temperature range from day to night. An open-building approach is more effective in warm and humid areas, where the temperature difference between day and night is relatively small.
- Consideration should be given to the use of fan-assisted cooling strategies. Andy Walker says, "Ceiling and whole-building fans can provide up to 9°F effective temperature drop at one tenth the electrical energy consumption of mechanical air-conditioning systems."
- When possible, provide ventilation to the attic space, as this greatly reduces heat transfer to conditioned rooms below. Ventilated attics have been found to be approximately 30°F cooler than unventilated attics.

• Maximize wind-induced ventilation by siting buildings so that summer wind obstructions are minimal.

7.3.6 Air Cleaners

There are a variety of types and sizes of air cleaners presently on the market, ranging from relatively inexpensive table-top models to larger, more sophisticated and expensive systems. Some air cleaners are highly effective at particle removal, while others, including most table-top models, are much less so. It should be noted that air cleaners are generally not designed to remove gaseous pollutants.

An air cleaner's effectiveness is expressed as a percentage efficiency rate. It depends on how well it collects pollutants from indoor air and how much air it draws through the cleaning or filtering element. The latter is expressed in cubic feet per minute. The fact needs noting that even an efficient collector that has a low air-circulation rate will not be effective, neither will an air cleaner with a high air-circulation rate but a less efficient collector. All being said, the long-term performance of any air cleaner depends largely on maintaining it in accordance with the manufacturer's directions.

Another critical factor in determining the effectiveness of an air cleaner is the level and strength of the pollutant source. Table-top air cleaners, in particular, may not be capable of adequately reducing amounts of pollutants from strong nearby sources. Persons who are sensitive to particular pollutant types may find that air cleaners are useful mostly when used in conjunction with collaborative efforts to remove the source.

The EPA does not currently recommend using air cleaners to reduce levels of radon and its decay products. The effectiveness of these devices is questionable because they only partially remove the radon decay products and do not diminish the amount of radon entering the home. EPA is planning to undertake additional research on whether air cleaners are, or could become, a viable means of reducing the health risk from radon.

For most indoor air quality problems, source control is the most effective solution approach. The EPA issued "Ozone Generators That Are Sold As Air Cleaners" to provide accurate information regarding the use of ozone-generating devices in indoor occupied spaces. This information is based on the most credible scientific evidence currently available. The EPA also issued the document "to assist" consumers. The document explains what air duct cleaning is, offers guidance to help consumers decide whether to have the service performed, as well as providing useful information on choosing a duct cleaner, determining if duct cleaning was done correctly, and how to prevent contamination of air ducts.

7.3.7 Ventilation Systems

Mechanical ventilation systems in large commercial buildings are normally designed and operated to heat and cool the air, as well as to draw in and circulate outdoor air. On the other hand, ventilation systems themselves can be a source of indoor pollution and contribute to indoor air problems if they are poorly designed, operated, or maintained. They can sometimes spread harmful biological contaminants that have steadily multiplied in cooling towers, humidifiers, dehumidifiers, air conditioners, or inside surfaces of ventilation ducts. For example, problems arise when, in an effort to save energy, ventilation systems are incorrectly programmed and bring in adequate amounts of outdoor air. Other examples of inadequate ventilation occur when the air supply and return vents within a space are blocked or placed in a manner that prevents the outdoor air to reach the breathing zone of building occupants. Improper location of outdoor air intake vents can also bring in contaminated air, particularly from automobile and truck exhaust, fumes from dumpsters, boiler emissions, or air vented from bathrooms and kitchens.

For mechanically ventilated spaces, designers should refer to ASHRAE Standard 55-1992, Addenda 1995, "Thermal Environmental Conditions for Human Occupancy." For naturally ventilated spaces, refer to California High Performance Schools (CHPS) Best Practices Manual Appendix C, "A Field Based Thermal Comfort Standard for Naturally Ventilated Buildings."

7.4 BUILDING MATERIALS AND FINISHES – EMITTANCE LEVELS

Several studies have over the years investigated the impact of pollution emitted by building materials on the indoor air quality and then related these to ventilation requirements. However, there has been a lack of systematic experiments, in which building materials are initially ranked according to their pollution strength, and then analyzing the impact on the indoor air quality of using these materials in real rooms. Such studies would allow us to quantify the extent to which using low-polluting building materials would reduce the energy used for ventilation of buildings without having to compromise indoor air quality. One of the primary objectives of an ongoing research project is to quantify this energy-saving potential based on the effects on the perceived air quality.

Healthy Building Network (HBN) identifies the primary building materials that are toxic and which have unacceptable high VOC emittance levels. These include the following:

Polyvinyl chloride (PVC): PVC has been the focal point of a controversial debate during much of the last two decades. A number of diverging scientific, technical, and economic opinions have been expressed on the question of PVC and its effects on human health and the environment. The Healthy Building Network (HBN), however, singled out PVC, or vinyl, for elimination because of its uniquely wide and potent range of chemical emissions throughout its life cycle, including many of the target chemicals listed below. It is virtually the only material that requires phthalate plasticizers, frequently includes heavy metals and emits large amounts of VOCs. In addition, during manufacture it produces a large quantity of highly toxic chemicals, including dioxins (the most potent carcinogens measured by man), vinyl chloride, ethylene dichloride, and PCBs among others. Moreover, when burned at the end of its useful life, whether in an incinerator or landfill fire, it releases hydrochloric acid and more dioxins. It is therefore prudent if not imperative to avoid products made with PVC.

Volatile organic compounds: VOCs essentially consist of thousands of different chemicals, such as formaldehyde and benzene, which evaporate readily into the air. Depending on the level of exposure, they can causing dizziness; headaches; eye, nose, and throat irritation; or asthma; and in some cases, can also cause cancer, induce longer term damage to the liver, kidney and nervous system, and stimulate higher sensitivity to other chemicals.

When dealing with wet products such as paints, adhesives, and other coatings, ensure that the products contain no or low VOC in them. Look for the Green Seal when using certified paints or paints with less than 20 g/L VOCs. For adhesives and coatings, make sure they are SCAQMD (South Coast Air Quality Management District) compliant.

For flooring and carpet, wall covering, ceiling tiles, and furniture, it is advisable to use only CA 01350–compliant products. A number of programs currently use the CA 01350 testing protocol to measure the actual levels of individual VOCs emitted from the material and compare it to allowable levels set by the state of California. These include CHPS, CRI's Green Label Plus, SCS's Indoor Advantage, RFCI's FloorScore®, and GREENGUARD's Schools & Children. Try to avoid flooring that requires waxing and stripping, a process that will release more VOCs than the original material. All composite wood products and insulation should have no added formaldehyde. The CA 01350 program has set limits on formaldehyde emissions; however, for these products, there are options that completely exclude the presence of added formaldehyde.

Phthalates: DEHP and other phthalates have attracted considerable adverse publicity for their use in PVC medical products and in toys, and concerns have been raised about their impact on the development of young children. Phthalates are, however, also used widely in flexible PVC building materials and have been linked to bronchial irritation and asthma. It is important therefore to avoid using products with phthalates (including PVC).

Heavy metals: Even though these metals are known to be hazardous to your health, they continue to be used for stabilizers or other additives in building materials. Lead, mercury, and organotins are all known potent neurotoxins, which is particularly damaging to the brains of fetuses and growing children. Cadmium is a carcinogen and can cause a variety of kidney, lung, and other damages. Look, therefore, for products that do not contain heavy metals.

Halogenated flame retardants (*HFRs*): The use of flame retardants in many fabrics, foams, and numerous plastics are known to have saved many lives over the years. However, the halogenated flame retardants (including PBDEs & other brominated flame retardants (BFRs), have been found to disrupt thyroid and estrogen hormones, which can cause developmental effects, such as permanent changes to the brain and to the reproductive systems. Being persistent and bioaccumulative, they are rapidly accumulating to dangerous levels in humans and have now become the subject of an increasing number of bans and phase-outs. All products that use halogenated flame retardants should be avoided.

Perfluorocarbons (*PFCs*): Numerous treatments for fabric and some building materials have been based on PFCs that, like the HFRs, are characteristically highly persistent and bioaccumulative and hence are concentrating at alarming levels in humans. PFOA, is a major component of treatment products such as Scotchguard, Stainmaster, Teflon, and Gore-Tex, and has been linked to a range of developmental and other adverse health effects. Thus, avoid all products that are treated with a PFC-based material.

Summary Checklist for Healthy Materials
- Avoid PVC (polyvinyl chloride, vinyl)
- Low or no VOC
- CA 01350 compliant
- Ensure *no* added formaldehyde
- Ensure *no* phthalates or heavy metals (lead, mercury, cadmium, organotins) are used

- Ensure *no* HFRs (PBDEs, BFRs, and other halogenated flame retardants) are used
- Ensure *no* PFCs (perfluorocarbons) are present

Resources for Locating Healthy Building Materials

Because of the great diversity of materials used in the construction and manufacturing industries, we are unable to produce a single building materials list or certification that covers all of the relevant health and environmental issues. For example, we find that of the various programs listed below that certify products to meet the CA 01350 VOC emissions standards, none also screen for phthalates and flame retardants but you will find many PVC/vinyl products on those lists. Furthermore, some 01350 VOC programs are managed by trade associations. It is prudent to always ask to see the actual lab certification of 01350 VOC emissions when first screening a material and ask about PVC/vinyl, heavy metals, HFRs, and PFCs.

HBN PVC alternatives database is a CSI-prepared listing of PVC-free alternatives for a wide range of different building materials. Website: www.healthybuilding.net/pvc/alternatives.html

GreenSeal certified products: Paints and coatings that meet the GreenSeal VOC content standards are listed; these materials do not contain certain excluded chemicals and meet typical LEED™ performance requirements.

EcoLogo: An environmental standard and certification mark, EcoLogo was founded in 1988 by the Government of Canada and is now recognized internationally. It was acquired by Underwriters Laboratories (UL) in 2010 and is currently the only North American eco-labeling program approved by the Global Ecolabelling Network (GEN), an association of third-party labeling organizations dedicated to improving transparency in environmental products and services. The EcoLogo Program provides customers – public, corporate, and consumer – with assurance that the products and services bearing the logo meet stringent standards of environmental leadership. The EcoLogo Program is a Type I eco-label, as defined by the International Organization for Standardization (ISO) and is one of two such programs in North America that has been successfully audited by the Global EcoLabelling Network (GEN) as meeting ISO 14024 standards for eco-labeling. The EcoLogo label helps manufacturers and consumers easily identify environmentally preferred products. It is used for a very wide variety of products, from building materials (including carpet, paint, and adhesives) and furniture to cleaning products, and paper products.

CHPS (collaborative for high-performance schools) low-emitting materials: CHPS maintains a table listing products that have been certified by the manufacturer and an independent laboratory to meet the CHPS Low-Emitting Materials criteria – Section 01350 – for use in typical classrooms. Certified materials include adhesives, sealants, concrete sealers, acoustical ceilings, wall panels, wood flooring, composite wood boards, resilient flooring, and carpet. Note: This list also includes paint listings, but CA 01350 is not yet a replacement for low VOC screening (www.chps.net/manual/lem_table.htm).

GreenLabel Plus: In 1992, the Carpet and Rug Institute (CRI) launched its Green Label program to test carpet, cushions, and adhesives to help specifiers identify products with very low emissions of VOCs. It means that the Carpet & Rug Institute (a trade association) assures customers that approved carpet products meet stringent requirements for low chemical emissions and furthermore, certifies that these carpets and adhesives meet CA 01350 VOC requirements. Furthermore, CRI recently launched its next series of improvements called Green Label Plus for carpet and adhesives. This is an enhanced program that sets an even higher standard for indoor air quality and ensures that customers are purchasing the very lowest emitting products currently on the market. Using scientifically established standards, the Green Label Plus program symbolizes the carpet industry's commitment to a better environment for living, working, learning, and healing. Thus, any architect, interior designer, government specifier, or facility administrator who is committed to using green building products just needs to look for the Green Label Plus logo as this signifies that the carpet product has been tested and certified by an independent laboratory and meets stringent criteria for low emissions.

FloorScore: Was developed by the Resilient Floor Covering Institute (RFCI) in conjunction with SCS Global Services (SCS). It tests and certifies hard surface flooring and flooring adhesive products for compliance with rigorous indoor air quality emissions requirements. Moreover, FloorScore conforms with US Green Building Council (USGBC) LEED criteria for EQ4.3 (carpet systems, flooring), Collaborative for High Performance Schools (CHPS) low-emitting products, and California 01350 Special Environmental Requirements. Scientific Certification Systems certifies for the Resilient Floor Covering Institute (a trade association) that resilient flooring meets CA 01350 VOC requirements. www.scscertified.com/iaq/floorscore_1.html. Among the benefits of attaining a FloorScore certification means meeting the requirements of programs, including the following:

• Qualify for leading building industry rating programs multiple LEED rating systems (NC, CI, CS, HC, EB, and Schools)

- GreenGuide for Healthcare
- NAHB Green Building Standard
- CHPS
- HBN Pharos Project
- EPA Tools for Schools

GREENGUARD *for Children & Schools*: Air Quality Sciences certifies for GREENGUARD that furniture and indoor finishes meet lower of CA 01350 VOC or 1/100 of TLV. The GREENGUARD Environmental Institute (GEI) has formulated performance-based standards to define goods with low chemical and particle emissions for use indoors; these primarily include building materials, interior furnishings, furniture, cleaning and maintenance products, electronic equipment, and personal care products. The standard establishes certification procedures including test methods, allowable emissions levels, product sample collection and handling, testing type and frequency, and program application processes and acceptance. GEI now certifies products across multiple industries.

Brominated flame-retardants, halogenated flame-retardants, and perfluorochemicals: No listings are yet screened for these emerging problem chemicals. The flame-retardants are added to plastics, particularly fabrics and foams. PBDEs are the most widely used. All halogen-based flame-retardants, however, are likely to remain problematic.

7.5 BEST PRACTICE FOR IEQ

Indoor environmental quality is a critical component of sustainable buildings. As we have already seen, numerous documented studies have confirmed the effect of the indoor environment on the health and productivity of building occupants. Ventilation, thermal comfort, air quality, and access to daylight and views are all important factors that play a significant role in determining indoor environmental quality.

The Architectural and Transportation Barriers Compliance Board (Access Board), which is an independent federal agency devoted to accessibility for people with disabilities, contracted with the National Institute of Building Sciences (NIBS) to establish an Indoor Environmental Quality Project as a first step in implementing an action plan. The NIBS issued a project report on Indoor Environmental Quality released in July 2005 which revealed that a growing number of people in the United States suffer a range of debilitating physical reactions caused by exposures to everyday materials and chemicals found in building products, floor coverings, cleaning products, fragrances, and others products. This condition is known as multiple chemical sensitivity (MCS). The range and severity of these reactions are very varied. In addition, the Access Board received numerous complaints from other people who reported adverse reactions from exposures to electrical devices and frequencies, a condition referred to as electromagnetic sensitivity (EMS) or electromagnetic hypersensitivity (EHS).

In response to these concerns, the Access Board sponsored a study on ways to tackle the problem of indoor environmental quality for persons with MCS and EMS as well as for the general population. In conducting this study for the Board, the NIBS brought together a number of interested parties to explore the relevant issues and to develop an appropriate action plan. Although the focus of the project was on commercial and public buildings, many of the issues addressed and recommendations offered can also be applied to residential settings. The report includes, among other things, recommendations on improving indoor environmental quality that addresses building products, materials, ventilation, and maintenance.

A steering committee was established for the project that included representation from MCS and EMS organizations, experts on indoor environmental quality, and representatives from the building industry. Committee members examined various methods and strategies for collecting and disseminating information, focusing on specific areas, increasing awareness of relevant issues, encouraging extended project participation, identifying potential partners for further study and outreach, and developing practical recommendations for best practices.

Below are some of the steps and best practices that can be applied to ensure good IEQ:

- Conduct a faculty-wide IEQ survey/inspection of facility, noting odors, unsanitary conditions, visible mold growth, staining, presence of moisture in inappropriate places, poorly maintained filters, personal air cleaners, hazardous chemicals, uneven temperatures, blocked vents.
- Determine operating schedule and design parameters for HVAC system and ensure adequate fresh air is provided to prevent the development of indoor air quality problems and to contribute to the comfort and well being of building occupants. Maintain complete and up-to-date ventilation system records.
- Ensure that appropriate preventive maintenance (PM) is performed on HVAC system, including but not limited to outside air intakes, inside of

air handling unit, distribution dampers, air filters, heating and cooling coils, fan motor and belts, air distribution ducts and VAV boxes, air humidification and controls, and cooling towers.

- Manage and review processes with potentially significant pollution sources, such as renovation and remodeling, painting, shipping and receiving, pest control, and smoking. Ensure adequate controls are instituted on all renovation and construction projects and evaluate control impacts on IEQ.
- Control environmental tobacco smoke by prohibiting smoking within buildings or near building entrances. Designate outdoor smoking areas at least 25 ft. from openings serving occupied spaces and air intakes.
- Control moisture inside buildings to inhibit mold growth, particularly in basements. Dehumidify when necessary and respond promptly to floods, leaks, and spills. Use of porous materials in basements should be monitored and restricted whenever possible.
- When mold growth is evidenced, immediate action should be taken to remediate.
- Choose low-emitting materials with minimal or no VOCs. This particularly applies to paints, sealants, adhesives, carpet and flooring, furniture, and composite wood products and insulation.
- Monitor CO_2 levels and install CO_2 and airflow sensors in order to provide occupants with adequate fresh air when required.
- To maintain occupants' thermal comfort, include adjustable features such as thermostats or operable windows.
- Window size, location, and glass type should be selected to provide adequate daylight levels in each space.
- Window sizes and positions in walls should be designed to take advantage of outward views and have high visible transmittance rates (greater than 50%) to ensure maximum outward visibility.
- Incorporate design strategies that maximize daylight and views for building occupants' visual comfort.
- Educate cleaning staff regarding use of appropriate methods and products, cleaning schedules, materials storage and use, and trash disposal.
- A process for complaint procedures should be established and IEQ complaints promptly respond to.
- Discuss with occupants how they can participate in maintaining acceptable indoor environmental quality.
- Permanent entryway systems such as grilles or grates should be installed to prevent occupant-borne contaminants from entering the building.

- Construction IAQ Management Plan should be in place so that during construction, materials are protected from moisture damage and particulates are controlled through the use of air filters.

There are a number of suggested IEQ-related recommendations that tenants should follow to ensure that a healthy indoor environment is maintained for all building occupants. These include the following:

- The use of air handlers during construction must be accompanied by the use of filtration media with a minimum efficiency reporting value (MERV) of 8 at each return grill as determined by ASHRAE 52.2-1999.
- Replace all filtration media immediately prior to occupancy and conduct, when possible, a minimum 2-week flush-out with new filtration media with 100% outside air after construction is completed and prior to occupancy of the affected space.
- Contractors to notify Property Manager 48 h prior to commencement of any work, which may cause objectionable noise or odors.
- Protect stored on-site materials and installed absorptive materials from moisture damage.
- All applied adhesives must meet or exceed the limits of the South Coast Air Quality Management District Rule #1168. Also, sealants used as fillers must meet or exceed Bay Area Air Quality Management District Reg. 8, Rule 51.
- Ensure that all paints and coatings meet or exceed the VOC and chemical component limits of GreenSeal requirements.
- Ensure that carpet systems meet or exceed the Carpet and Rug Institute Green Label Indoor Air Quality Test Program.
- Composite wood and agrifiber products should not contain any added urea-formaldehyde resins.
- Contractors should provide protection and barricades where needed to ensure personnel safety and should comply with OSHA at a minimum.

In conclusion, it should be remembered that the air quality in a building is one of the most important factors in maintaining employee productivity and health. Toward this end, IEQ monitoring will help minimize tenant complaints of BRI and SBS, and a cohesive proactive IEQ monitoring program is a powerful tool that can be used to achieve this goal.

It is also likely that in the coming years we will witness national and state regulations stipulating that designers, facility managers, and property owners meet specified indoor air quality standards. Several national and international organizations including the EPA, OSHA, ASHRAE, ASTM, GREENGUARD, USGBC, and others are currently in discussions concerning formulating new standards and updating and improving existing national indoor air quality standards. Thus, the EPA has set National Ambient Air Quality Standards for six principal pollutants, which are called "criteria" pollutants. Units of measure for the standards are parts per million (ppm) by volume, parts per billion (ppb) by volume, and micrograms per cubic meter of air ($\mu g/m^3$). Likewise, there are the GREENGUARD standards, which are among the most stringent indoor air quality standards currently available. They are based on available standards and guidelines from national and international public health agencies. However, not all GREENGUARD standards are ANSI standards, although all GREENGUARD standards are documented, publicly available (along with test protocols), and have been reviewed by public health/toxicology advisory groups.

There are also the ANSI/ASHRAE Standards 62.1 and 62.2, which are the recognized standards for ventilation system design and acceptable IAQ. In 2013, there was a revision of ASHRAE's standard for indoor air quality, which basically adds a number of important changes to remove inconsistencies and improve clarity. According to ASHRAE, these include

- modifying the zone air distribution effectiveness for underfloor air distribution systems that meet certain conditions;
- an increase in the required MERV rating for filters upstream of wetted cooling coils;
- the addition of a performance path for determining exhaust requirements;
- the addition of refrigerated warehouses to Table 6.1; and
- modification of the requirements for water used in humidification systems.

USGBC indoor air quality standards can be found in other sections of this handbook and also from the USGBC website (http://www.usgbc.org).

CHAPTER 8

Water Efficiency and Sanitary Waste

8.1 GENERAL ISSUES

Even though water is our most precious resource, it is often taken for granted. The *LA Times* on November 11, 2014, said, "Americans recently passed a milestone when federal officials reported that water use across the nation had reached its lowest level in more than 45 years: good news for the environment, great news in times of drought and a major victory for conservation." The current onslaught of sustainability in the building industry has helped make taking care of natural resources become part of our everyday culture. One of the more prominent issues facing us today is water conservation and using water more efficiently. Recent estimates place the amount of freshwater, that is, water needed for drinking, industry, and sanitation, at about 2.5% of the world's total. Roughly one-third of this is readily accessible to humans via lakes, streams, and rivers. Demand for freshwater continues to rise, and if current trends continue, experts project that demand for freshwater will double within the next three decades. Since 1950, the US population has increased by more than 90%. In that same time span, public demand for water increased by 209%. In 2010, Americans used an average of 89 gallons of water per person each day compared to 100 gpd in 2005. This increased demand has put tremendous stress on water supplies and distribution systems, threatening both human health and environment.

South Nevada Water Authority even has a Water Efficient Technologies program that offers financial incentives for capital expenditures when businesses retrofit existing equipment with additional water-efficient technologies. The Environmental Protection Agency (EPA) also has launched WaterSense, a water-oriented counterpart to the ENERGY STAR Program that promotes water efficiency and aims to boost the market for water-efficient products, programs, and practices.

There are many areas in the United States that already face water shortages, and the number of areas facing this issue has been growing. In drought-stricken areas, such as California and other states across the West, consumers have become accustomed to frequent warnings about the need

Copyright © 2016 Elsevier Inc.
All rights reserved.

to save water. In 2010, Californians consumed an estimated 38 billion gallons of water daily, compared to 46 billion a day in 2005. The reduced consumption of surface water accounted for the savings, but groundwater withdrawals actually increased because of the drought, which forced farmers to increase their reliance on irrigation. Overall in the United States, about 355 billion gallons of groundwater and surface water were used per day in 2010, compared with 410 billion a day in 2005.

Twelve states accounted for more than 50% of total water withdrawals. The largest was California, which accounted for 11% of the total withdrawals, followed by Texas, Florida, Idaho, Illinois, Arkansas, North Carolina, Colorado, Michigan, New York, Alabama, and Ohio. Moreover, the US Government Accountability Office reports that 40 of 50 state water managers expect water shortages under in some portion of their state over the next decade.

Unfortunately, local codes have not always kept pace with some of the new emerging technologies, which are not code compliant but are available in the marketplace. These include graywater systems, rainwater collection systems, high-efficiency irrigation systems, recirculating shower systems, regulations controlling hot water delivery, recirculation of hot water, insulation of hot water piping, demand-type tankless water heaters, water softeners, and drinking water treatment systems, all of which are being implemented through EPA WaterSense.

According to the EPA, toilets account for approximately 30% of the water used in residences, and Americans annually waste 900 billion gallons by the use of old, inefficient toilets. Recent advancements have allowed toilets to use up to 20% less water than the current federal standard, while still providing equal or superior performance. The WaterSense label is used on toilets that are certified by independent laboratory testing to meet rigorous criteria for both performance and efficiency. By replacing an older toilet with a WaterSense-labeled model, a family of four could reduce total indoor water use significantly and, depending on local water and sewer costs, save more than $90 annually. It is estimated that just by replacing one old toilet in every home with a WaterSense labeled High Efficiency Toilet, the total water savings could supply nearly 10 million US households with water for a year.

Moreover, water conservation translates into energy conservation and savings. By just one in every 10 homes in the United States installing WaterSense-labeled faucets or aerators in their bathrooms, in aggregate, this could result in a saving of about 6 billion gallons of water and more than

$50 million in the energy costs to supply, heat, and treat that water. The EPA also estimates that if the average home were retrofitted with water-efficient fixtures, there would be a savings of 30,000 gallons of water per year. If only 1 out of every 10 homes in the United States upgraded to water-efficient fixtures (including ENERGY STAR®-labeled clothes washers), the resultant savings could reach more than 300 billion gallons and nearly $2 billion annually. This could have a significant positive economic impact on small plumbing contractors and small businesses throughout various sectors.

Because of the increased demand and focus on water efficiency, the emerging water and energy conservation market has great potential to revitalize not only the plumbing industry but also traditional construction and small businesses across the country at a time when most small businessowners are suffering because of tough economic times and a lethargic construction industry.

8.2 WASTEWATER STRATEGY – WATER REUSE/RECYCLING

The US Department of Energy (DOE) estimates that commercial buildings consume 88% of the potable water in the United States. This offers facility managers a unique opportunity to make a huge impact on overall US water consumption. Benchmarking a facility's water use and implementing measures to improve overall efficiency will go a long way to achieving this goal.

Over the years, there has been an increased emphasis by LEED™ on water efficiency, and water-efficient design should be one of the main goals of any project, particularly because our nation's growing population is placing considerable stress on available water supplies. Although the US population has nearly doubled in the last five or six decades, public demand for water has more than tripled! This increased demand is putting additional stress on water supplies and distribution systems, and depleting reservoirs and groundwater can put our water supplies, human health, and our environment at serious risk. According to the EPA, lower water levels can contribute to higher concentrations of natural or human pollutants. Using water more efficiently helps maintain supplies at safe levels, protecting human health and the environment.

US Environmental Protection Agency estimates that an American family of four uses about 400 gallons of water per day. Roughly 30% of this is used outdoors for various purposes, including landscaping, cleaning sidewalks and driveways, washing cars, and maintaining swimming pools.

Nationally, landscape irrigation counts for almost one-third of all residential water use. That amounts to more than 7 billion gpd. Overall in the United States, about 355 billion gallons of groundwater and surface water were used per day in 2010, compared with 410 billion a day in 2005. Water efficiency is one of the principal categories of the LEED® Rating System and the number of WE credits available depend on the type of certification sought, for example, New Construction, Commercial Interiors, Schools, etc.

The main WE topics to know for LEED™ certification are:
- Water-efficient landscaping
  - Reduce by 50%
  - No potable use or no irrigation
- Innovative wastewater technologies
- Water use reduction
  - 20%
  - 30%

8.2.1 Water-Efficient Landscaping

Water-efficient landscapes using native and other climate-appropriate plant materials can reduce irrigation water use to better withstand drought, reduce drought loss or damage, and require less time and money to maintain. Water-efficient landscaping also includes maintenance techniques that create a landscape that requires less water. The purpose of water-efficient landscaping therefore is to reduce (by at least 50%) or eliminate the amount of potable water consumption and natural surfaces used for landscape irrigation. However, when landscape design strategies alone are unable of reaching a project's irrigation efficiency goals, attempts should be made to meeting efficiency demands through optimization of the irrigation system design. For example, use of high-efficiency drip, micro-, and subsurface systems can reduce the amount of water required to irrigate a given landscape. The US Green Building Council (USGBC) reports that drip systems alone can reduce water use by 30–50%. Climate-based controls, such as moisture sensors with rain shut-offs and weather-based evapotranspiration controllers, can further reduce demands by allowing naturally occurring rainfall to meet a portion of irrigation needs. For LEED certification, one point is awarded for a 50% reduction in water consumption for irrigation from a calculated midsummer baseline case, and a total of two points for a 100% water reduction (Figure 8.1). LEED stipulates that to achieve WE Credit 1.1, use only captured rainwater, recycled wastewater, recycled graywater, or water treated and conveyed by

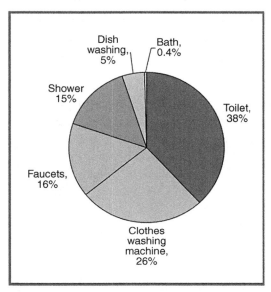

Figure 8.1 *Pie Chart Showing Proportion of Total Indoor Savings Available From Combining Efficiency and Curtailment Actions. Source: Benjamin Inskeep and Shahzeen Attari.*

a public agency specifically for nonpotable uses for irrigation. Alternatively, "Install landscaping that does not require permanent irrigation systems. Temporary irrigation systems used for plant establishment are allowed only if removed within one year of installation."

To assist in greening the supply, it is necessary to tap alternate water sources. LEED recognizes two alternate water sources: rainwater collection and wastewater recovery. Rainwater collection involves collecting and holding on-site rainfall in cisterns, underground tanks, or ponds during rainfall. This water can then be used during the dry periods by the irrigation system. Wastewater recovery can be achieved either on site or at the municipal scale. On-site systems capture graywater (which does not contain human or food processing waste) from the building and apply it to irrigation.

Reductions shall be attributed to any combination of the following approaches:

- Use a high-efficiency microirrigation system, such as drip, micromisters, and subsurface irrigation systems.
- Replace potable (drinking) water with captured rainwater, recycled wastewater (graywater), or treated water.
- Use of water treated and conveyed by a public agency that is specifically used for nonpotable purposes.

- Plant species factor – install landscaping that does not require permanent irrigation systems.
- Apply Xeriscape principles to all new development whenever possible.

When a landscaping design incorporates rainwater collection or wastewater recovery in particular, it is essential to assemble a team of experts and establish project roles at an early stage in the process. Rainwater collection and wastewater treatment systems stretch over multiple project disciplines, making it particularly important to clearly articulate responsibilities. Having an experienced landscape architect on board is pivotal for a water-efficient landscape and irrigation system design. To take advantage of the available LEED™ points for water-efficient landscaping credits, early planning is strongly advised.

Many of the LEED credits deal with graywater and blackwater. Graywater is typically considered to be untreated wastewater that has not come into contact with toilet waste, such as shower water, water from sinks (other than the kitchen), bathtubs, wash basins, and clothes washers. Graywater use includes indoor and outdoor reuse. When used outdoors, the graywater is usually filtered and then used in irrigation and for watering landscape. Indoor graywater use, on the other hand, consists of recycled water (nonpotable water) and is used mainly for flushing toilets. Graywater has other applications including construction activities, concrete mixing, and cooling water for power plants. In addition to graywater, nonpotable water can come from rainwater and municipally supplied reclaimed water. Blackwater lacks a specific definition that is accepted nationwide, but is generally considered to constitute toilet, urinal, dishwashers, and kitchen sink water (in most jurisdictions). However, depending on the jurisdiction, implementing graywater systems that reuse wastewater from showers and sinks for purposes like flushing of toilets or irrigation may encounter code compliance restrictions.

Reclaimed versus graywater systems: Recycling water and putting it back to use is commonly thought of in two different water usage strategies, reclaimed water and graywater, and it is important to distinguish between these systems (although some incorrectly use the terms reclaimed water and graywater interchangeably).

Simply put, reclaimed water is wastewater effluent/sewage that has been treated according to high standards at municipal treatment facilities and that meets the reclaimed water effluent criteria. Its treatment takes place offsite and is delivered to a facility. Reclaimed water is most commonly used for nonpotable purposes, such as landscaping, agriculture, dust control, soil compaction, and processes such as concrete production and cooling water for power plants.

The use of reclaimed water is increasing in popularity, especially in states like California, where openness to innovative, environmentally friendly concepts prevails, especially in the face of a very real and critical water crisis. Orange County, California, for example, started delivering purified wastewater last year, providing one of the first "toilet-to-tap" systems in the United States.

Graywater, on the other hand, is the product of domestic water use such as showers, washing machines, and sinks. It does not normally include wastewater from kitchen sinks, photo lab sinks, dishwashers, or laundry water from soiled diapers. These sources are typically considered to be blackwater producers because they contain serious contaminants and therefore cannot be reused. Graywater use is a point-of-source strategy, that is, graywater collected from a building will be reused in the same building.

8.2.2 Innovative Wastewater Technologies

The intent of the Innovative Wastewater Technologies credit is to reduce wastewater generation and potable water demand and increase the recharge of local aquifers. You can reduce potable water demand by using water-conserving fixtures, reusing nondrinking water for flushing, or reuse water treated on-site to tertiary standards (with the treated water infiltrated or used on site). Tertiary treatment is the final stage of treatment before water can be discharged back into the environment. If tertiary treatment is used, the water must be treated by biological systems, constructed wetlands, or a high-efficiency filtration system.

A Water Efficient Technologies program is now in place that offers financial incentives to commercial and multifamily property owners who install water-efficient devices and implement new, water-saving technologies. Examples of effective Water Efficient Technologies approaches include:

- Ultra high-efficiency toilets and efficient retrofits
- Use efficient showerheads and efficient retrofits
- Waterless and high-efficiency urinals
- Other ultralow water consumption products
- Converting a sports field from grass to an artificial surface
- Retrofitting standard cooling towers with qualifying, high-efficiency drift elimination technologies

Strategies for meeting one of Water Efficiency compliance requirements, reducing potable water use for sewage conveyance, falls into two categories that can be implemented either independently or in concert. As shown earlier, by simply meeting demands efficiently, the use of ultra high-efficiency plumbing fixtures can reduce the water required for sewage conveyance in excess of the 50% requirement. To use a typical example, composting toilets (not normally used in commercial facilities) and waterless urinals use no water. These two technologies alone can eliminate a facility's use of potable water for sewage conveyance, qualifying both for this credit's point, plus potentially a LEED™ Innovation in Design point for exemplary performance. Should the selected plumbing fixtures alone prove to be inadequate to reach the 50% reduction threshold, or if ultra high-efficiency plumbing fixtures are not selected, the water necessary for toilet and urinal flushing can be reduced by a minimum of 50%, or eliminated entirely, by applying rainwater collection or wastewater treatment strategies.

The Southface Eco Office in Atlanta provides an example of how this credit can be achieved (Figure 8.2). The facility, targeting LEED™

Figure 8.2 The new Eco Office in Atlanta, Georgia is a 10,000-sq. ft. facility seeking LEED™ Platinum rating, designed as a model for environmentally responsible commercial construction that is achievable utilizing existing off-the-shelf materials and technology. The Office provides a showcase of state-of-the-art energy and water conservation and waste-reducing features.

Platinum certification, completely eliminates the use of potable water for sewage conveyance using a variety of complementary strategies. Foam flush composting toilets and waterless urinals are used in the staff restrooms; composting toilets require only six ounces of water per use, which dramatically reduces the volume of water required for sewage conveyance. Water requirements in the public restrooms are reduced through the employment of a combination of dual-flush toilets, ultrahigh-efficiency toilets, and waterless urinals. The remaining reduced volume of water required for sewage conveyance is supplied by rainwater collected from a roof-mounted solar array and stored in a rooftop cistern, in addition to a supplemental in-ground storage tank.

A knowledgeable team and early involvement of local code officials are critical components for the successful design and implementation of nonpotable water supply systems. Furthermore, dual-plumbing lines for nonpotable water supply within the building are fairly easy to plan for during the design phase, but much more difficult to retrofit after construction is complete.

8.2.3 Water-Use Reduction

The intent of the Water Use Reduction credit according to LEED is to "maximize water efficiency within buildings to reduce the burden on municipal water supply and wastewater systems." One point is awarded for reducing water use by 20%, and this is increased to two points for reducing water use by 30%. The fixtures governed by this credit include water closets, urinals, lavatory faucets, showers, and kitchen sinks. For water-using fixtures and equipment, such as dishwashers, clothes washers, and mechanical equipment (nonregulated uses), that are not addressed by this credit may qualify for the LEED Innovation in Design point.

Employing proven, cost-effective technologies can facilitate achieving the 30% reduction necessary to earn both points for this credit. The use of low-flow lavatory faucets with automatic controls (0.5 gpm, 12 s per use) is normally sufficient to achieve a 20% reduction in water use, qualifying for 1 point. An additional 14% reduction can be achieved by the use of waterless urinals, which, when combined with low-flow faucets, should exceed the 30% reduction threshold, thereby earning both points.

Here too, the first step in the optimization process, reducing demands, does not apply. It is not possible to design away occupants' needs to use the restroom, wash their hands, or take a shower. Strategies for water-use reduction therefore fall into the same two categories identified for Innovative Wastewater Technologies – either meeting demands efficiently or fulfilling the demand in alternate, more environmentally appropriate means. The two credits complement one another, and water savings related to the Innovative Waste Water Technology credit will also contribute to the Water Use Reduction credit.

John Starr, AIA, and Jim Nicolow, AIA, of Lord, Aeck & Sargent, architects, state that among the LEED™ Water Efficiency credits,

Water Use Reduction can often be achieved without the early planning and design integration required by the other two credits. Most alternative plumbing fixtures use conventional plumbing supply and waste lines, allowing these fixtures to be substituted for less-efficient standard fixtures at any point in the design process, and even well into the construction process.

Recycling is a term usually reserved for waste like aluminum cans, glass bottles, and newspapers. Water can also be recycled and, indeed, through the natural water cycle, the earth has recycled and reused water for millions of years. Water recycling, though, generally refers to using technology to speed up these natural processes.

System approaches: As local municipalities and individual facilities continue to struggle to meet water needs in the face of dwindling water supplies, a variety of reclaimed water and graywater system approaches have emerged.

Reclaimed and graywater systems range in their size and complexity. Toward the high end are the multibuilding installations that draw wastewater from municipal sources, followed by the middle tier, which includes buildings that have installed storage tanks capable of collecting thousands of gallons of water from rainwater, sinks, and steam condensate, which is then treated and funneled to water reuse sources. There is also the more affordable undercounter systems that are simpler and on a smaller, yet significant, scale that carry out on-the-spot treatment of water that flows down sink drains, which is then pumped directly into toilet tanks.

The more complex systems should be built into new construction rather than to retrofit them later, whereas on-the-spot collection systems can be implemented at any time. It is important when specifying sustainable systems and technologies to remain within budget as a matter of setting goals and performing research up front to determine the additional value and payoff.

Graywater demonstrations: The amount of graywater produced in a particular building depends largely on the facility type. A typical office building, for example, may not yield as much graywater as a college dorm or multiuse retail and condominium building; the benefits are all about economies of scale and deriving value from the system, no matter how large or how small they may be.

Consider the amount of potable water that can be saved in a typical four-person household. On average, each person uses 80–100 gallons of water per day, with toilet flushing being the largest contributor to this use. The combined use of kitchen and bathroom sinks is only 15% of the water that comes into a home, which is significant considering that 100% of the water that comes into the home has been treated and made potable for drinking.

With the largest single source of fresh water in the home capable of using graywater instead of potable water, the household is able to make real gains on reusing water that is perfectly suitable for toilet flushing. For household and small commercial facilities, the most appropriate solution may be to use a graywater system that incorporates a reservoir, which is installed under the sink and attached to the toilet. These graywater systems are designed so that the toilet draws first from the collected water in the reservoir. However, the system remains connected to the fresh water pipes so that, should flushing deplete the amount of water stored in the reservoir, the toilet can then secondarily draw from outside water. Because toilets are the largest consumers of water in households, this type of system is able to save up to 5000 gpy.

Differing graywater policies and regulations between states have considerable impact on the extent to which facilities and homeowners can deploy graywater systems. The state of Arizona, for example, has graywater guidelines to educate residents on methods to build simple, efficient, and safe graywater irrigation systems. For those who follow these guidelines, their system falls under a general permit and automatically becomes "legal," which means that the residents do not have to apply or pay for any permits or inspections. Likewise, California has a graywater policy, but one that is very restrictive, which usually makes it difficult and unaffordable to install a permitted system. Many states have no graywater policy and do not issue permits at all, whereas other states issue experimental permits for systems on a case-by-case basis.

The recycling of water by whichever means provides substantial benefits, including reduction of stress on potable water resources, reduction of nutrient loading to waterways, reducing strain on failing septic tanks or treatment plants, using less energy and chemicals, and costing less than potable water. All of these benefits result in significant savings in both water and energy.

Long-term savings: It is scarcely believable that as little as 5–10 years ago, purchasing environmentally friendly building components that met LEED™ compliance standards could have added more than 10% to total building costs, whereas today, plumbers, engineers and other specifiers are now discovering that they can adopt higher sustainability standards without necessarily spending extra. And if they do have to spend extra, the long-term payoff more than compensates, when you factor in long-term operating costs, including water and wastewater utility bills, plus the energy it takes to heat water for faucets, showerheads, etc.

According to Flex Your Power, California's energy-efficiency marketing and outreach campaign, utilities account for about 30% of an office building's expenses. A 30% reduction in energy consumption can lower operating costs by $25,000 a year for every 50,000 ft.^{2} of office space.

The public is showing greater awareness and is taking greater notice of how companies and facilities expend water and energy, and both users and communities are holding building owners accountable for their use of precious local resources. Engineers need to stay abreast and monitor water and energy efficiency options in the restrooms and elsewhere in their facilities to minimize operating costs and help ensure that buildings meet LEED™ standards.

8.2.4 Construction Waste Management

Commercial construction typically generates between 2 pounds and 2.5 pounds of solid waste per square foot – the majority of which is recyclable. Salvaging and recycling construction and demolition waste can substantially reduce demand for virgin resources and the associated environmental impacts. Additionally, effective construction waste management, including appropriate handling of nonrecyclables, can reduce contamination from and extend the life of existing landfills. Whenever feasible, therefore, reducing initial waste generation is environmentally preferable to reuse or recycling.

The Construction Waste Management Plan should recognize project waste as an integral part of overall materials management. The premise being that waste management is a part of materials management, and the recognition that one project's wastes are materials available for another project, facilitates efficient and effective waste management. Moreover, waste management requirements should be included as a topic of discussion during both the preconstruction phase and at ongoing regular job meetings, to ensure that contractors and appropriate subcontractors are fully aware of the implications of these requirements on their work prior to and throughout construction.

Waste management should be coordinated with or part of a standard quality assurance program, and waste management requirements should

LEED v4 - BD+C		
Water Efficiency	Possible points:	11
Prereq 1	Outdoor water use reduction	Required
Prereq 2	Indoor water use reduction	Required
Prereq 3	Building level water metering	Required
Credit 1	Outdoor water use reduction	2
Credit 2	Indoor water use reduction	6
Credit 3	Cooling tower water use	2
Credit 4	Water metering	1

Figure 8.3 *Water Efficiency Credits – LEED v4 BD+C WE Category. Source: USGBC.*

be addressed regularly throughout the project. All topical applications of processed clean wood waste and ground gypsum board as a soil amendment must be implemented in accordance with local and state regulations. When possible, adherence to the plan would be facilitated by tying completion of recycling documentation to one of the payments for each trade contractor.

8.3 WATER FIXTURES AND REDUCING WATER STRATEGY

The LEED rating system can take credit for introducing highly efficient toilets, urinals, and entire irrigation systems to thousands of buildings – in the United States and globally. LEED v4, which is the latest LEED version, expands those water-savings targets to appliances, cooling towers, commercial kitchen equipment, and other areas that consume a vast proportion of water in many buildings (Figure 8.3).

The following is a list along with a summary description of the credits within the LEED® v4 BD+C WE category:

- WE Prerequisite: Outdoor Water Use Reduction (Mandatory): The intent of this prerequisite is to reduce outdoor water consumption using a variety of methods such as permeable pavement or landscaping that does not require irrigation.
- WE Prerequisite: Indoor Water Use Reduction (Mandatory): The intent of this prerequisite is to reduce indoor water use consumption, mainly through using plumbing fixtures with reduced flow rates.
- WE Prerequisite: Building Level Water Metering (Mandatory): The intent here is to monitor water consumption in order to track usage and identify water savings areas.
- WE Credit: Outdoor Water Use Reduction (up to 2 points – 18% of category total): The intent of this credit is to further reduce outdoor water usage beyond the prerequisite levels.

- WE Credit: Indoor Water Use Reduction (up to 6 points – 54% of category total): The intent of this credit is to further reduce indoor water usage beyond the prerequisite levels.
- WE Credit: Cooling Tower Water Use (up to 2 points – 18% of category total): The intent of this credit is to reduce the amount of water used for cooling tower makeup water.
- WE Credit: Water Metering (1 point – 10% of category total): The intent here is to monitor water consumption in order to facilitate tracking usage and identifying water savings areas. This goes beyond the prerequisite requirements to monitor water subsystems and the overall building system.

Today, there are thousands of plumbing fixtures and fittings in the mainstream market that can help save water, energy, and money. These include but are not limited to aerators, metering and electronic faucets, and prerinse spray valves. But when selecting energy-efficient equipment, it is vital to select quality products that meet conservation requirements without compromising performance. The product should deliver the consistent flow required, while maintaining the water and energy savings the industry demands. And with restroom fixtures accounting for most of a typical commercial building's water consumption, the best opportunities for increasing efficiency can be found there. The good news is that there is increased public awareness combined with higher-efficiency plumbing fixtures becoming more widely available.

Perhaps the best way to increase water efficiency in buildings is through plumbing-fixture replacement and implementation of new technologies. Replacing older, high-flow water closets and flush valves with models that meet current UPC and IPC requirements is important. While current codes require the lower flow rate for new fixtures, existing buildings often have older, high-flow flush valves. Despite the tremendous water savings available by updating the fixtures, facility managers often avoid the upgrade because of concerns about clogging. Solid waste removal must be 350 g or greater. Fixtures pass or fail based on whether the fixture can completely clear all test media in a single flush in at least four of five attempts. Toilets that pass qualify for the EPA WaterSense label. It should be noted that when the Energy Policy Act of 1992 was first enacted, many facility managers at the time experienced problems, with the low-flow fixtures clogging due to fixture-design issues. Those problems have long since been addressed and no longer present a problem (Table 8.1).

Selecting water-efficient fixtures will not only reduce sewer and water bills but efficient water use reduces the need for expensive water supply

Table 8.1 Illustrating a comparison of plumbing fixture water-flow rates

Plumbing fixture	Before 1992	EPA 1992	Current plumbing codes
Toilets	4–7 gpf	1.6 gpf	1.6 gpf
Urinals	3.5–5 gpf	1.0 gpf	1.0 gpf
Faucets*	5–7 gpm	2.5 gpm	0.5 gpm
Showerheads*	4.5–8 gpm	2.5 gpm	2.5 gpm

* At 80 psi flowing water pressure.
Source: Domestic Water Conservation Technologies, Federal Energy Management Program, US Department of Energy, Office of Energy Efficiency and Renewable Energy, National Renewable Energy Laboratory, October 2002.

and wastewater treatment facilities, and helps maintain healthy aquatic and riparian environments. Moreover, it reduces the energy needed to pump, treat, and heat water. Water is employed in a product's manufacture, during a product's use, and in cleaning, which means that water efficiency and pollution prevention can occur during several product life cycle stages.

Mark Sanders, product manager for Sloan Valve Company's AQUS Graywater System, says graywater and reclaimed water strategies make good use of water resources, especially when implemented in conjunction with efficient plumbing systems.

The maximum volume of water that may be discharged by the toilet, when field adjustment of the tank trim is set at its maximum water use setting, shall not exceed the following amounts:

• For single flush fixtures: 1.68 gpf
• For dual flush fixtures: 1.40 gpf in reduced flush mode and 2.00 gpf in full flush mode

The maximum volume of water discharged, using both original equipment tank trim and using after-market closure seals, shall be tested according to the protocol detailed on the WaterSense Web site. There are basically two approaches to measuring water volume: gallons per flush for toilets and urinals or gallons per minute (gpm) for flow-type fixtures such as lavatories, sinks, and showers. Metered faucets with controlled flow rates for preset time periods are measured in gallons per cubic yard.

For LEED™ purposes, baseline calculations should be computed by determining the number and gender of the users. As a default, LEED lets you assume that females use toilets three times per day, males once per day in addition to using the urinal two times per day. Both males and females will use the bathroom faucets three times each day and the kitchen sink once for 15 s each. The following section will discuss the various types of water-efficient fixtures on the market:

8.3.1 Toilet Types

By using water more efficiently, we can help preserve water supplies for future generations, save money and protect the environment.

High-efficiency toilets (HETs): The National Energy Policy Act was signed into law in 1994, requiring that toilets sold in the United States use no more than 1.6 gallons (6 L) per flush. This mandate to conserve has "nudged" manufacturers to produce a new generation of HETs that use technologies like pressure-assist, gravity flush, and dual flush to remove away waste using as little water as possible. Of these new technologies, the dual flush method has the advantage of intuitive flushing, where the operator can decide electively that less water is required and so uses 1 gallon (3 L) or less per flush instead of the 1.6-gallon maximum.

Two types of toilet fixtures dominate today's marketplace: (1) ultra-low flush toilets (ULFTs), aka "low flow" or "ultra low flow," and (2) HETs. ULFTs are defined by a flush volume between 1.28 gpf and 1.6 gpf. The HET is defined as a fixture that flushes at 20% below the 1.6 gpf maximum or less, equating to a maximum of 1.28 gpf. Dual flush fixtures are included in the HET category (Figure 8.4).

This 20% reduction threshold serves as a metric for water authorities and municipalities designing more aggressive toilet replacement programs and, in some cases, establishing an additional performance tier for their financial incentives such as rebate and voucher programs. It is also a part of the water-efficiency element of many green building programs that exist throughout the United States. Unfortunately, this standard currently applies only to tank-type toilets. Flushometer valve toilets have not been studied in the same way as tank types, and testing for flushometer valves needs to be performed on the flushometer valve with the various bowls on the market so that the pair can then be rated.

Although toilets purchased for new construction and retrofits are required to meet the new standards, millions of older inefficient toilets remain in use. As water and sewer costs keep rising, low-flow toilets are becoming increasingly attractive to the American consumer, while local and state governments use rebates and tax incentives to encourage households to convert to these new technologies.

The advantages of low-flow toilets in conserving water and thus reducing the demand on local water treatment facilities are obvious. According to the EPA, the elimination of inefficient old-style toilets would save the nation about 2 billion gallons of water a day. Having a growing population, an antiquated water treatment infrastructure, and the potential threat of

Figure 8.4 *Certified by WaterSense, the Brianne High-Efficiency Toilet From Gerber Plumbing Fixtures Uses Just 1.28 gpf.* The toilet also has a compact configuration, a FluidMaster 400A fill valve, a 3-in. flush valve for quick bowl cleaning, dual-fed siphon jets, and a 2-in. glazed trapway.

global warming contributing to uncertain weather, water conservation will continue to be a major concern to the public.

Dual-flush toilets: A dual–flush toilet (Figure 8.5) can help make bathrooms more environmentally friendly. They handle solid and liquid waste differently from standard American–style toilets, giving the user a choice of flushes. It contains an interactive toilet design that helps conserve water, which has become popular especially in countries where water is in short supply and in areas where water supply and treatment facilities are older or overtaxed. The EPA estimates that by the year 2013, a total of 36 states will possibly experience water shortages as a result of increased water usage and inefficient water management from aging regional infrastructures. Using less water to flush liquid waste while logical may face cultural biases in the United States that make accepting such an innovative approach to personal waste removal harder to accept.

How dual-flush toilets work: The simple biological process of handling bodily waste is apparently a delicate topic, indeed so much so that culture

Figure 8.5 *American Standard Dual-Flush High-Efficiency Toilet.* It meets EPA WaterSense® criteria and offers a full 1.6-gallon flush or a partial 1.0-gallon flush using 37.5% less water. Either way, the PowerWash® rim cleans the bowl well. An elongated bowl provides greater comfort, and its EverClean® surface stays cleaner, longer.

can be as much a factor in effecting change as necessity. For this reason, bodily functions are kept under wraps, and any changes in our approach to handling them may sometimes create culture shock and resistance. Interest in low-flow and dual-flush toilets is on the rise in the United States, partly due to increased government regulation and the rising cost of water, and the introduction of incentives in many States for making changes in the way we use the commode.

The way water is used to remove waste from the bowl impacts the amount of water needed to get the job done. Standard toilets use siphoning action, which basically employs a siphoning tube to discharge waste. A high volume of water enters the toilet bowl when the toilet is flushed fills the siphon tube and pulls the waste and water down the drain. Upon air entering the tube, the siphoning action stops. Dual-flush toilets employ a larger trapway (a hole at the bottom of the bowl) and a wash-down flushing design that pushes waste down the drain. Because no siphoning action is involved, the system requires less water per flush, and the larger trapway diameter facilitates the exit of waste from the bowl. Combined with the savings from using only half-flushes for liquid waste, a dual-flush toilet can save up to 68% more water than a conventional low-flow toilet.

Use of a larger-diameter trapway is the main reason a dual-flush toilet does not clog as often as a conventional toilet while requiring less water to flush efficiently and able to save more water than a low-flow toilet when flushing liquid waste. Among the downsides to buying dual-flush units is that they are a little more expensive than comparable low-flow toilet designs. In addition, there is the problem of aesthetics. Dual-flush toilets typically retain only a small amount of water in the bowl, and flushing doesn't always remove all the waste. Even in full flush mode, some occasional streaking will occur. With a dual-flush toilet, you'll probably need to use the toilet brush more often to keep the bowl clean.

Composting toilets: The first composting toilets were, in fact, outhouses. But composting toilets have come a long way in the last two decades and today are growing in popularity among off-grid home owners and seeking to reduce their burden on the environment by eliminating what they believe to be the unnecessary step of waste treatment plants and sewerage (Figure 8.6). Furthermore, public health professionals have started to appreciate composting toilets as an environmentally sound method for dealing with human wastes. By their very nature, they require little or no water to function effectively and are therefore particularly suitable (although not exclusively) for use in locations where mains water and sewerage connections are unavailable, or in locations where water consumption needs to be minimized to the greatest extent possible.

Advantages of composting toilet systems: One estimate shows that the average American uses 74 gallons (280 L) of water per day, one-third of which splashes down a flushing toilet. Although an older toilet may swallow up to 7 gallons (26.5 L) per flush, and federal law stipulates only 1.6 gallon (6.1 L) for low-flow models in new homes, a composting toilet can save more than 6,600 gallons (24,984 L) of water per person a year.

Not using water to flush a toilet also cuts out all the energy expended down the line, from the septic system to the treatment plant. That could

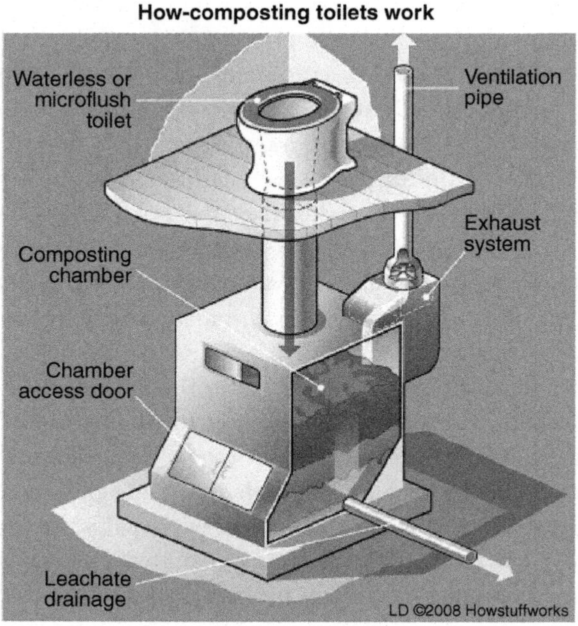

How-composting toilets work

Waterless or microflush toilet

Ventilation pipe

Composting chamber

Exhaust system

Chamber access door

Leachate drainage

LD ©2008 Howstuffworks

Figure 8.6 *Drawing Showing How a Composting Toilet Works.* Composting toilets use the natural processes of decomposition and evaporation to recycle human waste. The waste that enters the toilets is over 90% water, which is evaporated and carried back to the atmosphere through the vent system. The small amount of remaining solid material is converted to useful fertilizing soil by natural decomposition. *Source: HowStuffWorks, Inc.*

be beneficial to our waterways. An example of this is the Chesapeake Bay, which receives approximately 1.5 billion gallons (5.7 billion L) of waste-water flow every day from some 500 sewage treatment plants. This has caused much of the water to become a dead zone, incapable of keeping animals and plants alive.

For a self-contained composting toilet to work properly, appropriate ventilation is required that can both keep the smell out of your bathroom while providing enough oxygen for the compost to break down. Some toilets achieve this by employing fans and a heater powered by electricity (some models do not require electricity). The composter also has to be kept at a minimum temperature of 65°F (18.3°C), so for those living off the grid, a heater could potentially require more electricity than used in the rest of the house. The heater doesn't have to run all the time, however, and one model may only operate at a maximum level of 540 W for about 6 h a day. But because the self-contained models are relatively small, the power use for fans is fairly minimal. It may range from about 80 W to 150 W, which is the same amount of power used by a light bulb. A possible alternative would be to use solar panels to power the fans and heater.

8.3.2 Urinals

High-efficiency urinals (HEUs): HEUs are urinals that use 0.5 gpf or less – at least one-half of the amount of water used to flush the average urinal. The California Urban Water Conservation Council (CUWCC), in cooperation with water authorities and local agencies, defined them as fixtures that have an average flush volume lower than the mandated 1.0 gpf and zero water consumption urinals. Based on data from studies of actual usage, these urinals save 20,000 gallons of water per year with an estimated 20-year life. HEUs are making a significant difference in water usage, water bills, and our environment.

In addition to saving water and sewer cost, zero-water urinals are a significant improvement over traditional urinals in both maintenance and hygiene. These fixtures use a special trap with lightweight biodegradable oil that lets urine and water pass through but prevents odor from escaping into the restroom (Figure 8.7). Also, there are no valves to fail and no flooding.

As their name implies, Waterfree Urinals do not use water and therefore provide additional savings of water as well as sewage and water supply line costs. A biodegradable sealant liquid is contained within the cartridge.

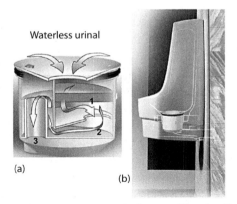

Figure 8.7 *Illustration of a Waterless Urinal Diagram Showing How it Functions.* The cartridge acts as a funnel directing flow through the liquid sealant, preventing any odors from escaping. The cartridge then collects uric sediment. The remaining liquid which is noncorrosive and free of hard water is allowed to pass freely down the drainage pipe. *Courtesy: Sloan Valve Company.*

A barrier is formed between the drain and the open air by the liquid sealant, thereby eliminating odors. Installation is easy whether in new or retrofit applications. The initial cost of a Waterfree Urinal is often less than conventional no-touch fixtures, lowering your initial investment. The urinal can be used to accumulate water-efficient LEED credits, including innovation points.

Zero-water and HEUs are part of the next generation of water-efficient plumbing products and contribute to USGBC LEED™ Credits for water use reduction.

Ultralow water urinals utilize only 1 pint (0.125 gallons) of water to flush. These systems combine the vitreous china fixture with either a manual or sensor-operated flush valve.

They provide effective, low-maintenance flushing in public restrooms while reducing water consumption by up to 88%.

8.3.3 Faucets

The EPAct includes specific requirements for faucet flow rates. For example, residential lavatory faucets must be regulated by an aerator to 2.2 gpm or less, kitchen faucets to 2.5 gpm or less. Commercial faucet requirements vary according to fixture type: handle-operated models are regulated by aerator to 0.5 gpm, whereas self-closing and sensor-operated models are limited to less than 0.25 gallons per cycle.

Again we find that the technology available greatly exceeds EPAct regulations. Although kitchen faucets may require about a 2.5 gpm flow rate in order to fill a pot in a timely fashion, studies have shown that residential lavatory faucets would be satisfactory for the user even when reduced to even a 0.5 gpm flow rate. Conservation-minded specifiers have started to recommend aerators that deliver that flow rate.

Electronic faucets: The electronic faucet is an easy way to save energy, and although it is more costly than a traditional faucet, it will pay for itself in water and energy savings in a short period of time. The electronic faucet has a sensor feature that prevents the faucet from being left on and from excess dripping. According to EnergyStar.gov, "hot water leaking at a rate of one drip per second from a single faucet can waste up to 1,661 gallons of water over the course of a year." An electronic faucet should come equipped with several standard features, including the choice of electric plugin (AC) and battery (DC) battery power options.

Metering faucets and aerators: The federal standard for faucets is currently 2.2 gpm, though WaterSense specification sets the maximum flow rate of faucets and aerators at 1.5 gpm (at 60 psi). Of note, WaterSense does not certify kitchen faucets. Metered faucets are less expensive than electronic faucets, yet can deliver similar energy-saving results. Metered faucets that are common in commercial washrooms are generally mechanically operated fixtures that deliver water (at no more than 0.25 gallons per cycle), and then self-close. The manual push feature prevents faucets from being left on after use and prevents unnecessary waste while scrubbing hands. The typical metering faucet's cycle time can be adjusted to deliver the desired amount of water per minute. Many of these devices are designed to allow the user to adjust the temperature before operation. However, in the majority of commercial washrooms we find that sensors are becoming the standard. But engineers appear to have reached the limit of water-efficiency for sensor models: 0.08 gallons per cycle. However, it is not user demand or engineering limitations that have determined this to be the limit; it is due to the fact that other environmental considerations come into play. For example, at less than 0.08 gallons per cycle, battery disposal (or power use) takes a greater toll on the environment than is represented by the water that might be saved.

Aerators are one of the most common faucet accessories; they add air into the water stream to increase the feeling of flow. Aerators are capable of controlling the flow to less than 1.5 gpm, and provide a simple and inexpensive low flow/energy solution. They come in a variety of models to provide the exact flow that complies with local plumbing codes.

8.3.4 Shower Heads

Flow-optimized showerheads: A 10-min shower can use between 25 gallons and 50 gallons of water because a typical high-flow showerhead uses between 6 gpm and 10 gpm. Flow-optimized showerheads have a flow of about 1.75 gpm and use 30% less water and can contribute toward maximizing LEED points. The flow-optimized single- and three-function showerheads outperform the standard 2.5 gpm flow rate without sacrificing performance.

This flow rate of 2.5 gpm is both the EPAct requirement and the LEED™ baseline. Attempts to reduce the flow rate still further are mostly met with very unhappy users. Some users even remove the flow restrictors from their fixtures, producing rates of 4–6 gpm, which is clearly not green by any standard. Likewise, flow rates below 2.5 gpm risk failure of certain types of thermostatic mixing valves, leading to scalding of the user. Before specifying valves and showerheads, it may be prudent to consult the manufacturer of the valve; the information may help alleviate this problem altogether. Of note, on December 22, 2010, the US DOE announced that Federal pre-emption of the states as to showerheads and other plumbing products have been waived. This denotes that states and local jurisdictions in the United States are now free to set their own showerhead performance requirements.

8.3.5 Baseline Water–Consumption Calculations

To achieve this LEED credit, one must first determine the baseline model for water usage in the building. The primary factors in determining this calculation are the types of fixtures in the building, the number of occupants, and the flow or flush rate for the specified fixtures. When evaluating a building's water-use efficiency, the USGBC offers a helpful method that allows one to benchmark annual water use and compare that use to current standards.

The first step is to establish water use based on past annual-use records or on estimates of building occupancy. This is followed by estimating a theoretical water-use baseline based on the types of fixtures in the building and the number of building occupants. To determine the number of occupants in the building, the number of Full-Time Equivalent (FTE) building occupants must be known (acquired from the LEED™ administrator). The FTE will typically be broken down 50/50 for men and women except in cases where the type of building is meant primarily for one gender, for example, a gym for women. In cases that do not adhere to a strict 50/50 split principle for male and female occupants, an explanation of the design case ration is recommended. This can be included in the narrative section of the LEED™ online template for this credit. The FTE should include the transient (visitors) building occupants the building is designed for, in addition to the primary occupants. In projects that will have both FTE and transient occupants, separate calculations are required for each type of occupancy.

In Table 8.2, we have an example used by the USGBC to illustrate the calculation process. It represents potable water calculation for sewage conveyance for a 2-storey office building with a capacity of 300 occupants. The calculations are based on a typical 8-h workday and a 50/50 male/female ratio. Male occupants are assumed to use water closets once and urinals twice in a typical work day (default), and females are assumed to use water closets three times (default). The reduced amount is the difference between the design case and the baseline case.

Table 8.2 Design case

Fixture type	Daily uses	Flow rate	Occupants (gpf)	Sewage generation (gal)
Low–flow water closet (male)	0	1.1	150	0
Low–flow water closet (female)	3	1.1	150	495
Composting toilet (male)	1	0.0	150	0
Composting toilet (female)	0	0.0	150	0
Waterless urinal (male)	2	0.0	150	0
Total daily volume (gal)				495
Annual workdays				260
Annual volume (gal)				128,700
Rainwater or graywater volume (gal)				36,000
Total annual volume (gal)				92,700

Source: USGBC.

Table 8.3 Baseline case

Fixture type	Daily uses	Flow rate	Occupants (gpf)	Sewage generation (gal)
Water closet (male)	1	1.6	150	240
Water closet (female)	3	1.6	150	720
Urinal (male)	2	1.0	150	300
Total daily volume (gal)				1,260
Annual workdays				260
Total annual volume (gal)				327,600

The USGBC requires that the baseline case must use the flow rates and flush volumes established by EPAct 1992.
Source: USGBC.

In Table 8.3, we show the baseline case being used in line with the Energy Policy Act of 1992 fixture flow rates. When undertaking these calculations, the number of occupants, number of workdays, and frequency data should remain the same. Furthermore, graywater or rainwater harvesting volumes should not be included. The baseline case in Table 8.3 estimates the amount of potable water per year used for sewage conveyance to be 327,600 gallons. This means that a reduction of 72% has been achieved in potable water volumes used for sewage conveyance. Using this strategy can earn one point in LEED's rating system.

Of note, the baseline calculation is based on the assumption that 100% of the building's indoor plumbing fixtures comply with the requirement of the 2006 Uniform Plumbing Code or the 2006 Intl. Plumbing Code fixture and fitting performance requirements. The *Uniform Plumbing Code* (UPC) is designated as an American National Standard and is a model code developed by the International Association of Plumbing and Mechanical Officials (IAPMO) to govern the installation and inspection of plumbing systems as a means of promoting the public's health, safety, and welfare. For building water use, confirm that calculations are up-to-date. The 2012 edition epitomizes the most recent approaches in the plumbing field and is the fourth edition developed under the ANSI consensus process. This edition is to be designated as an American National Standard; the latest UPC includes the following important changes:

- New alternate water sources for nonpotable applications and nonpotable rainwater catchment systems
- New Appendix L (sustainable practices)
- New minimum plumbing facilities table
- Water supply and drainage joint connection requirements (Source: Wikipedia)

Once the baseline has been established for the building, the actual use can be compared and measures can be implemented to reduce water use and increase overall water efficiency. Although this baseline methodology is specific to LEED, it can nevertheless be used in buildings that are not seeking LEED certification.

For calculation purposes in LEED™ projects, the precise number of fixtures is not important unless there are multiple types of the same fixture specified throughout the building. For example, if there are public restrooms with different water closets on the second floor, their use is to be accounted for as a percentage of the full time equivalent (FTE) in the LEED™ credit template calculations.

When the Energy Policy Act's fixture and flow rates are applied to FTE building occupants, the baseline quantity use can be established. FTE calculations for the project are used consistently throughout the baseline and design case calculations to determine the estimated use by the building occupants.

8.4 RETENTION PONDS, BIOSWALES, AND OTHER SYSTEMS

Stormwater runoff is generally generated when precipitation from rain and snowmelt events flows over land or impervious surfaces and is unable to percolate into the ground. As the runoff flows over the land or impervious surfaces (paved streets, parking lots, and building rooftops), it accumulates debris, chemicals, sediment, or other pollutants that could adversely affect water quality if the runoff is discharged untreated. The most appropriate method to control stormwater discharges is the application of best management practices (BMPs). Because stormwater discharges are normally considered point sources, they will require coverage under a National Pollutant Discharge Elimination System (NPDES) permit.

Utilizing rainwater collection systems such as cisterns, underground tanks and ponds, can substantially reduce or eliminate the amount of potable water used for irrigation. Rainwater can be collected from roofs, plazas, and paved areas and then filtered by a combination of graded screens and paper filters prior to its use in irrigation.

A retention pond is basically a body of water that is used to collect stormwater runoff for the purpose of controlling the release of this runoff. Retention ponds have no outlets or streams, creek ditches, etc. The water collects and then is released through atmospheric phenomenon such as evaporation or infiltration.

Retention ponds differ from detention ponds in that a detention pond has an outlet such as a pipe to discharge the water to a stream. A detention pond is defined as a body of water that is used to collect stormwater runoff for the purpose of controlling the release of this runoff. The pipe that a detention pond contains is sized to control the release rate of the stormwater runoff. Neighborhood ponds serve several purposes, but none of those purposes include swimming or wading.

Pond construction: Ponds must be of sufficient depth (at least 8–10 ft.) to prevent stagnation and algae growth, and to handle the amount of stormwater runoff that is expected to enter it.

Most ponds typically have a "safety ledge" at the edge to keep those who unintentionally enter the pond from getting into deep water immediately. This safety ledge is generally no wider than 10 ft., and leads directly to much deeper water. The slope off the safety ledge varies greatly, as does the depth of water it leads to.

One main problem that is encountered with ponds is the buildup of bacteria like *Escherichia coli* Because of the limited water flow and the tendency of wildlife like geese to gather around ponds, they can become breeding grounds for dangerous bacteria. With proper design and maintenance, ponds can be very attractive, but it may require more planning, and more land. However, without proper maintenance, these ponds can turn into a major liability for the owner.

A decade ago, most detention ponds were nothing more than an ugly hole in the ground hidden as far from view as possible. Today, most developers are attempting to incorporate their detention ponds as amenities, whether they have a permanent pool, or walking trails, picnic areas, playgrounds, etc. Ponds are today less a "waste of land" and more of a beneficial use of land.

Local requirements for rainwater harvest and wastewater treatment will vary greatly from location to location, which is why early involvement and input from local code officials is important. The owner should assemble an experienced team, including the architect, landscape architect, civil and plumbing engineers and rainwater system designer, early in the design process, if realistic efficiency cost effective goals are to be achieved.

Often developers will try to do away with retention ponds and replace them with pervious concrete pavement that although perhaps more expensive than typical concrete pavement, can partially or fully offset the cost by reducing or eliminating the need for drainage systems and retention ponds and their associated maintenance costs. In addition to the cost savings, elimination of retention ponds can also help meet the goal of reducing site disturbance found in LEED and therefore help earn additional LEED points.

The practicality and decision of whether to incorporate detention ponds or not will be dependent largely on site development, not large-scale land development. Costs of building a detention versus traditional detention should be studied, bearing in mind that the same storage volume will have to be provided. Then, after adding the cost for pumps, controls, additional storage, and thousands of linear feet of pipe, the decision needs to be made whether all that extra cost outweigh the cost of losing, say, 12–15% of your land for traditional detention.

In the past, this water was conveniently forced into the city storm drains or into retention ponds, thus becoming someone else's problem. Water from rainstorms and snow-melt needs to be carefully managed in order to conserve water in time of need, to better clean water before it starts its journey back to local aquifers, and to lessen the burden of excessive water runoff on municipal system drainage systems.

A system of interlocking, porous pavers resting on a multilayer bed of crushed stones and gravel of different sizes can be used for the parking area. This will allow water to diffuse through the surface of the parking lot, slowing the rush of water into the ground and permitting the surrounding landscaping to absorb the water while being diverted toward bioswales surrounding the property. Bioswales are gently sloped areas of the property designed to collect silt and other rainwater runoff while slowing down the speed with which water collects (Figure 8.8). The swales are shaped so that

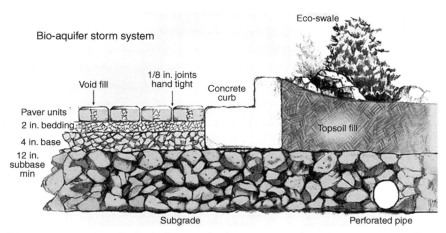

Figure 8.8 *Drawing of a Typical Bioswale.* These consist of gently sloped areas of the property designed to collect silt and other rainwater runoff – and slow down the speed with which water collects. The swales are shaped so that water is diverted, but not so sharply as to encourage erosion of the ground and soil. *Source: Other World Computing.*

water is diverted in a manner so as not to encourage erosion of the ground and soil.

Native vegetation can be planted in the bioswale to facilitate water absorption and lengthy root systems to prevent soil erosion while needing minimum maintenance. Native plants are hearty plants that can manage well during periods of dry, hot weather, yet manage to make use of and manage the flow of water from unexpected storms. The ability to combine nature with a well-planned surface system can make for an extremely efficient source of water management and filtering, in addition to an attractive design.

CHAPTER 9

Impact of Energy and Atmosphere

9.1 INTRODUCTION: LEED™ VERSION 4 – ENERGY AND ATMOSPHERE

With the threat of global economic collapse, it has become critically important to revise our thought patterns towards sustainability and green building and to make our buildings cost-effective and healthy place to live and work in. This can now be achieved through the use of integrated design processes that enable us to create high-performance buildings wherein all systems and components work together to produce overall functionality and environmental performance while meeting the needs of owner and tenant. Moreover, through integrated design we can now create "net zero energy buildings" (NZEB): buildings that, on an annual basis, draw from outside sources equal or less energy than produced on site from renewable energy sources.

This chapter will deal with the many aspects that impact the design and construction of creating high-performance, intelligent buildings that are both healthy and cost-effective. First, however, an overview of the new Leadership in Energy and Environmental Design (LEED™) Rating system is included as it applies to the Energy and Atmosphere category and also a brief discussion of some of the changes and new requirements for acquiring LEED™ credits, particularly in this category. It is important to note that many of the exam questions in the LEED™ exam tend to focus on the energy credits, especially strategies to optimize energy performance. It would therefore be especially prudent to pay particular attention to this category. For the latest updates relating to the LEED™ v4 tests and certification requirements, visit the GBCI and USGBC websites: www.gbci.org; www.usgbc.org.

The newly adopted LEED™ v4 requirements have changed from earlier versions, with an increased emphasis on sustainable sites, water efficiency, and energy and atmosphere. In terms of possible credits and points, Energy and Atmosphere must be considered the most important of the categories in the latest LEED™ v4 Rating System. For certification purposes, Energy and Atmosphere can earn up to 35 points (more or less) out of 100 + 10 (Tables 9.1 and 9.2). It should be stressed, however, that no individual product or system in itself can be LEED™ certified; they can only help contribute to the completion of LEED credits.

LEED v4 Practices, Certification, and Accreditation Handbook
http://dx.doi.org/10.1016/B978-0-12-803830-7.00009-8
Copyright © 2016 Elsevier Inc.
All rights reserved.

Table 9.1 Point allocation for different categories in LEED 2009 Rating System

Credit	NC	CI	EB	C&S	SCH	R	Homes
Sustainable Sites	26	21	26	28	24	26	22
Water Efficiency	10	11	14	10	11	10	15
Energy and Atmosphere	35	37	35	37	33	35	38
Materials/Resources	14	14	10	13	13	14	16
Indoor Environmental Quality	15	17	15	12	19	15	21
Innovation in Design	6	6	6	6	6	6	11
Regionalization	4	4	4	4	4	4	
Total	110	110	110	110	110	110 Points	

Table 9.2 Point distribution for New Construction (LEED 2009 and LEED v4 Rating Systems)

Credit categories	LEED 2009	LEED v4	Point gain/loss
Integrative Process[1]		1	+1
Location and Transportation[1,2]		16	−
Sustainable Sites[2]	26	10	−
Water Efficiency	10	11	+1
Energy and Atmosphere	35	33	−2
Materials and Resources	14	13	−1
Indoor Environmental Quality	15	16	+1
Innovation and Design Process[3]	6	6	−
Regional Priority[3]	4	4	−

[1] New category in LEED v4.
[2] Since the combined points in Location and Transportation and Sustainable Sites categories in LEED v4 are equal to the points in the Sustainable Site category in LEED 2009, we indicate "no change" in the points gain/loss column.
[3] Bonus points available toward maximum of 100 points total.

The significance of the dramatic changes to the LEED™ v4 scoring system cannot be overstated, particularly in how it relates to energy modeling. Energy and Atmosphere Prerequisite 2 (minimum energy performance) and Credit 1 (optimize energy performance) have changed significantly. Thus, the threshold for the prerequisite has changed from 14% to 10% and the points awarded in the optimize energy performance credit have increased from a 1-point to 10-point scale, to a possible 1–19-point scale, awarding basically 9 extra points for the same percentage improvement over the baseline building. But what is perhaps even more interesting is that the baseline itself has changed. LEED™ project teams are mandated to use the ASHRAE Standard referenced in the applicable Reference Guide and are permitted to use addenda within the most recent Supplement to that Standard. The LEED™ 2009 is largely governed by the 2007 update of ASHRAE 90.1 (as opposed to the previous version, ASHRAE 90.1 2004). The main modifications relating to LEED™ requirements include mandatory compliance with Appendix G of ASHRAE 90.1 2007 (ASHRAE 90.1 2007 has now been superseded by ASHRAE 90.1 2010).

For example, ASHRAE has made changes to the baseline heating, ventilation, and air-conditioning (HVAC) system types themselves. In the 2004 version, the cutoff in baseline criteria between commercial systems 3–4 (single-zone rooftop equipment) and 5–6 (variable air volume (VAV) rooftop equipment) was three floors or less, and larger than 75,000 ft.^{2} The new standard substantially reduces that square footage limit to 25,000 ft.^{2} So where a 60,000-ft.^{2} office building that used VAV HVAC could pick up 10–15% in fan energy savings alone using the old system, it will now be modeled against a similar baseline HVAC system.

Over the years, building construction values in the United States have become stricter. For example, in climate zone 3A, minimum compliance for roof insulation has increased from R-15 to R-20 and wall insulation has increased from R-13 to R-16.8. Although glass compliance has remained unchanged, different U-values have been introduced based on the type of glass, with the former assembly values of U-0.57 and SHGC-0.25 remaining consistent. Thus, using highly efficient glass remains an appropriate method for earning percentage points against the baseline. The significance of this change will be dramatically increased in building types where skin loads represent a large percentage of the peak HVAC load (i.e., office buildings), but be less significant in spaces where persons and ventilation loads dominate the sizing of the HVAC equipment (e.g., in assembly areas and schools).

Most state and local governments adopt commercial energy codes to establish minimum energy efficiency standards for the design and construction of buildings, and in the United States the majority of energy codes are based on ASHRAE 90.1 or IECC. It should be noted that several organizations have also produced standards for energy-efficient buildings, but ASHRAE and the US Green Building Council (USGBC) are perhaps the best known. ASHRAE 90.1-2010 is the new energy baseline, instead of 90.1-2007, *Energy Standard for Buildings Except Low-Rise Residential Buildings*, which was established by the USGBC as the commercial building reference standard for the new rating program, Leadership in Energy and Environmental Design (LEED) v4, which was launched on November 25, 2013, at the Greenbuild International conference and Expo in Philadelphia. The latest ASHRAE version is the last in a long succession to the original ASHRAE Standard 90-1975 standard, which is becoming increasingly more stringent.

LEED™ certification: In terms of qualifying for LEED certification, HVAC control systems and lighting control systems earn very few points on their own, perhaps 3 points or so. However, with the addition of the necessary sensors and building controls, the number of achievable points grows considerably – to as much as 25 points and even more with the use of fully integrated building systems. This is because with integrated systems, the building can earn multiple LEED™ points as well as cost savings by taking such measures as having a zone's occupancy sensor control both the lighting and HVAC.

A comprehensive building operation plan needs to be developed that addresses the heating system, cooling system, humidity control system, lighting system, and safety systems. Additionally, there is a need to develop a building automation control plan as well.

Energy conservation measures (ECMs) are recommended to ensure that the building will have the highest percentage of energy savings below the baseline building for the lowest upfront capital costs. Many recommended measures may involve minor upfront capital costs.

Other things to pay particular attention to in preparing for the LEED exam in the Energy and Atmosphere category include the following.

9.1.1 LEED™ EA Prerequisite 1: Fundamental Commissioning and Verification

Basically, the commissioning plan involves verification that the facility's energy-related systems are all installed, calibrated, and are performing according to the OPR (owner's project requirements) and BOD (basis of design). The building is to comply with the mandatory and prescriptive requirements of ASHRAE 90.1 2010, which is the new energy baseline, instead of 90.1-2007, in order to establish the minimum level of energy efficiency for the building type. The plans and data produced as a result of the building commissioning will lay the groundwork for later energy efficiency savings. This prerequisite is discussed in detail in Chapter 5. This is an extremely important prerequisite and should be completely understood. Several questions almost always turn up on the tests.

9.1.2 LEED™ EA Prerequisite 2: Minimum Energy Performance

The intent of this prerequisite is to achieve increasing levels of energy performance beyond the prerequisite standard to reduce environmental and economic harms associated with excessive energy use. LEED requirements are to establish an energy performance target no later than the schematic design phase. The target must be established as kBtu per square foot-year (kW per square meter-year) of source energy use. The important thing to remember here is to comply with both the mandatory and prescriptive provisions of ASHRAE 90.1-2010 or State Codes, whichever is more stringent. Of note, there are many mandatory lighting control requirements in ASHRAE 90.1-2010 that need to be addressed.

9.1.3 LEED™ EA Prerequisite 3: Building-Level Energy Metering

The intent is to support energy management and identify opportunities for additional energy savings by tracking facility-level use. This is a new EA prerequisite and impacts a broad range of facility types under the Building Design and Construction (BD + C) and Existing Buildings – Operations and Maintenance (EBOM) rating systems. It requires permanently installed metering capable of aggregating whole-facility energy use, including electricity, gas, water, steam, chilled water, BTU, and more; this new prerequisite requires minimum monthly or utility-billing-period interval data of consumption (kilowatt-hours) and demand (kilowatts). One of the requirements is that the data must be shared with the USGBC for 5 years, with certain caveats applying. It is applicable to BD+C and EBOM, the latter system also accepts manual or remote meter readings and requires monthly and annual summary data reports.

9.1.4 LEED™ EA Prerequisite 4: Fundamental Refrigerant Management

The intent is to reduce ozone depletion. This can be achieved by zero use of chlorofluorocarbon (CFC)–based refrigerants in new HVAC and refrigeration (HVAC&R) systems. For existing construction, a comprehensive CFC phase-out conversion prior to project completion is required if reusing existing HVAC equipment (Montreal Protocol–1995). Additionally, if CFC-based refrigerants are maintained in the building, reduce annual leakage to 5% or less using the procedures in the Clean Air Act, Title VI, Rule 608, governing refrigerant management and reporting (or a local equivalent for projects outside the United States), and reduce the total leakage over the remaining life of the unit to less than 30% of its refrigerant charge. Existing small HVAC&R units (defined as containing less than 0.5 pound (225 g) of refrigerant) and other equipment, such as standard refrigerators, small water coolers, and any other equipment that contains less than 0.5 pound (225 g) of refrigerant, are exempt.

9.1.5 LEED™ EA Credit 1: Enhanced Commissioning (3–4 Points)

This credit is discussed in Chapter 5. The basic intent is to begin the commissioning process early in the design process and implement additional activities after systems performance verification.

- Path 1: Enhanced commissioning (3 points) or
- Path 2: Enhanced and monitoring-based commissioning (4 points) (visit http://www.usgbc.org/credits/new-construction/v4 for latest updates)

Exemplary performance: enhanced commissioning – For NC, CS and Schools, projects that conduct comprehensive envelope commissioning may be considered for an innovative credit. These projects will need to demonstrate the standards and protocol by which the envelope was commissioned.

9.1.6 LEED™ EA Credit 2: Optimize Energy Performance (1–18 Points for NC and 1–16 for Schools and 1–20 Points for Healthcare)

The intent is to increase levels of energy performance in comparison to prerequisite standards. *Option 1.* Whole Building Energy Simulation using an approved energy-modeling program (1–18 points for NC, 1–16 for schools, and 1–20 points for healthcare). Demonstrate an improvement of 5% for new construction, 3% for major renovations, or 2% for core and shell projects in the proposed building performance rating compared with the baseline building performance rating. Calculate the baseline building performance according to ANSI/ASHRAE/IESNA Standard 90.1–2010, Appendix G, with errata (or a USGBC-approved equivalent standard for projects outside the United States), using a simulation model. Projects must meet the minimum percentage savings before taking credit for renewable energy systems.

Option 2. Prescriptive Compliance Path – Comply with ASHRAE's 50% Advanced Energy Design Guide (6 points under LEED v4 and 1 point under LEED™ v2009) appropriate to the project scope; facility must be 20,000 ft.² (1,860 m²) or less and must be office occupancy or retail occupancy.

According to LEED v4, they must

Comply with the mandatory and prescriptive provisions of ANSI/ASHRAE/IESNA Standard 90.1–2010, with errata (or a USGBC-approved equivalent standard for projects outside the U.S.).

Comply with the HVAC and service water heating requirements, including equipment efficiency, economizers, ventilation, and ducts and dampers, in Chapter 4, Design Strategies and Recommendations by Climate Zone, for the appropriate ASHRAE 50% Advanced Energy Design Guide and climate zone:

- *ASHRAE 50% Advanced Energy Design Guide for Small to Medium Office Buildings, for office buildings smaller than 100,000 square feet (9290 square meters);*
- *ASHRAE 50% Advanced Energy Design Guide for Medium to Large Box Retail Buildings, for retail buildings with 20,000 to 100,000 square feet (1860 to 9290 square meters);*
- *ASHRAE 50% Advanced Energy Design Guide for K–12 School Buildings; or*
- *ASHRAE 50% Advanced Energy Design Guide for Large Hospitals. Over 100,000 square feet (9290 square meters)*

For projects outside the U.S., consult ASHRAE/ASHRAE/IESNA Standard 90.1–2010, Appendixes B and D, to determine the appropriate climate zone.

Option 3. Prescriptive Compliance Path: Advanced Buildings "Core Performance" Guide (1–3 Points available only under LEED v2009). Moreover, you are required to,

Comply with the mandatory and prescriptive provisions of ANSI/ASHRAE/IESNA Standard 90.1-2010, with errata (or USGBC approved equivalent standard for projects outside the U.S.).

Comply with Section 1: Design Process Strategies, Section 2: Core Performance Requirements, and the following three strategies from Section 3: Enhanced Performance Strategies, as applicable. Where standards conflict, follow the more stringent of the two. For projects outside the U.S., consult ASHRAE/ASHRAE/IESNA Standard 90.1-2010, Appendixes B and D, to determine the appropriate climate zone.

3.5 Supply Air Temperature Reset (VAV)

3.9 Premium Economizer Performance

3.10 Variable Speed Control

To be eligible for Option 3, the project must be less than 100,000 square feet (9290 square meters). (Source: LEED v4.)

Note: Healthcare, Warehouse, or Laboratory projects are ineligible for Option 3.

Exemplary performance: For project teams pursuing Option 1, new construction must exceed **ASHRAE 90.1 2010 Appendix G** baseline performance rating by 50% (previously 45.5% for NC) and for existing buildings by 46% (previously 38.5% for NC) to be considered under the Innovation in Design category.

9.1.7 LEED™ EA Credit 3: Advanced Energy Metering (Possible Credits: 1–2)

To support energy management and identify opportunities for additional energy savings by tracking building-level and system-level energy use. This credit applies to Commercial Interiors (1–2 points), Retail (1–2 points), and Hospitality (1–2 points). In addition to whole-building metering, any particular energy use representing 10% or more of the facility total must also be metered. Also intended to track energy use at the system-level (HVAC) to identify additional energy savings opportunities.

The advanced energy metering must have the following characteristics. The Advanced Metering credit requires the permanent installation of electricity meters capable of recording 60-min or less consumption (kilowatt-hours) and demand (kilowatts), interval data, remote communications and interface capability to a LAN, building automation system, wireless network or other advanced communication infrastructure. The system must be capable of storing all meter data storage for at least 18 months. The data must be remotely accessible. In addition to remote data access, the system must also be capable of reporting hourly, daily, monthly, and annual energy use.

9.1.8 LEED™ EA Credit 4: Demand Response

Programs encourage building owners and managers to reduce their energy use during peak hours in return for payment and/or other economic incentives. Preliminary findings show that effective Demand Response strategies can lead to load sheds, from 5% to 18% (WFP%). The system is capable of limiting demand to reduce peak demand values. Intended to increase participation in Demand Response technologies and programs that make energy generation and distribution systems more efficient, increase grid reliability, and reduce environmental impacts and greenhouse gas (GHG) emissions. Impacting the BD+C and EBOM rating systems, points are given based on level of implementation: not available but has infrastructure in place (1–3 points). For the v4 credit: 10% or more of the estimated peak electricity demand is required to be shown. The system will adjust the duty cycle to minimize the number of devices on at the same time.

9.1.9 LEED™ EA Credit 5: On-Site Renewable Energy – (1 Possible Point)

1–7 points for NC and schools and 1–4 for CS – The intent is to encourage increase of renewable energy self-supply and reduce impacts associated with fossil fuel energy use. For the minimum renewable energy percentage for each point threshold see Table 9.3 and the reference guide. LEED v4 says,

Table 9.3 The minimum renewable energy percentage for each point threshold

Percentage renewable energy	Points
1	1
3	2
10	5
20	6
30	7
40	8

Use on-site renewable energy systems to offset building energy costs. Calculate project performance by expressing the energy produced by the renewable systems as a percentage of the building's annual energy cost and use the table below to determine the number of points achieved. Use the building annual energy cost calculated in EA Credit 1: Optimize Energy Performance or the U.S. Department of Energy's Commercial Buildings Energy Consumption Survey database to determine the estimated electricity use.

Exemplary performance: on-site renewable energy – For NC and Schools, projects can earn credit for exemplary performance by showing that on-site renewable energy accounts for at least 15% of annual building energy cost. For the CS category, the on-site renewable energy must account for at least 5% of the annual building energy cost to earn an exemplary performance credit.

9.1.10 LEED™ EA Credit 6: Enhanced Refrigerant Management

(1 Point each for New Construction, Core and Shell, Schools, Retail, Data Centers, Warehouses and Distribution Centers, Hospitality, Healthcare) – The intent is to reduce ozone depletion while complying with Montreal Protocol while minimizing direct contributions to climate change. *Option 1.* Do not use refrigerants or use only refrigerants (naturally occurring or synthetic) that have an ozone depletion potential (ODP) of zero and a global warming potential (GWP) of less than 50. *Option 2.* Select refrigerants and HVAC&R that minimize or eliminate the emission of compounds that contribute to ozone depletion and global warming. Meet or exceed requirements set by the maximum threshold for the combined contributions to ozone depletion and global warming potential.

9.1.11 LEED™ EA Credit 7: Green Power and Carbon Offsets (Possible 1–2 Points)

Applies to BD+C 1–2 points. This credit applies to New Construction (1–2 points), Core and Shell (1–2 points), Schools (1–2 points), Retail (1–2 points), Data Centers (1–2 points), Warehouses and Distribution

Centers (1–2 points), Hospitality (1–2 points), and Healthcare (1–2 points). The intent is to encourage the reduction of greenhouse gas emissions through the use of grid-source, renewable energy technologies, and carbon mitigation projects. Green power and RECs must be Green-e Energy certified or the equivalent. RECs can only be used to mitigate the effects of Scope 2, electricity use. Carbon offsets may be used to mitigate Scope 1 or Scope 2 emissions on a metric ton of carbon dioxide–equivalent basis and must be Green-e Climate certified, or the equivalent. For US projects, the offsets must be from greenhouse gas emissions reduction projects within the United States.

Option 1. Use the annual electricity consumption results of EA Prerequisite Credit 1: Optimization Energy Performance to determine baseline electricity use and cost. *Option 2.* Determine the baseline electricity consumption and cost by using the DOE Commercial Buildings Energy Consumption Survey (CBECS) database. Renewable Energy Certificates (RECs) provide the renewable attributes associated with green power. RECs can be provided at a much more competitive cost than from local utilities. A facility is not required to switch its current utility in order to procure off-site renewable energy for this credit.

Exemplary performance: green power (possible 1 point) – For NC, CS, and Schools, projects that purchase 100% of their electricity from renewable sources may be considered for an innovation design credit. Credit based on total building energy usage.

Air filters are discussed in Section 9.3.6. It suffices to say that a building's HVAC air filtration system provides measurable and discernable ways to improve indoor air quality (IAQ) and energy efficiency – two main tenets of the LEED™ program. In addition, proper filtration systems can contribute to the completion of LEED credits and prerequisites.

Finally, with respect to fire protection systems, these should never be compromised as they serve the purpose of life safety. However, just like other building systems, they should be designed, sourced, installed, and maintained in a manner that is environmental friendly and reduces their impacts on the environment as discussed in Section 9.6.

9.2 INTELLIGENT ENERGY MANAGEMENT SYSTEMS

Building automation is becoming one of the landmarks of today's society; it basically consists of a programmed, computerized, "intelligent" network of electronic devices that monitor and control the mechanical and lighting systems in a building. In fact, more and more buildings are incorporating central communications systems and the Computer Integrated Building has not only become a reality but has become an integral part of mainstream America. The intent is to create an intelligent sustainable building and reduce energy and maintenance costs. In addition, increasing consumer demand for clean renewable energy and the deregulation of the utilities industry have spurred growth in green power – solar, wind, geothermal steam, biomass, and small-scale hydroelectric sources of power. Small commercial solar power plants are beginning to emerge and have started to serve some energy markets.

9.2.1 Building Automation and Intelligent Buildings

There have been various definitions of intelligent buildings and sustainability. Regarding sustainability, the *ASHRAE GreenGuide* defines it as "providing for the needs of the present without detracting from the ability to fulfill the needs of the future." An intelligent building on the other hand can be said to be one that provides a productive and cost-effective environment through optimization of its basic elements: structure, systems, services, and management and the interrelationships between them. Thomas Hartman, PE, a building automation expert on the other hand, believes there are three cardinal elements of an intelligent building. These are as follows:

1. *The occupants*: An intelligent building is one that provides easy access, keeps people comfortable, environmentally satisfied, secure, and provides services to keep the occupants productive for their purpose in the building.
2. *Structure and systems*: An intelligent building is one that at a bare minimum significantly reduces environmental disruption, degradation, or depletion associated with the building while ensuring a long-term useful functional capacity for the building.
3. *Advanced technologies*: An intelligent building is one that because of its climate and/or use is challenged to meet elements (1) and (2), and succeeds in meeting those challenges through the use of appropriate advanced technologies.

Most engineers today, however, understand an intelligent building to be a building that incorporates computer programs to coordinate many building subsystems to regulate the interior temperatures HVAC and providing power. The goal is usually to reduce the operating cost of the building while maintaining the desired environment for the occupants (Figure 9.1). Many people fail to realize that it is really about the use of advanced technologies

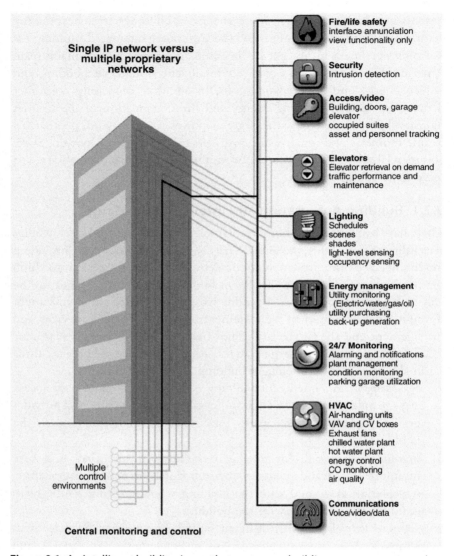

Figure 9.1 An intelligent building is one that can merge building management requirements with IT systems to achieve optimized system performance as well as simplifying general facility operations. In this illustration, a single IP network is compared with a multiple propriety network. *Source: BPG Properties Ltd.*

to dramatically improve the comfort, environment, and performance of its occupants while minimizing the external environmental impact of its structure and systems. The key phrase here is "comfort of its occupants" – which is what it is all about. In the final analysis, intelligent buildings help property owners and developers as well as tenants to achieve their objectives in the areas of comfort, cost, safety, long-term flexibility, and marketability.

There are a number of commercial-off-the-shelf Building Automation Systems (BAS) now on the market, and the majority of facility and building managers recognize the potential value of such systems as a powerful energy saving tool, and if it was not for the initial costs involved, there would be no hesitation in employing them. For example, one basic BAS that is readily available saves energy by widening temperature ranges and reducing lighting in unoccupied spaces and reduces costs for electricity by shedding loads when electricity is higher-priced.

Kristin Kamm, a senior research associate at E Source notes that some of the most common strategies that BASs employ to cut energy use include:

- *Scheduling*: Scheduling turns equipment on or off depending on time of day, day of the week, day type, or other variables such as outdoor air conditions.
- *Lockouts*: Lockouts ensure that equipment does not turn on unless it is necessary. For example, a chiller and its associated pumps can be locked out according to calendar date, when the outdoor air falls below a certain temperature, or when building cooling requirements are below a minimum.
- *Resets*: When equipment operates at greater capacity than necessary to meet building loads, it wastes energy. A BAS can ensure equipment operates at the minimum needed capacity by automatically resetting operating parameters to match current weather conditions. For example, as the outdoor air temperature decreases, the chilled water temperature can be reset to a higher value.
- *Diagnostics*: Building operators who use a BAS to monitor information such as temperatures, flows, pressures, and actuator positions may use that data to determine whether equipment is operating incorrectly or inefficiently, and to troubleshoot problems. Some systems also the use the data to automatically provide maintenance bulletins.

Customized building automation can be complex depending on the needs of the client. Some intelligent buildings have the capability to detect and report faults in the mechanical and electrical systems, especially critical systems. Many also have the ability to track individual occupants to adapt building systems to the individual's wants and needs (e.g., setting a room's temperature and lighting levels automatically when a homeowner enters), as well as anticipate forecasted weather, utility costs, or electrical demand. There are other nonenergy uses for automation in a building such as scheduling preventive maintenance for the building, monitoring security, monitoring rent or consumables charges based on actual usage, and even giving directions within the building.

Some of the typical elements and components that are frequently employed in building automation include the following:

Controller: Today's controllers allow users to take advantage of networking and gain real-time access to information from multiple resource segments in a building's network, creating an "Intelligent Building." These controllers come in a wide range of sizes and capabilities to control devices that are common in buildings. Usually the primary and secondary buses are chosen based on what the controllers provide.

Occupancy sensors: There are different types of these devices such as infrared, ultrasonic and dual tech sensors that are designed to meet a wide range of applications. Occupancy is usually based on time of day schedules, but override is possible through different means. Some buildings can sense occupancy in their internal spaces by an override switch or sensor. Sensors can be either ceiling mounted or wall mounted, depending on the type and application (Figure 9.2).

Lighting: With a building automation system lighting can be turned on and off based on time of day, or the occupancy sensors and timers. One

Figure 9.2 *A Multicriteria Intelligent Sensor, Acclimate™.* This sensor incorporates both thermal and photoelectric technologies that interact to maximize detection. Acclimate is a fire sensor with an onboard microprocessor with advanced software that makes adjustments to reduce false alarms. *Source: System Sensor.*

typical example is to turn the lights in a space on for a half hour since the last motion was sensed. A photocell placed outside a building can sense darkness, and the time of day, and modulate lights in outer offices and the parking lot.

Air handlers: With most air handlers less temperature change is required because they typically mix the return and outside air. This in turn can save money by using less chilled or heated water (not all AHUs use chilled/hot water circuits). Some external air needs to be introduced to keep the building's air quality healthy. The supply fan (and return if applicable) is started and stopped based on either time of day, temperatures, building pressures or a combination of all three.

Constant volume air-handling unit (CAV): This is a less efficient type of air-handler because the fans do not have variable-speed controls. Instead, CAVs open and close dampers and water-supply valves to maintain temperatures in the building's spaces. They heat or cool the spaces by opening or closing chilled or hot water valves that feed their internal heat exchangers. Generally one CAV serves several spaces, but in larger buildings may incorporate many CAVs.

Variable volume air-handling unit (VAV): The VAV is a more efficient unit than the CAV. VAVs supply pressurized air to VAV boxes, usually one box per room or area. A VAV air handler can change the pressure to the VAV boxes by changing the speed of a fan or blower with a variable frequency drive or (less efficiently) by moving inlet guide vanes to a fixed-speed fan. The amount of air is determined by the needs of the spaces served by the VAV boxes.

VAV hybrid systems: This is a variation between VAV and CAV systems. In this system the interior zones operate as in a VAV system but the outer zones differ in that the heating is supplied by a heating fan in a central location usually with a heating coil fed by the building boiler. The heated air is ducted to the exterior dual duct mixing boxes and dampers controlled by the zone thermostat calling for either cooled or heated air as needed.

Central plant: The function of a central plant is to supply the air-handling units with water. It may supply a chilled water system, hot water system and a condenser water system, as well as transformers and an auxiliary power unit for emergency power. If well managed, these can often help each other. For example, some plants generate electric power at periods with peak demand, using a gas turbine, and then use the turbine's hot exhaust to heat water or power an absorptive chiller.

Chilled water system: This is typically used to cool a building's air and equipment. Chilled water systems usually incorporate chiller(s) and pumps. Analog temperature sensors are used to measure the chilled water supply and return lines. The chiller(s) are sequenced on and off to ensure that the water supply is chilled.

Condenser water system: Cool condenser water is supplied to the chillers through the use of cooling tower(s) and pumps. In order to ensure that the condenser water supply to the chillers is constant, speed drives are commonly employed on the cooling tower fans to control temperature. Proper cooling tower temperature ensures the proper refrigerant head pressure in the chiller. The cooling tower set point used depends upon the refrigerant being used. Analog temperature sensors measure the condenser water supply and return lines.

Hot water system: The hot water system supplies heat to the building's air-handling units or VAV boxes. The hot water system will have a boiler(s) and pumps. Analog temperature sensors are placed in the hot water supply and return lines. Some type of mixing valve is typically incorporated to control the heating water loop temperature. To maintain supply, the boiler(s) and pumps are sequenced on and off.

Alarms and security: Most building automation systems today incorporate some alarm capabilities. If an alarm is detected, it can be programmed to notify someone. Notification can be implemented via a computer, pager, cellular phone, or audible alarm. Security systems can also be interlocked to a building automation system. If occupancy sensors are present, they can also be used as burglar alarms. This is discussed in greater detail in Section 9.6.

There are a large number of propriety protocols and industry standards on the market, including ASHRAE, BACnet, DALI, DSI, Dynet, ENERGY STAR, KNX standard, LonTalk, and ZigBee. Details of these systems are outside our scope.

9.3 ACTIVE MECHANICAL SYSTEMS – ZONING AND CONTROL SYSTEMS

9.3.1 General

The majority of people living today in urban American cities now take for granted that the buildings they live and work in will have appropriate mechanical HVAC systems in place. It is understood that these systems are designed to provide air at comfortable temperature and humidity levels,

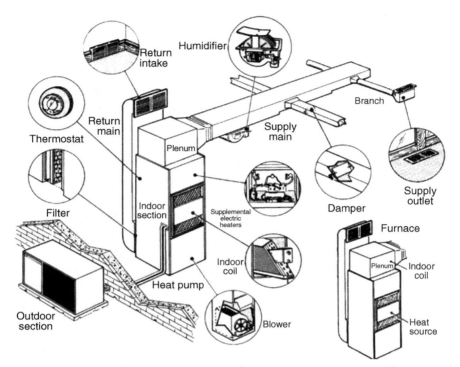

Figure 9.3 *Diagram Illustrating the Basics of HVAC Systems.* Source: Southface Energy Institute.

free of harmful concentrations of air pollutants (Figure 9.3). The continuous development of air-conditioning systems has brought about fundamental changes in the way we design projects because it has allowed investors to build larger, higher, and more efficient buildings than was previously possible. But even as buildings today are being designed with increasingly sophisticated energy management and control systems (EMCS) for monitoring and controlling the conditions of a building's interior space, we nevertheless frequently discover that a building's HVAC equipment routinely fails to satisfy the performance expectations of its designers and owners. What is surprising, such failures often go unnoticed for extended periods.

New technologies and recent developments in computers and electronics equipment have made it possible to create HVAC systems that are smarter, smaller, and more efficient. These advancements have reshaped how the systems are installed, how they are maintained, and how they operate. Other important recent developments in HVAC equipment design include the introduction of VAV. This involves a technique for controlling the capacity of an HVAC system. This means that with these systems, persons who have conditioned air circulating in, on, or around them can control the temperature in their own particular personal space. For example, if two individuals are on the same system and one seeks to increase the temperature, the system can heat that person's space and cool the other's. Another advantage is that VAV can change the volume of air delivered to the space and also damper off a space that is not used or occupied, thereby increasing efficiency. The fan capacity control, especially with modern electronic variable speed drives, reduces the energy consumed by fans, which can be a substantial part of the total cooling energy requirements of a building. Dehumidification with VAV systems is also greater than it is with constant-volume systems, which modulate the discharge air temperature to attain part load cooling capacity.

Some estimates suggest that buildings in the United States annually consume about 42% of America's energy and 68% of its electricity. Of this, HVAC systems consume a significant percentage. Energy sources that provide power to an HVAC system are usually gas, solid fuels, oil, or electricity, the conducting medium usually being water, steam, or gas. The heating and cooling source equipment comprise components that use the energy source to heat or cool the conducting medium. The heating and cooling units (such as air conditioners and air handling units) are the components of the system that are instrumental in modifying the air temperatures in the interior spaces.

Scientists and others have known for many years that physical comfort is critical to work effectiveness, satisfaction, and physical and mental well-being. As discussed in Chapter 7, Indoor Environmental Quality, we know that uncomfortable conditions in the workplace such as noise, inadequate lighting, uncomfortable temperature, high humidity, ergonomics, and other physiological stressors invariably restrict the ability of people to function to their full capacity, leading in many cases to lower job satisfaction and increases in building-related illness (BRI) symptoms. And because humans generally spend most of their time indoors, health, well-being, and comfort in buildings are crucial issues.

9.3.2 Choosing Refrigerants

A refrigerant is chemical compound that is used as the heat carrier, which changes from gas to liquid and then back to gas in the refrigeration cycle. Refrigerants are used primarily in refrigerators/freezers, air-conditioning, and fire suppression systems. According to Mike Opitz, Certification Manager, LEED™ for Existing Buildings (USGBC),

Chemical refrigerants are the heart of a large majority of building HVAC and refrigeration equipment. These manufactured fluids provide enormous benefits to society, but in recent decades have been found to have harmful consequences when released to the atmosphere: all refrigerants in common use until the 1990s caused significant damage to the protective ozone layer in the earth's upper atmosphere, and most also enhanced the greenhouse effect, leading to accelerated global warming.

For example, in the EA Credit 4 (Enhanced Refrigerant Management) for NC, Schools, and CS, points can be earned by either not using refrigerants (Option 1) or by selecting environmentally friendly refrigerants and HVAC&R that minimize or eliminate the emission of compounds that contribute to ozone depletion and global warming (Option 2).

To address these and other relevant issues, on September 21, 2007, parties to the Montreal Protocol, including the United States, overwhelmingly agreed to accelerate the phase-out of hydrochlorofluorocarbons (HCFCs) to protect the ozone and combat climate change with adjustments beginning in 2010 to production and consumption allowances for developed and developing countries. This refrigerant phase out of CFCs (production of CFCs ceased in 1995) and HCFCs will have a significant impact on proposed real estate purchases that still utilize this equipment. This means that owners and administrators need to take the long view when making decisions about their capital investments.

The 2007 Montreal Protocol definitely energized the green building movement and equipment manufacturers to undergo changes in the types of refrigerants used in certain equipment because of general environmental concerns and to seek suitable more environmental alternatives. It is no longer a question of whether facilities managers will upgrade their HVAC and other equipment, but when and in what way.

The Environmental Protection Agency (EPA), in accordance with the Montreal Protocol, is now obligated to phase out hydrochlorofluorocarbon (HCFC) refrigerants used in heat pump and air-conditioning systems because of their impact on ozone depletion. Chlorofluorocarbon (CFC) refrigerants manufacture has been banned in the United States since 1995. To date, the main alternatives are hydrofluorocarbons (HFCs) and HFC blends, although there are several potential non-HFC alternatives as well. DuPont has produced a complete family of easy-to-use, nonozone-depleting HFC retrofit refrigerants for CFC and HCFC equipment. But while (HFCs) may be suitable as short to medium-term replacements, they may not be suitable for long-term use because of their high global warming potential (GWP).

In some categories of the LEED Rating System, *Fundamental Refrigerant Management* is included as a prerequisite, the *intent* being to reduce ozone depletion.

The *LEED requirements* being to have zero use of CFC-based refrigerants in new HVAC&R systems. When reusing the existing base building HVAC, a comprehensive CFC phase-out conversion must be conducted prior to project completion. Also, various categories of the USGBC LEED programs award one credit point for using non-ozone-depleting, HFC refrigerants.

9.3.3 Types of HVAC Systems

Although there are a wide variety of HVAC systems in use in today's real estate, no system is right for every application. In order to service the various needs, there are a number of different types of HVAC systems available (e.g., single zone/multiple zone, constant volume/variable air volume). The most common classification of HVAC systems is by the carrying mediums used to heat or cool a building. The two main transfer mediums for this purpose are air and water, which take them to emitters. On smaller projects, electricity is often used for heating although some systems now use a combination of transfer media. HVAC systems range in complexity from stand-alone units that serve individual rooms or zones to large, centrally controlled systems serving multiple zones in a building.

Heating systems: This mechanism is used to regulate temperature in commercial and public offices, residences, and other facilities. Heating systems can be either central or local. The most commonly used setup is the central heating system where the heating is concentrated in a single central location from where it is then circulated for various heating processes and applications. Some of the more common heating systems currently in use include:

Electric heating: This is a device that transforms electrical energy into heat. Every electric heater contains an electric resistor, which acts as its heating element. The practice of using electricity for heating is becoming increasingly popular in both residences and public buildings. Electric heating generally costs more than energy obtained from combustion of a fuel, but the convenience, cleanliness, and reduced space needs of electric heat have often justified its use. The heat can be provided from electric coils or strips used in varying patterns, such as convectors in or on the walls, under windows, or as baseboard radiation in part or all of a room. By the incorporation of a heat pump, the overall cost of electric heating can be substantially reduced.

Electric baseboard heaters: These systems are a fairly common heat source, heating the room by using a process call electric resistance. These types of heaters are zonal heaters controlled by thermostats located within each room. Inside baseboard heaters are electric cables, and it is these that warm the air that passes through it. Electric baseboard heaters are typically installed along the lower part of outside walls to provide perimeter heating. Room air heated by the resistance element rises and is replaced by cooler room air, establishing a continuous convective flow of warm air while in operation.

Central heating: The vast majority of modern commercial buildings including office buildings, high-rise residential apartments, hotels, and shopping malls are today provided with some form of central heat. There are many different types of central heating systems on the market, most of which comprise a central boiler (which is actually a heat generator because the water is not "boiled," it peaks at 82–90°C) or furnace to heat water, pipes to distribute the heated water, and heat exchangers or radiators to conduct this heat to the air. However, there is no such thing as a standard central heating system, and each project requires the system to be tailored to meet its own requirements; and with advanced controls, a correctly programmed central heating system when optimized will be able to constantly monitor and automatically adjust the system basically on its own.

Central heating differs from local heating in that the heat generation occurs in one place, such as a furnace room in a house or a mechanical room in a large building. The most common method to generate heat involves the combustion of fossil fuel in a furnace or boiler. The resultant heat then gets distributed: typically by forced-air through ductwork, by water circulating through pipes, or by steam fed through pipes. Increasingly, buildings utilize solar-powered heat sources, in which case the distribution system normally uses water circulation.

Most modern systems have a pump to circulate the water and ensure an equal supply of heat to all the radiators. The heated water is often fed through another heat exchanger inside a storage cylinder to provide hot running water. Forced air systems send air through ductwork, which can be reused for air-conditioning, and the air can be filtered or put through air cleaners. The heating elements (radiators or vents) are ideally located in the coldest part of the room, typically next to the windows. A central heating system provides warmth to the whole interior of a building (or portion of a building) from one point to multiple rooms.

Furnace: This is a heating system component designed to heat air for distribution to various building spaces. Furnaces can be used for residential and small commercial heating systems. Furnaces use natural gas, fuel oil, and electricity for the heat source as well as on-site energy collection, and heat transfer. Natural gas furnaces are available in condensing and noncondensing models. The cooling can be packaged within the system, or a cooling coil can be added. When direct expansion systems with coils are used, the condenser can be part of the package or remote. The efficiency of new furnaces is measured by the annual fuel utilization efficiency (AFUE), a measure of seasonal performance. Today's furnaces are designed to be between 78% AFUE and 96% AFUE. Traditional "power combustion" furnaces are 80–82% AFUE. Above percent AFUE, a furnace is "condensing," which generally means it recaptures some of the heat wasted in traditional systems by condensing escaping water vapor.

Radiant heating is increasing in popularity because it is clean, quiet, efficient, dependable, and invisible. The heat is provided in part by radiation in all forms of direct heating, but the term is usually applied to systems in which floors, walls, or ceilings are used as the radiating units. Steam or hot water pipes are placed in the walls or floors during the construction process, and radiant heating systems circulate warm water through continuous loops of tubing. The tubing system transfers the heat into the floor and upward into virtually any surface including carpeting, hardwood, parquet, quarry and ceramic tile, vinyl flooring, or concrete. If electricity is used for heating, the panels containing heating elements are mounted on a wall, baseboard, or the ceiling of the room. Radiant heating provides uniform heat and is both efficient and relatively inexpensive to operate.

Warm air systems: There are basically two types of warm air heating systems: gravity and forced air. The gravity system operates by air convection and is based on the principle that when air is heated it expands, becomes lighter, and rises. Cooler air is dense and therefore falls. The difference in air temperature creates the convection or motivation for air movement. The return of a gravity system must be unrestricted and even a filter is considered too restrictive. This is necessary to develop positive convection and better distribution. The furnace consists of a burner compartment (firebox) and a heat exchanger. The heat exchanger is the medium used to transfer heat from the flame to the air, which moves via ducts to the various rooms. Besides being the medium of heat transfer, the heat exchanger keeps the burned fuels separate from the air. Often the furnace is arranged so the warm air passes over a water pan in the furnace for humidification before circulating through the building. As the air is heated,

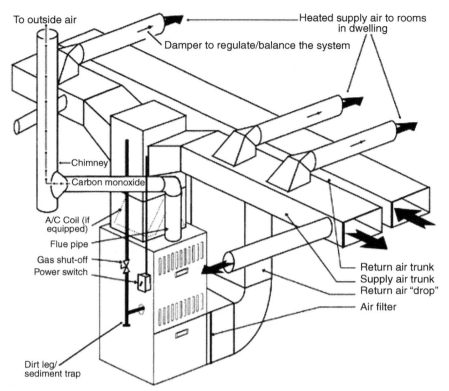

To outside air

Heated supply air to rooms in dwelling

Damper to regulate/balance the system

—Chimney

—Carbon monoxide

A/C Coil (if equipped)

Flue pipe

Gas shut-off

Power switch

Return air trunk
Supply air trunk
Return air "drop"
Air filter

Dirt leg/ sediment trap

Figure 9.4 *Diagram Describing How a Forced Warm Air System Operates. Source: warmair.cominc.*

it passes through the ducts to individual grills or registers in each room (which may be opened or closed to control the temperature) of the upper floors.

As with gravity warm air-heating systems, the heat exchanger is the medium of heat transfer and separates the burned fuel from the air that moves through the building. Forced-circulation systems typically have a fan or blower placed in the furnace casing that blows air through an evaporator coil, which cools the air (Figure 9.4). This cool air is routed throughout the intended space by means of a series of air ducts, thus ensuring the circulation of a large amount of air even under unfavorable conditions. When combined with cooling, humidifying, and dehumidifying units, forced-circulation systems may be effectively used for heating and cooling. The ability to utilize the same equipment to provide air-conditioning throughout the year has given added impetus to the use of forced-circulation warm air systems in residential installations.

Hot water systems: Such systems typically have a central boiler, in which water is heated to a temperature of from 140°F to 180°F (60–83°C), and which is then circulated by means of pipes to some type of coil units, such as radiators, located in the various rooms. Circulation of the hot water can be accomplished by pressure and gravity, but forced circulation using a pump is more efficient because it provides flexibility and control. In the rooms, the emitters give out the heat from their surfaces by radiation and convection. The cooled water is then returned to the boiler. In addition, there are combination systems that use ducts for supplying air from the central air-handling unit, and water to heat the air before it is transferred into the conditioned space. Combination boiler heating systems are the most commonly used in central heating systems. Running on mains pressure water eliminates both the need for tanks to be placed in the loft and the need for a hot water cylinder as the water is instantly heated when needed. Although hot water circulating systems generally provide convenience and save water, they can also expend large amounts of energy and may not be cost-effective.

In boiler hydronic systems, there are different approaches to arrange the piping depending on the budget at installation time and the efficiency level required. Generally speaking, hot water systems use either a one-pipe or a two-pipe system that are used to circulate the heated water. The one-pipe system uses fewer pipes than the two-pipe arrangement, which is why it is less expensive to install. However, it is also less efficient because larger radiators or longer baseboards are required at the end of the loop as this part of the loop gets less heat. The operation of a one-pipe system is fairly simple: water enters each radiator from the supply side of the main pipe, circulates through the radiator, and flows back into the same pipe.

Most modern layouts use a two-pipe layout, in which radiators are all supplied with hot water at the same temperature from a single supply pipe, and the water from these radiators flows back to the furnace to be reheated through a common return pipe. Although the two-pipe system requires more pipe work, it is more efficient and easier to control than the one-pipe system. Another advantage of the two-pipe direct return and reverse return loop over the one-pipe series loop is that it can be zoned. Zoning offers additional control over where and when heat is required, which in turn can reduce heating costs. As with many hydronic loop systems, the two-pipe direct return needs balancing valves, and in both systems an expansion tank is required to compensate for variations in the volume of water in the system. Another system that is sometimes used is the sealed hot water system. This is basically a closed system that does not need water tanks because the hot water is supplied direct from the mains.

Steam systems: Steam heating systems are similar to heating water systems, except steam is used as the heating medium instead of water. Steam is often used to carry heat from a boiler to consumers as heat exchangers, process equipment, etc. Sometimes steam is also used for heating purposes in buildings. Steam heating systems closely resemble their hydronic systems counterpart except that steam rather than hot water is circulated through the pipes to the radiators and no circulating pumps are required. The steam condenses in the radiators and/or baseboards, giving up its latent heat. Both one-pipe and two-pipe arrangements are employed for circulating the steam and for returning to the boiler the water formed by condensation.

Each heating unit in a one-pipe gravity-flow system has a single-pipe connection through which it simultaneously receives steam and releases condensate. All heating units and the end of the supply main are sufficiently above the boiler water line so that condensate is able to flow back to the boiler by gravity. In a two-pipe system, steam supply to the heating units and condensate return from heating units are through separate pipes. Air accumulation in piping and heating units discharges from the system through the open vent on the condensate pump receiver. Piping and heating units must be installed with proper pitch to provide gravity flow of all condensate to the pump receiver. The three main types of steam systems generally employed are air-vent systems, vacuum, or mechanical-pump systems and vapor systems.

Heat pumps are actually air conditioners that run in reverse to bring heat from outdoors into the interior. This works by heating up a piped refrigerant in the outdoor air, then pumping the heat that is generated by the warmed refrigerant inside to warm the indoor air. This type of system works best in moderate climates and becomes less efficient in very cold winter temperatures, when electrical heat is needed for auxiliary heating demands. This is because as the outdoor temperature begins to drop, the heat loss of a space becomes greater, requiring the heat pump to operate for longer stretches of time for it to be able to maintain a constant indoor temperature. As with furnaces, heat pumps are usually controlled by thermostats. Figure 9.5 illustrates a typical residential application of a water pump system.

Reverse cycle chiller (RCC) is a recently introduced heat pump variant (although the technology is not new) that heats or cools an insulated tank of water and then distributes the heating or cooling either through fans and ducts or radiant floor systems. A fan and coil system can then pump hot

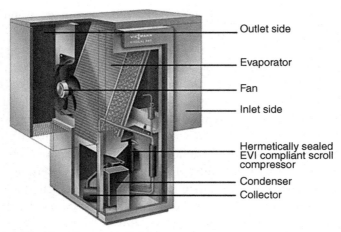

Figure 9.5 *The Vitocal 350A-G Air/Water Heat Pump.* Heating output is 10.6–18.5 kW. *Source: Viessmann Werke.*

water or cooled air away from the tank and through the ductwork to one or more heating zones. RCC normally uses a highly efficient heat exchanger to transfer energy to a water line, attaining temperatures up to 120° in the heating mode, and down to 50° in cooling mode. Some in the industry imply that the RCC system is superior because it uses the hot water from the tank to defrost the coils, so no backup burner is needed. This means that the system will never blow cold air when it should not – the result being that you stay nice and warm.

The need for auxiliary electric heating coils and defrosting cycles to prevent icing of the refrigerant is eliminated, thus making these systems more suitable for cold climates. Newer models also now offer solar-powered hot water heating for the unit. These systems still require an exterior condenser unit similar to traditional HVAC systems.

Geothermal heat pump (GHP): This is a relatively new technology that is gaining wide acceptance for both residential and commercial buildings. Studies show that approximately 70% of the energy used in a geothermal heat pump system is renewable energy from the ground. This system utilizes the relatively constant temperature of the ground or water several feet below the earth's surface as source of heating and cooling. The earth's constant temperature is what makes geothermal heat pumps one of the most efficient, comfortable, and quiet heating and cooling technologies available today. Geothermal heat pumps are appropriate for retrofit or new facilities, where both heating and cooling are desired, and business owners

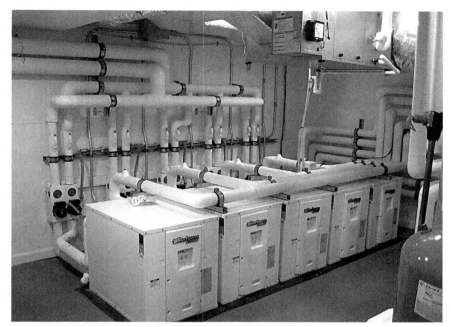

Figure 9.6 *A Residential Geothermal Heat Pump System.* *Source: Climate Heating and Cooling, Inc.*

around the United States are now installing geothermal heat pumps to heat and cool their buildings. These systems can also be located indoors because there is no need to exchange heat with the outdoor air. Although this technology may be more expensive to install than traditional HVAC systems, geothermal systems will greatly reduce gas or electric bills through reduced energy, operation, and maintenance costs, thus allowing for relatively short payback periods (Figure 9.6). Conventional ductwork is generally used to distribute heated or cooled air from the geothermal heat pump throughout the building.

When selecting ground-source heat pumps, models should be chosen that qualify for the ENERGY STAR® label, or that meet the recommended levels of COP (coefficient of performance) and EER (energy efficiency ratio). Efficiency is measured by the amount of heat a system can produce or remove using a given amount of electricity. A common measurement of this performance is the seasonal energy efficiency ratio (SEER). The Federal Appliance Standards, which took effect on January 23, 2006, will require the new standards for central air conditioners be a minimum of 13 SEER. Most manufacturers now offer SEER 10, 11, 12, and 13 and SEER

14 models. This translates into five separate efficiency options, with model numbers usually keyed to the SEER numbers, so they are easy to recognize. The most efficient models, however, generally involve dual compressor systems and increased heat exchange area, and thus cost significantly more.

Heat pump performance continues to improve with the ongoing development of new technologies and innovations. The introduction of two-speed compressors allows heat pumps to operate close to the heating or cooling capacity that is needed at any particular moment. This saves large amounts of electrical energy and reduces compressor wear. Some heat pumps are equipped with *variable-speed* or *dual-speed motors* on their indoor fans, outdoor fans, or both. The variable-speed controls for these fans attempt to keep the air moving at a comfortable velocity, minimizing cool drafts and maximizing electrical savings. Another advance in heat pump technology is the introduction of a device called a *scroll compressor*, which compresses the air or refrigerant by forcing it into increasingly smaller areas. This device uses two interleaved scrolls to pump, compress, or pressurize fluids such as liquids and gases.

Vapor compression refrigeration unit: This refrigeration system uses a circulating liquid refrigerant as the medium that absorbs and removes heat from the space to be cooled and subsequently rejects that heat elsewhere. It is the most common refrigeration system used today for air-conditioning of large public buildings, private residences, hotels, hospitals, theaters, and restaurants. These systems have four essential components: a gas compressor, an evaporator, a condenser (heat transfer), and an expansion valve (also called a valve).

Solar thermal collectors: Solar collectors are considered to be one of the renewable energy technologies with the best economics. A solar collector specifically is intended to collect heat, that is, to absorb sunlight to provide heat, and may be used to heat air or water for building heating purposes. A solar collector operates on a very simple basis. The radiation from the sun heats a liquid, which goes to a hot water tank. The liquid heats the water and flows back to the solar collector. Water-heating collectors may replace or supplement a boiler in a water-based heating system. Air-heating collectors may replace or supplement a furnace. Solar collectors have an estimated lifetime of 25–30 years or more and require very little maintenance, only control of antifreeze and pressure in the system.

There are various types of solar systems, which can be either active or passive. The terms passive and active in solar thermal systems refer to whether the systems rely on pumps or only thermodynamics to circulate water through

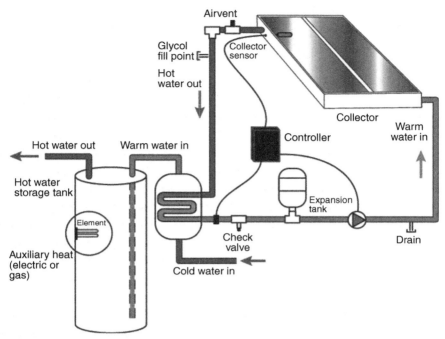

Figure 9.7 *Diagram of an Indirect Active Solar System.* This type of system is preferred in climates with extended periods of below-freezing temperatures. *Courtesy: Southface Energy Institute.*

the systems. As solar energy in an active solar system is typically collected at a location remote from the spaces requiring heat, solar collectors are normally associated with central systems. Solar water-heating collectors may also provide heated water that can be used for space cooling in conjunction with an absorption refrigeration system. Since the sun provides free energy, a saving of up to 70% can be made of the energy that would otherwise be used for heating the water. Besides the economical reward, there is also a significant environmental advantage. By using solar collectors for heating of water, an average family can save up to 1 ton of CO_2 per year. Figure 9.7 illustrates an example of an indirect active solar system.

Air-conditioning systems: According to ASHRAE, an air-conditioning system is a system that accomplishes four specific objectives simultaneously: (1) control air temperature; (2) control air humidity; (3) control air circulation; (4) control air quality. The fact is that air-conditioning systems are typically designed to provide heating, cooling, ventilation, and humidity control for a building or facility. It is much easier to incorporate air-conditioning systems into new modern offices and public buildings under construction than to retrofit existing buildings because of the bulky air ducts that are required.

There are many types of air conditioner systems; some of them use direct expansion coils for cooling such as window units, package unit air conditioners, split system air conditioners, packaged terminal air conditioners like the air conditioners used in hotels, and mini-split ductless air conditioners. Other types of air conditioners utilized chilled water for air-conditioning, and these are typically commercial air conditioners for large commercial buildings. But whichever type of air conditioner is used, the coils in the air conditioner system are required to be brought to a temperature colder than the air.

When operating properly, air conditioners use the direct expansion coils or chilled water coils to remove the heat from the air as air is blown across the coils. The evaporator coil in an air conditioner system is responsible for absorbing heat. The evaporator and condenser both comprise tubing, usually copper, surrounded by aluminum fins. As air (or water in a chiller) passes over the evaporator coils a heat exchange process takes place between the air and the refrigerant. The refrigerant absorbs the heat and evaporates in the indoor evaporator coils, and draws the heat out of the air and cools the facility. Finally, the hot refrigerant gas is pumped (compressor) into the outdoor condenser unit where it returns to liquid form. Some of the different types of systems that are currently available are:

Wall or window air-conditioning units: Window and through-the-wall electric air-conditioning units are often used in single zone applications that do not have central air-conditioning installed such as small buildings and trailers. They are also used in retrofit situations in conjunction with an existing system. Basically, they are small ductless units with casings extending through the wall, and are generally noisy and designed to cool small areas, though some larger units may be able to cool larger spaces. The primary advantage of a wall unit over a window unit is that it does not occupy window space. When a window unit is installed, the window becomes unusable.

Removal of the cover of an unplugged window unit will show that it is comprised of a compressor, an expansion valve, a hot coil (on the outside), a chilled coil (on the inside), two fans, and a control unit (Figure 9.8). The wall unit air conditioner works by removing hot air from the room into the unit, and the hot air that enters the unit is brought over the air-conditioning condenser and cooled. The cooled air is then pushed back into the room. Many of the newer units incorporate significant innovations such as

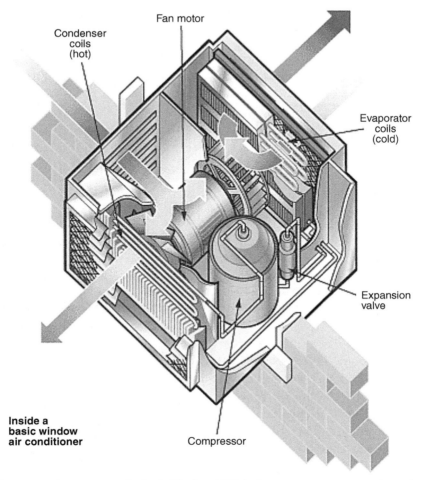

Figure 9.8 *Components of a Basic Window AC Unit.* A compressor, an expansion valve, a hot coil (on the outside), a chilled coil (on the inside), two fans, and a control unit. *Source: HowStuffWorks.com*

electronic touchpad controls, energy saver settings, and digital temperature readouts. Some units also incorporate timers that allow you to set the AC to cycle on and off at certain times of the day.

Central air-conditioning systems: Air-conditioning is one of those amenities that is frequently taken for granted, and often, particularly in relatively warm climates, we find it has become more the rule than the exception. In addition to cooling, A/C systems dehumidify and filter air, making it more comfortable and cleaner.

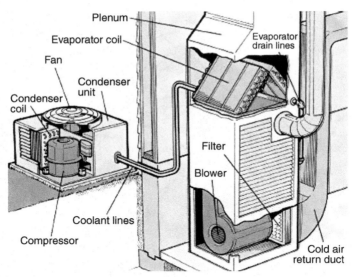

Figure 9.9 Domestic central air conditioners are made up of two basic components: the condenser unit, located outside the house on a concrete slab, and the evaporator coil above the furnace. These components in turn also comprise several elements. *Source: HowStuffWorks.com*

Many central HVAC systems today tend to serve one or more thermal zones with its main components being located outside of the zone or zones being served – usually at a convenient central location in, on, or near the building. Central air-conditioning systems are extensively installed in offices, public buildings, theaters, stores, restaurants, and other building types. Although centralized air-conditioning systems provide fully controlled heating, cooling, and ventilation, they need to be installed during construction. In recent years, these systems have increasingly become automated by computer technology for purposes of energy conservation. In older buildings, indoor spaces may be equipped with a refrigerating unit, blowers, air ducts, and a plenum chamber in which air from the interior of the building is mixed with air from the exterior. Such installations are used for cooling and dehumidifying during the summer months, and the regular heating system is used during the winter. Figure 9.9 shows the general components of a small central air-conditioning system that can be used in small commercial buildings or residences.

Central air-conditioning systems consist of three main components: the outdoor unit (condenser and compressor), the indoor unit (blower coil or evaporator), and the indoor thermostat to regulate the temperature (Figure 9.8). The success of a central air system is dependent upon these three systems appropriately functioning together. Likewise, the design of an air-conditioning system depends among other things, on the type of structure in which the system is to be placed, the amount of space to be cooled, the function of that space, and the number of occupants using it. For example, a room or building with large windows exposed to the sun, or an indoor office space with many heat-producing lights and fixtures, requires a system with a larger cooling capacity than a space with minimal windows in which cool fluorescent lighting is used. Also, a space in which the occupants are allowed to smoke will require greater air circulation than a space of equal capacity in which there is no smoking. Air-conditioned buildings often have sealed windows because open windows would disrupt the attempts of the HVAC system to maintain constant indoor air conditions.

Split system: A split AC system uses at least two pieces of hardware and generally implies that the condenser and compressor are located in an outdoor cabinet and the refrigerant metering device and the evaporator stored in an indoor cabinet. With many split-system air conditioners, the indoor cabinet also houses a furnace or a part of the heat pump. The cabinet or main supply duct of this furnace or heat pump also houses the air conditioner's evaporator coil. For reverse cycle applications, the heat exchangers can swap roles, with the heat exchanger exposed to outside air becoming the evaporator and the inside heat exchanger becoming the condenser (Figure 9.10). Split systems may have a variety of configurations. The four basic components of the vapor compression refrigeration cycle – compressor, condenser, refrigerant metering device, and evaporator – can be grouped in several ways. The grouping of components is based on practical considerations, such as available space, ease of installation, keeping noise outside occupied spaces, etc. Split systems are generally more expensive to purchase, but potentially less expensive to install, and a ductless system is practical for homes that do not already have ductwork. The big advantage of a ductless mini-split system is that you can adjust temperature levels for individual rooms or areas. However, split systems require more maintenance than single-box central air conditioners.

The condensing unit can sometimes be quite massive and normally located on the roof in the case of department stores, businesses, malls, warehouses, etc. An alternative is to have many smaller units on the roof, each attached inside to a small air handler that cools a specific zone within the building. The split-system approach may not always be suitable for larger buildings and particularly in multistory buildings because problems start to appear; for example, the distance between the condenser and the air handler

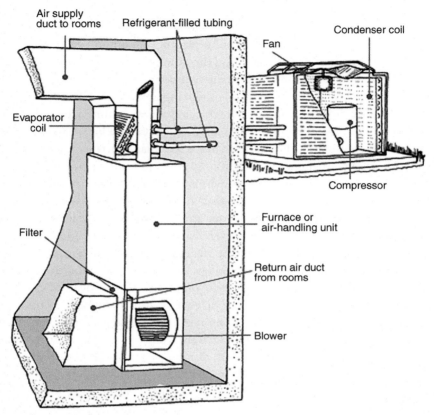

Figure 9.10 *A Drawing Illustrating How a Split System Works.* With a typical "split system," the condenser and the compressor are located in an outdoor unit; the evaporator is mounted in the air-handling unit, which is often a forced-air furnace. With a "package system," all of the components are located in a single outdoor unit that may be located on the ground or on the roof of the house. *Source: HomeTips.com*

exceeds pipe distance limitations, or the amount of duct work and the length of ducts cease to be viable.

Packaged systems: Packaged air conditioner models are used for medium sized halls and multiple rooms on the same floor. They are usually used for applications where air–conditioning of more than 5 tons is required The main difference between a package unit and a split–system is that a split system uses indoor and outdoor components to provide a complete comfort system whereas a package unit or self-contained unit requires no external coils, air handlers, or heating units. Packaged units commonly use electricity to cool and gas to heat. Packaged systems typically have all their components located in a single outdoor unit located either on the ground or roof. A packaged rooftop unit is a self-contained air handling unit, typically used in low-rise buildings, and mounted directly onto roof curbs, discharging conditioned air into the building's air duct distribution system. Air handling units come in many capacities, from units of just over one ton to systems of several hundred tons that contain multiple compressors and designed for single or multiple-zone application.

All-air systems: These systems transfer cooled or heated air from a central plant via ducting, distributing air through a series of grilles or diffusers to the room or rooms being served. They represent the majority of systems currently in operation. The overall energy used to cool buildings with all-air systems includes the energy necessary to power the fans that transport cool air through the ducts. Because the fans are usually placed in the air stream, fan movement heats the conditioned air (Figure 9.11a), thus adding to the thermal cooling peak load.

All-water systems: Robert McDowall, author of Fundamentals of HVAC systems, says that "when the ventilation is provided through natural ventilation, by opening windows, or other means, there is no need to duct ventilation air to the zones from a central plant. This allows all processes other than ventilation to be provided by local equipment supplied with hot and chilled water from a central plant. These systems are grouped under the name 'all-water systems.' The largest group of all-water systems are heating systems." McDowall also notes that "both the air-and-water and all-water systems rely on a central supply of hot water for heating and chilled water for cooling."

The conditioning effect in these systems is distributed from a central plant to conditioned spaces via heated or cooled water. Water is an effective heat transfer medium; thus, distribution pipes are often of relatively small volume (compared to air ducts). On the other hand, water cannot be directly dumped into a space through a diffuser, but requires a more sophisticated delivery device. All-water heating-only systems employ a variety of delivery devices, such as baseboard radiators, convectors, unit heaters, and radiant floors. All-water cooling-only systems are rare, with valance units being the most common delivery device for such systems. When full air-conditioning is contemplated, the most appropriate delivery device may be the fan-coil unit. All-water systems are generally the most expensive to install and own, and are classed as the least energy efficient in terms of transfer of energy (Figure 9.11b).

Air–water systems: A category of central HVAC systems that distribute conditioning effect by means of heated or chilled water and heated or cooled air (Figure 9.11c).

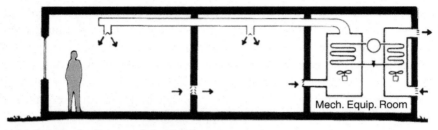

(a) Schematic diagram of an all-air system

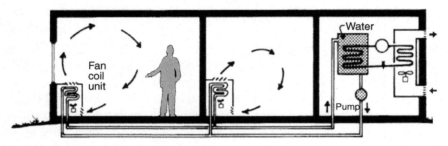

(b) Schematic diagram of an all-water system

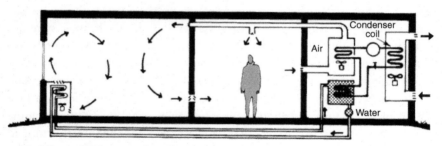

(c) Schematic diagram of an air-water system

Figure 9.11 *Diagram Illustrating Three Basic HVAC Systems.* (a) All-air system. (b) All-water system. (c) Air–water system. *Courtesy: N. Lechner.*

9.3.4 HVAC System Requirements

Building spaces such as cavities between walls can support platforms for air handlers, and plenums defined or constructed with materials other than sealed sheet metal, duct board, or flexible duct must not be used for conveying conditioned air including return air and supply air. Care should be taken to ensure that ducts installed in cavities and support platforms are not compressed in a manner that would cause reductions in the cross-sectional area of the ducts. Connections between metal ducts and the inner core of flexible ducts must be mechanically fastened, and openings must be sealed with mastic, tape, or other duct closure systems that meet the codes and standards of local jurisdictions.

National building codes generally stipulate that access be provided to certain components of mechanical and electrical systems. This is usually required for maintenance and repair, and includes such elements as valves, fire dampers, heating coils, mechanical equipment, and electrical junction boxes. Commercial construction usually takes advantage of ceiling plenums to run horizontal ducts while vertical ducts are contained within their own chases. Depending on the type of structure and depth of the plenum, large ducts may occupy much of this depth, leaving little if any space for recessed light fixtures. Where the plenum is used as a return air space, most local and national building codes prohibit the use of combustible materials such as wood or exposed wire within the space in commercial building projects.

Access flooring (typically in computer room applications), which consists of a false floor of individual panels raised by pedestals above the structural floor, is sometimes used in commercial construction. This is designed to provide sufficient space to run electrical and communication wiring as well as HVAC ductwork. Sometimes small pipes are designed to run within a wall system, whereas larger pipes may need deeper walls or even chase walls to accommodate the pipes. Fan systems that exhaust air from the building to the outside must be provided with back draft or automatic dampers. Gravity ventilating systems must have an automatic or readily accessible, manually operated damper in all openings to the exterior, except combustion inlet and outlet air openings and elevator shaft vents.

9.3.5 Common HVAC Deficiencies

HVAC deficiencies are usually maintenance related. When maintenance of the equipment is deferred or performed by unqualified personnel, the system will increasingly experience problems. When properly maintained, a building's HVAC system can enjoy a substantial life span. HVAC deficiencies fall into two main categories: issues that are fairly simple to address, such as filter or belt replacement, and complex issues requiring the attention of specialized personnel, such as pump or boiler replacement. Another deficiency often encountered is inadequacy of the system for the size of the facility. Most designers of mechanical systems now utilize computer analysis software to determine heating and cooling loads, but a general rule of thumb in an assessment is to compare the actual tonnage of the unit to the standard design tonnage using the following formulas:

BTU of unit ÷ 12,000 = actual tonnage of the unit

Square footage of the building + 350 = design tonnage

Deficiencies and other issues requiring to be checked for conformance when conducting HVAC evaluations include the following:

- Evidence of abnormal component vibrations or excessive noise
- Evidence of unsafe equipment conditions, including instability or absence of safety equipment (guards, grills, or signage)
- Identify the location of the thermostats
- Does the building have an exhaust system and are the toilets vented independently or mixed with the common area venting system?
- Insufficient air movement to reach all parts of the room or space being cooled or, alternatively, the presence of drafts in the room or space being cooled
- Return air at the return registers is not at least 10–15°F warmer than the supply air
- Evidence of leaking caused by inadequate seals
- Evidence of fan alignment deficiencies, deterioration, corrosion, or scaling
- Is there a fresh air make-up system in the building?

9.3.6 HVAC Components and Systems

The principal function of an HVAC system is to provide building occupants with healthy and comfortable interior thermal conditions. HVAC systems generally include a number of active mechanical and electrical systems employed to provide thermal control in buildings. Control of the thermal environment is one of the key objectives of virtually all occupied buildings. Numerous systems and components are used in combination to provide fresh air, as well as temperature and humidity control in both residential and commercial properties (Figure 9.12).

HVAC system components may best be grouped into three functional categories: *source components, distribution components,* and *delivery components*:

- *Source components* generally provide or remove heat or moisture. There are four basic types of heat sources employed in buildings: (1) on-site combustion (fuel such as natural gas or coal); (2) electric resistance (converting electricity to heat); (3) solar collector on roof to furnace; and (4) heat pump in furnace. Choosing a heat source for a given building

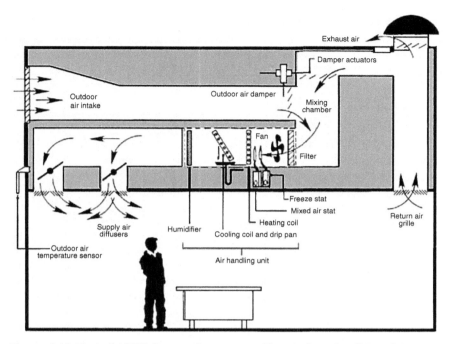

Figure 9.12 *Typical HVAC System Components That Deliver Conditioned Air to a Building or Space to Maintain Thermal Comfort and Indoor Air Quality.* Source: Terry Brennan, Camroden Associates.

depends on several factors such as source availability, fuel costs, required system capacity, and equipment costs.

- *Distribution components* are used to convey a heating or cooling medium from a source location to portions of a building that require conditioning. Central systems produce a heating and/or cooling effect in a single location, which is then transmitted to the various spaces in a building that require conditioning. Three transmission media are commonly used in central systems: air, water, and steam. Hot air can be used as a heating medium, cold air as a cooling medium. Hot water and steam can be used as heating media, whereas cold water is a common cooling medium. When a central system is used, it will always require distribution components to convey the heating or cooling effect from the source to the conditioned locations.
- *Delivery components* basically serve as an interface between the distribution system and occupied spaces. The heating or cooling effect produced at a source and distributed by a central system to all the spaces within a

building has to be properly delivered to each space to promote comfort and well-being. In air-based systems, the dumping of heated or cooled air into each space does not provide the control over air distribution required of an air-conditioning system. Likewise, with water-based systems, the heated or cooled media (water or steam) cannot just be dumped into a space. Some means of transferring the conditioning effect from the media to the space is required. Devices designed to provide the interface between occupied building spaces and distribution components are collectively termed delivery devices.

It should be noted that compact systems that only serve a single zone of a building frequently incorporate all three functions in a single piece of equipment, whereas systems that are intended to condition multiple spaces in a building (central systems) usually employ distinctly different equipment elements for each function. Furthermore, for each commercial property type, some systems will perform better than others. However, from a lender's perspective, performance is judged by how well the needs of tenants and owners are met regarding comfort, operating costs, aesthetics, reliability, and flexibility.

Ductwork: The primary objective of duct design is to provide an efficient distribution network of conditioned air to the various spaces within a building. To achieve this, ducts must be designed to facilitate airflow and minimize friction, turbulence, and heat loss and gain. Optimal air distribution systems have correctly sized ducts with minimal runs, smooth interior surfaces, and minimum direction and size changes. Ducts that are badly designed and installed can result in poor air distribution, poor indoor air quality (IAQ), occupant discomfort, additional heat losses or gains, increased noise levels, and increased energy consumption. Duct system design requirements and construction can be impacted by the design of the building envelope.

A duct system's overall performance can also be impacted by the materials used. Fiber glass insulation products are currently used in the majority of duct systems installed in the United States, and serve as key components of well-designed, well-operated, and well-maintained HVAC systems that provide both thermal and acoustical benefits for the life of the building. Other materials commonly used for low-pressure duct construction include sheet metal (galvanized steel), black carbon steel, aluminum, stainless steel, fiberglass-reinforced plastic, polyvinyl steel, concrete, and copper.

It is strongly advised to put in place a preventive maintenance inspection program that will help identify system breaches that are typically more prevalent at duct intersections and flexible connections. Supply and return air may both utilize ductwork that may be located in the ceiling cavity or below the floor slab, depending on the configuration of the system. In single-duct systems, both cool air and hot air utilize the same duct, and in double-duct systems, separate ducts are used for cooling and heating. It should be noted that air flowing through a duct system will encounter friction losses through contact with the duct walls and in passing through the various devices such as dampers, diffusers, filters, and coils. To overcome these friction losses, a fan is utilized that provides both the energy input required to overcome friction losses and also to circulate air through the system. Depending on the size of the ductwork, a central HVAC system may require the use of several fans for air supply and return, and for exhaust air.

Where ducts penetrate a firewall, the use of fire dampers will be required. Modern fire dampers contain a fusible link that melts and separates when a particular temperature is reached, causing it to slam shut in the event of a fire. Another major factor that impacts the fabrication and design of duct systems and that has to be considered is that of acoustics, and unless the ducting system is properly designed and constructed, it can act as large speaker tubes transmitting unwanted noise throughout the building. Splitters and turning vanes are often used to reduce the noise that is generated within the ducts while at the same time mitigate friction losses by reducing turbulence within the ductwork.

Grills, registers, and diffusers: These are employed in conjunction with ductwork and assist in controlling the return, collection, and supply of conditioned air in HVAC systems. A *grille* is basically a decorative cover for return air inlets; it does not have an attached damper and in most cases, has no moving parts. However, a grille can be used for both supply air and return air. The same is not true for a register or diffuser. Grills are also used to block sightlines to prevent persons from seeing directly into return air openings.

A *diffuser* is an airflow device designed primarily to discharge supply air into a space, mix the supply air within the room air, and to minimize unwelcome drafts. The location of supply air diffusers and return air grills should be integrated with other ceiling elements, including lights, sprinkler heads, smoke detectors, speakers, and the like, so that the ceiling is aesthetically pleasing and well planned. To assist in this integration, these components are usually connected to the main ductwork with flexible ducting to allow some adjustability in their placement.

Registers are adjustable grill-like devices that cover the opening of a duct in a heating or cooling system, providing an outlet for heated or cooled air to be released into a room. They are similar to diffusers except that they are designed and used for floor or sidewall air supply applications or sometimes as return air inlets. Supply air registers are essentially grills equipped with double-deflection adjustable vanes at the face and a damper behind the face for balance and to control the direction of flow and/or flow rate. A diffuser performs the task of diffusing the gas. The grills, registers, and diffusers introduce and blend the fresh air with the air of another location. When fitted together, the grills take in fresh air, the registers blow the air out, and the diffusers scatter the air in the required space.

Thermostats: These are devices whose main function is to control the operation of HVAC systems, turning on heating or cooling as required. Boilers and heating systems use thermostats to prevent overheating and also control the temperature of the circulating water. Thermostats are fitted to hot water cylinders, boilers, radiators, and in rooms. In addition to providing the basic function of maintaining comfortable indoor temperatures, modern programmable thermostats are capable of being programmed to automatically raise or lower the temperature of a facility according to predefined schedules and allow input of weekday and weekend schedules. More sophisticated thermostats will also allow humidity control, outdoor air ventilation, and even signal when the system filters need changing. Some modern thermostats can also include a communications link and demand management features that can be used to reduce air-conditioning system energy use during periods of peak electrical demand or high electricity costs. Thermostat location should be coordinated with light switches, dimmers, and other visible control devices. They are typically placed 48–60 in. above the floor and away from exterior walls and heat sources.

Zoning and high-efficiency equipment can substantially increase the overall energy performance of a home or office while maintaining rising energy costs to a manageable level. Zoning systems can automatically direct the flow of the conditioned air to those zones needing it and at the same time automatically switch over and provide the opposite mode to other zones, eliminating the need for constant balance and outlet adjustments based on continuously changing indoor conditions. But zoning represents just one of several actions that are designed to improve HVAC performance and which gives building occupants personal climate control throughout their work environment.

HVAC systems can have single-zone or multizone capabilities. In a single-zone system, the entire building is considered one area, whereas in a multizone system, the building is divided into various zones, allowing specific control of each area. In fact, many of today's newest HVAC systems are designed to incorporate individually controlled temperature zones to improve occupant comfort and provide the ability to manage the heating or cooling of individual rooms or spaces by use. Additionally, it allows us to adjust individual room temperatures for individual preferences, and to close off airflow to areas that are rarely used. A zoned HVAC system is typically provided with a series of dampers. The use of zone dampers can save on the installation cost of multiple-unit systems. But whether zoning with one HVAC unit using zone dampers or using multiple HVAC units, zoning can save on utility and maintenance costs. More importantly, perhaps, the design of the duct system for today's zoning is an important factor to a comfortable and efficient zoning system.

Boilers: This is a heating system component designed to generate steam or hot water for distribution to various building spaces (Figure 9.13). As water cannot be used to directly heat a space, boilers are only used in central systems where hot water is circulated to delivery devices (such as baseboard radiators, unit heaters, convectors, or air-handling units). Once the delivery device is heated with the hot water, the water is returned back to the boiler to be reheated, and the water circulation loop continues. Generally speaking, hot water boilers are more efficient than steam boilers for several reasons. For example, there is less heat loss throughout the hot water piping and the shell of the boiler because a hot water boiler operates at a lower temperature than a steam boiler. This means there is less heat loss throughout the entire

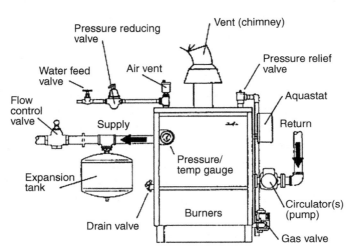

Figure 9.13 *Drawing of a Gas-Fired Hot Water Boiler Showing Main Components. Courtesy: Home-Cost.com*

boiler and piping system. Also, because a hot water boiler operates at a lower temperature, it requires less fuel or energy to convert into heat.

An on-site solar energy collection system may serve in lieu of a boiler. Heat transfer systems (heat pumps) likewise may serve as a substitute for a boiler. Constructed of cast iron or steel, and occasionally copper, boilers can be fired by various fuel sources, including natural or propane gas, electricity, coal, oil, steam or hot water, and wood.

Chilled water systems: These are cooling systems that remove heat from one element (water) and move it into another element (ambient air or water). They are a key component of air-conditioning systems for large buildings although they typically use more energy than any other piece of equipment in large buildings. It is similar to an air-conditioning system in that it is compressor based, but a chiller cools liquid whereas an air-conditioning system cools air. Other components are a reservoir, recirculating pump, evaporator, condenser, and a temperature controller. Chillers vary in terms of condenser cooling method, cooling specifications, and process pump specifications. The cooling fluid used is usually a mix of ethylene glycol and water.

Chillers can either be *air-cooled, water-cooled,* or *evaporatively cooled.* Water-cooled chillers incorporate the use of cooling towers, which improve the chillers' thermodynamic effectiveness as compared to air-cooled chillers. This is due to heat rejection at or near the air's wet-bulb temperature rather than the higher, sometimes much higher, dry-bulb temperature. Evaporatively cooled chillers offer efficiencies better than air cooled, but lower than water-cooled. Air-cooled chillers are usually located outside and consist of condenser coils cooled by fan-driven air. Water-cooled chillers are typically located inside a building, and the heat from these chillers carried by recirculating water to outdoor cooling towers. Evaporatively cooled chillers are basically water-cooled chillers in a box. These packaged units cool the air by humidifying it and then evaporating the moisture.

Chilled water systems are mainly employed in modern commercial and industrial cooling applications although there are some residential and light commercial HVAC chilled water systems in use. One of the reasons behind the popularity of chilled water systems is because they use water as a refrigerant. Water is much less expensive than refrigerant, which makes them cost-effective especially in commercial HVAC air-conditioning applications. Thus, instead of running refrigerant lines over a large area of the building, water pipes are run throughout the building and to evaporator coils in air handlers for HVAC air-conditioning systems. The chilled water is pumped through these pipes from a chiller where the evaporator coil absorbs heat and returns it to the chiller to reject the heat. Maintaining them well and operating them smartly can yield significant energy savings.

There are two main types of chillers commonly used today: the *compression chiller* and the *absorption chiller*. Absorption chillers use a heat source such as natural gas or district steam to create a refrigeration cycle that does not use mechanical compression. During the compression cycle, the refrigerant passes through four major components within the chiller: the evaporator, the compressor, the condenser, and a flow-metering device such as an expansion valve. The evaporator is the low-temperature (cooling) side of the system, and the condenser is the high-temperature (heat-rejection) side of the system. Compression chillers, depending on the size and load, use different types of compressors for the compression process. Mechanical compression chillers, for example, are classified by compressor type: (1) reciprocating, (2) rotary screw, (3) centrifugal, and (4) frictionless centrifugal.

Important factors that impact the choice of a water-cooled chiller or an air-cooled unit include whether a cooling tower is available or not. The water chiller option is often preferred over the air-cooled unit because it costs less, has a higher cooling capacity per horsepower, and consumes less energy per horsepower. Compared to water, air is a poor conductor of heat, making the air-cooled chiller much larger and less efficient, which is why it is less frequently used unless it is not possible to construct a water-cooling tower.

Cooling towers: Their main function is removing heat from the water discharged from the condenser so that the water can be discharged to the environment or recirculated and reused. Cooling towers are used in conjunction only with water-cooled chillers and vary in size from small rooftop units to very large hyperboloid structures. Cooling towers are also characterized by the means by which air is moved. *Mechanical draft cooling* towers are the most widely used in buildings, and rely on power-driven fans to draw or force the air through the tower. They are normally located outside the building.

Mechanical draft towers can be one of two types that are common to the HVAC industry: *induced draft* and *forced draft*. Induced draft towers have a large propeller fan at the top of the tower (discharge end) to draw air upward through the tower while warm condenser water spills down. This type requires much smaller fan motors for the same capacity than forced draft towers (Figure 9.14a,b). Forced draft towers utilize a fan at the bottom or side of the structure. Air is forced through the water spill area and discharged out the top of the structure. After the water has been cooled

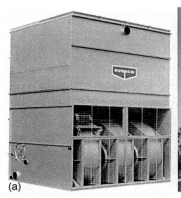

Figure 9.14 (a) Forced draft towers with fans on the air inlet to push air either counterflow or cross-flow to the movement of the water. (b) Induced draft towers have a large propeller fan at the top of the tower to draw air counterflow to the water. Induced draft towers are considered to be less susceptible to recirculation, which can result in reduced performance. *Source: McQuay International.*

in the cooling tower, it is pumped to a heat exchanger or condenser in the refrigeration unit where it picks up heat again and is returned to the tower.

Condensers: A condenser is one of the critical components used in air-conditioning systems to cool and condense the refrigerant gas that becomes hot during the evaporation stage of the cooling process. There are two common condenser types: air-cooled and cooling tower. Achieving the cooling process is accomplished through the use of air, water, or both.

Air filters: The primary function of air filters is to remove particles from the air. They are a critical component of the air-conditioning system; without them, air-conditioning systems would become dirty and the interior environment would be filled with pollutants and become unhealthy (Figure 9.15). Although no individual product or system in itself can be LEED-certified, proper employment of air filtration systems provide tangible ways to improve indoor air quality (IAQ) and energy efficiency and can contribute to the completion of LEED prerequisites and credits. This is why the right filter media strategy is important and can help buildings become "greener" and meet LEED and other green building rating system criteria.

The latest LEED™ *Reference Guide* (depending on the LEED certification targeted) should be consulted for more detailed information relating to achievable credits. For example, for EQ Credit 3.1 (Construction IAQ Management Plan: During Construction: "If permanently installed air handlers are used during construction, filtration media with a Minimum Efficiency ReportingValue (MERV) of 8 shall be used at each return air grille,

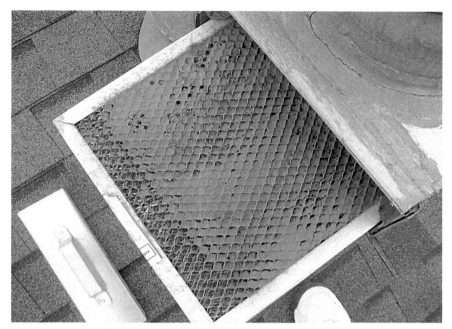

Figure 9.15 *A Photograph Showing a Clogged Filter that has been Removed From an AC Unit.* It is critically important to change filters periodically to minimize pollution and improve indoor air quality.

as determined by ASHRAE 52.2-1999." Upon completion of construction and prior to occupancy (after all interior finishes completed), new MERV 13 filters should be installed followed by a 2-week building flush-out by supplying a total air volume of 14,000 ft.^{3} of outdoor air per square feet of floor area. Also for NC EQ Credit 5: Indoor Chemical and Pollutant Source Control, the reference guide states, "In mechanically ventilated buildings, provide regularly occupied areas of the building with new air filtration media prior to occupancy that provides a Minimum Efficiency Reporting Value (MERV) of 13 or better. Filtration should be applied to process both return and outside air that is to be delivered as supply air."

An understanding of ASHRAE 52.1 and 52.2 is pivotal in identifying what to look for when selecting the right filter to meet IAQ and energy efficiency requirements to help achieve green building standards. On this point, Dave Matela of Kimberly-Clark Filtration Products, says,

One of the biggest determining factors is filtration efficiency, which defines how well the filter will remove contaminants from air passing through the HVAC system. Initial and sustained efficiency are the primary performance indicators for HVAC

filters. Initial efficiency refers to the filter's efficiency out of the box or immediately after installation. Sustained efficiency refers to efficiency levels maintained throughout the service life of the filter. Some filters have lower initial efficiency and do not achieve high efficiency until a "dirt cake" builds up on the filter. Other filters offer high initial, as well as sustained, efficiency, meaning they achieve an ideal performance level early and maintain that level.

There is a variety of air-conditioning filters, the most common types are conventional fiberglass disposable filters (1-in. and 2-in.), pleated fiberglass disposable filters (1-in. and 2-in.), electrostatic filters, electronic filters, and carbon filters. Most air-conditioning filters are sized 1.5–2 ft.^{2} for each ton of capacity for a home or commercial property. Applying MERV rating is a good way to help evaluate the effectiveness of a filter. MERV means minimum efficiency reporting value, which was developed by the American Society of Heating, Refrigeration and Air Conditioner Engineers – ASHRAE. MERV values vary from 1 to 16, and the higher the MERV value is, the more efficient the filter will be in trapping airborne particles. Another consideration is airflow through the HVAC system. Leaving a dirty air filter in place or using a filter that is too restrictive may result in low airflow and possibly cause the system to malfunction. Of note, there are various types of filters with a MERV 13, each having different design requirements and pressure drop. It would be prudent, therefore, for building owners and designers to consult with a Certified Air Filter Specialist (CAFS) to obtain the best information on the optimum filters and prefilters to obtain LEEDS certification. Filters should be selected for their ability to protect both the HVAC system components and general indoor air quality.

9.4 ELECTRICAL POWER AND LIGHTING SYSTEMS

9.4.1 General

Electrical systems that provide a facility with accessible energy for heating, cooling, lighting and equipment (telecommunication devices, personal computers, networks, copiers, printers, etc.) and appliance operation (e.g., refrigerators and dishwashers) has witnessed dramatic developments in the past few decades, comprising the fastest-growing energy load within a building. More than ever, facilities today need electrical systems to provide power with which most of the vital building systems operate. These systems control the energy required in the building and distributes it to the location utilizing it. Most frequently, distribution line voltage carried at utility poles is delivered at 2400/4160V. Transformers step down this voltage to predefined levels for use within buildings. In an electric power distribution grid, the most common form of electric service is through the use of overhead wires known as a *service drop*, which is an electrical line running from a utility pole to a customer's building or other premises. It is the point where electric utilities provide power to their customers.

In residential installations in North America and countries that use their system, a service drop comprises of two 120-V lines and a neutral line. When these lines are insulated and twisted together, they are referred to as a *triplex* cable. In order for these lines to enter a customer's premises, they must usually first pass through an electric meter and then the main service panel, which will usually contain a "main" fuse or circuit breaker. This circuit breaker controls all of the electrical current entering the building at once, and a number of smaller fuses/breakers, which protect individual branch circuits. There is always a main shutoff switch to turn off all power; when circuit breakers are used, this is provided by the main circuit breaker. The neutral line from the pole is connected to an earth ground near the service panel — often a conductive rod driven into the earth.

In residential applications, the service drop provides the building with two separate 120-V lines of opposite phase, so 240 V can be obtained by connecting a circuit between the two 120-V conductors, whereas 120-V circuits are connected between either of the two 120-V lines and the neutral line. In addition, 240-V circuits are used for high-power devices and major appliances, such as air conditioners, clothes dryers, ovens, and boilers, whereas 120-V circuits are used for lighting and ordinary small appliances. It should be noted that these are "nominal" numbers, meaning that the actual voltage may vary.

In Europe and many other countries, a three-phase 416Y/230 system is used. The service drop consists of three 240-V wires, or phases, and a neutral wire, which is grounded. Each phase wire provides 240 V to loads connected between it and the neutral. Each of the phase wires carries 50-Hz alternating current, which is 120° out of phase with the other two. The higher voltages, combined with the economical three-phase transmission scheme, allow a service drop to be longer than in the North American system, and allow a single drop to service several customers.

For commercial and industrial service drops, which are usually much larger and more complex, a three-phase system is used. In the United States, common services consist of 120Y/208 (three 120-V circuits 120° out of phase, with 208 V line to line), 240-V three-phase, and 480-V three-phase. In Canada, 575-V three-phase is common, and 380–415V or 690-V three-phase

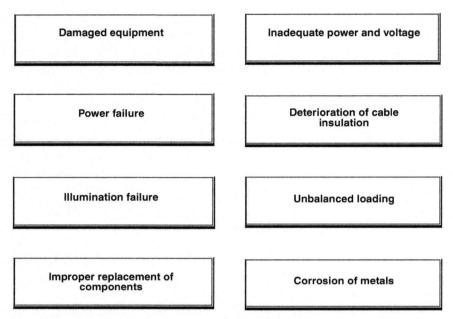

Damaged equipment	Inadequate power and voltage
Power failure	Deterioration of cable insulation
Illumination failure	Unbalanced loading
Improper replacement of components	Corrosion of metals

Figure 9.16 *Diagram Showing Typical Deficiencies Found in Electrical Systems.*

is found in many other countries. Generally, higher voltages are used for heavy industrial loads, and lower voltages for commercial applications.

The difference between commercial and residential electrical installations can be quite significant, particularly with large installations. Although the electrical needs of a commercial building can be simple, consisting of a few lights for some small structures, they are often quite complex, with transformers and heavy industrial equipment. When electrical or lighting system deficiencies become evident and need attention, they are usually measurable and include power surges, tripped circuit breakers, noisy ballasts, and other more obvious conditions such as inoperative electrical receptacles or lighting fixtures that are frequently discovered or observed during a review of the system. As illustrated in Figures 9.16 and 9.17, there are a number of typical deficiencies found in both the electrical and the lighting systems.

In many commercial buildings, the major load placed on a given electrical system comes from the lighting requirements; therefore, the distribution and management of electrical and lighting loads must always be monitored on a regular basis. Lighting management should also be periodically checked because building space uses change and users relocate

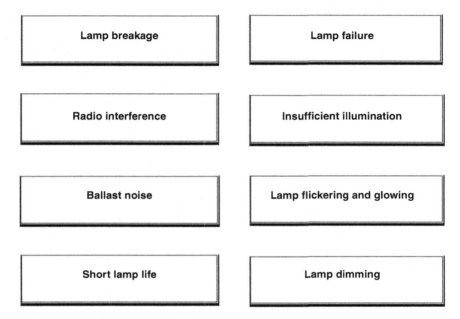

Figure 9.17 *Diagram Showing Typical Deficiencies Found in Lighting Systems.*

within the building. It is also highly advisable for the lighting system to be integrated with the electrical system in the facility. Lighting systems are designed to ensure adequate visibility for both the interior and exterior of a facility and are composed of an energy source, and distribution elements normally consisting of wiring and light–emitting equipment.

There are several different electrical codes today being enforced in various jurisdictions throughout the United States. Some of the larger cities, such as New York and Los Angeles, have created and adopted their own electrical codes. The National Electrical Code (NEC) and the National Fire Protection Code (NFPC), published by the National Fire Protection Association (NFPA), cover almost all electrical system components. The NEC is commonly adopted in whole or in part by municipalities. Inspection of the electrical and lighting system should include a determination of general compliance with these codes at the facility.

9.4.2 Basic Definitions: Amperes, Volts, and Watts

Amperage (A or Amp): This is a unit of electrical current. The Amperage or Amps provided by an electrical service is the flow rate of "electrical current" that is available. Appliance will typically have an Amp rating or, if only a wattage is quoted, Amps = Wattage/Voltage. Practically speaking, the voltage level provided by an electrical service, combined with the ampacity rating of the service panel determines the electrical load or capacity. Branch circuit wire sizes and fusing or circuit breakers used to typically set the limit on the total electrical load or the number of electrical devices that can be run at once on a given circuit. Thus, for example, if you have a 100-A current flow rate in place, you may be able to run approximately ten 10-A electric heaters simultaneously. If you have only 60 A available, you won't be able to run more than 6 such heaters without risking overheating the wiring, tripping a circuit breaker, blowing a fuse, or causing a fire.

The service ampacity and voltage must be known to be able to determine the amount of electrical service a facility receives. The safe and proper service amperage available at a property is set by the smallest of the service conductors, the main disconnect fuse or switch, or the rated capacity of the electric panel itself. The main fuse/circuit breaker (CB) is the only component that actively limits amperage at a property by shutting off loads, drawing more than the main fuse rating. The main breakers or fuses are allowed to have lower overcurrent protection than the capability of the service equipment (panel) and conductors (entrance cable).

Volt (*V*): A volt can be defined as the potential difference across a conductor when a current of one ampere dissipates one watt of power. Practically speaking, a volt is a measure of the strength of an electrical source at a given current or amperage level. If we bring 100 A into a building at 240 V, we have twice as much power available as if we bring in 100 A at 120 V. However, if we exceed the current rating of a wire it will get hot, risking a fire. This is why fuse devices are employed to limit the current flow on electrical conductors to a safe level and thereby prevent overheating and potential fires.

As mentioned earlier, a "240-V" circuit is a nominal rating, which implies that the actual voltage level will vary. In many countries, the actual voltage level varies around the nominal delivered "voltage rating" and in fact depending on the quality of electrical power delivered on a particular service, voltage will also vary continuously around its actual rating. Most electrical power systems are prone to slight variations in voltage due to demand or other factors. Generally, this difference is inconsequential, as most appliances are built to tolerate current a certain percentage above or below the rated voltage. However, severe variations in current can damage electrical equipment. It is always advisable to install a voltage stabilizer where sensitive electronic equipment is used.

Watts: In electricity, a watt is equal to current (in amperes) multiplied by voltage (in volts). The formula watts = volts × amps basically describes this relationship. In buildings, the unit of electricity consumption measure is the watt-hour, which is usually in thousands, called kilowatt-hours. In larger buildings, not only is the total consumption rate measured but the peak demand as well.

9.4.3 Components to be Evaluated

Service connection: This is basically a device that provides a connection between the power company service and the facility and also measures the amount of electricity a facility uses. From here a meter either feeds a disconnect switch or a main breaker or fuse panel. The connection can be located either overhead or underground. The service connection should be checked for type (i.e., voltage, amperage), and general condition and whether the total power adequately serves the facility's requirements. The equipment should be clean and free from debris or overgrown planting.

Switchgear and switchboards: The function of switching equipment is to control the power supply in the facility and all the services arriving on the site (service drop). Switchgear is used to interrupt or reestablish the flow of electricity in a circuit. It is generally used in combination with metering, a disconnect switch, protective, and regulating equipment to protect and control motors, generators, transformers, and transmission and distribution lines. The main service switch is the system disconnect for the entire electrical service. To avoid excessive voltage drop and flicker, the distance from the transformer to the meter should not exceed 150 ft.

Switchgears are typically concentrated at points where electrical systems make significant changes in power, current, or routing, such as electrical supply substations and control centers. Switchgear assemblies range in size from smaller, ground-mounted units to large walk-in installations and can be classified as outdoor or indoor units. Commercial and industrial assemblies are usually indoors, while utilities and cogeneration facilities are more likely to have outdoor gear. Manufactured for a variety of functions and power levels, all switchgear conforms to standards set by the Institute of Electrical and Electronic Engineers (IEEE), the American National Standards Institute (ANSI), or the National Electrical Manufacturers Association (NEMA).

A switchboard is composed of one or more panels with various switches and indicators that are used to route electricity and operate circuits. The main switchboard controls and protects the main feeder lines of the system.

Switchgear and switchboards should be readily accessible, in good condition, and have protective panels and doors. Moreover, they should be checked for evidence of overloading or burn marks. Switchboard covers should not normally be removed.

Meters: Measurement of electric consumption in a building can be accomplished in one of two ways. In residential applications, only the total electric consumption is measured. In larger facilities, both the total consumption rate peak demand is measured. This is because large peaks require the utility company to build more power-generating capacity to meet the peak. Commercial services of up to 200 A single-phase may have service panels similar to those found in residences. Larger services may require standalone switchboards with one or more meters.

Panelboards: Electrical panel boards and their cabinets house an assembly of circuit breakers and from where power generation can be monitored and the power generated can be distributed. Panelboards are designed to control and protect the branch circuits in addition to providing a central distributing point for the branch circuits for a building, a floor, or part of a floor. Each breaker serves a single circuit, and the overload protection is based on the size and current-carrying capacity of the wiring in that circuit. A building may have a number of panelboards and a main panel, with a disconnect switch for the entire building.

To estimate the electric service panel ampacity: Evidence of a tag (normally paper) or embossed rating on fuse pull-outs on the panel itself and often includes the amperage rating of the panel. This information is usually present in newer panels on a panel side, or on the panel cover. Actual dimensions of an electric panel are not a reliable determinant of ampacity. For example, many larger panels can be fitted with a variety of bus-bar and main switch assemblies of varying ampacity.

Aluminum wiring: This type of wiring was used in some residences mainly between the 1965 and the early 1973, but this application has been largely discontinued because aluminum-wired connections have been found to have a very high probability of overheating compared with copper-wired connections and were therefore a potential fire hazard. A large number of connection burnouts have occurred in aluminum-wired homes, and according to the US Consumer Product Safety Commission, many fires have resulted, some involving injury and death. The main problem with aluminum wiring is a phenomenon known as "cold creep." When aluminum wiring warms up, it expands, and when it cools down, it contracts. However, unlike copper, when aluminum goes through a number of warm/cool cycles it begins to lose some of its tightness. To add to the problem, aluminum oxidizes, or corrodes when in contact with certain types of metal, so the resistance of the connection increases. This causes it to heat up and corrode/oxidize still more and until eventually the wire may start getting very hot and melt the insulation or fixture it's attached to, and possibly even cause a fire without ever tripping the circuit breaker.

Aluminum wire "alloys" were introduced in the early 1970s but this did not address most of the connection failure problems. Aluminum wiring is still permitted and used for certain applications, including residential service entrance wiring and single-purpose higher amperage circuits such as 240 V air-conditioning or electric range circuits. Although the fire risk from single-purpose circuits is much less than for branch circuits, field reports indicate that these connections remain a potential fire hazard.

A simple method of identifying aluminum wiring is to examine the wire sheathing for the word aluminum. If you can't find the word aluminum embossed in the wire sheathing then look for silver-colored wire instead of the copper-colored wire used in modern wiring. Without opening any electrical panels or other devices, it is possible to still look for printed or embossed letters on the plastic wire jacket where wiring is visible at the electric panel. Some aluminum wire has the word "Aluminum" or a specific brand name such as "Kaiser," "Alcan," "Aluminum," or "AL/2" plainly marked on the plastic wire jacket. Some white-colored plastic wire jackets are inked in red; others have embossed letters without ink and are hard to read. Shining a light along the wire may make it easier to identify. Of note, the fact that no aluminum wiring was evident in the panel does not necessarily mean that none is present. Aluminum may have been used for part of circuits or for some but not other circuits in the building.

Service outlets and receptacles: Service outlets include convenience receptacles, motors, lights, and appliances. Receptacles are commonly known as *outlets* or sometimes erroneously as wall plugs (a plug is what actually goes into the outlet). It is preferable for outlets to be three-prong, where the third prong is grounded. For large spaces or areas, all of the outlets should not be on the same circuit so that when a fuse or circuit breaker trips due to an overload, the space will not be plunged into complete darkness.

Grounding: The grounding of a service to earth is basically a safety precaution and is necessary mainly to protect against lightning strikes or other high-voltage line strikes. Earth grounding in a commercial building might be to a grounding rod inside a switchboard, to a steel cold water pipe in the plumbing system, or to the steel frame of a building. Other methods of grounding are also used depending on the equipment or system to be grounded. Ground wires are typically covered with green insulation or sometimes may be without cover.

Motors: Are devices that convert any form of energy into mechanical energy, especially an internal-combustion engine or an arrangement of coils and magnets that convert electric current into mechanical power. There are basically four types of motors in general use. (1) The *DC motor* is a rotating electric machine designed to operate from a direct voltage source. It is used for small-scale applications and for elevators, where continuous and smooth acceleration to a high speed is important. (2) *Stepper/switched reluctance (SR) motors*: These are brushless, synchronous electric motors that can divide a full rotation into a large number of steps. The motor's position can be controlled precisely, without any feedback mechanism. Stepper motors are basically similar to SR Motors – in fact, the latter are very large stepping motors with a reduced pole count, and generally are closed-loop commutated. The main advantage of stepper motors is that they can achieve accurate position control without the requirement for position feedback. Stepper motors operate differently from normal DC motors, which rotate when voltage is applied to their terminals. (3) *Induction motors* can be either 3-Phase AC or single-Phase AC. The 3-Phase AC induction motor is a rotating electric machine designed to operate from a three-phase source of alternating voltage and usually applies to larger motors. These motors are characterized by extreme reliability and remain constant in rpm, unless heavily overloaded. The single-phase motor is a rotating machine that has both main and auxiliary windings and a squirrel-cage rotor. (4) The fourth type of motor in general use is the *Universal motor*, which is a rotating electric machine similar to a DC motor but designed to operate either from direct current or single-phase alternating current, and which varies in speed based on the load. The universal motor is usually found in mixers, hand drills, and similar appliances. Motors should always be protected against overload by *thermal relays* that shut off the power when any part of the motor or housing overheats.

Switches and controls: Switches and controls are devices that direct the flow of power service to the electrical equipment. Safety switches are installed in locations where service cut-off is available in case of emergencies. These include toggle switches, dials, and levers.

Emergency power and emergency lighting: Large facilities often have standby power with which to ensure continued electrical service when a shutdown of the standard power service takes place. Emergency power is required for life support systems, fire and life safety circuits, elevators, exit and emergency lighting. Facilities, which require full operation during emergencies or disasters, such as hospitals and shelters, always have back up power. Computer facilities, to ensure continued storage and survival of the data, also commonly have emergency power. For major equipment, a diesel engine generator with an automatic starting switch and an automatic transfer switch is often provided for emergency power (Figure 9.18) while for lighting, battery units are installed. The typical AC power frequency in the United States is 60 cycles per second or 60 Hz, whereas in Europe, 50 Hz is the standard.

Having an emergency lighting system in place is necessary in the event of a power failure or other emergency; emergency lighting in a facility enables the occupants to exit safely. Emergency lighting can consist of individual battery units placed in all corridors and areas that may require sufficient lighting for building user exiting, in interior and some exterior exitways. These batteries are continuously recharged while power is on, and which take over when power is lost. Alternatively, the lighting can be powered by a central battery unit. Fluorescent lamps will require some method of power conversion as batteries are typically 12 V.

Transformers: It is a device that converts an alternating current (A/C) circuit of a certain voltage to a higher or lower value, without change of frequency, by electromagnetic induction. A "step up" transformer receives a low voltage and converts it into a higher voltage, and a "step down"

Figure 9.18 *Photo of an Amtrak Generator for Emergency Backup Power Inside a Building.*

transformer does just the reverse. Transformers are often used to step up voltage in order to transmit power over long distances without excessive losses, and subsequently step down voltage to more useable levels. While a transformer changes the voltage in a circuit, it has practically no effect on the total power in the circuit.

There are two distinct types of transformers: Wet or dry type. There are also subcategories of each main type. For wet or liquid-filled transformers, the cooling medium can be conventional mineral oil. Some wet-type transformers use less flammable liquids, such as high fire point hydrocarbons and silicones. Wet transformers are typically more efficient than dry-types, and usually have a longer life expectancy. There are some drawbacks, however. For example, fire prevention is more important with liquid-type units because the liquid cooling medium used may catch fire (although dry-type transformers are also susceptible) or even explode. Wet-type transformers typically contain a type of fire-resistive fluid or mineral oil such as polychlorinated biphenyls, and, depending on the application, wet transformers may require a containment trough for protection against possible leaks of the fluid, which is why they are preferred predominantly when placed outdoors.

For lower voltage type indoor-installed distribution transformers of 600 V and less, the dry-type transformer is preferred even though they have minimal requirements for insulation and avenues for ventilation of heat generated by voltage changes. Dry-type transformers come in enclosures that have louvers or are sealed. Location of transformers should be carefully considered, and there should be clear access surrounding exterior transformers and adequate ventilation and access for interior transformers, which should be inside a fireproof vault. On-site transformers in parking lots may require bollards or other protection (Figure 9.19). Transformers should be analyzed for polychlorinated biphenyls and their registration number noted. In addition, transformers tend to make a certain amount of noise (hum), and it is desirable to address this.

9.4.4 Lighting Systems

The main objectives of a good lighting system are to provide the visibility required based on the task to be performed while achieving the desired economic objectives. Another primary objective must be to minimize energy usage while achieving the visibility, quality, and aesthetic objectives. The quality and quantity of lighting affects the ambience, security, and function of a facility as well as the performance of its employees. Divergent artificial

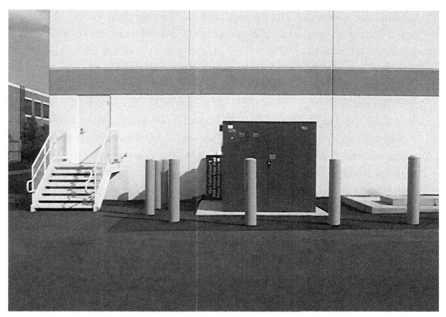

Figure 9.19 *Illustration of a Transformer Located Outside a Facility and Protected by Bollards.*

sources produce different kinds of light and vary significantly in their efficiency, which is the calculated lumen output per watt input. Regrettably, US lighting design does not readily translate overseas – not when different regions have their own voltage, product standards, construction methods, and conceptions about what light is meant to achieve.

Interior lighting: It goes without saying that without a light source we cannot see, and without surfaces to reflect light, there is nothing to see. To understand this relationship between a source of light, the surfaces that reflect light, and how we see light, we need to have a common comprehensive lighting language. This is particularly important because lighting typically accounts for a significant percentage of the annual commercial business and residential electric bills. Advances in lighting technology can make significant reductions in the amount of money that is spent for lighting a facility.

Today we find many types of interior lighting systems that enable us to make full use of a facility around the clock. The most common categories are the following:

Fluorescent lamps: This type of fixture has long been preferable to incandescent lighting, in terms of energy efficiency. Fluorescent lighting is far more efficient than incandescent bulbs (3–5 times as efficient as standard incandescent lamps) and has an average life of 10–20 times longer (fluorescent lamps last up to 20,000 h of use), and use roughly one-third as much electricity as incandescent bulbs with comparable output. Compact fluorescent lamps (CFLs) are similar in operation to standard fluorescent lamps but are manufactured to produce colors similar to incandescent lamps. New developments with fluorescent technology, including the high-efficacy T-5, T-8, and T-10 lamps, have pushed the energy efficiency envelope further. Recently, attention has also been paid to the mercury content of fluorescents and the consequences of mercury releases into the environment. As with all resource use and pollution issues, reduction is the best way to limit the problem. Even with low-mercury lamps, however, recycling of old lamps remains a high priority.

A fluorescent fixture typically consists of the lamp and associated *ballast*, which controls the voltage and the current to the lamp. Replacing standard incandescent light bulbs with CFLs will reportedly slash electrical consumption in homes and offices where incandescent lighting is widely used. By reducing the amount of electricity used, corresponding emissions of associated carbon dioxide, sulfur dioxide, and nitrous oxide are reduced. CFL technology continues to develop and evolve, and is now capable of replacing most of the light fixtures that were originally designed for incandescent light bulbs. Types of fluorescent lamps include

- *Linear fluorescent lamps* T5HO lamps are increasingly being used in many high-bay applications in place of high-intensity discharge (HID) lamps. These smaller-diameter lamps have replaced the T12 lamps that have dominated the market for several decades. These latest lamps work best in luminaires that provide the general ambient lighting for a space. The long and diffuse nature characterized by these lamps provides superior surface lighting, and the smaller lamp diameters make for enhanced optical performance in many luminaires. Typical applications include indirect/direct linear fluorescent pendants and wall-mounted uplights.

- *Compact fluorescent lamps* (CFL) are typically used as basic substitutes for incandescent lamps due to their significantly longer life and better energy efficiency. Of note, one disadvantage is that retrofit lamps cannot be dimmed. Also, the performance of screw-in lamps is usually not as good as the separate lamp and ballast combination. Their small size encourages CFL lamps to be used in recessed luminaires, wall and ceiling mounted fixtures, as well as track lighting and task lighting. The diffuse nature of the fluorescent lamp makes the CFL lamp an appropriate choice for downlighting and wall lighting.

- *Inductive fluorescent lamps* are white light sources with superior color rendering and color temperature properties. These lamps, like their fluorescent counterparts, are energy efficient and offer extremely long life (over 100,000 h), good lumen maintenance characteristics, and instant-on capability. The lamp is coated on the inside with phosphor, and is powered by a small generator (about the size of a fluorescent ballast) attached to the lamp via a short fixed-length cable. The generator prompts a current in the lamp, which causes it to glow – there are no electrodes to wear out. Because of their larger size and diffuse nature, these sources are considered to be ideal for lighting larger volumes and surfaces. They are often used in place of low- to medium-wattage high-intensity discharge sources because of the instant-on capability and reduced maintenance typically associated with the longer lamp life.
- *Low-mercury fluorescent lamps* can be disposed of in landfills in a number of states. In these states, lamps that have sufficiently low levels must pass the testing procedure known as the Toxic Characteristic Leaching Procedure test (see EPA SW-846, "Test Methods for Evaluating Solid Waste (Physical/Chemical Methods)." It should be noted that many states have pending legislation that would not allow the disposal of products containing mercury in a landfill. Specifying a low-mercury product and then recycling that lamp at the end of its life offers the best environmental solution to disposal of mercury-containing lamps.

Incandescent lamps: Incandescent lamps have relatively short lives (typically 1000–2000 h of use) and are the least efficient of common light sources. In fact, only about 15% of the energy they use comes out as light and the rest becomes heat. Nevertheless, they remain popular because they produce a pleasant color that is similar to natural sunlight and they are the least expensive to buy. Incandescent lamps come in various shapes and sizes with different characteristics. The most common incandescent outdoor lighting options are metal halide, and high-pressure sodium.

Environmental issues include lamp efficacy (lumens per watt), luminaire efficiency, controllability of the light source, potential for PV power, and control of light pollution. To control light pollution, full-cutoff luminaires should be specified. It also makes very good sense to use whenever possible environmentally friendly, commercial outdoor lighting systems. For example, ENERGY STAR lights consume only about 20% of the energy consumed by traditional lighting products. This provides substantial savings in money spent, energy consumed and reducing greenhouse gas emissions.

Tungsten halogen lamps are a type of incandescent lamp that has become increasingly popular in recent years. They produce a whiter, more intense light than standard incandescent lamps and are typically used for decorative, display, or accent lighting. They are about twice as efficient as regular incandescent lamps and last two to four times longer than most incandescent lamps.

High-intensity discharge: This category of high-output light source consists of a lamp within a lamp that runs at a very high voltage. HID lamps are considered one of the best performing and most efficient lamps for lighting large areas or great distances. There are basically four types of HID lamps: (1) high-pressure sodium lamp, (2) mercury vapor lamps, (3) metal halide gas, and (4) low-pressure sodium lamps. Metal halide (white light) lamps are replacing high-pressure sodium lamps in many outdoor applications because white light sources have proved to be 2–30 times more effective in peripheral visual detection than yellow-orange sources, including high-pressure sodium. Pulse initiated, or "pulse-start" metal halide lamps provide improved color stability and longer life than earlier technologies. As with fluorescent lights, HID lights require a ballast for proper lamp operation. HID lamps do not normally work well with occupancy sensors because most HID lamps take a long time to start each time they are switched off. The efficiency of HID sources varies widely from mercury vapor, with a low efficiency (almost as low as incandescent), to low-pressure sodium, which is an extremely efficient light source. Color rendering also varies widely from the bluish cast of mercury vapor lamps to the distinctly yellow light of low-pressure sodium.

Fiber optics: This is a technology that is increasing in popularity, providing an alternative that is superior to conventional interior and exterior lighting systems. The technology is based on the use of hair-thin, transparent fibers to transmit light or infrared signals. The fibers are flexible and consist of a core of optically transparent glass or plastic, surrounded by a glass or plastic cladding that reflects the light signals back into the core. Light signals can be modulated to carry almost any other sort of signal, including sounds, electrical signals, and computer data, and a single fiber can carry hundreds of such signals simultaneously, literally at the speed of light. A typical fiber optic lighting system can be broken down into two basic components: a light source, which generates the light, and the fiber optics, which will deliver the light.

Although fiber optic lighting offers unique flexibility compared to conventional lighting, it does have its limitations. Areas of high ambient light should be avoided, as they tend to "wash out" the color. However, often fiber optics can be installed in areas not accessible to conventional lighting. Good ventilation is very necessary for all illuminators. Light-colored reflective surfaces are preferable for end light or sidelight applications. Dark surfaces absorb light and should only be used to provide contrast. Typical applications include cove lighting, walkway lighting, and entertainment illumination (Figure 9.20).

Interior lighting should meet minimum illumination levels (Table 9.4). It is important to determine the amount of light required for the activity that will take place in a space. Typically, the light needed for visibility and perception increases as the size of details decreases, as contrast between details and their backgrounds is reduced, and as task reflectance is reduced. However, interior lighting must not exceed allowed power limits. Interior lighting includes all permanently installed general and task lighting shown on the plans but does not include specialized lighting for medical, dental, or research purposes and display lighting for exhibits in galleries, monuments, and museums.

Figure 9.20 *Photo Showing Glass Block Walls – Optical Fiber Interior. Source: Lumenyte International Corp.*

Table 9.4 Recommended illumination levels for various functions

Area	Footcandles
Building surrounds	1
Parking area	5
Exterior entrance	5
Exterior shipping area	20
Exterior loading platforms	20
Office corridors and stairways	20
Elevators and escalators	20
Reception rooms	30
Reading or writing areas	70
General office work areas	100
Accounting/bookkeeping areas	150
Detailed drafting areas	200

The useful life of a lighting installation becomes progressively less efficient during its operation as a result of dirt accumulation on the surface and aging of the equipment. The rate of reduction is influenced by the equipment choice and the environmental and operating conditions. In lighting scheme design, we must take account of this deficiency by the use of a maintenance factor and plan suitable maintenance schedules to limit the decay.

Exterior lighting systems: Outdoor lighting is common around buildings and there are many innovative, energy-efficient lighting solutions for outdoor applications. Exterior lighting should be carefully designed and sufficient thought given to its placement, intensity, timing, duration, and color and should meet the requirements of the Illuminating Engineering Society of North America (IES or IESNA). Outdoor lighting is used to illuminate statues, signs, flags, or other objects mounted on a pole, pedestal, or platform; spotlighting or floodlighting used for architectural or landscape purposes; must use full cut-off or directionally shielded lighting fixtures that are aimed and controlled so that the directed light is substantially confined to the object intended to be illuminated. Facility evaluations are often required to identify inadequate exterior lighting conditions. Full cut-off lighting fixtures are required for all outdoor walkway, parking lot, canopy, and building/wall-mounted lighting, and all lighting fixtures located within those portions of open-sided parking structures that are above ground. An open-sided parking structure is a parking structure that contains exterior walls that are not fully enclosed between the floor and ceiling.

To meet code requirements, automatic controls are typically required for all exterior lights. The control may be a directional photocell, an astronomical time switch, or a building automation system with astronomical time switch capabilities. The control should automatically turn off exterior lighting when daylight is available. Lights in parking garages, tunnels, and other large covered areas that are required to be on during daylight hours are exempt from this requirement. Incandescent and high-intensity discharge is the most common type for exterior lighting. Illumination levels should be adequate and in good condition.

Emergency lighting: Emergency lights and exit signs are required for most buildings and occupancies. Emergency lights are intended to illuminate the quickest and safest pathway to the exit door, and exit signs show us either the direction to or the exact location of the exit door. The requirements under OSHA 1910.37 and NFPA 101 show that the lights shall be tested for 30 s monthly and 90 min annually. Moreover, NFPA 101 2006 stipulates that emergency illumination (when required) must be provided for a minimum period of 1.5 h to compensate for the possible failure of normal lighting. NFPA also requires emergency lighting to be arranged to provide initial illumination of not less than an average of 1 footcandle and a minimum at any point of 0.1 footcandle measured along the path of egress at floor level. In all cases, an emergency lighting system must be designed to provide illumination automatically in the event of any interruption of normal lighting (NFPA 101 2006 7.9.2.3). Emergency lighting and LED signs typically use relatively small amounts of energy and have a long life expectancy. Although LED fixtures may cost more than incandescent fixtures, reduced energy costs and labor savings will often quickly make up the difference.

9.4.5 Harmonics Distortion

Loads connected to electricity supply systems may be broadly categorized as either linear or nonlinear. There was a time when almost all electrical loads were linear – those that weren't made up such a small portion of the total that they had little effect on electrical system operation. All that changed, however, with the arrival of the solid-state electronic revolution. Today, we are immersed in an environment rich in nonlinear loads, such as

- UPS equipment,
- Industrial equipment (welding machines, arc furnaces, induction furnaces, rectifiers),
- Office equipment (PCs, photocopy machines, fax machines, etc.),
- Household appliances (television sets, microwave ovens, fluorescent lighting, etc.),

- Inverters,
- Induction motors,
- Variable-speed drives for asynchronous and DC motors

Operation of these devices represents a double-edged sword. While they provide greater efficiency, they can also cause serious consequences to power distribution systems – in the form of harmonic distortion. In reality, total harmonic distortion is hardly perceptible to the human ear, and even though the voltage distortion caused by the increasing penetration of nonlinear loads is often accommodated without serious consequences, power quality is compromised in other cases unless steps are taken to address this phenomenon.

While the majority of loads connected to the electricity supply system draw power that is a linear (or near linear) function of the voltage and current supplied to it, these linear loads do not normally cause disturbance to other users of the supply system. Some types of loads, however, cause a distortion of the supply voltage/current waveform because of their nonlinear impedance. Harmonic distortion can surface in electric supply systems through the presence of nonlinear loads of sufficient size and quantity. The severity of problems depends upon the local and regional supply characteristics, the size of the loads, the quantity of these loads, and how the loads interact with each other. Utility companies are clearly concerned about emerging problems caused by increasing concentrations of nonlinear loads resulting from the growing proliferation of electronic equipment, particularly computers and their AC to DC power supply converters, and electronic controllers.

Harmonic reduction: Reducing harmonic voltage and current distortion from nonlinear distribution loads such as adjustable frequency drives can be achieved through the use of several basic approaches. However, in the presence of excessive harmonic distortion, it is highly recommended to bring in a specialized consultant to correct the issue. Some of the methods used by harmonic specialists to reduce harmonic distortion may include the use of a DC choke, line reactors, 12-pulse converters, 12-pulse distribution, harmonic trap filters, broadband filters, and active filters. Whatever approach is applied to achieve reduction, it must meet the guidelines of the Institute of Electrical and Electronics Engineers, Inc. (IEEE).

9.5 SOLAR ENERGY SYSTEMS

There are two types of solar energy systems: active and passive.

9.5.1 Active Solar Energy Systems

9.5.1.1 Solar Photovoltaics: Overview

Photovoltaic (PV) materials convert sunlight into useful, clean electricity. By adding PV to your home or office, you can generate renewable energy, reduce your own environmental impact, enjoy protection from rising utility costs, and reduce greenhouse gas emissions. Electricity is only one of many uses for solar energy. The sun of course is essential to your garden, and it can heat water very cost-effectively, but the most fundamental use of solar energy is in overall building design. Good design uses solar radiation to passively and/or actively heat your building and to help keep it cool. Solar energy is also increasingly being used for street lighting (Figure 9.21).

Building-integrated photovoltaic (BIPV) systems offer additional design options, allowing electricity to be generated by windows, shades and awnings, roofing shingles, and PV-laminated metal roofing, for example. BIPV options can be used in retrofits or new construction.

Solar energy is a renewable resource that is environmentally friendly. Unlike fossil fuels, solar energy is available in abundance and is free, and immune to rising energy prices. Solar energy can be used in many ways – to provide heat, lighting, mechanical power, and electricity. It helps minimize the impact of pollution from energy generation, which is considered to be the single largest contributor to global warming. Renewable energy could clean the air, stave off global warming, and help eliminate our nation's dependence on fossil fuels from overseas. The recent upsurge in consumer demand for clean renewable energy and the deregulation of the utilities industry have spurred growth in green power – solar, wind, geothermal steam, biomass, and small-scale hydroelectric sources of power. Small commercial solar power plants are emerging and starting to serve some of this energy demand.

For decades, solar technologies in the United States have used the sun's energy and light to provide heat, light, hot water, electricity, and even cooling, for homes, businesses, and industry. The types of renewable technologies available for a particular facility depend largely on the application and what sort of energy is required, as well as a building's design and access to the renewable energy source. Building facilities can use renewable energy for space heating, water heating, air-conditioning, lighting, and refrigeration. Commercial facilities include assembly and meeting spaces, educational facilities, food sales, food service, health care, lodging, stores and service businesses, offices, and warehouses.

Figure 9.21 *Example of Solar Power LED Street Lighting With Automatic On–Off, Lasts for 4–5 Nights After Fully Charged.* Source: Hankey Asia Ltd.

In the LEED rating system, the on–site renewable energy credits are not always easily achieved, particularly in urban locations. Essentially you need to generate 2.5–7% of the building's electricity from wind, water, or solar energy, which because of the many site constraints in a city environment, leaves only solar energy to focus on.

9.5.1.2 Solar Electric System Basics

Solar electric systems, also known as photovoltaic systems, convert sunlight into electricity. When interconnected solar cells convert sunlight directly into electricity, they form a solar panel or "module," and several modules connected together electrically form an array. Most people picture a solar electric system as simply the solar array, but a complete system consists of several other components. The working of a solar collector is very simple (Figure 9.22). The energy in sunlight takes the form of electromagnetic radiation from the infrared (long) to the ultraviolet (short) wavelengths. This radiation from the sun heats a liquid, which goes to a hot water tank.

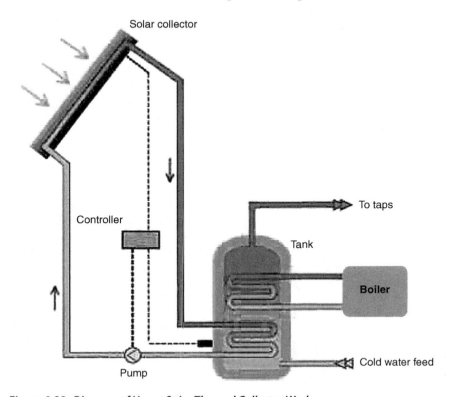

Figure 9.22 *Diagram of How a Solar Thermal Collector Works.*

The liquid heats the water and flows back to the solar collector. The solar energy that strikes the earth's surface at any particular time largely depends on weather conditions, as well as location and orientation of the surface, but overall, it averages approximately 1000 W/10 ft.^{2} (equivalent to one square meter) under clear skies with the surface directly perpendicular to the sun's rays. This solar thermal heat is able to provide hot water for an entire family during the summer. The collector size needed per person is just over 16 ft.^{2} (1.5 m^2). An average family of four people therefore needs a collector of about 65 ft.^{2} (6 m^2).

System components: Understanding the basic components of PV systems and how they function is not particularly difficult. The most common component equipment generally used in on-grid and off-grid solar-electric systems is listed as follows (of note, systems vary and not all equipment is necessary for every system type):

Solar electric panels: The technical word for solar panels that create electricity is *photovoltaic*. PV material, most commonly utilizing highly purified silicon, converts sunlight directly into electricity. When sunlight strikes the material, electrons are dislodged, creating an electrical current that can be captured and harnessed. The PV materials can consist of several individual solar cells or a single thin layer, which make up a larger solar panel. Panels are usually mounted on either a stationary rack or a tracking rack that follows the movement of the sun (Figure 9.23). Life expectancy of a typical system is 40–50 years. Panels are generally warranted for 20–25 years. Also, as they have no moving parts, solar electric panels operate silently.

Over the recent years, PV has been making significant inroads as supplementary power for utility customers already served by the electric grid. In fact, grid-connected solar systems now comprise a larger market share than off-grid applications. However, compared to most conventional fuel options, PV remains a very small percentage of the energy makeup both within the United States and globally. Still, with increasing concerns of global warming, more and more individuals, companies, and communities are choosing PV for a variety of reasons, including environmental, economic, emergency backup, and fuel and risk diversification. The economics of a PV system for your home or business is not just the solar resource, but rather a combination of the solar resource, electricity prices, and local/national incentives.

Siting is a critical aspect of solar design. For example, panels like full sun, facing within 30° of south and tilting within 30° of the site's latitude. A 1-kW system requires about 80 ft.^{2} of solar electric panels. Stationary racks can roof or pole-mounted. Tracking racks are pole-mounted.

Figure 9.23 *Nellis Solar Power Plant, the Largest Photovoltaic Power Plant in North America.* The 70,000 solar panels sit on 140 acres of unused land on the Nellis Air Force Base, Nevada, forming part of a solar photovoltaic array that will generate in excess of 25 million kW-h of electricity annually and supply more than 25% of the power used at the base. *Source: Wikimedia Commons; Photo taken by Airman 1st Class Nadine Barclay.*

Inverter: An *inverter* converts the direct current (DC) electricity produced by the PV modules into usable 120-V alternating current (AC) electricity, which is the most common type for powering lights, appliances, and other needs. Grid-tied inverters are utilized to synchronize the electricity they produce with the grid's utility grade AC electricity, allowing the system to feed solar-made electricity to the utility grid. Inverters are typically warranted for 5–10 years.

Array mounting rack: Mounting racks provide a secure platform on which to anchor the PV panels, ensuring that they are fixed in place and correctly oriented. Panels can be mounted on a rooftop, on top of a steel pole set in concrete, or at ground level. A PV array is the complete power-generating unit, comprising one or more solar PV modules (solar panels) that convert sunlight into clean solar electricity. The solar modules need to be mounted facing the sun and avoiding shade for best results. Solar panels generate DC power, which can be converted to AC power with an inverter.

Wiring: Selecting the correct size and type of wire will enhance the performance and reliability of the PV system. The size of the wire must be sufficiently large to carry the maximum current expected without undue voltage losses.

Battery bank: *Batteries* are used to store solar-produced electricity for evening or emergency backup power. Batteries may be required in locations that have limited access to power lines, as in some remote or rural areas. If batteries are part of the system, a *charge controller* may be required to protect the batteries from being overcharged or drawn down too low. Depending on the current and voltages for certain applications, the batteries are wired in series and/or parallel.

Charge controller: The main function of a charge controller is to protect the battery bank from overcharging. This is done by monitoring the battery bank, and when the bank is fully charged, the controller interrupts the flow of electricity from the PV panels. Modern charge controllers usually incorporate maximum power point tracking (MPPT), which optimizes the PV array's output, thereby increasing the energy it produces.

System meter: They are used to measure and display several different aspects of a solar-electric system's performance and status, tracking how full the battery bank is; how much electricity the solar panels are producing or have produced; and how much electricity is in use.

Array DC disconnect: The DC disconnect is used to safely interrupt the flow of electricity from the PV array. It is an essential component when system maintenance or troubleshooting is required. The disconnect enclosure houses an electrical switch rated for use in DC circuits, and if required, may also integrate either circuit breakers or fuses.

Main DC disconnect: *Disconnect switches* are required in battery-based systems to allow the power from a solar electric system to be turned off for safety purposes during maintenance or emergencies. It also protects the inverter-to-battery wiring against electrical fires. A disconnect typically consists of a large, DC-rated breaker mounted in a sheet metal enclosure.

AC breaker panel: The AC breaker panel is the point at which all of a property's electrical wiring meets with the provider of the electricity, whether that is the grid or a solar-electric system. The AC breaker panel typically consists of a wall-mounted panel or box that is normally installed in a utility room, basement, garage, or on the exterior of the building. It contains a number of labeled circuit breakers that route electricity to the various spaces throughout a structure. These breakers allow electricity to be disconnected for servicing, and also protect the building's wiring against electrical fires.

Kilowatt-hour meter. Homes and businesses with a grid-tied solar-electric system will often have AC electricity both coming from and going to the electric utility grid. A bidirectional kilowatt-hour meter is able to simultaneously keep track of how much electricity flows in each direction, which tells you how much electricity is being used and how much the solar-electric system is producing. Intertied-capable meters are typically provided by the utility company free of charge.

Backup generator. Off-grid solar-electric systems can be sized to provide electricity during cloudy periods when the sun doesn't shine. But sizing a system to cover a worst-case scenario, like several cloudy weeks during the winter, can result in an unduly large system that will rarely be used to its full capacity. Engine generators can be fueled with biodiesel, petroleum diesel, gasoline, or propane, depending on the design. These generators produce AC electricity that a battery charger (either standalone or incorporated into an inverter) converts to DC energy, which is stored in batteries.

Solar electric systems are attracting increasing attention because they are environmentally friendly and do not generate emissions of greenhouse gasses or other pollutants, thereby reducing global climate impacts. Solar panels reflect visible demonstration of concern for the environment, community education, and proactive forward thinking.

9.5.1.3 Types of Solar Energy Systems

The three most widely used types of solar-electric systems are grid-tied, grid-intertied with battery backup, and off-grid (stand-alone). Each has distinct applications and component needs.

Grid-tied solar system (alternating current) – also known as on-grid, or grid-intertied: This type of solar-electric system does not require storage equipment (i.e., batteries) because it generates solar electricity and routes it to the electric utility grid, offsetting a home's or business's electrical consumption and, in some instances, even turning the electric meter backwards. Living with a grid-connected solar-electric system does not really differ from living with grid power, except that some or all of the electricity used comes from the sun. The crucial issue relative to the PV panel systems is the technical aspects of tying into the electricity grid.

Applications of this type require the use of grid-tied inverters that meet the requirements of the utilities. They must not emit "noise," which can interfere with the reception of equipment (e.g., televisions), switch off in the case of a grid failure, and retain acceptable levels of harmonic distortion (i.e., quality of voltage and current output waveforms). This type of system tends to be an optimum configuration from an economic viewpoint because all the electricity is utilized by the owner during the day and any surplus is exported to the grid. Meanwhile, the cost of storage to meet nighttime needs is avoided because the owner simply draws on the grid in the usual way. Also, with access to the grid, the system does not need to be sized to meet peak loads. This arrangement is termed net metering or net billing. The specific terms of net metering laws and regulations vary from state to state and utility-to-utility, which is why the local electricity provider or state regulatory agency should be consulted for their guidelines.

Stand-alone grid-tied solar system with battery backup (alternating current): This type of solar energy system is the same as the grid–tied system except that battery storage (battery bank or generator backup) is added to enable power to be generated even when the electricity grid fails. Incorporating batteries into the system requires more components, is more expensive, and lowers the system's overall efficiency. But for homeowners and businesses that regularly experience utility outages or have critical electrical loads, having a backup energy source is invaluable. The additional cost to the customer can be quantified against the value of knowing that their power supply will not be interrupted.

Stand-alone off-grid solar-electric system without energy storage (direct current): This configuration (i.e., without any energy storage device) consists of a PV system whose output is dependent upon the intensity of the sun. In this system, the electricity generated is used immediately and, therefore, the application must be capable of work on both direct current (DC) and variable power output. Stand-alone off-grid electric systems are most common in remote locations where there is no utility grid service. These systems operate independently from the grid to provide all the electricity required by a household or small business. The choice to live off-grid may be because of the prohibitive cost of bringing utility lines to remote locations, the appeal of an independent lifestyle, or the general reliability a solar-electric system provides. However, those who choose to live off-grid often need to make adjustments to when and how they use electricity, to allow them to live and work within the limitations of the system's capabilities.

To meet the greatest power needs in an off-grid location, the PV system may need to be configured with a small diesel generator. This increases the capability of the PV system as it no longer has to be sized to cope with the worst sunlight conditions available during the year. The diesel generator can also provide the back-up power, but its use is minimized during the rest of the year of the PV system, to keep fuel and maintenance costs to a minimum.

For any module with a defined peak power, the actual amount of electricity in kilowatt-hours that it generates will depend primarily on the amount of sunlight it receives. The electrical power output of a PV module is the current that it generates (dependent on its surface area) multiplied by the voltage at which it operates. The larger the module, or the solar array – the number of modules connected together, the more power is generated. A linear current booster can be added to convert excess voltage into amperage to keep a pump running in low light conditions. A linear current booster can boost pump output by 40% or more. For safety considerations, PV arrays are normally earthed.

Energy production: Each kilowatt of unshaded stationary solar electric panels generates about 1200 kWh of electricity per year. A 1-kW, dual-axis tracking system will generate about 1600 kWh/year. Power is generated during peak daylight hours. Solar power exhibits a very good peak coincidence with commercial building electrical loads. Dual axis tracking systems, where the panels follow the sun, will require periodic maintenance.

9.5.2 Passive Solar Energy Systems

The sun's energy arrives on earth in the primary form of heat and light. Passive solar heating and cooling represents an important strategy for displacing traditional energy sources in buildings and is an effective method of heating and cooling through utilization of sunlight. To be successful, building designs must carefully balance their energy requirements with the building's site and window orientation. The term *passive* indicates that no additional mechanical equipment is used, other than the normal building elements. Solar gains are generally introduced through windows, and minimum use is made of pumps or fans to distribute heat or effect cooling. Passive cooling minimizes the effects of solar radiation through shading or generating airflows with convection ventilation.

Correct building orientation, thermal mass, and insulation are specified in conjunction with careful placement of windows and shading. The thermal mass absorbs heat during the day and radiates it back into the space at night. To do this, passive solar techniques make use of building elements such as walls, windows, floors, and roofs, in addition to exterior building elements and landscaping, to control heat generated by solar radiation. Solar heating designs collect and store thermal energy from direct sunlight. The effect is a quiet, comfortable, energy efficient space with stable year-round temperatures.

Another solar concept is daylighting design, which optimizes the use of natural daylight and contributes greatly to energy efficiency. The quantity and quality of light around us helps determine how well we see, work, and play. Light impacts our health, safety, comfort, morale, and productivity. Whether at home or in the office, it is possible to save energy and still maintain good light quantity and quality.

The benefits of using passive solar techniques include simplicity, price, and the design elegance of fulfilling one's needs with materials at hand. Some of the advantages of passive solar designs are as follows:

- At little or no cost, passive solar design can easily be designed into new construction, and can in some cases be retrofitted into existing buildings.
- It pays dividends over the life of the building through reduced or eliminated heating and cooling costs.
- Indoor air quality is improved through elimination of forced air systems.
- Retrofitting is rarely as effective as initially designing for this method.
- Sites with good southern exposure are most suitable.

LEED offers credits in its Indoor Environmental Quality section – Daylight and Views. The *intent* of the credits are to reduce electric lighting, increase productivity, and provide building occupants with a connection between indoor and outdoor spaces by incorporation of daylight and views into regularly occupied spaces.

LEED requirements for the credits is to achieve daylight (through computer simulations) in a minimum of 75% or 90% of regularly occupied spaces, and achieve a daylight illuminance level of a minimum of 25 footcandles and a maximum of 500 footcandles in a clear sky condition on September 21 at 9:00 a.m. and 3:00 p.m. A combination of side-lighting and/or top-lighting may be used to achieve the total daylighting zone required, which is at least 75% of all the regularly occupied spaces. Sunlight redirection and/or glare control devices may be provided to ensure daylight effectiveness. The provision of daylight redirection and/or glare control devices to avoid high-contrast situations should be provided to avoid impeding of visual tasks. Exceptions for areas where tasks would be hindered by daylight will be considered on their merits.

It should be stressed that the USGBC Reference Guide or website should be consulted for the latest updated requirements, including possible exemplary performance credits.

9.5.3 Federal Tax Credits for Energy Efficiency

Source: ENERGY STAR – US EPA – US Department of Energy.

Tax Credits for Consumers: Home improvements that fall under the Non-business Energy Property Credit and the Residential Energy

Efficient Property Credit can be applied towards certain energy property expenditures to help make energy-saving retrofits more affordable for homeowners. Tax credits extended for home improvements made in 2014 include:

- Windows, Skylights and Doors (Tax credit: 10% of the cost, up to $500, but windows are capped at $200)
- Insulation (Tax credit: 10% of the cost, up to $500. Does not include installation but you can install the insulation/home sealing yourself and get the credit)
- Roofs – Metal and Asphalt (Tax credit: 10% of the cost, not including installation, up to $500.)
- HVAC (For full details, see the ENERGY STAR website. The following are the proposed improvements:
  - Central air-conditioning, $300 tax credit.
  - Advanced main air circulating fan, $50 tax credit.
  - Air source heat pumps, $300 tax credit.
  - Gas, propane, or oil hot water boiler. $150 tax credit, including installation costs.
  - Natural gas, propane, or oil furnace, $150 tax credit)
- Non-Solar Water Heaters (Tax credit: $300)
- Biomass Stoves (Tax credit: $300)

For homeowners, tax credits are available at 30% of the cost of alternative energy equipment installed in or on their homes (for existing homes and new construction) as listed as follows:

- Geothermal heat pumps
- Solar panels
- Solar water heaters
- Small wind energy systems
- Fuel cells

The Residential Energy Efficient Property Credit is valid until 2016 and has no dollar limit for most types of property. If your credit exceeds the tax owed, you can carry the unused portion forward to the following year's tax return (except for fuel cells, which is limited to $500 for each 0.5 kW of capacity of the property).

9.5.3.1 Tax Credits for Consumers

Congress and President Obama extended more than 50 tax breaks before the end of 2014. More than 50 of these tax breaks did not expire but were extended at the last minute by Congress and the president with the Tax

Increase Prevention Act, which was signed December 19, 2014. Taxpayers who upgrade their homes to improve energy efficiency or make use of renewable energy may be eligible for tax credits to offset some of the costs. The 50-plus tax breaks apply to both businesses and individuals; they are scheduled to expire at the end of 2015 – unless they are extended again. As of the 2014 tax year, the federal government offers two credits: (1) the Residential Energy Efficiency Property Credit, which is good through 2016, and (2) the Nonbusiness Energy Property Credit. To claim either credit, file Form 5695 with your tax return.

(For latest updates visit: https://www.energystar.gov/about/federal_tax_credits)

9.5.3.2 Efficient Cars
Fuel-efficient vehicles and energy-efficient appliances and products, mitigate the adverse impacts on our environment, as well as provide many other benefits including better gas mileage (and therefore lower gasoline costs), fewer emissions, and lower energy bills, increased indoor comfort, and reduced air pollution.

Starting January 1, 2009, there has been a new tax credit for plug-in hybrid electric vehicles, starting at $2500 and capped at $7500 for cars and trucks (the credit is based on the capacity of the battery system). The first 250,000 vehicles sold get the full tax credit (then it phases out like the hybrid vehicle tax credits).

Tax credits are available to buyers of hybrid gasoline-electric, diesel, battery-electric, alternative fuel, and fuel cell vehicles. The tax credit amount is based on a formula determined by vehicle weight, technology, and fuel economy compared to base year models. These credits are available for vehicles placed in service starting January 1, 2006. For hybrid and diesel vehicles made by each manufacturer, the credit will be phased out over 15 months starting after that manufacturer has sold 60,000 eligible vehicles. (Check the IRS Website for updated information.)

9.5.3.3 Tax Credits for Home Builders
Energy saving requirements: The federal Energy Policy Act of 2005 established tax credits of up to $2000 for home builders of new energy-efficient homes, including manufactured homes constructed in accordance with the Federal Manufactured Homes Construction and Safety Standards. In essence, a project must achieve 50% energy savings for heating and cooling relative to the International Energy Conservation Code (IECC) and supplements.

At least one-fifth of the energy savings must come from building envelope improvements. It must also meet minimum efficiency standards established by the Department of Energy. The building envelope component improvements must account for at least one-third of the reduction in energy consumption.

Manufactured homes qualify for $1000 credit if they conform to Federal Manufactured Home Construction and Safety Standards and reduce energy consumption by 30% relative to IECC 2006. Alternatively, manufactured homes can also qualify for a $1000 credit if they meet ENERGY STAR Labeled Home requirements.

A home qualifies for a tax credit if
- it is located in the United States;
- its construction is substantially completed before December 31, 2014;
- it meets the energy saving requirements outlined in the statute; and
- it is acquired from the eligible contractor after December 31, 2013, and before January 1, 2015, for use as a residence.

For latest Tax Credit updates, visit: https://www.energystar.gov/about/federal_tax_credits.

Please note that with the exception of the tax credit for an ENERGY STAR®qualified manufactured home, these tax credits are not directly linked to ENERGY STAR. Therefore, a builder of an ENERGY STAR qualified home may be eligible for a tax credit but it is not guaranteed.

9.5.3.4 Tax Deductions for Commercial Buildings

A tax deduction of up to $1.80 ft.$^{-2}$ is available to owners or designers of new or existing commercial buildings that save at least 50% of the heating and cooling energy of a building that meets ASHRAE Standard 90.1-2001. This standard expounds minimum requirements for the energy-efficient proposals of new buildings so that they may be constructed, operated, and maintained in a manner that reflects minimum energy use without constraining the building function or the comfort and productivity of the occupants. The total deduction may not exceed the amount equal to the cost of energy-efficient commercial building property placed in service during the taxable year. As the deduction is based on the square footage, 179D is more beneficial to large buildings.

Additionally, partial deductions of up to $.60 ft.$^{-2}$ can be taken for measures affecting any one of three building systems: the building envelope, lighting, or heating and cooling systems, where the overall building does not meet the 50% energy savings threshold. In these instances, the individual systems for which the deduction is claimed must meet various reductions in energy usage.

9.6 FIRE SUPPRESSION SYSTEMS

9.6.1 General

As more and more high-rise, high–performance buildings are built both nationwide and globally, the planning for fire protection has taken on a real urgency. Fire suppression design requires an integrated approach in which system designers need to analyze building components as a total package. As with other aspects of sustainable design, to achieve the most beneficial symbiosis between these components, an experienced system designer, such as a fire-protection engineer, should be involved early in the planning and design process and should be an integral part of the project team. Moreover, moving forward, we should start seeking out sustainable environmentally friendly fire-suppression approaches to reduce the environmental impacts during design and testing and also to help a project earn LEED credits.

Fire protection systems play an important role in overall building design and construction and should never be comprised because they serve the purpose of life safety. Indeed, it is frequently argued that the life safety system is the most important system to be evaluated in a facility, particularly when it comes to high-rise structures. Furthermore, like any other building system, green concepts and specifications can be applied to their design, installation, and maintenance in a manner that reduces their harmful impacts on the environment. Moreover there have been significant advances recently in fire-detection technology and fire-suppression systems in addition to an ongoing development of international and national codes and standards, all of which have made possible the "greening" of facility fire-safety systems and which is taking on increasing importance for building owners and property developers.

For maximum efficiency, the various components of modern fire-protection systems should work together to detect, contain, control, and/or extinguish a fire in its early stages – and to survive during the fire. And the installation of environmentally friendly fire-protection technology can help earn credits under the USGBC's LEED™ Green Building Rating System for new or retrofitted buildings.

A facility's type, size, and function will generally determine the complexity of the life safety system used. In some of the smaller structures, the system may comprise only smoke detectors and fire extinguishers. In

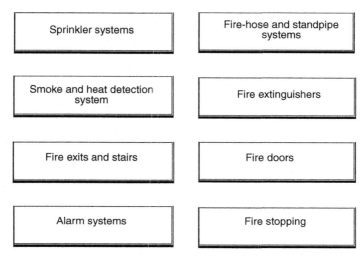

Figure 9.24 *Typical Fire Suppression System Components.*

other larger more complex buildings, a complete fire suppression system such as fire sprinklers is installed throughout the facility. An important aspect in the assessment of any life safety system includes verification that periodic maintenance, inspection, and testing of the main components of the system is being conducted.

As illustrated in Figure 9.24, there are several types of life safety systems normally employed to address fire safety requirements. Each of these gives rise to their own set of issues that need to be taken into account in facility surveys. The extent of a life safety system survey and the expertise required to perform such an evaluation varies greatly from facility to facility.

Fortunately, fire detection and prevention technologies have become increasingly sophisticated, intelligent, and powerful in recent years. Frank Monikowski and Terry Victor of SimplexGrinnell list some of the advances and emerging technologies that can be found in today's Life/Safety systems:

- *Control mode sprinklers* – standard manufactured sprinklers that limit fire spread and stunt high heat release rather than extinguish a fire; they also "pre-wet" adjacent combustibles.
- *Suppression sprinklers* – operate quickly for high-challenge fires, and are expected to extinguish a fire by releasing a high density of water directly to the base of the fire.
- *Fast-response sprinklers* – provide quicker response and are now required for all light–hazard installations.

- *Residential sprinklers* – designed specifically to increase the survivability of an individual who is in the room where a fire originates.
- *Extended coverage sprinklers* – designed to reduce the number of sprinklers needed to protect a given area. These come in quick-response, residential, and standard-response types and are also available for both light- and ordinary-hazard occupancies.
- *Special sprinklers, such as early suppression fast response (ESFR)* – designed for high-challenge rack storage and high-pile storage fires. In most cases, these sprinklers can eliminate the expense and resources needed to install in-rack sprinkler heads.
- *Low-pressure sprinklers* – provide needed water coverage in multistory buildings, where pressure may be reduced. These low-pressure sprinklers bring a number of benefits: reduced pipe size, reduction or elimination of a fire pump, and overall cost savings.
- *Low-profile, decorator, and concealed sprinklers* – designed to be more aesthetically pleasing.
- *Sprinkler system valves* that are smaller, lighter, and easier to install and maintain and, therefore, less costly.
- *A Fluid Delivery Time* computer program that simulates water flowing through a dry system in order to accurately predict critical "water-to-fire" delivery time for dry-pipe systems.
- *The use of cost-efficient CPVC piping* for light-hazard and residential sprinkler systems.
- *Advanced coatings on steel pipes*, designed to resist or reduce microbiologically influenced corrosion (MIC) and enhance sprinkler system life.
- *Corrosion monitoring devices* to alert users of potential problems.
- *More efficient coordination in evaluating building sprinkler system need* – including site surveys, accurate measurements, and the use of CAD and hydraulics software to ensure that fire sprinkler system designs respond to the specific risks and the physical layout of the premises.

The National Fire Protection Association (NFPA) developed and issued a new Emergency Evacuation Planning Guide for People with Disabilities. This document provides general information to assist designers in identifying the needs of people with disabilities related to emergency evacuation planning. This guide covers the following five general categories of disabilities: *mobility impairments, visual impairments, hearing impairments, speech impairments,* and *cognitive impairments.* The four elements of evacuation information needed by occupants are *notification, way finding, use of way,* and *assistance.*

LEED™ contributions: Fire-suppression systems are only indirectly referenced in LEED certification documents. For example, LEED™ for New Construction (LEED-NC) V3 Energy and Atmosphere (EA) Credit 4, Enhanced Refrigerant Management, and LEED™ for Existing Buildings: Operations and Maintenance (LEED EBOM) V4 EA Credit 5, Refrigerant Management, the intent being reducing ozone depletion, supporting compliance with the Montreal Protocol and minimizing direct contributions to global warming. It appears that credits can be earned with the installation/operation of fire-suppression systems that do not contain ozone-depleting substances such as chlorofluorocarbons (CFCs), hydrochlorofluorocarbons (HCFCs), and halons.

Likewise, LEED credits in the Innovation in Design category can also be obtained for fire-suppression systems. The LEED *Reference Guide* in the relevant category should be consulted, but generally, to earn those points, it is necessary to document and substantiate the innovation and design processes used.

9.6.2 Components to be Evaluated

9.6.2.1 Sprinkler System Types

It is best to have protection in commercial and residential buildings in case of fire or smoke. Installing an appropriate sprinkler system is a good preventative measure to take. There are numerous types of fire sprinklers, and various systems are described as follows.

Automatic sprinkler systems: Optimized sprinkler system designs offer an effective means of addressing environmental impact and sustainability. They are the most common, widely specified, and most effective fire suppression system in commercial facilities – particularly in occupied spaces. Furthermore, automatic sprinkler systems are now not only required in new high-rise office buildings, but in many American cities it is mandated by code that existing high-rises be retrofitted with automatic sprinkler systems.

For commercial facilities, sprinklers are the most widely specified and most effective fire suppression system, particularly in occupied spaces. There are several types of sprinkler systems that are commonly used, these include wet- and dry-pipe, preaction, deluge, and fire cycle systems. Of these, wet-pipe and dry-pipe are the most common. In a wet-pipe system, the sprinklers are connected to a water supply, enabling immediate discharge of water at sprinkler heads triggered by the heat of the fire. In a dry pipe system, the sprinklers are under air pressure which, when the pressure is eased by the opening of the sprinkler heads, fills the system with water.

With optimized sprinkler system designs, effective use of the available water source is made, requiring the minimal necessary number of components. It also employs techniques and technologies that make it adaptable to future building modifications; plus a well-designed valve or pump room can maximize the life of the system and facilitate any modifications that may be required in the future. By minimizing variations in piping, construction waste can be reduced and a more efficient installation achieved. In some cases, the need for a fire pump may be eliminated altogether, thus reducing water waste.

In the design of automatic sprinkler systems, careful attention should be given to proper connections for flow and flow testing. Likewise, flexible connections and arm-overs may be employed to provide a means for facilitating the relocation of sprinklers with minimal need for additional materials if the system designer incorporates appropriate flow restrictions due to friction losses.

Sometimes, when sprinklers are not feasible because of special considerations (e.g., water from sprinklers would damage sensitive equipment or inventory), alternative fire-suppression systems such as gaseous/chemical suppression may be considered. But in the final analysis, the type of sprinkler system decided upon depends mainly on a building's function. Of note, the majority of today's fire sprinklers incorporate the latest advances in design and engineering technologies, thereby providing a very high level of life-safety and property protection. The features and benefits now available are making fire sprinkler systems more efficient, reliable, and cost-effective.

As the benefits of sprinkler systems become better understood and more obvious, and the cost more affordable, their installation in residential structures is becoming more common. However, these sprinkler systems typically fall under a residential classification and not a commercial one. The main difference between commercial and residential sprinkler systems is that a commercial system is designed to protect the structure and the occupants from a fire, whereas most residential systems are primarily designed to suppress a fire in a manner that allows for the safe escape of the building occupants. While these systems will often also protect the structure from major fire damage, this consideration nevertheless remains of secondary importance.

In residential structures, sprinklers are typically omitted from closets, bathrooms, balconies, and attics because a fire in these areas would not normally impact an occupant's escape route. When a system is operating as intended, fire sprinkler systems are highly reliable, but like any other

Figure 9.25 *Photo of a Fire Sprinkler Control Valve Assembly Installation. Source: Wikipedia.*

mechanical system, sprinkler systems require periodic maintenance and inspection in order to sustain proper operation. In the rare event a sprinkler system does fail to control a fire, the root cause of failure has often been found to be the lack of proper maintenance. Figure 9.25 is an example of a fire sprinkler control valve assembly including pressure switches and valve monitors.

Wet-pipe systems: Wet-pipe sprinkler systems are by far the most widely used and have the highest reliability. Wet systems are typically used in buildings where there is no risk of freezing. The systems are simple, with the

Figure 9.26 *Example of Ceiling-Mounted Sprinkler Head.* *Source: Sujay Fire & Safty Equipment*

only operating component being the automatic sprinkler. A water supply provides pressure to the piping, and all of the piping is filled with water adjacent to the sprinklers. The water is held back by the automatic sprinklers (Figure 9.26) until activated. When one or more of the automatic sprinklers is exposed to sufficient heat, the heat-sensitive element releases, allowing water to flow from that sprinkler. Each sprinkler operates individually. Sprinklers are manufactured to react to a specific range of temperatures, and only sprinklers subjected to a temperature at or above their specific temperature rating will operate. Figure 9.27 shows a drawing of a typical wet-pipe sprinkler system.

The principal disadvantage of wet-pipe sprinkler systems is that they are not suited for subfreezing environments. Another potential concern is

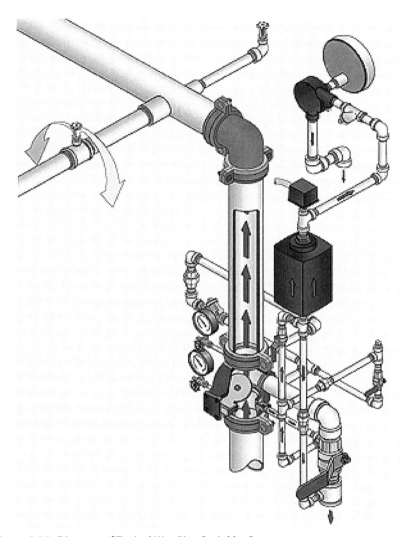

Figure 9.27 *Diagram of Typical Wet-Pipe Sprinkler System.*

where piping is subject to severe impact damage and could consequently leak, for example, in warehouses.

Dry-pipe systems: After the wet-pipe system, this system is the most widely used. A dry-pipe sprinkler system is one in which pipes are filled with pressurized air or nitrogen, rather than water. This air holds a remote valve, known as a dry-pipe valve, in a closed position. The dry-pipe valve is located in a heated space and prevents water from entering the pipe until a fire causes one or more sprinklers to be activated. Once this happens, the air escapes and the dry-pipe valve releases. To prevent the larger water supply pressure from forcing water into the piping system, the design of the dry-pipe valve intentionally includes a larger valve clapper area exposed to the specified air pressure, as compared to the water pressure. Water then enters the pipe, flowing through open sprinklers onto the fire. However, regulations (NFPA 13 2007 ed. Sections 7-2 and A7-2) typically stipulate that these systems only be used in spaces in which the ambient temperature can be cold enough to freeze the water in a wet-pipe system, thus rendering it inoperable. For this reason, we often find dry-pipe systems used in refrigerated coolers and in unheated buildings.

The system is activated (becomes operational) when one or more of the automatic sprinklers are exposed to sufficient heat, allowing the maintenance air to vent from that sprinkler. Each sprinkler operates individually. As the air pressure in the piping drops, the pressure differential across the dry pipe valve changes, allowing water to enter the piping system. Delays can be experienced in dry-pipe systems, since the air pressure must drop before the water can enter the pipes and suppress the fire. Dry-pipe systems are therefore not as effective as wet-pipe systems in fire control during the initial stages of the fire, although aiding in faster activation, and dry-pipe valves may employ quick opening devices connected to them. Other important characteristics of dry-pipe systems include the following:

- In a dry-pipe system, the fire will continue to grow after the activation of a sprinkler. As the water travels the pipework in the sprinkler system, the fire size grows. The dry-pipe system should therefore be designed to be as small as possible.
- Dry-pipe systems are not intended to limit accidental damage from broken or open sprinklers.
- Dry-pipe sprinkler systems cannot be gridded.
- Sprinklers may only be installed in the upright position unless specifically listed dry-pipe sprinklers are utilized or horizontal sidewall sprinklers installed so that water is not trapped.
- The location of the dry-pipe riser is required to be heated to a minimum of 40°F.
- Piping must be pitched back to the riser so that water is not trapped.

Deluge systems: Less common are the "deluge system" and the "preaction system." These systems are used in environments that require special sprinkler protection and are activated by smoke or heat fire detection systems. These systems represent only a small percentage of market share of sprinkler systems in operation.

A deluge system is similar to a preaction system, except the sprinkler heads are open and the pipe is not pressurized with air. This means that in a Deluge system, the heat-sensing operating element is removed during installation, so that all sprinklers connected to the water piping system remain open by the operation of a smoke or heat detection system. These detection systems are normally installed in the same area as the sprinklers, so that when the detection system is activated water readily discharges through all of the sprinkler heads in the system.

These systems provide a simultaneous application of water over the entire hazard and are typically used in high hazards areas where rapid fire spread is a major concern such as power plants, aircraft hangars, and chemical storage facilities. Water is not present in the piping until the system operates. Because the sprinkler orifices are open, the piping is at ambient air pressure. To prevent the water supply pressure from forcing water into the piping, a *deluge valve* is used in the water supply connection, which is a mechanically latched nonresetting valve that stays open once tripped.

Because the heat-sensing elements present in the automatic sprinklers have been removed (resulting in open sprinklers), the deluge valve must be opened as signaled by a specialized fire alarm system. The type of fire alarm activation device used is based largely on the hazard (e.g., smoke detectors or heat detectors). The activation/initiation device signals the fire alarm panel, which in turn signals the deluge valve to open. Activation can also be achieved manually, depending on the system goals. Manual activation is usually via an electric or pneumatic fire alarm pull station, which signals the fire alarm panel, which in turn signals the deluge valve to open, allowing water to enter the piping system. Water flow effectively takes place from all sprinklers simultaneously.

Deluge fire sprinkler systems are typically installed in buildings such as industrial parks and buildings with many tanks.

Preaction systems: Preaction sprinkler systems employ the basic concept of a dry-pipe system in that water is not normally contained within the pipes. It differs from the dry-pipe system, however, in that water is held from entering the piping by an electrically operated valve, known as a preaction valve. Valve operation is controlled by independent flame, heat, or smoke detection. This type of system thus requires two triggers to start water flow. It helps greatly that the preaction fire sprinkler can be set to prevent water from spouting in case of a false alarm or a mechanical failure. A preaction system is generally used in locations where it is not acceptable to have the pipes full of water until there is a fire, for example, data centers, or where accidental activation is undesired, such as in museums with rare art works, manuscripts, or libraries. Preaction systems are hybrids of wet, dry, and deluge systems depending on the exact system goal.

There are two subtypes of preaction systems: single interlock and double interlock. The operation of single interlock systems are similar to dry systems except that these systems require that a "preceding" and supervised event (typically the activation of a heat or smoke detector) takes place prior to the "action" of water introduction into the system's piping due to opening of the preaction valve (i.e., a mechanically latched valve). The operation of double interlock systems is similar to a deluge system except that automatic sprinklers are used. Upon detection of the fire by the fire alarm system, it basically converts from a dry system into a wet system.

Water mist: These systems consist of environmentally friendly systems that normally force water and pressurized gas together through stainless steel tubes that are much narrower in diameter than pipes used in traditional sprinkler systems. The water mist system produces a fine mist with a large surface area that absorbs heat efficiently through vaporization. Because water mist systems utilize water as the extinguishing medium, they are totally safe for humans.

These systems suppress fire with three main mechanisms:

1. As the water droplets contact the fire they convert to steam. This process absorbs energy from the surface of the burning material.
2. As the water turns into steam it expands greatly. This removes heat and lowers the temperature of the fire and the air surrounding it.
3. The water and the steam act to block the radiant heat and prevent the oxygen from reaching the fire (thus starving it of oxygen) so the fire smothers.

These systems can be useful for suppressing fires in gas turbine enclosures, and machinery spaces and are FM (Factory Mutual) approved for such applications. Water mist systems are ideally suited for cultural heritage buildings where large amounts of water can potentially cause unacceptable damage to irreplaceable items and in retro-fits where space is often limited. Water mist systems are also often used to protect passenger cruise ships, where the system's excellent performance and low total system weight have made them very popular.

Foam water sprinkler systems: This type of sprinkler system is essentially a special application type system, discharging a mixture of water and low expansion foam concentrate, resulting in a foam spray from the sprinkler. These systems are generally more economical than a water-only system, when evaluated for the same risk, and provide for actual extinguishment of the fire and a lower water demand. The conversion assists in the reduction of property loss, loss of life, and in many cases the reduction of insurance rates.

Foam concentrates and expanded foams that are generally considered to be safe with regard to exposure to humans but, unless specifically indicated, can adversely impact the environment if allowed to flow freely into watershed areas. The base properties of typical foaming agents include nitrates, phosphorus, and organic carbon. It should be noted that the use of halons in fire suppression systems was phased out in the early 1990s to comply with the Montreal Protocol because they were determined to cause significant damage to the ozone layer. Moreover, halons have a long life in the atmosphere and a high global warming potential (GWP).

One of the characteristics of this system is that almost any sprinkler system – wet, dry, deluge, or preaction, can be readily adapted to include the injection of AFFF foam concentrate in order to combat high-risks situations. These systems are typically used with special hazards occupancies associated with high-challenge fires, such as flammable liquids and airport hangars. Added components to the sprinkler system riser include bladder tanks to hold the foam concentrate, concentrate control valves to isolate the sprinkler system from the concentrate until activation, and proportioners for mixing the appropriate amount of foam concentrate with the system supply water. The main standard that delineates the minimum requirements for the design, installation, and maintenance of foam-water sprinkler and spray system is: NFPA 16: Standard for the installation of Foam–Water Sprinkler and Foam–Water spray systems.

The following checklist is provided by the New York Property Insurance Underwriting Association to help identify general problems that may arise in typical sprinkler systems. This checklist is intended to identify what is required to be done and to ensure that the sprinkler system is properly maintained:

- Are sprinkler heads free of paint, dust, and grease?
- Are the sprinkler heads obstructed by stored material? There should be no less than 18 in. of clearance at each head. Obstructions will diminish the operation of the head.
- Are the sprinkler pipes used to support lighting or other objects?
- Are there extra sprinkler heads and wrenches located at the control area for maintenance purposes?
- Is the OS&Y valve chained in an open position to avoid disabling of the system?

- Are the sprinkler heads directed properly for their location?
- Is there a sprinkler contractor that supervises and inspects the system as required by NFPA and ISO? Is a service log maintained?
- Are the sprinkler alarms activated to protect your property in the event of accidental discharge or fire?
- Has the occupancy classification of the material in the building changed since its installation so that the sprinkler system is now ineffective?
- Is the heat supply in the premises adequate for the operation of a wet-pipe system?

9.6.2.2 Fire-Hose and Standpipe Systems

Michael O'Brian, president of Code Savvy Consultants says,

Standpipes are a critical tool that requires preplanning on first responding apparatus in order to be used effectively. The initial approval process for these systems is critical, and the fire prevention bureau can assist responding crews by ensuring proper installation and maintenance of these systems. Standpipe systems vary in design, use, and location. These factors vary based on the adopted code; the use, size, and type of building they are installed in. Typically, model codes refer to NFPA 14, Standpipe and Hose Systems for the design, installation, and maintenance of these systems.

Standpipe systems consist of piping, valves, outlets, and related equipment designed to provide water at specified pressures and installed exclusively for the fire department or trained occupant use for the fighting of fires. The system is used in conjunction with sprinklers or hoses and basically consists of a water pipe riser running vertically through the building, although sometimes a building is provided with only piping for the standpipe system. Standpipe systems can be wet or dry. Dry systems are normally empty and are not connected to a water source. A Siamese fitting is located at the bottom end of the pipe, allowing the fire department to pump water into the system. In a wet-type system, the pipe is filled with water and attached to a tank or pump. This type also contains Siamese fittings for the fire department's use.

O'Brian says,

Many buildings are required to have an Automatic Class I standpipe system with a design pressure of 100 psi. Based on friction loss, municipal water supply, and pressure loss for the height of the standpipe a fire pump may need to be designed into the system. Due to the pressure requirements standpipes are limited to a maximum height of 275 feet. Those buildings over 275 feet in height will require the standpipe systems to be split in different pressure zones.

Model fire and building codes stipulate among other things, the requirements for the installation of standpipe systems. The specific type of system is based on the occupancy classification and building height. Standpipe systems have three common classifications:

- Class I standpipe signifies that it is equipped with a 2.5-in. fire hose connection for fire department use and those trained in handling heavy fire streams. These connections must match the hose thread utilized by the fire department and are typically found in stairwells of buildings.
- Class II standpipe system is one directly connected to a water supply and serves a 1.5-in. fire hose connection which provides a means for the control or extinguishment of incipient-stage fires. They are typically found in cabinets, and are intended for trained occupant use and are spaced according to the hose length. The hose length and connection spacing is intended for all spaces of the building.
- Class III standpipe system is a combined standpipe system (have both Class I and II connections) directly connected to a water supply and is for the use of in-house personnel capable of furnishing effective water discharge during the more advanced stages of fire in the interior of workplaces. Many times, these connections will include a 2.5-in. reducer to a 1.5-in. connection.

When a standpipe system *control valve* is located within a stairwell, the maximum length of hose should not exceed 100 ft. If the control valve is located in areas other than the stairwell, the length of hose should not exceed 75 ft. Code requires that fire hose on Class II and Class III standpipe systems be equipped with a shut-off-type nozzle.

9.6.3 Hand-Held Fire Extinguishers

There are several different classifications of fire extinguishers, each of which extinguishes specific types of fire (Figure 9.28). Newer fire extinguishers use a picture/labeling system to designate which types of fires they are to be used on, whereas older fire extinguishers are labeled with colored geometrical shapes with letter designations (Figure 9.29).

9.6.3.1 Classification of Hand-Held Fire Extinguisher Ratings

Class A fire extinguishers are designed to put out fires caused by organic solids and ordinary combustibles like wood, textiles, paper, some plastic, and rubber. The numerical rating for this class of fire extinguisher refers to the amount of water the fire extinguisher holds and the amount of fire it will extinguish. To extinguish a Class A fire, extinguishers utilize either the heat-absorbing effects of water or the coating effects of certain dry chemicals. *Class A* fire

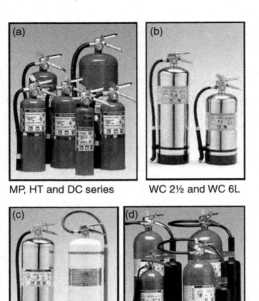

Figure 9.28 *Various Types of Fires Extinguishers in Common Use.* (a) MP series multipurpose dry chemical, DC series regular dry chemical, HT series Halotron I; (b) WC series wet chemical; (c) WM series water mist; (d) CD series carbon dioxide. *Source: Larsen's Manufacturing Co.*

extinguishers should be clearly marked with a triangle containing the letter "A." If in color, the triangle should be green.

Class B fire extinguishers are used to put out fires involving flammable and combustible liquids and gases. They work by starving the fire of oxygen and interrupting the fire chain by inhibiting the release of combustible vapors. *Class B* fires include gasoline, oil, and paraffin. The numerical rating for this class of fire extinguisher states the approximate number of square feet of a flammable liquid fire that a nonexpert person can expect to extinguish. This includes all hydrocarbon- and alcohol-based liquids and gases that will support combustion. *Class B* fire extinguishers should be clearly marked with a square containing the letter "B." If in color, the square should be red.

Class C fire extinguishers are most effective for use on fires that involve live electrical equipment where a nonconducting material is required. This class of fire extinguishers does not have a numerical rating, but the presence of the letter "C" indicates that the extinguishing agent is nonconductive.

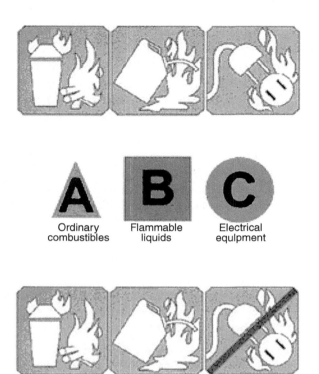

Figure 9.29 *New- and Old-Style Labeling Systems Indicating Suitability for Use on Class A, B, and C Fire Extinguishers.*

Class C fire extinguishers should be clearly marked by a circle containing the letter "C." If in color, the circle should be blue.

Class D fire extinguishers are special types designed and approved for specific combustible materials (metals) such as magnesium, titanium, zirconium, potassium, sodium, etc., which require an extinguishing medium that does not react with the burning metal. *Class D* fire extinguishers should be clearly marked by a five-point painted star containing the letter "D." If in color, the star should be colored yellow. These extinguishers generally have no rating nor are they given a multipurpose rating for use on other types of fires.

Class K fire extinguishers are effective for fighting fires involving cooking fats, grease, oils, etc., in commercial cooking environments. These fire extinguishers work on the principle of saponification, which takes place when alkaline mixtures such as potassium acetate, potassium citrate, or potassium carbonate are applied to burning cooking oil or fat. The alkaline mixture combined with the fatty acid creates soapy foam on the surface, which holds in the vapors and steam and extinguishes the fire. Class K fire extinguishers should be clearly marked with the letter "K."

Labeling: If a multipurpose extinguisher is being used and in order for users to be able to quickly identify the classification of a fire extinguisher in the event of an emergency, each unit should be clearly labeled. The approved marking system combines pictographs of both recommended and unacceptable extinguisher types on a single identification label. Many extinguishers available today can be used on different types of fires and will be labeled with more than one designator, for example, A–B, B–C, or A-B-C. It should also be noted that British Standards and classifications differ slightly from American Standards.

9.6.3.2 Types of Fire Extinguishers

Dry chemical extinguishers come in a variety of types and are usually rated for multiple purpose use (class A, B, and C fires). They are filled with a foam or powder extinguishing agent and use a compressed, nonflammable gas as a propellant. One advantage a dry chemical extinguisher has over a CO_2 extinguisher is that it leaves a nonflammable substance on the extinguished material, reducing the likelihood of reignition.

Water extinguishers or APW extinguishers (air-pressurized water) are the cheapest and most widely used fire extinguishers. They are filled with water and pressurized with oxygen. AWP extinguishers should only be used on Class A (ordinary combustibles) fires and never on grease fires, electrical fires, Class B (Liquid) fires, or Class D fires – the flames will only spread and likely make the fire bigger.

Carbon dioxide (CO_2) extinguishers contain carbon dioxide, a nonflammable gas, and are highly pressurized. They are most effective on Class B and C (liquids and electrical) fires. Since the gas disperses quickly, these extinguishers are only effective from 3 ft. to 8 ft. The carbon dioxide is stored as a compressed liquid in the extinguisher; as it expands, it cools the surrounding air. The cooling will often cause ice to form around the "horn" where the gas is expelled from the extinguisher. However, they don't work very well on Class A fires because they may not be able to displace enough oxygen to put the fire out, causing it to reignite. The advantage CO_2 extinguishers have over dry chemical extinguishers is that they do not leave a harmful residue and may therefore be a good choice for an electrical fire on a computer or other favorite electronic device such as a stereo or TV.

Halon extinguishers contain a gas that interrupts the chemical reaction that takes place when fuels burn. Halon is an odorless, colorless gas that can cause asphyxiation, and halon extinguishers have a limited range, usually 4–6 ft. An advantage of halon is that it is a clean agent because it leaves no corrosive or abrasive residue after release, minimizing cleanup, which makes it more suitable for valuable electrical equipment, computer rooms, telecommunication areas, theaters, etc. However, pressurized fire suppression system cylinders can be hazardous and if not handled properly are capable of violent discharge. Moreover, the cylinder can act as projectile, potentially causing injury or death. Halon has been banned from new production, except for military use, since January 1, 1994, because it contributes to ozone depletion and has a long atmospheric lifetime, usually 400 years. However, Halon reuse is still permitted in the United States.

NFPA Code 10 addresses all the issues pertaining to portable fire extinguishers and contains the clear, widely accepted rules for distribution and placement, maintenance, operation, inspection, testing, and recharging. It is recognized as a first line of defense against fires, portable extinguishers, when maintained and operated properly on a small containable fire, and can prevent fires from spreading beyond its point of origin. NFPA Code 10 requires owners of extinguishers to have monthly inspections performed and to maintain records of the inspections.

9.6.3.3 Fire Detection – Smoke and Heat Detection Systems

Fire-detection systems play a pivotal role in green buildings. Kate Houghton, director of marketing for Kidde Fire Systems says,

> By detecting a fire quickly and accurately (i.e., by not sacrificing speed or causing false alarms) and providing early warning notification, a fire-detection system can limit the emission of toxic products created by combustion, as well as global-warming gases produced by the fire itself. These environmental effects often are overlooked, but undoubtedly occur in all fire scenarios. Therefore, reducing the likelihood of a fire is an important part of designing a green building.

A *smoke detector* or *smoke alarm* is a device that detects smoke and issues an alarm to alert nearby people of the threat of a potential fire. A household smoke detector will typically be mounted in a disk shaped plastic enclosure about 150 mm in diameter and 25 mm thick, but the shape can vary by manufacturer (Figure 9.30). Because smoke rises, most detectors are mounted on the ceiling or on a wall near the ceiling. It is imperative that smoke detectors are regularly maintained and checked that they operate properly. This will ensure early warning to allow emergency responses to occur well

Figure 9.30 *Drawing of Ceiling-Mounted Smoke Detector. Source: Scott Easton.*

before a fire causes serious damage. It is not uncommon for modern types of systems to detect smoldering cables or overheating circuit boards. Early detection can save lives and help limit damage and downtime. Laws governing the installation of smoke detectors differ depending on the jurisdiction.

Smoke detectors are typically powered by one or more batteries but some can be connected directly to a building's wiring. Often the smoke detectors that are directly connected to the main wiring system also have a battery as a power supply backup in case the facility's wiring goes out. Batteries should be checked and replaced periodically to ensure appropriate protection.

The majority of smoke detectors work either by *optical detection* or by *ionization*, and in some cases both detection methods are used to increase sensitivity to smoke. A complete fire-protection system will typically include spot smoke detectors that can signal a fire control panel to deploy a fire-suppression system. Smoke detectors can either operate alone or be interconnected to cause all detectors in an area to sound an alarm if one is triggered, or be integrated into a fire alarm or security system. Smoke detectors with flashing lights are also available for the deaf or hearing impaired. Smoke detector cannot detect carbon monoxide to prevent carbon monoxide poisoning unless they come with integrated carbon monoxide detectors.

Aspirating smoke-detectors (ASDs) can detect the early stages of combustion and are 1000 times more sensitive than conventional smoke detectors, giving early warning to building occupants and owners. An aspirating smoke detector generally consists of a central detection unit that sucks air through a network of pipes to detect smoke, and in most cases it requires a fan unit to draw in a representative sample of air from the protected area through its network of pipes. Aspirating smoke detectors are extremely sensitive, and are capable of detecting smoke before it is even visible to the human eye. However, their use is not recommended in unstable environments because of the wide range of particle sizes that are detected.

Optical smoke detectors are light sensors. When used as a smoke detector it includes a light source (infra-red LED), a lens to collimate the light into a beam like a laser, and a photodiode or other photoelectric sensor at right angles to the beam as a light detector. Under normal conditions (i.e., in the absence of smoke), the sensor device detects no light signal and therefore produces no output. The source and the sensor device are arranged so that there is no direct "line of sight" between them. When smoke enters the optical chamber into the path of the light beam, some light is scattered by the smoke particles, and some of the scattered light is detected by the sensor. An increased input of light into the sensor sets off the alarm.

Projected *beam detectors* are employed mainly in large interior spaces, such as gymnasia and auditoria. A unit on the wall transmits a beam, which is either received by a receiver, or reflected back via a mirror. When the beam is less visible to the "eye" of the sensor, it sends an alarm signal to the Fire alarm control panel. Optical smoke detectors are generally quick in detecting slow-burning, smoky fires.

Ionization detectors are sometimes known as an ionization chamber smoke detector (ICSD), which is capable of quickly sensing flaming fires that produce little smoke. It employs a radioactive material to ionize the air in a sensing chamber; the presence of smoke affects the flow of the ions between a pair of electrodes, which triggers the alarm. Although more than 80% of the smoke detectors in American homes are of this type and although ionization detectors are less expensive than optical detectors, they are frequently rejected for projects seeking LEED certification for environmental reasons. The majority of residential models are self-contained units that operate on a 9-V battery, but construction codes in some parts of the country now require installations in new homes to be connected to the house wiring, with a battery backup in case of a power failure.

A *heat detector* is a device that can detect heat and can be either electrical or mechanical in operation. Most heat detectors are designed to trigger alarms and notification systems before smoke even becomes a factor.

*Conventional and intelligent heat detectors:*These types of detectors are set to alarm when ambient temperatures reach a fixed point, typically indicating a fire; fixed-temperature heat detectors are a highly cost-effective solution for many property protection applications. If rapid response to fire is vital, rate-of-rise heat detectors are an ideal solution where rapid temperature increases would only be caused by a fire emergency. Combination heat detectors provide both fixed and rate-of-rise detection.This enables the heat detector to communicate an alarm to the central control panel prior to reaching its fixed set point for high rates of rise, providing a timely response to both rapid and slow temperature increases. The main benefit of good detection (beyond triggering the alarm system) is that, in many cases, there is a chance to extinguish a small, early blaze with a fire extinguisher. Also, intelligent smoke detectors can differentiate between different alarm thresholds.These systems typically have remote detectors located throughout the facility, which are connected to a central alarm station.

9.6.3.4 Fire Doors

These are basically doors made of fire-resistant material that can be closed to prevent the spread of fire and are designed to provide extra fire-spread protection for certain areas of a building (Figure 9.31). The fire rating classification of the wall into which a door is installed dictates the required fire rating of the door.The location of the wall in the building and prevailing building code establish the wall's fire rating.

Fire doors are commonly installed at staircases from corridors or rooms, cross-corridor partition, to laboratories, plant rooms, workshops, storerooms, machine rooms, service ducts, and kitchens as well as to defined fire compartments.They are also employed in circulation areas, which extend the escape route from the stair to a final exit or to a place of safety, and entrances and lobbies; at routes leading onto external fire escapes, and corridors that are protected from adjoining accommodation by fire-resistant construction.

According to the National Fire Protection Association (NFPA), doors are rated with respect to the number of hours they can be expected to withstand fire before burning through. There are 20-, 30-, 45-, 60-, and 90-min rated fire doors as well as 2HR and 4HR rated fire doors that are certified by an approved laboratory such as Underwriters Laboratories.The certification only applies if all parts of the installation are correctly specified

Interior fire-rated doors and glass lites

Class	Fire rating	Location and use	Glass lite size allowed		
			Area	Height	Width
A	3 h	Fire walls separating buildings or various fire areas within a building. 3–4 h walls	None allowed	None allowed	None allowed
B	1½ h (H.M.) 1 h (other)	Vertical shafts and enclosures such as stairwells, elevators, and garbage chutes. 2 h walls	100 in.	33 in.	10–12* in.
B	1 h	Vertical shafts in low-rise buildings and discharge corridors. 1–1½ h walls	100 in.	33 in.	10–12* in.
C	¾ h	Exit access corridors and exitway enclosures. 1 h walls	1296 in.	54 in.	54 in.
N/A	20-min (⅓ h)	Exit access corridors and room partitions. 1 h walls	No limit	No limit	No limit

*Final size depends on code publication used

Figure 9.31 *Fire-Rated Door and Glass Lites Classification.*

and installed. For example, fitting the wrong kind of glazing may severely reduce the door's fire resistance rating.

Because Fire doors are rated physical fire barriers that protect wall openings from the spread of fire, they are required to provide automatic closing in the event of fire detection. Fire doors should usually be kept closed at all times, although some fire doors are designed to stay open under normal circumstances, and are designed to close automatically or manually in the event of a fire. Fire door release devices are electromechanical devices that enable automatic closing fire doors to respond to alarm signals from detection devices such as smoke detectors, heat detectors, and central alarm systems. This permits closing the door before high temperatures melt the fusible link. Fusible links should always be used as a backup to the releasing device.

Fire-rated door assemblies: Assemblies complying with NFPA 80 that are listed and labeled by Underwriters Laboratories, for fire ratings indicated, based on testing according to NFPA 252. Assemblies must be factory-welded or come complete with factory-installed mechanical joints and must not require job site fabrication.

9.6.3.5 Fire Exits and Stairs

OSHA defines an exit route as a continuous and unobstructed path of exit travel from any point within a workplace to a place of safety. Every building has fire exits that enable users to exit safely in the event of an emergency.

Well-designed emergency exit signs are necessary for emergency exits to be effective. In the United States, fire escape signs often display the word "EXIT" in large, well-lit, green, or red letters. An exit route must be permanent and must be separated by fire-resistant materials. Construction materials used to separate an exit from other parts of the workplace must have a 1-hour fire resistance rating if the exit connects three or fewer stories and a 2-hour fire resistance rating if the exit connects four or more stories.

Exit routes: Unless otherwise stipulated by code, at least two exit routes must be provided in a workplace to permit prompt evacuation of employees and other building occupants during an emergency. The exit routes must be located as far away as practical from each other so that if one exit route is blocked by fire or smoke, employees can evacuate using the second exit route. More than two exit routes must be available in a workplace if the number of employees, the size of the building, its occupancy, or the arrangement of the workplace is such that all employees would not be able to evacuate safely during an emergency. Likewise, a single exit route is permitted where the number of employees, the size of the building, its occupancy, or the arrangement of the workplace is such that all employees would be able to evacuate safely during an emergency.

Exit routes must be arranged so that employees are not required to travel toward a high-hazard area, unless the path of travel is appropriately shielded from high-hazard areas by suitable partitions or other physical barriers. Exit routes must be free and unobstructed. No materials or equipment may be placed, either permanently or temporarily, along the exit route. The exit access must not go through a room that can be locked, such as a bathroom, to reach an exit or exit discharge, nor may it lead into a dead-end corridor. Stairs or ramps are required where the exit route is not substantially level.

Exit discharge: OSHA stipulates that each exit discharge must lead directly to the exterior or to a street, walkway, refuge area, public way, or open space with access to the outside. The street, walkway, refuge area, public way, or open space to which an exit discharge leads must be large enough to accommodate the building occupants likely to use the exit route. Exit stairs that continue beyond the level on which the exit discharge is located must be interrupted at that level by doors, partitions, or other effective means that clearly indicate the direction of travel leading to the exit discharge.

Exit door access: Exit doors must not be locked from the inside, and each doorway or passage along an exit access that could be mistaken for an exit (such as a closet) must be marked "Not an Exit" or similar designation, or be identified by a sign indicating its actual use. Furthermore, exit route doors must be free of decorations or signs that obscure the visibility of the exit route door, and employees must be able to readily open an exit route door from the inside at all times without keys, tools, or special knowledge. A device such as a panic bar that locks only from the outside is permitted on exit discharge doors. Exit route doors may be locked from the inside only in mental, penal, or correctional facilities and then only if supervisory personnel are continuously on duty and the employer has a plan to remove occupants from the facility during an emergency.

Outdoor exit routes: When an outdoor exit route is used, it must have guardrails to protect unenclosed sides where a fall hazard exists. The outdoor exit route must be covered if snow or ice is likely to accumulate along the route, unless it can be demonstrated that any snow or ice accumulation will be removed before it presents a slipping hazard. The outdoor exit route must be reasonably straight and have smooth, solid, substantially level walkways, and the outdoor exit route must not have a dead end that is longer than 20 ft. (6.2 m).

9.6.3.6 Fire Stopping

The protection of people and property during building fires requires the employment of three essential design elements: (1) Alarms to provide early warnings, (2) automatic sprinklers or other suppression systems, and (3) fireproof compartments to contain flames and smoke. These elements work together to give occupants time to escape and firefighters time to arrive. Eliminating any one of the three fire protection elements – detection, suppression, or compartmentation would compromise the integrity of the building.

Building regulations in most jurisdictions stipulate that large buildings need to be divided into compartments and that these fire compartmentations must be maintained should a fire occur. In order to do this, there are a range of fire stopping products and methods available offering between 30 min and 240 min fire compartmentation protection for construction of movement joints and service penetrations.

Compartmentation: Provisions for fire safety may require dividing a building into fire-isolated compartments, which restricts a fire to an area of a building until it can be extinguished. But to be effective, the walls, floor, and ceiling need to contain flames and smoke within the compartment. These components must also provide sufficient insulation to prevent excessive heat radiating outside the compartment.

A fire compartment can therefore be defined as a space within a building extending over one or several floors that is enclosed by separating members such that the fire spread beyond the compartment is prevented during the relevant fire exposure. Fire compartments are sometimes referred to as *fire zones*. Compartmentation is critical to preventing fire to spread into large spaces or into the whole building. It involves the specification of fire-rated walls and floors sealed with firestop systems, fire doors, and fire dampers, etc.

The division of the building into discrete fire zones offers perhaps the most effective means of limiting fire damage. Compartmentation techniques are designed to contain the fire to within the zone of origin, by limiting vertical and horizontal fire spread. Compartmentation also provides at least some protection for the rest of the building and its occupants even if first-aid firefighting measures are used and fail. It also provides protection for inventory and business operations and delays the spread of fire prior to the arrival of the fire brigade.

Determining the required fire resistance for a compartment depends largely upon its intended purpose and on the expected fire. Either the separating members enclosing the compartment shall resist the maximum expected fire or contain the fire until occupants are evacuated. The load-bearing elements in the compartment must always resist the complete fire process or be classified to a certain resistance measured in terms of periods of time, which is equal or longer than the requirement of the separating members. The most important elements to be upgraded are the doors, floors, and walls, penetrations through floors and walls, and cavity barriers in the roof spaces. Halls and landings should typically be separated from staircases to prevent a fire from traveling vertically up or down the stairwell to the other floors. However, creation of new lobbies can have an unacceptable negative impact on the character of a fine historic interior. To be effective, compartmentation needs to be correctly planned and implemented.

The main function of fire stopping is to stop the spread of fire between the floors of a building. Flame-retardant material is installed around floor openings designed to contain conduit and piping. A fire stop is a product that when properly installed, impedes the passage of fire, smoke, and toxic gases from one side of a fire-rated wall or floor assembly to another. Typical fire stop products include sealants, sprays, mechanical devices (fire stop collar), foam blocks, or pillows. These products are installed primarily in two applications: (1) around penetrations that are made in fire-resistive construction for the passage of pipes, cables, or HVAC systems, and (2) where two assemblies meet, forming a expansion joint such as the top of a wall, curtain wall (edge of slab), or floor-to-floor joints. Typical opening types include electrical through-penetrations, mechanical through-penetrations, structural through-penetrations, nonpenetrated openings (e.g., openings for future use), re-entries of existing fire stops, control or sway joints within fire-resistance rated wall or floor assemblies, junctions between fire-resistance rated wall or floor assemblies and "head-of-wall" (HOW) joints, where nonload-bearing wall assemblies meet floor assemblies.

Compliance with all applicable laws and regulations relating to a building is the owner's responsibility, including the adopted and enforced fire code within a specific jurisdiction. Fire codes govern the construction, protection, and occupancy details that affect the fire safety of buildings throughout their life span. Numerous different fire codes have been adopted throughout the United States – the vast majority of which are similar and based on one of the model codes available today or in the past. One requirement in all of these model codes is that fire safety features incorporated into a building at the time of its construction must be maintained throughout a building's life. Therefore, this would require any fire resistance-rated construction to be maintained (Figure 9.32).

9.6.3.7 Alarm Systems and Notification Systems

Fire alarm systems are essential to any facility, particularly in large buildings where there may be visitors or personnel who are unfamiliar with their surroundings. Bruce Johnson, Regional Manager for Fire Service Activities with the International Code Council says,

> Fire alarm systems and smoke alarms are life safety systems that save countless lives each year, both civilians and firefighters. The International Residential Code requires interconnected, hardwired smoke alarms in all new construction (Section R313) and the International Building Code and International Fire Code (Section 907.2) call for manual or automatic fire alarm systems in most commercial buildings with high life occupancy or other hazards. In addition to new construction, the International Fire Code also has provisions for fire alarm systems and smoke alarms in existing structures (Section 907.3).

Fire alarms alert building occupants of a fire *and* alarms that alert emergency public responders (police and fire) through a central station link to initiate an appropriate response.

Mass notification systems (MNS) are invaluable in the protection of a wide range of facilities, and MNS use both audible and visible means to distribute potential lifesaving messages. An MNS is much more than an alarm system. By using the technologies based on fire alarm codes and standards, fire system manufacturers are able to produce a robust life safety and security system.

The safest test method

(a)

(b)

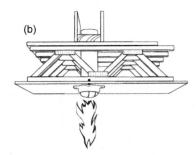

Smoke seals or firestops?

(c)

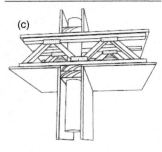

(d)

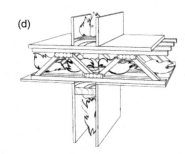

Shutting off the path of a fire

(e)

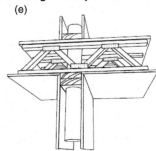

(f)

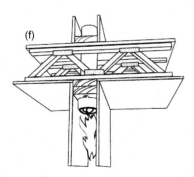

Figure 9.32 *Drawing Showing Various Fire-Stopping Systems Used in Construction.* (a) Chase wall testing does not evaluate performance where the fire enters or starts within the cavity. (b) Systems tested as through-penetrations provide a worst-case evaluation, indicating performance of the firestop regardless of the wall. (c) Caulk only systems provide a seal but cannot shut down a burning pipe. (d) A fire in the wall cavity can quickly spread to the ceiling and wall cavities above. (e,f) Intumescent systems like the firestop collar installed below the top plate shut down burning pipes helping to confine the fire to its point of origin, prolonging structural integrity and buying time for occupants to safely exit the building.

Figure 9.33 *A Siemens MXL Fire Alarm Control Panel (top) and Graphic Annunciator (bottom) for Potomac Hall, at James Madison University. Source: Wikipedia, Photograph by Ben Schumin.*

Fire alarm control panels (FACPs), or fire alarm control units (FACUs), comprise electric panels that function as the controlling components of a fire alarm system (Figure 9.33). The FACP receives information from environmental sensors designed to detect any changes associated with fire. It also monitors their operational integrity and provides for automatic control of equipment and transmission of information necessary to prepare the facility for fire based on a predetermined sequence. An FACP may also supply electrical energy to operate any associated sensor, control, transmitter, or relay. There are currently four basic types of FACP on the market: coded panels, conventional panels, addressable panels, and multiplex systems.

With the introduction of increasingly sophisticated technology, today's alarm systems have the ability to provide more information to the fire department and first responders. In many cases, they can do more than just tell them that there has been an alarm in the building; they can be directed by the kind of alarm and where the alarm is. Moreover, many modern systems now include speakers that provide alerts in place of (or in addition to) traditional bell-type alarms. These speakers also can be used in emergencies other than fires to instruct and inform occupants of the situation.

These voice-actuated systems can include prerecorded or live messages that play in the event of fire or another emergency. Typical prerecorded messages tell occupants that an alarm has been sounded and that they should remain in their designated area for further instruction. Building management can then manually use the system to deliver additional information and prepare occupants for an evacuation, if necessary. Alert systems can also close fire doors, recall elevators, and interface and monitor the installed suppression systems, such as sprinklers. It should be noted that when fire alarms systems are properly installed and maintained, they perform very well. But when they are not, the public and fire service may be subject to unnecessary "false alarms" that puts everyone at risk.

Alarm systems can also connect with a building's ventilation, smoke-management, and stairwell-pressurization systems – all of which are critical to life safety. Again, these features are dependent on the building in which the system is installed. In addition, the marriage of MNS and fire alarm control systems is a growing trend that it is hoped will continue and be applied in larger varieties of facilities and multibuilding properties, including schools, high-rise buildings, mass transit hubs, and even public gathering places such as theatres, restaurants, and places of worship.

An annunciator panel is sometimes employed to monitor the status of the different areas in a designated fire zone, theft protection, and control of a facility's alarm devices. There may be several fire zones in a building. Each fire zone is clearly marked on the panel. The annunciator panel identifies the different zones and their specific security status. Should a fire occur, an indicator light flashes on the panel and identifies the fire's location. For example, the light on the panel might indicate that a fire has occurred in Fire Zone 4. This information allows the Fire Department to quickly locate the fire.

9.6.4 Codes and Standards

Code compliance is the first objective in any design. There are a number of relevant national codes that relate to green-building fire-protection systems that are published by the National Fire Protection Association (NFPA). It should be noted that fire codes can vary substantially from one jurisdiction to another, and although these codes are not mandatory in all jurisdictions, they should nevertheless be adhered to whenever possible because they provide maximum safety for property and personnel and can help guide system design and installation:

NFPA 72 – National Fire Alarm Code: Governs the design, installation, operation, and maintenance of fire-detection and alarm systems. It includes requirements for detector spacing, occupant notification, and control-panel functionality.

NFPA 750 – Standard on Water Mist Fire Protection Systems: Governs water-mist-system classification and incorporates requirements for water-mist-system design, installation, operation, and maintenance.

NFPA 2001 – Standard on Clean Agent Fire Extinguishing Systems: Governs the design, installation, operation, and maintenance of clean-agent systems. It additionally includes requirements for assessing design concentrations, safe personnel-exposure levels, and system-discharge times. The standard also stipulates that an agent be included on the US EPA's Significant New Alternatives Policy list.

Finally, green buildings today have numerous fire protection system options that can be employed. Careful consideration of a building and its anticipated hazards will help determine which areas require protection. Because of the recent advances in technology, fire detection and suppression systems can now adequately support and sustain a modern green-building philosophy. The methodical selection of a clean-agent or water-mist system can also help contribute to LEED certification credits for building owners and developers.

CHAPTER 10

Economics of Green Design

10.1 GENERAL OVERVIEW

Developing an understanding of the future is one of the critical aspects of the construction planning process. Construction is a very diverse industry that is heavily interconnected with the economy as a whole. Additionally, the construction industry is invariably impacted by global events and trends. A February 2015 USGBC article, *The Business Case for Green Building*, says,

> Commercial building owners and managers will invest an estimated $960 billion globally between now and 2023 on greening their existing built infrastructure. Major priority areas include more energy-efficient heating, ventilation and air conditioning, windows, lighting, plumbing fixtures, and other key technologies. As the industry standard, and the recognized leader in innovative design, LEED for Existing Buildings and Maintenance is poised to expand exponentially as a result of this near trillion-dollar investment in sustainable design.

Like most industries, the US construction industry continues to evolve and witness rapid advancements in science and technology, especially as it relates to green building and sustainability. Unlike most industries, however, the construction industry has until recently been dependent on age-old techniques. The current financial crisis and global economic recession put an intolerable strain on the nation's construction industry. Peter Morris, principal of the construction consultancy Davis Langdon, believes that the drastic reduction in construction activity in recent years encouraged increased competition among bidders and lower escalation pressure on projects to the extent that in many projects, cost trends have become negative, leading in part to moderate construction price deflation. But one of the biggest causes of concern according to Morris is the issue of contractor financing and working capital. Many contractors are finding it increasingly difficult to maintain adequate cash flow for operations, and none have the resources to manage significant expansion of working capital. This has caused considerable concern and obliged bidders to be very cautious and judicious in selection of projects, with a strong preference toward projects that have good-quality cash flows.

Looking at the upside, there has emerged an increasingly broad awareness of the significant benefits that green buildings have to offer. For example,

LEED v4 Practices, Certification, and Accreditation Handbook
http://dx.doi.org/10.1016/B978-0-12-803830-7.00010-4
Copyright © 2016 Elsevier Inc.
All rights reserved.
519

the US military including the US Air Force and Navy now require that their new buildings be Leadership in Energy and Environmental Design (LEED™) green buildings. Boston Mayor Thomas M. Menino clearly put it, "High performance green building is good for your wallet. It is good for the environment. And it is good for people."

Benefits such as reduced energy bills and reduced potable water consumption are easily quantified, whereas other benefits like green design's impact on occupant health or security are usually much more difficult to quantify. Incisive Media's "2008 Green Survey: Existing Buildings" found that nearly 70% of commercial building projects in the United States have already incorporated some kind of energy monitoring system. The survey also found that energy conservation is the most widely implemented green program in commercial buildings; this is followed by recycling and water conservation. Moreover, approximately 65% of building owners who have implemented green building features claim their investments have already resulted in a positive return on their investment.

A recent study released by Lawrence Livermore National Laboratory in California and the US Department of Energy indicates that the United States used more energy but produced fewer carbon emissions in 2014 compared to 2010. The recently released analysis implies that the use of wind and natural gas as energy sources account for part of the improvement. Moreover, the emissions reductions measured in the study are the result of wind and solar energy development, and a greater acceptance of natural gas for electricity generation and industrial applications.

Likewise, according to Turner Construction Company's Green Building Barometer, roughly 84% of respondents maintain that their green buildings have resulted in lower energy costs, and 68% recorded lower overall operating costs. Additionally, "75% of executives said that recent developments in the credit markets would not make their companies less likely to construct Green buildings." In fact, the survey contends that 83% would be "extremely" or "very" likely to seek LEED™ certification for buildings they are planning to build within the next 3 years. In the same survey, executives reported that Green buildings have better financial performance than non-Green buildings, especially in the following sectors:

- Higher building values (72%)
- Higher asking rents (65%)
- Greater return on investment (52%)
- Higher occupancy rates (49%)

According to the 2014 Turner Green Building Marketing Barometer, there is an increasing Focus on Benefits to Employees and Occupants.

In deciding whether to incorporate Green features, financial considerations were most often rated as extremely or very important such as energy efficiency (81%), asking rents (81%), ongoing operations and maintenance costs (79%), and occupancy rates (78%). However, several non-financial factors were rated almost as highly including health and well-being of occupants (78%), indoor air quality (78%), employ productivity (74%), impact on brand/reputation (73%), and satisfaction of employees/occupants (71%). Recognizing the importance of an organization's reputation for sustainability in its ability to attract and retain talented employees, employee hiring/retention was rated as extremely or very important by 62% of executives, up from 49% in the 2012 survey.

In a comparative study by Langdon (2006) in which the construction costs of 221 buildings were analyzed, it was found that 83 buildings were constructed with the intent of achieving LEED certification and 138 that did not have any sustainable design intentions. The study found that a majority of the buildings analyzed were able to achieve LEED certification without increased funding. In another investigation conducted by Davis Langdon of a wide and diverse range of studies by other organizations, it was found that the average construction cost premium required to achieve a moderate level of Green features, equivalent to a Silver LEED certification, was roughly between 1% and 2%. However, Davis Langdon also found that often half or more of the Green projects in these studies revealed no increase in construction costs at all.

Yet, despite the effusive praise and increased awareness for sustainable design, property owners and developers were not always quick to embrace green building practices. This appears to be the prevailing sentiment of CB Richard Ellis's *Green Downtown Office Markets: A Future Reality*, a report depicting the general progress of the green building movement. The report looks at the obstacles preventing a broad-based acceptance of sustainable design in office construction. Perhaps the main obstacle to embracing design sustainability is the perception of initial outlay compared to long-term benefits; even though an increasing number of studies similar to the one conducted by Davis Langdon in 2006 clearly show that there is no significant difference in average costs for green buildings as compared to conventionally constructed buildings. The other hurdle — which has now been addressed — was the lack of data on development, construction costs, and time needed to recoup costs, making education the most crucial tool for sustainable design. However, green building is now growing fast in the United States and is expected to represent more than half of all commercial and institutional construction by 2016.

However, increased awareness and education has had a very positive effect on moving forward with sustainable development. For example, Alex Wilson, president of BuildingGreen, LLC, notes, "Given that we manage 1.7 billion square feet globally, we have a real opportunity to help our clients understand how they can make changes in operations in their own facilities to reduce greenhouse gas emissions and be more energy efficient."

A rough estimate of green building research is currently put at about $193 million per year or roughly 0.2% of federally funded research. This approximates only 0.02% of the estimated $1 trillion value of annual US building construction, despite the fact that the building construction industry represents approximately 9% of the US GDP. The construction industry is currently reinvesting only 0.6% of sales back into research, which is markedly less than the average for other US industries and private sector construction research investments in other countries.

The US Green Building Council (USGBC) suggests that we need to move decisively toward increasing and improving green building practices, to avoid being confronted with a dramatic backlash in adverse impact of the built environment on human and environmental health. Building operations today are estimated to account for 38% of US carbon dioxide emissions, 71% of electricity use and 40% of total energy use. If the energy required in the manufacture of building materials and constructing buildings is included, this number then goes up to an estimated 48%. Buildings also consume roughly 12% of the country's water in addition to rapidly increasing amounts of land. Moreover, construction and remodeling of buildings account for 3 billion tons, or 40%, of raw material used globally each year, which in turn has a negative impact on human health; in fact, up to 30% of new and remodeled buildings may experience acute indoor air quality problems.

10.2 COSTS AND BENEFITS OF GREEN DESIGN

Peter Morris opines that "clearly there can be no single, across the-board answer to the question, 'what does green cost?' On the other hand, any astute design or construction professional recognizes that it is not difficult to estimate the costs to go green for a specific project. Furthermore, when green building concepts and features are incorporated early in the design process, it greatly increases the ability to construct a certified green building

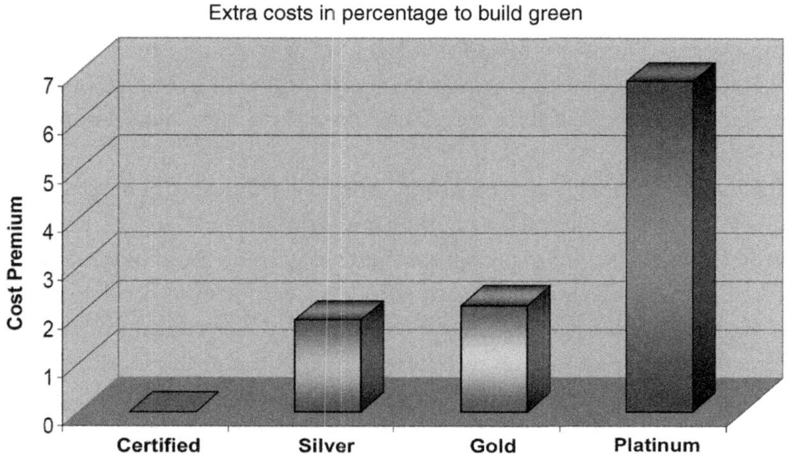

Figure 10.1 *Diagram Showing One Study's Estimated Cost of Building Green for the Various LEED™ Rating Systems. Source: USGBC.*

at a cost comparable to a code-compliant one. This means that it is possible today to construct green buildings or buildings that meet the USGBC's LEED® third-party certification process with minimal if any increase in initial costs (Figure 10.1).

Davis Langdon also suggests that to be successful in building green and to keep the costs of sustainable design under control, three critical factors must be understood and implemented. These are,

First, clear goals are critical for managing the cost. It is not enough to simply state, "we want our project to be green"; the values should be determined and articulated as early in the design process as possible. Second, once the sustainability goals have been defined, it is essential to integrate them into the design and to integrate the design team so that the building elements can work together to achieve those goals. Buildings can no longer be broken down and designed as an assemblage of isolated components. This is the major difference between traditional building techniques and the new sustainable design process. Third, integrating the construction team into the project team is critical. Many sustainable design features can be defeated or diminished by poor construction practices. Such problems can be eliminated by engaging the construction team, including sub contractors and site operatives in the design and procurement process.

Another noteworthy study that represents the largest international research of its kind, Greening Buildings and Communities: Costs and Benefits, is based on extensive financial and technical analysis of 150 green buildings across the United States and in 10 other countries and provides the most detailed findings to date on the costs and financial benefits of building green. The study found that benefits of building green consistently outweigh any potential cost premium. Among the main conclusions arrived at in the Greening Buildings and Communities: Costs and Benefits study are the following:

Conclusion 1: Most green buildings cost 0%–4% more than conventional buildings, with the largest concentration of reported "green premiums" between 0% and 1%. Green premiums increase with the level of greenness but most LEED™ buildings, up through gold level, can be built for the same cost as conventional buildings. This stands in contrast to a common misperception that green buildings are much more expensive than conventional buildings.

Conclusion 2: Energy savings alone make green building cost-effective. Energy savings alone outweigh the initial cost premium in most green buildings. The present value of 20 years of energy savings in a typical green office ranges from \$7 ft.^{2} (certified) to \$14 ft.^{2} (platinum), more than the average additional cost of \$3–\$8 ft.^{2} for building green.

Conclusion 3: Green building design goals are associated with improved health and with enhanced student and worker performance. Health and productivity benefits remain a major motivating factor for green building owners, but are difficult to quantify. Occupant surveys generally demonstrate greater comfort and productivity in green buildings.

Conclusion 4: Green buildings create jobs by shifting spending from fossil fuel–based energy to domestic energy efficiency, construction, renewable energy, and other green jobs. A typical green office creates roughly one-third of a permanent job per year, equal to \$1 ft.^{2} of value in increased employment, compared to a similar nongreen building.

Conclusion 5: Green buildings are seeing increased market value (higher sales/rental rates, increased occupancy and lower turnover) compared to comparable conventional buildings. CoStar, for example, reports an average increased sales price from building green of more than \$20 ft.^{2} providing a strong incentive to build green even for speculative builders.

Conclusion 6: Roughly 50% of green buildings in the study's data set see the initial "green premium" paid back by energy and water savings in 5 years or less. Significant health and productivity benefits mean that more than 90% of green buildings pay back an initial investment in 5 years or less.

Conclusion 7: Green community design (e.g., LEED™-ND) provides a distinct set of benefits to owners, residents and municipalities, including reduced infrastructure costs, transportation and health savings and increased property value. Green communities and neighborhoods have a greater diversity of uses, housing types, job types, and transportation options and appear to better retain value in the market downturn than conventional sprawl.

Conclusion 8: Annual gas savings in walkable communities can be as much as $1000 per household. Annual health savings (from increased physical activity) can be more than $200 per household. CO_2 emissions can be reduced by 10–25%.

Conclusion 9: Up-front infrastructure development costs in conservation developments can be reduced by 25%, approximately $10,000 per home.

Conclusion 10: Religious and faith groups build green for ethical and moral reasons. Financial benefits are not the main motivating factor for many places of worship, religious educational institutions, and faith-based nonprofits. A survey of faith groups building green found that the financial cost-effectiveness of green building makes it a practical way to enact the ethical/moral imperative to care for the Earth and communities. Building green has also been found to energize and galvanize faith communities.

Even when a green building costs more up front as a result of inefficient planning and execution, the costs are quickly recouped through lower operating costs over the life of the building. The green building approach is an analytical method that applies to a project's life cycle cost analysis for determining the appropriate upfront expenditure.

10.2.1 The Economic Benefits of Green Buildings

The numerous savings in costs can only be fully realized when they are incorporated at the project's conceptual design phase and with the assistance of an integrated team of professionals. Using an integrated systems approach ensures that the building is designed in a holistic manner as one system rather than a number of stand-alone systems.

Some of the benefits of building green are not easy to quantify; how do you measure improving occupant health, comfort, productivity, or pollution reduction? This is one of the main reasons why they are excluded from being adequately considered in cost analysis. It would be prudent therefore to consider setting aside a small portion of the building budget (e.g., as a contingency) to cover differential costs associated with less tangible green building benefits or to cover the cost of researching and analyzing green building options. Even when experiencing difficult times, many green building measures can be incorporated into a project with minimal or zero increased up-front costs but which are capable of yielding enormous savings (Figure 10.2).

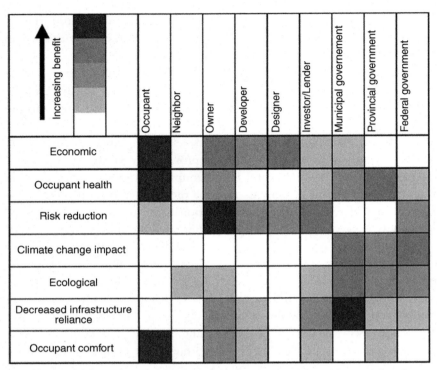

Figure 10.2 *Matrix Illustrating Various Green Building Stakeholder Benefits. Source: Report – A Business Case for Green Buildings in Canada.*

What does "green" cost is one of the first questions often asked by owners and developers regarding sustainable design; typical translation: does it cost more? This raises the question: more than what? For example, is the question more than the building would have cost without the sustainable design features than that of comparable buildings, or more than available funds? The answers to these questions have until recently been largely elusive because of the lack of hard data. However, because of the considerable research undertaken in recent years, we now have substantial data on building costs that allows us to compare the cost of green buildings and nonsustainable buildings with comparable programs.

When considering sustainable design and how it relates to construction costs, it is important to analyze the costs and benefits using a holistic approach. This means including evaluation of operations and maintenance costs, user productivity and health, and design and documentation fees, among other financial measurements. This is because empirical experience has repeatedly shown that it is the construction cost implications that have the greatest impact to fundamentally drive decisions about sustainable design. Assisting teams comprehend the actual construction costs on real projects of achieving green, and by providing a methodology that will allow teams to viably manage these construction costs can go a long way to facilitate a team's ability to get past the question of whether to green or not.

From such an analysis, we can conclude that many projects are able to achieve sustainable design within the initial budget, or with minimal supplemental funding. This suggests that developers continue to find ways to incorporate project goals and values, regardless of budget, by making choices. However, every building project is unique and should be considered as such when addressing the cost and feasibility of LEED certification. There is no one-size-fits-all answer, and benchmarking with other comparable projects can be valuable and informative, but not predictive. Any estimate of cost relating to sustainable design for a specific building must be made with reference to that building, its particular circumstances, and goals.

The cost of green design has dropped significantly in recent years as the number of green buildings has increased. The trend of declining costs associated with increased experience in green building construction has manifested itself in a number of states throughout the country. Moreover, studies have shown that LEED™-certified buildings constructed at the Certified or Silver level have often been built without an increase in preliminary costs.

In fact, Davis Langdon explored this topic in a report issued in 2004, *Costing Green: A Comprehensive Cost Database and Budgeting Methodology*. The report concluded that many LEED™-rated projects cost the same as projects with no stated green goals. The study actually found that many projects with LEED™-level ratings often cost less than non-LEED™ projects. Davis Langdon also discovered that the opposite was often true, in that no direct correlation was found between higher costs for going green and lower costs for conventional building.

Some recent studies, however, suggest that gold- and platinum-level projects indicate an increase of 1–5%, with 2% being the average increased additional first cost. But even in cases that show a small increase in upfront costs, these costs are usually recouped within a short period, that is, 1–2 years during the life cycle of the building, through tax incentives and operational savings resulting from decreased energy and water use.

A recent study by Greg Kats of Capital E Analysis provides a summary of financial benefits from going green, as shown in Figure 10.3. The report concluded that financial benefits for building green are estimated to be

Category	20-year net present value
Energy saving	$5.80
Emissions savings	$1.20
Water savings	$0.50
Operations and maintenance savings	$8.50
Productivity and health value	$36.90 to $55.30
Subtotal	**$52.90 to $71.30**
Average extra cost of building green	(–3.00 to –$5.00)
Total 20-year net benefit	**$50 to $65**

Figure 10.3 *Financial Benefits of Green Buildings – Summary of Findings (Per Square Foot).* The financial benefits of going green are related mainly to productivity. *Source: Capital E. Analysis.*

between $50 and $70 ft.^{2} in a LEED building; this represents more than 10 times the additional cost associated with building green. These financial benefits come in the form of lower energy, waste and water costs, lower environmental and emissions costs, and lower operational and maintenance costs, lower absenteeism, increased productivity and health, greater retail sales, and easier reconfiguration of space, resulting in less downtime and lower costs. Cost estimates that are based on a sample of 33 office and school buildings suggested only 0.6% greater costs for LEED certification, 1.9% for silver, 2.2% for gold, and 6.8% for platinum certification. Although these estimates are direct costs, they nevertheless closely reflect those provided by the USGBC.

Impact of green building on long-term value: The principal drivers of green building appear to be (1) increasing energy costs (75%); (2) government regulations/tax incentives (40%); and (3) global influences (26%).

The absence of sufficient clear up-to-date data pertaining to development, construction costs and time required to recoup costs are contributing factors to the industry's slow acceptance of green construction, which is why education has become the most important tool in promoting the green construction initiative. The main obstacles measured in the study include

- Too multidisciplinary – 41%
- Not convinced of increased return on investment (ROI) – 37%
- Lack of understanding benefits – 26%
- Lack of service providers – 20%
- Too difficult – 17%
- Greenwashing – 16%
- Lack of shareholder support – 10%

Most building owners and developers build to sell, in the sense that they construct or revamp an office building, lease its space, and with the hope of selling the asset within a 3- to 5-year time frame to repay debt and ensure a profit. The speed with which the process is completed impacts the amount of profit generated upon consummation of the sale of the building. Uninformed developers that are under the perception that green construction costs more spur fears as they are already concerned about the cost of short-term debt and conventional building materials.

Typical benefits for building owners and tenants alike as deduced by Davis Langdon include the following:

- Reduced risk of building obsolescence
- Ability to command higher lease rates, thereby increasing profits
- Potential higher occupancy rates
- Higher future capital value
- Less need for refurbishment in the future
- Higher demand from institutional investors
- Lower operating costs
- Mandatory for government tenants
- Lower tenant turnover
- Enhance occupant comfort and health
- Improve air/thermal/acoustic environment, thereby improving employee productivity and satisfaction

10.2.2 Cost Considerations

The economic considerations with respect to green buildings can normally be divided into (1) Direct Capital Costs and (2) Direct Operating Costs as shown in the following:

1. *Direct capital costs*: These are costs associated with the original design and construction of the building and generally include interest during construction (IDC). The general perception of some building stakeholders is that the capital costs of constructing green buildings are significantly higher than those of conventional buildings, whereas many others within the green building field believe that green buildings actually cost less or no more than conventional buildings. The green building industry believes that savings are achieved by the downsizing of systems through better design and the elimination of unnecessary systems that will offset any increased costs caused by implementing more advanced systems.

   Capital and operational costs are normally relatively easy to measure because the required data are readily quantifiable and available.

Productivity effects on the other hand are difficult to quantify, yet are nevertheless important to consider because of their potential impact. Other indirect and external effects can be wide-reaching, and may also be difficult to quantify.

2. *Direct operating costs*: These costs consist of all necessary expenditures required to operate and maintain a building over its full life. They include the total costs of building operation, such as energy use, water use, insurance, maintenance, waste, property taxes etc. over the entire building life. The primary costs are those associated with heating and cooling and maintenance activities (painting, roof repairs, and replacement, etc.). Included in this cost category are less obvious items such as churn (the costs of reconfiguring space and services to accommodate occupant moves). Excluded from this category are costs relating to major renovations, cyclical renewal, and residual value or demolitions costs, which are considered to be direct capital investments.

Insurance rates: Insurance is a direct operating cost, and green buildings have many tangible benefits that reduce or mitigate a variety of risks, and which should be reflected in the insurance rates for the building. Also, the fact that green buildings generally provide a healthier environment for occupants should be reflected in health insurance premiums. Indeed, the general attributes of green buildings (e.g., the incorporation of natural light, off-grid electricity, and commissioning) should reduce a broad range of liabilities, and the general site locations also potentially reduces risks of property loss due to natural disasters.

A fully integrated design of a building will typically reduce the risk of inappropriate systems or materials being employed, which could have a positive impact on other insurable risks. In some cases, insurance companies offer premium reductions for certain green features, such as commissioning or reduced reliance on fossil fuel–based heating systems (resulting in reduced fire threat). The list of premium reductions will undoubtedly increase with further education and awareness and as the broad range of benefits are more fully recognized.

Churn rates: Churn rate reflects the frequency with which building occupants are moved, either internally or externally, including occupants who move but remain within a company, and those who leave a company and are replaced. Statistics show that green buildings typically have lower churn rates than conventional buildings because of increased occupant comfort and satisfaction.

10.2.3 Life-Cycle Costing

This is basically a method of combining both capital and operating costs to determine the net economic effect of an investment, and to evaluate the economic performance of additional investments that may be required for green buildings. It is based on discounting all future costs and benefits to dollars of a specific reference year that are referred to as present value (PV) dollars. This makes it feasible to intelligibly quantify costs and benefits and compare alternatives based on the same economic criterion or reference dollar.

Peter Morris, a principal with Davis Langdon, says, "Perhaps a measure of the success of the LEED™ system, which was developed to provide a common basis for measurement, is the recent proliferation of alternative systems, each seeking to address some perceived imbalance or inadequacy of the LEED™ system, such as the amount of paperwork, the lack of weighting of credits, or the lack of focus on specific issues." Of note, many of these criticisms have now been addressed with the latest version of LEED™ v4. Peter Morris also says that "among these alternative measures are broad-based approaches, such as Green Globes, and more narrowly focused measures, such as calculations of a building's carbon footprint or measurements of a building's energy efficiency (the ENERGY STAR® rating). All these systems are valid measures of sustainable design, but each reflects a different mix of environmental values, and each will have a different cost impact."

By effectively integrating sustainable design elements into a project during its early project development and design phases, one can reduce building costs. On the other hand, if green design elements are not considered until late in the design process or during construction, and when parts of the project or the entire project has to be redesigned, then a significant increase in costs can be contemplated. However, according to a recent study by the World Business Council for Sustainable Development (WBCSD), key players in real estate and construction unfortunately often misjudge the costs and benefits of "green" buildings.

First cost: In documented research, the most often cited reason for not incorporating green elements into building designs is the assumed increase in first cost. Some aspects of design have little or no first cost, including site orientation and window and overhang placement. Other sustainable systems that incorporate additional costs in the design phase, such as an insulated shell, can be offset, for example, by the reduced cost of a smaller mechanical system. Material costs can be reduced during the construction phase of a project by the use of dimensional planning and other material efficiency strategies. Such strategies can reduce the amount of building materials needed and cut construction costs but they require forethought on the part of designers to ensure a building that creates less construction waste solely on its dimensions and structural design. An example of dimensional planning is designing rooms of 4-ft. multiples, since wallboard and plywood sheets come in 4- and 8-ft. lengths. Furthermore, one dimension of a room can be designed in 6- or 12-ft. multiples to correspond with the length of carpet and linoleum rolls, and whenever possible, rooms should be designed with 2-ft. incremental dimensions.

Nevertheless, construction projects typically have initial or upfront costs, which may include capital investment costs related to land acquisition, construction, or renovation and for the equipment needed to operate a facility. Land acquisition costs are normally included in the initial cost estimate if they differ among design alternatives. A typical example of this would be when comparing the cost of renovating an existing facility with new construction on purchased land.

A 2007 study by Davis Langdon updating an earlier study states, "It is clear from the substantial weight of evidence in the marketplace that reasonable levels of sustainable design can be incorporated into most building types at little or no additional cost. In addition, sustainable materials and systems are becoming more affordable, sustainable design elements are becoming widely accepted in the mainstream of project design, and building owners and tenants are beginning to demand and value those features." He follows this up in a 2010 study saying, "Green design strategies have become 'the norm' for higher quality buildings, and the perception of 'extra' cost has diminished."

Ashley Katz, a communications coordinator for the USGBC, says, "Costs associated with building commissioning, energy modeling and additional professional services typically turn out to be a risk mitigation strategy for owners. While these aspects might add on to the project budget, they will end up saving projects money in the long run, and are also best practices for building design and construction."

Many experienced users of LEED have found it possible to build at Silver and Gold LEED™ levels for the same cost as conventional buildings, although some studies have concluded that certification adds roughly 1–2% of the overall budget to the cost of construction. LEED™ registration and certification fees are not substantial, averaging about $2000, depending on a project's size. In any case, registration is required for projects pursuing LEED certification and provides access to a variety of online resources, including

LEED™ Online, which is an online project management tool that project teams use to prepare their documentation for LEED certification.

10.2.4 Life-Cycle Cost Method

Life-cycle cost analysis (LCCA) is a method used to estimate the total cost of facility ownership. It takes into consideration all costs relating to acquiring, owning, and disposing of a building or building system. Sieglinde Fuller of the National Institute of Standards and Technology (NIST) says, "LCCA is especially useful when project alternatives that fulfill the same performance requirements, but differ with respect to initial costs and operating costs, have to be compared in order to select the one that maximizes net savings. For example, LCCA will help determine whether the incorporation of a high-performance HVAC or glazing system, which may increase initial cost but result in dramatically reduced operating and maintenance costs, is cost-effective or not." However, LCCA is not useful when it comes to budget allocation.

Despite the general consensus on the valid reliable basis for adopting a life cycle approach, many building stakeholders still prefer to focus on minimizing direct costs or, at best, applying short time frame payback periods. Other developers, building owners, and other stakeholders hold the view that basing opinions on anything other than a reduced direct cost approach is fiscally irresponsible, when in reality the opposite is often the case. This lack of adoption is largely due to the typical corporate structure that dissociates direct and operating costs and with most constructers often lacking the mandate to reduce operating costs, although they are mandated to reduce construction cost. This unfortunate reality is also evidenced by owner/developers who oversee construction of buildings for their own use. Other causal factors identified with this issue include the following:
- Not comprehending the life cycle concept;
- Cash-flow constraints;
- Arduousness in calculating performance as opposed to the relative ease in calculating direct cost; and
- Lack of adequate support from lending institutions.

The main objective of an LCCA is to calculate the overall costs of project alternatives and to select the design that safeguards the ability of the facility to provide the lowest overall cost of ownership in line with its quality and function. The LCCA should be performed early in the design process to allow any needed design refinements or modifications to take place before finalization to optimize the life-cycle costs (LCCs).

Perhaps the most important and challenging task of an LCCA (or any economic evaluation method for that matter) is to evaluate and determine the economic effects of alternative designs of buildings and building systems and to be able to quantify these effects and depict them in dollar amounts.

Lowest LCC provides a straightforward and easy-to-interpret measure of economic evaluation. Other commonly used methods include net savings (or net benefits), savings-to-investment ratio (or savings benefit-to-cost ratio), internal rate of return, and payback period. Sieglinde Fuller sees them as being consistent with the lowest LCC measure of evaluation if they use the same parameters and length of study period. Almost identical approaches can be made to making cost-effective choices for building-related projects irrespective of whether it is called cost estimating, value engineering, or economic analysis.

Sustainable buildings can be assessed as cost-effective through the life-cycle cost method, a way of assessing total building cost over time. After all costs by year and amount are identified and then discounted to present value, they are added to arrive at total life-cycle costs for each alternative: It consists of

- Initial design and construction costs.
- Operating costs, including energy, water/sewage, waste, recycling, and other utilities.
- Maintenance, repair, and replacement costs.
- Other environmental or social costs/benefits including but not limited to impacts on transportation, solid waste, water, energy, infrastructure, worker productivity, and outdoor air emissions, etc.

Because sustainable buildings are also considered healthy buildings, they can have a positive impact on worker illness costs.

Supplementary measures: Sieglinde Fuller says,

Supplementary measures of economic evaluation are Net Savings (NS), Savings-to-Investment Ratio (SIR), Adjusted Internal Rate of Return (AIRR), and Simple Payback (SPB) or Discounted Payback (DPB). They are sometimes needed to meet specific regulatory requirements. For example, the FEMP LCC rules (10 CFR 436A) require the use of either the SIR or AIRR for ranking independent projects competing for limited funding. Some federal programs require a Payback Period to be computed as a screening measure in project evaluation. NS, SIR, and AIRR are consistent with the lowest LCC of an alternative if computed and applied correctly, with the same time-adjusted input values and assumptions. Payback measures, either SPB or DPB, are only consistent with LCCA if they are calculated over the entire study period, not only for the years of the payback period.

Supplementary measures are considered to be relative measures; that is, they are computed for an alternative relative to a base case.

The design of buildings requires the integration and incorporation of different kinds of data into a synthetic whole. This integrated process, or "whole building" design approach requires the active and full participation of many players – users, code officials, building technologists, cost consultants, specifications specialists, architects, civil engineers, mechanical and electrical engineers, structural engineers, and consultants from many specialized fields. The best buildings produced through active, deliberate, and full collaboration among all players.

Building-related investments typically involve a great deal of uncertainty relating to their costs and potential savings. The performing of an LCCA greatly increases the ability and likelihood of deciding on a project that can save money in the long run. Yet this does not alleviate some of the potential uncertainty associated with the LCC results, mainly because LCCAs are typically conducted early in the design process when only estimates of costs and savings are available, rather than specific dollar amounts. This uncertainty in input values means that actual results may differ from estimated outcomes.

Deterministic techniques, such as sensitivity analysis and breakeven analysis, are easily conducted without the need for additional resources or information. They will produce a single-point estimate highlighting how uncertain input data affect the analysis outcome. However, the basis of the probabilistic method is that of quantifying risk exposure by deriving probabilities of achieving different values of economic worth from probability distributions for input values that are uncertain. Probabilistic techniques possess greater informational and technical requirements than do deterministic ones. The final choice for using one or the other method depends on several factors, mainly the size of the project, its significance, and the available resources. Both sensitivity analysis and break-even analysis are approaches that are simple to perform and should both be an integral component of every LCCA.

10.2.5 Increased Productivity

One of the many pivotal benefits of green buildings is their positive effect on productivity. Recent research and numerous studies clearly illustrate that green buildings dramatically affect productivity. However, because many of these studies are often broad in nature and rarely focus on unique or specific green building attributes, they need to be supplemented by other thorough, accurate, and statistically sound research to fully comprehend the effects of green buildings on occupant productivity, performance, and sales.

It does seem prudent, however, that any productivity gains attributable to a green building should be included in the LCCA, particularly for an owner-occupied building. The difficulty of properly attributing such gains like reduced absenteeism and staff turnover rates appears to have made this the exception rather than the rule.

Some of the key features of green buildings that relate to increased productivity include controllability of systems relating to ventilation, temperature, and lighting; daylighting and views; natural and mechanical ventilation; pollution-free environments; and vegetation. It is not always clear why these features produce improved productivity, although healthier employees typically mean happier employees, which studies show almost always produce increased worker satisfaction, improved morale, reduced absenteeism, and increased productivity.

One study by Lawrence Berkeley National Laboratory concluded that improvements to indoor environments such as commonly found in green buildings could reduce health care costs and work losses as follows:

- From communicable respiratory diseases by 9–20%
- From reduced allergies and asthma by 18–25%
- From nonspecific health and discomfort effects by 20–50%

Hannah Carmalt, a Project Analyst with Energy Market Innovations, says, "The most intuitive explanation is that productivity increases due to better occupant health and therefore decreased absenteeism. When workers are less stressed, less congested, or do not have headaches, they are more likely to perform better." High-performance buildings have many potential benefits, including increased market value, lower operating and maintenance costs, improved occupancy for commercial buildings, and increased employee satisfaction and productivity for owner-occupied buildings. Occupants of green buildings have consistently recorded increased productivity and healthier working and living environments.

In a William Fisk study, green buildings were found to add $20–$160 billion in increased worker productivity annually. This is due to the fact that LEED-certified buildings were found to yield significant productivity and health benefits, such as heightened employee productivity and satisfaction, fewer sick days, and fewer turnovers. Moreover, other independent studies have shown that better climate control and improved air quality can increase employee productivity by an average of 11–15% annually. However, it should be noted that in commercial and institutional buildings, payroll costs generally significantly overshadow all other costs, including those involved in the design, construction, and operation of a building.

Elements affecting productivity: While theoretical explanations may help explain why productivity increases have been documented to occur in green buildings, it is still necessary to define the particular elements of green buildings that are directly related to productivity. Sound control, for example, while recognized as increasing productivity, is often excluded from green-related studies, mainly because it is not necessarily a green building feature. Likewise, the presence of biological pollutants, such as molds, are also associated with decreased productivity but are also excluded because typical green buildings do not automatically eliminate the presence of such materials (although their presence is reduced by improved ventilation in green buildings).

As mentioned previously, among the most significant elements of green buildings that affect productivity and thermal comfort is occupant control over temperature, ventilation, and lighting. Green buildings usually try to incorporate this feature because it can noticeably decrease energy use by ensuring areas are not heated, cooled, or lit more than is required. These measures are decisive to maintaining energy efficiency and occupant satisfaction within a building. The Urban Land Institute and the Building Owners and Managers Association conducted a survey, which found that occupants rated air temperature (95%) and air quality (94%) most crucial in terms of tenant comfort. The study also determined that 75% of buildings did not have the option or capability to adjust features and that many individuals were willing to pay higher rents in order to obtain such measures.

Various studies have associated increased productivity with increased emotional well-being. In a study conducted by the Heschong Mahone Group (HMG), it was found that higher test scores in daylight classrooms were achieved because of students being happier. HMG also found that when teachers were able to control the amount of daylighting in classrooms, students progressed 19–20% faster than students in classrooms that lacked controllability. Similar studies that were performed in office settings clearly showed that increases in individual control over temperature, ventilation, and lighting produced significant increases in productivity.

Office space with views to the outdoors is another common feature of green buildings that is associated with productivity. The 2003 office study conducted by the consultants HMG mentioned previously also found correlations between productivity and access to outdoor views: test scores were generally 10–15% higher and calling performance increased by 7–12%. This reinforces HMG's earlier 1999 study findings of schools where children in classrooms of the Capistrano School District progressed

15% faster in math and 23% faster in reading when they were located in the classrooms with the largest windows. Furthermore, along with happier students and workers, substantial financial and human-performance benefits have been associated with increased daylight.

Ventilation too is a feature of green buildings that is associated with productivity. The importance of ventilation is that it facilitates the introduction of fresh air to cycle through the building and removing stale or pollutant air from the interior. Germs, molds, and various VOCs, such as those emitted by paints, carpets, and adhesives, can often be found within buildings lacking adequate ventilation, causing sick building syndrome (SBS). Typical symptoms include inflammation, asthma, and allergic reaction. Ventilation can play a critical role in worker productivity, as evidenced by the extensive research that has been conducted to address these issues. This is also why it is imperative to minimize the use of toxic materials inside the building. Many of the products used in conventional office buildings such as carpets and copying machines contain toxic materials, and minimizing them decreases the potential hazards associated with them and their disposal. Furthermore, because these materials are known to leak pollutants into the indoor air, they can have an adverse impact on worker productivity if the building is not properly ventilated.

Productivity can therefore be influenced by many factors. Even the simple inclusion of vegetation inside and around the building has been shown to have a positive impact on productivity. Plants inside buildings help control indoor air pollution whereas outdoor vegetation aids in outdoor air pollution and also helps improve water quality, controlling erosion, and provide more natural habitats.

The fact that green buildings improve job satisfaction resulting in increased productivity is undeniable. Additionally, job satisfaction generates less staff turnover, thereby improving the overall productivity of a firm. Less time spent on job training allows more time to be spent on being productive work. Staff retention is one of the decisive factors why many firms take the decision to green their office buildings.

As for speculative or leased facilities, it is more problematic to commit a market value to occupant productivity gains and have them accurately reflected in the business case at the decision-making point. Nevertheless, there is now adequate data and evidence quantifying the effects to support taking them into account on some basis. And while an owner of a leased facility may not financially benefit directly from increased user productivity, there could be indirect benefits in the form of increased rental fees and occupancy

Figure 10.4 *LEED Certified Buildings Provide Healthier Work and Living Environments.* Photo of interior space in Ohlone College, Newark Center for Health Sciences and Technology, Newark, CA, which received LEED™ Platinum. *Source: Green Building Market Barometer – Turner 2008.*

rates. For the majority of commercial buildings, the use of a conservative estimate for the potential reduction in salary costs and productivity gains will loom large in any calculation, as indicated by various case studies.

10.2.6 Improved Tenant/Employee Health

LEED™ standards are intended to create improved indoor air quality and reduce potential health risks, especially allergies and other sensitivities. Moreover, LEED-certified buildings use critical resources more efficiently when compared to conventional buildings that are only built to meet code requirements. LEED-certified buildings have healthier work and living environments, which contributes to higher productivity and improved employee health and comfort (Figure 10.4).

Green buildings are typically expected to incorporate superior air quality, abundant natural light, access to views, and effective noise control. Each of these qualities is for the benefit of building occupants, making these buildings better places to work or live. Building occupants are increasingly seeking many green building features, such as superior air quality, control of air temperatures, and views. An extensive North American study focusing on office building tenant satisfaction determined that tenants highly value comfort in office buildings. Respondents reportedly attributed the highest importance to comfortable air temperature (94%) and indoor air quality (94%). These two features were the only ones that were considered "most important" and on the list with which tenants were least satisfied. This study also determined that the principal reasons why tenants move out are due to heating or cooling problems.

Improved indoor environmental quality is one of green buildings' core components. Building occupants understand that natural light, clean air, and thermal comfort are required elements to stay healthy and productive, in addition to providing an enjoyable living and workspace. The number of credible studies demonstrating an intrinsic connection between green building strategies and occupant health and well-being is endless. In office settings and learning environments, this directly translates into increased productivity and reduced absenteeism. Studies continue to show that companies are losing billions of dollars annually because of employee absenteeism.

By building to LEED™ green standards as outlined earlier, tenants recognize the true value of a healthy indoor environment as it relates to enhanced employee productivity and decreased susceptibility to the VOCs, molds, and maladies that contribute to "sick building syndrome (SBS)." This becomes a significant attribute when tenants take into account how much time their employees spend at work. The health benefits of leasing in a LEED-certified building may even contribute to a firm's bottom line.

For most tenants, the enhanced productivity and health benefits provided by a LEED-certified building add to cost savings generated by the operating efficiencies previously mentioned. This becomes viable as owners are able to pass utilities and maintenance savings on to leasing tenants – a feature that tenants of conventional, nongreen buildings do not enjoy.

William Fisk's 2002 study "How IEQ Affects Health, Productivity" estimates that 16–37 million cases of colds and flu could be avoided by improving indoor environmental quality. This translates into a $6–$14 billion annual savings in the United States and at the same time reducing sick building syndrome (SBS) symptoms (a condition whereby occupants become temporarily ill), by 20–50%, resulting in $10–$30 billion annual savings in the United States.

10.2.7 Enhancement of Property Value and Marketability

Enhancement of property value is a key factor for speculative developers who fail to directly achieve operating cost and productivity savings. To date, there remains a lack of thorough and reliable studies that have been conducted on the intrinsic relationship between property values and green buildings, even though this is an important aspect that should be quantified and included in the economic calculations. It is an element of particular relevance to speculative developers who intend to either sell or lease a new building, although it can also have a bearing on the decision process in general, including developers who intend to occupy a building while maintaining an eye on the market value of the asset.

There are many factors that will or could increase property values for green buildings. Much of the real estate industry, unfortunately, still does not fully comprehend the benefits of green buildings and therefore are unable to correctly convey these benefits to prospective purchasers. Moreover, the benefits may not be properly reflected in the selling prices or lease rates.

Jerry Yudelson maintains that increased annual energy savings have been found to promote higher building values. As an example, a 75,000-ft.^{2} building saves $37,500 per year in energy costs versus a comparable building built to code. (This savings might result from saving of 50 cents/ft.^{2}/year.) At capitalization rates of 6%, typical today in commercial real estate, green-building standards would add $625,000 ($8.33 ft.^{2}) to the value of the building. This means that for a small upfront investment, an owner can reap benefits that typically offer a payback of 3 years or less and a rate of return exceeding 20%.

High-performance buildings can offer owners many important benefits ranging from higher market value to more satisfied and productive employee occupants. In many cases, the benefits of green and high-performance buildings are not evident in higher base rents in some local markets. The primary reason for this is that majority of the benefits accrue to tenants, and tenants usually need proof before they are willing to participate in the cost of investments that are perceived will help them be more productive or save money. It is only very recently, due mainly to increased awareness, that tenants have begun to fully appreciate the benefits of cleaner air, more natural lighting, and easier-to-modify spaces.

Resale values and rental rates: Several recent American studies have provided considerable evidence that green buildings, particularly those with good-quality natural lighting, can have a dramatic effect on sales in commercial buildings. Furthermore, without exception, these studies report that there is a sound economic basis for green buildings, but only when operational costs are included into the equation. More specifically, whole building studies conclude that the net present values (NET) for pursuing green buildings as opposed to conventional buildings ranges from $50 ft.^{2} to $400 ft.^{2} ($540–$4300 m^2). The NET depends on a building's length of time analyzed (e.g., 20–60 years) and the degree to which the buildings implement green strategies. One of the main conclusions from these studies is that, generally, the greener the building, the higher the NET.

With regard to rental sales, a CoStar study found that LEED buildings can command rent premiums of about $11.24 ft.^{2} versus their conventional peers in addition to a 3.8% increase in occupancy rate. The study also found that rental rates in ENERGY STAR® buildings can boast a $2.38 ft.^{2} premium versus comparable non–ENERGY STAR buildings in addition to a 3.6% greater occupancy rate. However, what is perhaps more remarkable and what may prove to be a trend that could signal greater attention from institutional investors is that ENERGY STAR buildings are commanding an average of $61 ft.^{2} more than their conventional counterparts, and LEED buildings are commanding a surprising $171 more per square foot.

This is quite extraordinary because most leasing arrangements, particularly in the office/commercial sectors, provide little incentive to undertake changes that might be construed as being beneficial to the environment. For example, leases often have fixed rates with no regard to energy or water consumption, even though the lessees have control over most energy- and water-consuming devices. This can prove very frustrating for owners and lessees because consumption patterns are often difficult to determine as a result of a lack of detailed metering by space.

10.2.8 Other Indirect Benefits

There are numerous other indirect benefits that include such features as improved image, risk reduction, future proofing, and self-reliance. These and other similar benefits of green buildings may be captured by investors and should not be discarded in decision economics considerations. Although they may not be easily quantified and in some cases even be intangible, they should nevertheless be factored into the business case because they are intrinsically connected to sustainable design and they can significantly affect the value of a green building.

Improved image: Even if we disregard the financial benefits, green buildings are generally perceived by the public as modern, dynamic, and altruistic. Another key message conveyed by sustainable buildings is concern for the environment. Green buildings can therefore send a strong "symbolic message" of an owner's commitment to sustainability. Some of the benefits that companies can enjoy from these perceptions include employee pride, satisfaction, and well-being, which translate into reduced turnover, advantages in recruitment of employees, and improved morale. These powerful images can be an important consideration in a company's decision to pursue occupancy in a green building.

Risk reduction: Peter Morris, Principal with Davis Langdon, commenting on the downturn in the US economy, tells us that "risk remains a serious concern for construction projects. Delay and cancellation of projects, even projects under construction, is a growing trend." Morris also proposes a key theme for project owners in the current market turmoil, which is that "the successful adoption of a competitive procurement strategy, in order to secure lower costs, will depend on active steps by the owner and the project team to ensure that the contractor is in a position to provide a realistic and binding bid and that contractors' bidding costs are minimized."

However, with that in mind, many risks can be mitigated through the application of green building principles. In this regard, the Environmental Protection Agency in the United States currently classifies indoor air quality as one of the top five environmental health risks. The "Sick Building Syndrome" (SBS) and "Building Related Illness" (BRI), both of which are discussed in Chapter 7, are among the issues of major concern and which often end up being resolved in the courts. Business owners and operators are increasingly facing legal action from building tenants blaming the building for their health problems.

The base cause of SBS and BRI is poor building design and/or construction, particularly with respect to the building envelope and mechanical systems. Green buildings should accentuate and promote not only safe, but also exceptional, air quality, and no functioning green building should ever have to suffer from SBS or BRI. A similar argument may be made for mold-related issues, as this too is facing increased litigation today.

Future proofing: According to Davis Langdon, "Going green is 'future-proofing' your asset." Green buildings are inherently efficient and safe, and as such help ensure that they will not be at a competitive disadvantage in the future. In this respect, there are a number of potential future risks that are significantly mitigated in green buildings, including the following:

- Energy conservation protects against future energy price increases.
- Water conservation shields against water fee increases.
- Occupants of green buildings are generally more comfortable and contented, so it can be assumed that they will generally be less likely to be litigious.
- A documented effort to build or occupy a healthy green building demonstrates a level of due diligence that could stand as an important defense against future lawsuits or changes in legislation, even when faced with currently unknown problems.

Self-reliance: The fact that green buildings often incorporate natural lighting and ventilation, and internal energy and water generation, makes them less likely to rely on external grids and less likely to be effected by grid-related problems or failures such as blackouts, water shortages, or contaminated water. This element is acquiring increasing importance globally because of the potential risk of terrorism.

10.2.9 External Economic Effects

External effects generally consist of costs or benefits of a project that accrue to society and are not readily captured by the private investor. Examples of this are reduced reliance on infrastructure (sewers, roads, etc.), reduced greenhouse gases, and reduced health costs, etc. The extent to which these benefits can be factored to a business case relies on the extent to which they can be converted from the external to the internal sides of the ledger. This constitutes a vital factor in any assessment of the costs and benefits of green buildings. Thus the costs of green vegetated roofs are borne by the developer or investor, while much of the benefit accrues at a broader societal level such as reduced heat island effects and reduced storm water runoff.

Where the investor is a government agency, or where a private developer is compensated for including features that produce benefits at a societal level, then the business case can encompass the much broader range of effects. For example, there are many jurisdictions, such as in the state of Oregon, that offer tax incentives for green building, thereby providing a direct business case payoff to the investor. Another example is Arlington, Virginia, which allows higher floor space to land coverage ratios for green buildings.

Infrastructure costs: Costs such as those for water use and disposal are typically provided by governments and are rarely cost-effective or even cost-neutral, and in many instances, governments are required to heavily subsidize water use and treatment.

External environmental costs: These consist mainly of pollutants in the form of emissions to air, water, and land and the general degradation of the ambient environment. Buildings can be singled out to have the largest indirect environmental impact on human health. Other perhaps less critical impacts, such as damage to ecosystems, crops, structures/monuments, and resource depletion, should also be considered even though they do not have a large associated indirect cost relative to human health.

Job creation: There are significant environmental impacts associated with the transportation of materials for the construction industry. Accordingly, LEED™ rating systems promote the use of local or regional materials, which in turn promote local or regional job creation. In addition to the aforementioned, green building attributes are often labor intensive, rather than material or technology intensive.

International recognition and export opportunities: Green building can also have economic ramifications on a much broader scale as a result of increased international recognition and related export sales. The 2005 Environmental Sustainability Index prepared by Yale and Columbia Universities benchmarked the ability of nations to protect the environment by integrating data sets, including natural resource endowments, pollution levels, and environmental management efforts into a smaller set of indicators of environmental sustainability. This study ranked the United States 45th.

10.2.10 Increased Recruitment and Retention

Providing a healthy and pleasant work environment increases employee satisfaction, productivity, and retention. It also increases the ability to compete for the most qualified employees as well as for business. Statistical data and other evidence indicate that high-performance green buildings can increase a company's ability to recruit and retain employees because buildings with good air quality, abundant amounts of natural light, and better circulated heat and air conditioning are more pleasant, healthier, and more productive places to work in.

Willingness to join an organization and the ability to retain them are aspects often overlooked when considering how green buildings affect employees. It is estimated that the cost of losing a single good employee is roughly between $50,000 and $150,000 and many organizations experience a 10–20% annual turnover, some of it from persons they would have really liked to retain. In some cases, people decide to leave due to poor physical and working environments. In a workforce of say 100 people, turnover at this level implies 10–20 people leaving per year.

If green buildings are able to reduce turnover by only 10%, for example, that is, one to two people out of 10–20, that value would range from $50,000 to possibly as much as $300,000, which is more than enough to justify the costs of certifying a building project. Figure 10.5 compares the difference in occupancy rates between ENERGY STAR and non–ENERGY STAR buildings.

10.2.11 Tax Benefits

Many states now offer tax incentives for green buildings. States like New York and Oregon offer state tax credits, whereas others, such as Nevada, offer property- and sales-tax abatements. The federal government also offers tax credits. The state of Oregon credits vary and are based on building area and LEED™ certification level. At the Platinum level, for example, a 100,000-ft.^{2} building in Oregon can expect to receive a net-present-value tax credit of up to $2 ft.^{2}, which is transferrable from public or nonprofit entities to private companies (e.g., contractors or benefactors), making it even more attractive than a credit that applies only to private owners.

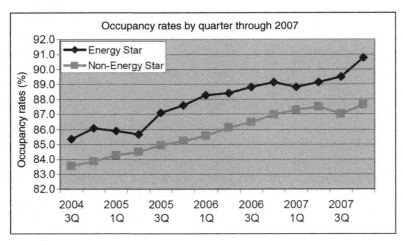

Figure 10.5 *Diagram Comparing the Occupancy Rates of Two Types of Buildings – ENERGY STAR and Non-ENERGY STAR. Source: Does Green Pay Off? by Norm Miller, Jay Spivey and Andy Florance.*

The state of New York has a tax credit that allows builders who meet energy goals and use environmentally preferable materials to apply up to $3.75 ft.² for interior work and $7.50 ft.² for exterior work against their state tax bill. In order to qualify for this credit, a building needs to be certified by a licensed architect or engineer in addition to meeting specific requirements for energy use, water use, indoor-air quality, waste disposal, and materials selection. This means that the energy used in new buildings must not exceed 65% of that allowed under the New York energy code, and in rehabilitated buildings, energy use cannot exceed 75%.

The state of Nevada offers a property-tax abatement of up to 35% for up to 10 years to private development projects that achieve LEED™ Silver certification. This means that if the property tax represents 1% of value, it could be worth as much as 5% of the building cost, which translates to much more than the actual cost of achieving LEED Silver on a large project. This has encouraged a large number of Nevada projects to pursue LEED certification, including the $7-billion, 17-million-ft.² Project City Center in Las Vegas, which is one of the world's largest private development projects to date. The state of Nevada also provides for sales-tax abatement for green materials used in LEED™ Silver–certified buildings.

In addition to the state tax incentives, there are federal tax incentives such as the 2005 federal Energy Policy Act that offers two major tax incentives for differing aspects of green buildings: (1) a tax credit of 30% on use of both solar thermal and electric systems and (2) a tax deduction of up to $1.80 ft.[2] for projects that reduce energy use for lighting, HVAC, and water-heating systems by at least 50% compared with the 2001 baseline standard. These tax deductions may be taken by the design team leader (typically the architect) when applied to government projects. Readers should check with the USGBC and the jurisdictions in question for the latest updates as tax incentives can vary from year to year.

10.3 MISCELLANEOUS OTHER GREEN BUILDING COSTS

10.3.1 Energy and Water Costs

Operational expenses for energy, water, and other utilities depend to a large extent on consumption, current rates, and price projections. But since energy and to a lesser extent water consumption, building configuration, and building envelope are interdependent elements, energy and water costs are usually assessed for the building as a whole rather than for individual building systems or components.

Energy costs: To accurately predict energy costs during the preliminary design phase of a project is rarely simple. Assumptions must be made regarding use profiles, occupancy rates, and schedules, all of which can have a dramatic impact on energy consumption. There are numerous computer programs currently on the market such as Energy-10 and eQuest that can provide the required data regarding assumptions on the amount of energy consumption for a building. Alternatively the information and data can come from engineering analysis. Other software packages, such ENERGY PLUS (DOE), DOE-2.1E, and BLAST are excellent programs but which require more detailed input not normally available until later in the design process.

Before selecting a program, it is important to determine whether you require annual, monthly, or hourly energy consumption estimates and whether the program is capable of adequately tracking savings in energy consumption even when design changes take place or when different efficiency levels are simulated. Figure 10.6 is an example of the typical costs incurred by an HVAC System over 30 years (useful life).

Energy represents a substantial and widely recognized cost of building operations that can be reduced through energy efficiency and related measures that are part of green building design. Although estimates vary slightly, the consensus is that green buildings on average, use 30% less energy than conventional buildings. A detailed survey of 60 LEED™-rated

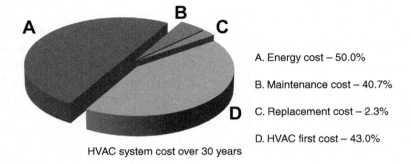

A. Energy cost – 50.0%

B. Maintenance cost – 40.7%

C. Replacement cost – 2.3%

D. HVAC first cost – 43.0%

HVAC system cost over 30 years

Figure 10.6 *Pie Diagram Illustrating Typical Costs (in Percentage) Incurred by an HVAC System Over 30 Years, Which Represents its Useful Life.* Source: Washington State Department of General Administration.

buildings demonstrates that green buildings, when compared to conventional buildings, reaffirms that they are:

- On average more energy efficient by approximately 25–30%.
- Characterized by lower electricity peak consumption.
- More likely to generate renewable energy on-site.
- More likely to purchase grid power generated from renewable energy sources (green power and/or tradable renewable certificates).

Green building energy savings come mainly from reduced electricity purchases and secondarily from reduced peak energy demand. On average, green buildings are estimated to be 28% more efficient than conventional buildings and on average generate 2% of their power on-site from photovoltaics (PV). The financial benefits that accrue from a 30% reduced consumption at an electricity price of $0.08 kWh[1] comes to about $0.30/ft^2/year, with a 20-year NPV (Net Present Value) of more than $5 ft^2, equal to or more than the average additional cost associated with building green.

Jerry Yudelson, author of *The Green Building Revolution*, says that

Many green buildings are designed to use 25- to 40-percent less energy than current codes require; some buildings achieve even higher efficiency levels. Translated to an operating cost of $1.60 to $2.50 per square foot for electricity (the most common energy source for building), this energy savings could reduce utility operating costs by 40 cents to $1 per square foot per year. Often, these savings are achieved for an added investment of just $1 to $3 per square foot. With building costs reaching $150 to $300 per square foot, many developers and building owners are seeing that it is a wise business decision to invest 1 to 2 percent of capital cost to secure long-term savings, particularly with a payback of less than three years. In an 80,000-sq-ft building, the owner's savings translates into $32,000 to $80,000 per year, year after year, at today's prices.

Table 10.1 Reduced energy use in green buildings as compared with conventional buildings (in percentage)

	Certified	Silver	Gold	Average
Energy efficiency (above standard code)	18	30	37	28
On-site renewable energy	0	0	4	2
Green power	10	0	7	6
Total	28	30	48	36

Source: USGBC, Capital E Analysis.

The environmental and health costs associated with air pollution caused by nonrenewable electric power generation and on-site fossil fuel use are generally excluded when making investment decisions. Table 10.1 highlights the reduced energy used in green buildings as compared with conventional buildings.

10.3.2 Operation, Maintenance, and Repair Costs

Scores of studies on sustainability have generally shown that LEED-certified buildings typically both cost less and are easier to operate and maintain over the life of the structure than conventional buildings. This puts them in a position to command higher lease rates than conventional buildings in their markets.

However, maintenance and repair (OM&R) costs and nonfuel operating costs, are often more difficult to estimate than other building expenditures. Operating schedules and maintenance standards will vary from one building to the next; the variation in these costs is significant even when the buildings are of the same type and age. It is therefore important to use common sense and good judgment in estimating these costs.

Supplier quotes and published estimating guides can sometimes provide relevant information on maintenance and repair costs. Some of the data estimation guides derive their cost data from sources such as Means and BOMA, which typically report, for example, average owning and operating costs per square foot, by age of the building, its geographic location, number of stories, and number of square feet in the building.

Once the project is operational, buildings tend to recoup any added costs within the first 1–2 years of the life cycle of the building. LEED-certified buildings use considerably less energy and water than a conventional building. Green buildings typically use 30–50% less energy and water than a conventional building.

than its conventional counterpart, yielding significant savings in operational costs. The New Buildings Institute (NBI) recently released a new research study indicating that new buildings certified under the USGBC LEED certification system are, on average, performing 25–30% better than non–LEED-certified buildings in terms of energy use. The study also suggests that Gold and Platinum LEED™ buildings have average energy savings approaching 50%.

10.3.3 Replacement Costs

According to Sieglinde Fuller, the number and timing of capital replacements of building systems depend largely on the estimated life of the system and the length of the study period. He further states that the same sources providing the cost estimates for the initial investments should be used to obtain estimates of replacement costs and expected useful lives, and a good starting point for estimating future replacement costs is to use their cost as of the base date. The LCCA method is designed to escalate base-year amounts to their future time of occurrence.

Residual values: The residual value of a system (or component) is basically the value it will have after it has depreciated, that is, its remaining value at the end of the study period, or at the time it is replaced during the study period. According to Fuller, residual values can be based on value in place, resale value, salvage value, or scrap value, net of any selling, conversion, or disposal costs. For rule of thumb calculations, the residual value of a system with remaining useful life in place can be determined by linearly prorating its initial costs.

10.3.4 Other Costs

Finance charges: Neither finance charges nor taxes usually apply to federal projects. However, finance charges and other payments do apply if a project is financed through an Energy Savings Performance Contract (ESPC) or Utility Energy Services Contract (UESC). These charges are normally included in the contract payments negotiated with the Energy Service Company (ESCO) or the utility.

Nonmonetary benefits or costs: These are project-related features for which there is no meaningful way of assigning a dollar value, and despite efforts to develop quantitative measures of benefits, there are situations that simply do not lend themselves to such an analysis. For example, projects may provide certain benefits such as improved quality of the working environment, preservation of cultural and historical resources, or other similar qualitative advantages. By their nature, these benefits are external to the LCCA and difficult to assess, but if considered significant they should be taken into account in the final investment decision and portrayed in the project documentation and included in an LCCA.

The analytical hierarchy process (AHP) can be used to formalize the inclusion of nonmonetary costs or benefits in the decision-making process. It is one of a set of multiattribute decision analysis (MADA) methods that are used when considering qualitative and quantitative nonmonetary attributes in addition to common economic evaluation measures when evaluating project alternatives. The ASTM E 1765 Standard Practice for Applying Analytical Hierarchy Process (AHP) to Multi-attribute Decision Analysis of Investments Related to Buildings and Building Systems that is published by ASTM International presents a general procedure for calculating and interpreting AHP scores of a project's total overall desirability when making building-related capital investment decisions. The WBDG Productive Branch is an excellent source of information for estimating productivity costs.

10.4 DESIGN AND ANALYSIS TOOLS AND METHODS

The federal government is the nation's largest owner and operator of built facilities, and during the energy crisis of the 1970s followed by the 1980s crisis, it was faced with increasing initial construction costs and ongoing operational and maintenance expenses. As a result, facility planners and designers decided to use economic analysis to evaluate alternative construction materials, assemblies, and building services with the objective of lowering costs. In today's difficult economic climate, building owners wishing to reduce expenses or increase profits are again employing economic analysis to improve their decision making during the course of planning, designing, and constructing a building. Moreover, federal, state, and municipal entities have all enacted legislative mandates (in varying degrees) requiring the use of building economic analysis to determine the most economically efficient or cost-effective choice among building alternatives. Figure 10.7 illustrates the general steps taken in an economic analysis process.

10.4.1 Present-Value Analysis

This is a standard method for using the time value of money to appraise long-term projects. The basic concept of present-value analysis is that the value of a dollar profit today is greater than the value of a dollar profit next year. How much greater is determined by what is called the "Discount Rate," as

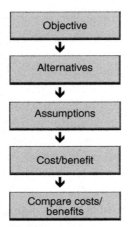

Figure 10.7 *Diagram Illustrating the Economic Analysis Process. Source: Whole Building Design Guide.*

in "how much of a discount would you expect if you were buying a dollar's worth of next year's profit." Net present value (NPV) allows decision makers to compare various alternatives on a similar time scale by converting the various options to current dollar figures. A project is generally considered acceptable if the net present value is positive over the expected lifetime of the project.

As an example, let us take a building that is considering having its lighting changed from traditional incandescent bulbs to fluorescents. The initial investment to change the lights themselves is estimated to be $40,000. After the initial investment, it is estimated to cost $2,000 to operate the lighting system, but which will yield $15,000 in savings each year. This thus produces an annual cash flow of $13,000 every year after the initial investment. If, for simplicity, a discount rate of 10% is assumed and it is calculated that the lighting system will be utilized over a 5-year time period, this scenario would have the following NPV calculations:

$t = 0$ NPV $= (-40,000)/(1 + .10)\ 0 = -40,000.00$
$t = 1$ NPV $= (13,000)/(1.10)\ 1 = 11,818.18$
$t = 2$ NPV $= (13,000)/(1.10)\ 2 = 10,743.80$
$t = 3$ NPV $= (13,000)/(1.10)\ 3 = 9,767.09$
$t = 4$ NPV $= (13,000)/(1.10)\ 4 = 8,879.17$
$t = 5$ NPV $= (13,000)/(1.10)\ 5 = 8,071.98$

Based on this information, the total NPV over the lifetime of the project would come to $9,280.22.

Discount rate: Discounting adjusts costs and benefits to a common point in time. Thus, in order to be able to add and compare cash flows that are incurred at different times during the life cycle of a building, they need to be made time-equivalent. To make cash flows time equivalent, the LCC method converts them to present values by discounting them to a common point in time, which is usually the base date. The interest rate used for discounting essentially represents the investor's minimum acceptable rate of return. To some extent, the selection of the discount rate is dependent on the use to which it will be put.

The discount rate for federal energy and water conservation projects is determined annually by the DOE's Federal Energy Management Program (FEMP); for other federal projects, those not primarily concerned with energy or water conservation, the discount rate is determined by the Office of Management and Budget (OMB). These discount rates, however, represent real discount rates that do not include the general rate of inflation.

Length of study period: The study period begins with the base date, which is the date to which all cash flows are discounted. The study period includes any planning, construction, and implementation periods as well as the service or occupancy period. The study period remains the same for all of the alternatives considered.

Service period: The service period essentially begins when the completed building is occupied or when a system is taken into service. This is the period over which operational costs and benefits are evaluated. In FEMP analyses, the service period cannot exceed 25 years.

Contract period: The contract period in ESPC and UESC projects lies within the study period, starting when the project is formally accepted, energy savings begin to accrue, and contract payments begin to be due. The contract period generally ends when the loan is paid off.

Discounting convention: In OMB and FEMP studies, annually recurring cash flows such as operational costs are normally discounted from the end of the year in which they are incurred. In MILCON studies, they are typically discounted from the middle of the year. All single amounts such as replacement costs and residual values are discounted from their dates of occurrence.

Application: With regard to the application of LCCA, Sieglinde Fuller states that it can be applied to any capital investment decision in which relatively higher initial costs are traded for reduced future cost obligations. Fuller also maintains that "it is particularly suitable for the evaluation of building design alternatives that satisfy a required level of building performance but may have different initial investment costs, different operating and maintenance and repair costs, and possibly different lives." LCCA is an approach that provides a much better assessment of the long-term cost-effectiveness of a project than alternative economic methods that mainly focus on first costs or on operating-related costs in the short run.

According to Fuller, LCCA can be performed at various levels of complexity, but its scope could vary from a "back-of-the envelope" study to a detailed analysis with thoroughly researched input data, supplementary measures of economic evaluation, complex uncertainty assessment, and extensive documentation. The comprehensiveness of the exercise should be tailored to the requirements of the project.

Treatment of inflation: An LCCA can be performed in either constant dollars or current dollars. Both methods of calculation produce identical present-value life-cycle costs. However, a constant-dollar analysis does not include the general rate of inflation, which means it has the advantage of not requiring an estimate of the rate of inflation for the years in the study period. A current-dollar analysis, on the other hand, does include the rate of general inflation in all dollar amounts, discount rates, and price escalation rates.

Constant-dollar analysis is generally recommended for federal projects, except for projects financed by the private sector such as through the Energy Savings Performance Contracting (ESPC) and the Utility Energy Services Contract (UESC). There are several alternative financing studies available and that are usually performed in current dollars if the analyst wants to compare contract payments with actual operational or energy cost savings from year to year.

10.4.2 Sensitivity Analysis

Sensitivity analysis is a technique recommended by FEMP for energy and water conservation projects. Critical assumptions should be varied and net present value and other outcomes recomputed to determine how sensitive outcomes are to changes in the assumptions. The assumptions that deserve the greatest attention will rely on the dominant benefit and cost elements and the areas of greatest uncertainty of the program being analyzed. In general, a sensitivity analysis is used for estimates of (1) benefits and costs; (2) the discount rate; (3) the general inflation rate; and (4) distributional assumptions. Models used in the analysis should be well documented and, where possible, available to facilitate independent review. Sensitivity analysis is useful for

- identifying which of a number of uncertain input values has the greatest impact on a specific measure of economic evaluation,

- determining how variability in the input value affects the range of a measure of economic evaluation, and
- testing different scenarios to answer "what if" questions.

10.4.3 Break-Even Analysis

Break-even analysis focuses on the relationship between fixed cost, variable cost, and profit. It is mostly used when decision makers want to know the maximum cost of an input that will allow the project to still break even, or conversely, what minimum benefit a project can produce and still cover the cost of the investment. To perform a break-even analysis, benefits and costs are set equal, all variables are specified, and the break-even variable is solved mathematically.

10.4.4 Computer Programs

The use of computer programs can considerably reduce the time and effort spent on formulating the LCCA, performing the computations, and documenting the study. There are a large number of LCCA-related software programs available, all of which can be found on the Internet. The following are some of the more popular packages:

- Building Life-Cycle Cost (BLCC) Program, version 5.3-08, is a program and economic analysis tool developed by the National Institute of Standards and Technology (NIST) for the US Department of Energy Federal Energy Management Program (FEMP). It is designed to provide computational support for the analysis of capital investments in buildings.
- ECONPACK (Economic Analysis Package) for Windows is a comprehensive economic analysis computer package incorporating economic analysis calculations, documentation and reporting capabilities. It is structured to permit it being used by noneconomists to prepare complete, properly documented economic analysis (EA) in support of DoD funding requests. The program was developed by the US Army Corps of Engineers.
- ENERGY-10™ is a cost-estimating program tool that assists architects, builders, and engineers to rapidly (within 20 min) identify the most cost-effective, energy-saving measures to employ in designing a low-energy building. Using the software at the early phases of a design can reportedly result in energy savings of 40–70%, with little or no increase in construction cost. The software is available through the Sustainable Buildings Industry Council (SBIC).

- Life-Cycle Cost in Design WinLCCID Program was originally developed for MILCON analyses by the Construction Engineering Research Laboratory of the US Army Corps of Engineers. The program is a life cycle costing tool that is used to evaluate and rank design alternatives for new and existing buildings and carry out "what if" analyses based on variables such as present and future costs and/or maintenance and repair costs.
- Success Estimator Estimating and Cost Management System is a cost estimating tool available from US Cost that gives estimators, project managers, and owners real-time, simultaneous access to their cost data and estimating projects from any Internet-connected computer.

10.4.5 Relevant Codes and Standards

There are many standards that are relevant to green building. These include the
- 10 CFR 436 Subpart A – Federal Energy Management and Planning Programs, Methodology and Procedures for Life-Cycle Cost Analyses
- *ICC 700-2012: 2012 National Green Building Standard (ICC 700)*
- *International Code Council's 2012 International Green Construction Code (IgCC)*
- ASTM E2432 – Standard Guide for the General Principles of Sustainability Relative to Building
- *ANSI/ASHRAE/USGBC/IES Standard 189.1-2011: Standard for the Design of High-Performance Green Buildings Except Low-Rise Residential Buildings (ASHRAE 189.1)*
- Circular No. A-94 Revised – Guidelines and Discount Rates for Benefit-Cost Analysis of Federal Programs
- Energy Policy Act of 2005
- Executive Order 13123 – Greening the Government through Efficient Energy Management, DOE Guidance on Life-Cycle Cost Analysis Required by Executive Order 13123
- Facilities Standard for the Public Buildings Service, P100 (GSA), Chapter 1.8 – Life-Cycle Costing
- NAVFAC P-442 Economic Analysis Handbook
- Standards on Building Economics, 6th ed. ASTM, 2007.
- Sustainable Building Technical Manual (DOE/EPA)
- *US Green Building Council's Leadership in Energy and Environmental Design (LEED)*
- Green Globes™

10.5 LIABILITY ISSUES

10.5.1 General Overview

Liability can be an extremely complex matter, and this section cannot therefore possibly address the many concerns and legal matters that may arise with regard to liability issues, and builders, manufacturers, and designers are strongly advised to consult their attorneys and professional liability insurance carriers for advice on these matters. While building owners and managers are not expected to guarantee the safety or well being of their tenants, visitors, and guests, they are required to exercise reasonable care to protect them from foreseeable events. The number of liability lawsuits filed against American companies has increased dramatically over the last decade.

Thus, while the green building movement is having a very significant and positive impact on the construction industry, there are aspects such as the risk of liability that are causing considerable concern and which need to be addressed. For example, green buildings are generally more efficient users of energy and materials, resulting in reduced safety factors for the different systems. But green buildings are sometimes prone to using nonstandard materials and systems that may result in an increased risk of failure of the affected materials or systems incorporated in green buildings. To minimize these risks, qualified designers should be employed to ensure that the design process is correctly implemented. In this respect, Ward Hubbell, previous President, Green Building Initiative, says, "One of our most pressing issues is the fact that some buildings designed to be green fail to live up to expectations. And in business, as we all know, where there are failed expectations there are lawsuits. All practices and/or products that could possibly result in a firm's exposure and liability should be clearly identified. The good news is that this period of increased legal action, or the threat thereof, will in fact motivate the kind of clarity and measurement that both reduces liability risks and results in better buildings."

Nevertheless, acronyms and phrases such as IAQ (Indoor Air Quality), IEQ (Indoor Environmental Quality), Sick Building Syndrome (SBS) and Building Related Illness (BRI) are tossed around to the degree that building owners and managers just shrug them off. This is unfortunate since recent studies indicate that the incidence of commercial buildings with poor IAQ and the frequency of litigation over the effects of poor IAQ is increasing substantially. These increases will have obvious ramifications for insurance carriers, which pay for many of the costs of health care and general commercial liability.

Some of this increased concern toward liability issues may be due to the enormous surge in interest in green buildings, which has resulted in many misconceptions and exaggerations put forth by owners, designers, manufactures, and distributors. "Greenwash" is the general term used within the industry for this form of misconception. It can apply to building materials, systems, buildings, or companies among other things. Greenwash can ultimately discredit the entire green building industry in addition to being the source of numerous lawsuits because the ultimate goals of green buildings are not achieved through their use. Attorneys often categorize these claims into two basic groups: (1) Materials and (2) Performance.

1. *Material claims*: Because of the lack of preciseness in what constitutes a green building, material or system providers will frequently find a material property with limited green characteristics, and market this property, and the material, as being "green." As an example, a material that uses high-recycled content might also contain unacceptable amounts of urea–formaldehyde in its production; this material is erroneously marketed as green even though its overall impact on the environment is negative. Such false claims have also occurred when material or system providers base their claims on unreliable and inaccurate information. But as the green building field matures, and as processes such as life cycle analyses become more mainstream, these risks should decline. The employment of a reliable material rating system would also reduce the plethora of false material or system claims.

2. *Performance Claims*: A not uncommon phenomenon within the green building industry is the misrepresentation of a person or building stakeholder company's knowledge and expertise regarding green building. When building owners and other building stakeholders rely on this expertise, the result can be a dismal failure of the green building to achieve its ultimate goal. Misconceptions of this type permeate many building stakeholder groups.

Problems often arise when building owners, designers, and builders differ in their interpretation of what constitutes a successful green building, and particularly when building owners fail to explicitly communicate their thoughts at the commencement of the project. Issues of this kind are compounded when the parties are relatively new to the concepts of the green building process. The two main areas that typically need to be addressed are (1) a building's failure to achieve a promised level of green building certification and (2) a building's operational performance.

The first area may be problematic in that it could impact the building owner's ability to qualify for a grant or tax incentive, on which the owner may have relied on to assist in offsetting the project's initial costs. As for public buildings, certification may indeed be required by law.

In considering a building's operational performance, there is some expectation that green buildings will, in addition to reducing environmental impacts, reduce energy and water costs, require less maintenance, and afford other long-term benefits to the building owner. The point is, however, that while the design may incorporate a wide range of green features, there are numerous variables between a building's design and occupancy that can invariably impact its operational performance. These potential areas of misunderstanding can be minimized by following good business practices that facilitate clear communication and common expectations between building owners, designers, and rating organizations.

An action may be brought against the building owner, the builder, the architect or engineer, or the product manufacturer. In such cases, an expert(s) will likely be required to give an opinion as to whether there has been negligence in the design, execution, or performance of duty. But often the investigation involves much more than expert opinion; for example, laboratory and other tests may be recommended to help determine the cause of failed performance. The role that experts are required to play will therefore vary depending on the case. Sometimes, an expert will serve solely as a consultant to the lawyer, and remain in the background, without his or her name ever being known to the other side. At other times, an expert may be used in the pretrial stages, perhaps to give an affidavit supporting one or more issues of the case. In other cases, the expert may serve solely as an expert witness at trial. Sometimes, an expert will play a combination of these roles.

Herbert Leon Macdonell, author of *The Evidence Never Lies*, rightly said, "You can lead a jury to the truth, but you can't make them believe it," which is why good field notes and photographs are necessary as they are the basis of solid documentation. Field notes should be written in a clear, legible, and articulate manner as they provide the firsthand recorded observations and are irreplaceable. Moreover, they should be self-explanatory in addition to being accurate. Photographs provide a visual record and are cardinal to forensic investigations, particularly with issues such as mold, filters, etc. Photographs should be of the highest quality and taken from different positions so as to get a comprehensive overview of the scene in question. Photographs should be well annotated and filed appropriately. Digital photographs may be stored on the computer or on CD. The forensic expert may also decide to supplement documentation of the project or scene with video photography. This will depend largely on the circumstances prevailing at the time.

Assigning culpability for green disputes often boil down to a matter of negligence, ignorance, or incompetence. American courts often require that qualified experts testify to the standard of care that is applicable to the case in dispute, and that qualified experts testify to the professional's performance as measured by that applicable standard of care. The principle of standard of care may not apply to building contractors, since they are not deemed to be "professionals" in the sense of making independent evaluations and judgments based on learning and skill. However, builders are held to a "Duty to Perform," that is, to strictly follow the plans, specifications, and provisions of their contracts.

10.5.2 Traditional Litigation – Pretrial and Trial Procedures

Traditional civil litigation is expensive and continues to escalate; it is complex and can drag on for years. The traditional litigation process requires the observance of specific protocols regarding rules of evidence and procedures for reports, pretrial discovery techniques, motion practices, interrogatories, depositions, hearsay rules, direct and cross examination, and redirect and recross examination. The consultant, building owner, or product manufacturer need not be fully conversant with the procedural details of civil lawsuits, although a basic understanding of the different stages is desirable. Litigation procedures are typically governed by statute in each jurisdiction. If we take New York as an example, procedure is governed by the *Civil Practice Law and Rules*; in federal courts, it is governed by the *Federal Rules of Civil Procedure*.

Lawsuits are typically divided into two basic phases: (1) pretrial and (2) trial.

1. *Pretrial*: In pretrial procedures, the parties to a lawsuit are identified and the issues in dispute are clarified. In addition, each party is offered an opportunity to learn about the other party's witnesses and potential evidence.

2. *The trial*: Before a trial can actually begin, a jury has to be selected. The method of jury selection depends on the jurisdiction of the trial. It can either be accomplished by the attorneys themselves or by the judge, depending on the rules of the court where the case is being heard. Upon jury selection, the trial can start with the attorneys of the plaintiff and defendant giving opening statements. These statements typically outline the strategy to be followed by the respective attorneys to prove their case in the trial.

10.5.3 Alternative Dispute Resolution

General concern and dissatisfaction with the present state of traditional civil litigation has provided impetus to the development of various alternative dispute resolution (ADR) techniques for the construction industry, including sustainable construction and manufacturing industries. ADR techniques such as arbitration, mediation, negotiation, and other out-of-court settlement procedures are now employed in the vast majority of such disputes. They are essentially based on the premise that disputes can best be resolved through negotiation or mediation immediately after a conflict comes to light rather than through the tedious, costly, and time-consuming route of traditional civil litigation. When one party to the dispute insists on litigation, it may be because legal precedents have been shown to be favorable to that party.

In the United States, construction disputes are basically resolved in one of several ways (Table 10.2):

- Settlement discussion among the parties: this often requires some compromise by all parties to the dispute.
- Mediation: this is usually nonbinding, and where an impartial mediator that is mutually agreed upon by the disputing parties is employed in the hope of bringing about an acceptable resolution through diplomacy and conciliation. They can be successful and satisfying in its outcome, provided the parties act in good faith.
- Arbitration: this can be binding or nonbinding, where one or more arbitrators are agreed upon to hear the disputing parties' arguments, take and record testimony and evidence, ask probing questions, and finally arrive at a decision on a resolution through a quasi-judicial process.
- Trial in court – before a judge and usually a six-member jury that results in a binding verdict. However, only a small percentage of cases of failure incidents end up in "trial" in court in the United States.
- Settlement discussions. These are the best, least bruising, most private, and least expensive ways of resolving disputes.

Arbitration and mediation are viable, cost-effective alternatives to litigation. According to the American Arbitration Association (AAA), "Arbitration is the submission of a dispute to one or more impartial persons for a final and binding decision, known as an 'award.' Awards are made in writing and are generally final and binding on the parties in the case. Mediation, on the other hand, is a process in which an impartial third party facilitates communication and negotiation and promotes voluntary decision making by the parties to the dispute. This process can be effective for resolving disputes prior to arbitration or litigation."

Table 10.2 The main advantages and disadvantages of different forms of resolution

Resolution process	Advantages	Disadvantages
Negotiation/assisted negotiation	• Parties have control • Confidential	• No structure • Entrenched bargaining positions likely
Mediation	• Structured • Skilled mediator helps avoid entrenched positions • Control and resolution lies with parties • Helps maintain future commercial relationship for parties • Costs less than litigation • Quick result • Confidential	• No decision if parties do not agree • A resolution may not be reached
Arbitration	• Structured • Can be quick, timetable controlled by parties • Costs may be less than litigation • Confidential	• Parties do not have control • Imposed decision • May jeopardize future relationship of parties
Litigation (Court action)	• Structured	• Timetable controlled by Court • Costs may be significant • Parties do not have control • Imposed decision • May jeopardize future relationship of parties • Long waiting times • Goes on public record (no confidentiality)

A number of jurisdictions have now made it mandatory to use ADR methods prior to accepting a case, and then only if ADR methods have failed to resolve the dispute. Most attorneys, however, advise caution in choosing ADR methods over traditional litigation, and when ADR is chosen, preference is usually given to methods that are voluntary and nonbinding.

ACRONYMS AND ABBREVIATIONS

1 ORGANIZATIONS AND AGENCIES

ACEEE	American Council for an Energy Efficient Economy
AFA	American Forestry Association
AIA	American Institute of Architects
ANSI	American National Standards Institute
APCA	Air Pollution Control Association
ASAE	American Society of Architectural Engineers
ASCE	American Society of Civil Engineers
ASHRAE	American Society of Heating, Refrigeration, and Air-Conditioning Engineers
ASID	American Society of Interior Designers
ASME	American Society of Mechanical Engineers
ASNT	American Society for Nondestructive Testing
ASPE	American Society of Plumbing Engineers
ASTM	American Society for Testing Materials
AWEA	American Wind Energy Association
BBRS	Board for Building Regulations and Standards
BCDC	Bay Conservation and Development Commission
BIFMA	Business and Institutional Furniture Manufacturer's Association
BLM	Bureau of Land Management
BOCA	Building Officials and Code Administrators
BREEM	Building Research Establishment Environmental Assessment Method
CEC	California Energy Commission
CFR	Code Federal Regulation
CIBSE	Chartered Institution of Building Services Engineers
CRI	Carpet and Rug Institute
CRS	Center for Resource Solutions
CSI	Construction Specifications Institute; Construction Standards Institute
CUWCC	California Urban Water Conservation Council
DoD	Department of Defense
DOE	US Department of Energy
DPW	Directorate of Public Works
DWR	Department of Water Resources (CA)
EIA	Energy Information Administration
EPA	US Environmental Protection Agency
ERDC	USACE Engineer Research and Development Center
ESI	European Standards Institute
FEMA	US Federal Emergency Management Agency
FSC	Forest Stewardship Council
GBCI	Green Building Certification Institute
HUD	US Department of Housing and Urban Development
IEA	International Energy Agency

Copyright © 2016 Elsevier Inc.
All rights reserved.

IEEE	Institute of Electrical and Electronics Engineers, Inc.
IESNA	Illuminating Engineering Society of North America
IPMVP	International Performance Measurement and Verification Protocol
ISO	International Organization for Standardization
NAE	National Academy of Engineering
NAHB	National Association of Home Builders
NAS	National Academy of Sciences
NBI	New Building Institute
NCARB	National Council of Architectural Registration Boards
NFRC	National Fenestration Rating Council
NIST	National Institute of Standards and Technology
OEE	Office of Energy Efficiency
OSHA	Occupational Safety and Health Administration (or Act)
OSWER	US EPA Office of Solid Waste and Emergency Response
SBIC	Sustainable Building Industry Council
SCAQMD	South Coast Air Quality Management District
SEC	Securities and Exchange Commission
SMACNA	Sheet Metal and Air Conditioning Contractors' National Association
UL	Underwriters Laboratories
USACE	US Army Corps of Engineers
USDA	United States Department of Agriculture
USEPA	US Environmental Protection Agency
USFS	US Forest Service
USGBC	United States Green Building Council

2 REFERENCED STANDARDS AND LEGISLATION

ADA	Americans With Disabilities Act
ASHRAE 52.2	Standardized method of testing building ventilation filters for removal efficiency by particle size
ASHRAE 55	Standard describing thermal and humidity conditions for human occupancy of buildings
ASHRAE 62	Standard that defines minimum levels of ventilation performance for acceptable indoor air quality
ASHRAE 90.1	Building energy standard covering design, construction, operation and maintenance
ASHRAE 192	Standard for measuring air-change effectiveness
ASTM E408	Standard of inspection-meter test methods for normal emittance of surfaces
ASTM E903	Standard of integrated-sphered test method for solar absorptance, reflectance, and transmittance
CAA	Clean Air Act; Compliance Assurance Agreement
CASBEE	Comprehensive Assessment for Building Environmental
CERCLA	Comprehensive Environmental Response, Compensation, and Liability Act
CERL	Construction Engineering Research Lab – part of USACE
CWA	Clean Water Act (aka FWPCA)
EISA	Energy Independence and Security Act of 2007

EPAct	Energy Policy Act of 2005
EPAct	US Energy Policy Act of 1992
FCAA	Federal Clean Air Act
FFHA	Federal Fair Housing Act
FOIA	Freedom of Information Act and similar state statutes
GS	Green Seal
MERV	Minimum Efficiency Reporting Value
NFPA	National Fire Protection Association
PURPA	Federal Public Utilities Regulatory Policy Act of 1978
RCRA	Resource Conservation and Recovery Act
TSCA	Toxic Substances Control Act
UBC	Uniform Building Code: the International Conference of Building Officials Model Building Code

3 ABBREVIATED GENERAL TERMINOLOGY

A	Area
A or AMP	Ampere
AAQS	Ambient air quality standards
AC	Alternating current
A/C	Air-conditioning unit
ACH	Air change per hour
ACM	Asbestos-containing material
ACT AGE	Actual age
ADAAG	ADA architectural guidelines
AE	Architect-engineer firm (typically contracted for design services)
AEI	Advanced energy initiative
AEO	Annual energy outlook, DOE/EIA publication
AF	Acre-foot (of water)
AFC	Application for certification
AFV	Alternative-fueled vehicle
AFY	Acre-feet per year
AGMBC	LEED-NC application guide for multiple buildings and on-campus building projects (USGBC document)
AHM	Acutely hazardous materials
AHU	Air-handling unit
AIB	Air infiltration barrier
AIRR	Adjusted internal rate of return
AL	Aluminum
APPA	America Public Power Association
AQMD	Air quality management district
AQMP	Air quality management plan
ARB	Air Resources Board (CA)
ATC	Acoustical tile ceiling
A/V	Audiovisual
BAAQMD	Bay Area Air Quality Management District
BAS	Building automation system

BC	Building code
Bcfd	Billion cubic feet per day
BEA	US Bureau of Economic Affairs
BEEP	BOMA Energy Efficiency Program
BEES	Building for environmental and economic sustainability support
BG	Biomass gasification
BIM	Building information model
BIPV	Building integrated photovoltaics
BL	Building line
BM	Benchmark
BMP	Best management practices
BOD	Basis of design
BOD	Beneficial occupancy date
BRI	Building-related illness
BT	Building technologies
BTU	British thermal unit
BTUH	British thermal unit per hour
BUR	Built-up roofing
CAA	US Clean Air Act
CAAQS	California Ambient Air Quality Standards
CalEPA	California Environmental Protection Agency
CAPM	Capital asset pricing model
CARB	California Air Resources Board
CBC	California Building Code
CBECS	Commercial Building Energy Consumption Survey
CD	Construction division
CDVR	Corrected design ventilation rate
CEERT	Coalition for energy efficiency and renewable technologies
CEU	Continuing Education Unit
CFCs	Chlorofluorocarbons
CFM	Cubic feet per minute
CFR	Code of Federal Regulations
CFS	Cubic feet per second
CHPS	Collaborative for High Performance Schools
CO_2	Carbon dioxide
CO	Carbon monoxide
COC	Chain-of-custody: proper accounting of materials flows, as used by the FSC
COS	Center of standardization
CMBS	Commercial Mortgage-Backed Securities
CMU	Concrete masonry unit
CPG	Comprehensive procurement guidelines
CSA	Canadian Standards Association
CWA	Clean Water Act
CxA	Commissioning Authority
dB	Decibel
DB	Design-build single contract
DBB	Design-bid-build

DC	Direct current
DEC	Design energy cost
DHWH	Domestic hot water heater/water heater
DOC	Determination of compliance
DOR	Designer of record
DPB	Discounted payback
DPTN	Demountable partitions
DS	Daylight sensing control; disconnect switch; downspout
DW	Drinking water; drywall
E	Each; east; modulus of elasticity
E&C	Engineering and construction
ECB	Engineering and Construction Bulletin
ECBEMS	Energy management system
ECMs	Energy conservation measures
EER	Energy efficiency ratio
EFF AGE	Effective age
EIA	Energy information administration
EIFS	Exterior insulation and finish system
EIR	Environmental impact report
EIS	Environmental impact statement
EL	Easement line; elbow
EL, ELEV	Elevation
EMCS	Energy monitoring and control system
EMF	Electromagnetic fields
EMP	LEED Energy Modeling Protocol
EPCA	Energy Policy and Conservation Act
EPDM	Ethylene propylene diene monomer
EQP, EQUIP	Equipment
ESA	Environmental site assessment
ESC	Erosion and sedimentation control plan
ESP	Energy service providers
ETS	Environmental tobacco smoke
EUL	Expected useful life
FAU	Forced air unit
FEMP	Federal Energy Management Program
FERC	Federal Energy Regulatory Commission
FF&E	Finishes, furniture (fixtures) and equipment
FFL	Finished floor line
FIO	For information only
FIX	Fixture
FOIL	Freedom of information letter
FS	Full scale; full size; federal specification
FSC	Forest Stewardship Council
ft.²/ft.³	Square feet(foot)/cubic feet(foot)
FTC	Federal Trade Commission
FTE	Full-time employee
FTE	Full-time equivalent
FTG	Footing

FY	Fiscal year
GBI	Green Building Initiative
GC	General contractor
GDP	Gross domestic product
GEP	Good engineering practice
GF	Glazing factor
GHG	Greenhouse gases
GIS	Geographic information system
gpd	Gallons per day
gpf	Gallons per flush
gpm	Gallons per minute
GRD, GD, G	Grade
GW	Gigawatt
GW(h)	Gigawatt (hour) = 1 billion watts
GWP	Global warming potential
GYP	Gypsum
GYP BD	Gypsum board
H₂S	Hydrogen sulfide
HAZMAT	Hazardous materials
HBD, HDB	Hardboard
HCFCs	Hydrochlorofluorocarbons
HFCs	Hydrofluorocarbons
HP	Horse power
HRA	Health risk assessment
HT	Height
HV	High voltage
HVAC	Heating, ventilating and air-conditioning
HVAC&R	Heating, ventilation, air-conditioning, and refrigerants
HWD	Hardwood
Hz	Hertz
IAQ	Indoor air quality
IDG	Installation design guide
IEPR	Integrated energy policy report
IEQ	Indoor Environmental Quality – one of the six LEED credit categories
IFMA	International Facilities Management Association
in.	Inch(es)
INFO	Information
INSUL	Insulate (ion)
IPCC	Intergovernmental Panel on Climate Change
IPLV	Integrated Part Load Value
IRR	Internal rate of return
ISO	Independent system operator
kg	Kilogram
KIT	Kitchen
km	Kilometer
kV	Kilovolt
kVA	Kilovolt-ampere (transformer size rating)
kVAR	Kilovolt-ampere reactive

kW	Kilowatt = 1000 watts
LADWP	Los Angeles Department of Water and Power
LAN	Local area network
LAV	Lavatory
LBNL	Lawrence Berkeley National Laboratory
lbs	Pounds
lbs/h	Pounds per hour
LCA	Life cycle assessment
LCC	Life cycle cost
LCCA	Life cycle cost analysis
LCGWP	Life-cycle global warming potential
LCODP	Life-cycle ozone depletion potential
LD BRG	Load bearing
LEED	Leadership in Energy and Environmental Design (USGBC)
LEED AP	LEED Accredited Professional
LEED-EB	LEED tool for Existing Buildings
LEED Homes	LEED tool for homes
LEED-NC	LEED tool for New Construction and Major Renovations
LEED ND	LEED tool for Neighborhood Development
LE/FE	Low-emission/fuel-efficient vehicle
LID	Low-impact development
LL	Live load
LPG	Liquified petroleum gas (propane and butane)
LQHC	Low quality hydrocarbons (i.e., tar sands and oil shale)
LR	Living room
LTV	Loan-to-value
LV	Low voltage
LVL	Level
LW	Lightweight
LZ	Lighting zone
M, m	Meter, million, mega, milli or thousand
MAIN	Maintenance
max	Maximum
MDF	Medium-density fiberboard
MEP	Mechanical, electrical, and plumbing
MERV	Minimum efficiency reporting value
M/F ratio	Male/female ratio
MMT	Million metric tons
MOU	Memorandum of understanding
MR	Moisture resistant
m/s	Meters per second
MSDS	Material safety data sheet
MV	Megavolt
MVA	Megavolt-amperes
MW	Megawatt = 1 million watts
MWD	Metropolitan Water District
MWh	Megawatt-hour
N	North

NAAQS	National Ambient Air Quality Standards
NC	New Construction
NCPA	Northern California Power Agency
NEMA	National Electrical Manufacturers Association
NEPA	National Environmental Policy Act (federal "equivalent" of CEQA) of 1969
NES	National Energy Savings
NG	Natural gas
NO$_2$	Nitrogen dioxide
NO	Nitrogen oxide
NOM	Nominal
NO$_x$	Nitrogen oxides
NPDES	National Pollutant Discharge Elimination System
NPV	Net present value
NS	Net savings
NTS	Not to scale
O$_3$	Ozone
O&M	Operation and maintenance
OC, O/C	On center
ODP	Ozone-depleting potential
OH, OVHD	Overhead
OPR	Owner's Project Requirements Document
OSA	Outside air
OSB	Oversight Board
oz	Ounce
PBD	Particleboard
PBP	Payback period
PCA	Property condition assessment
PCC	Precast concrete
PDT	Project development team
PFL	Pounds per lineal foot
PL	Property line
PM	Project manager
PML	Probable maximum loss
PMO	Project management oversight
POC	Point of contact
ppm	Parts per million
ppt	Parts per thousand
PRM	Performance rating method
psf	Pounds per square foot
psi	Pounds per square inch
PTO	Permit to operate
PU	Per unit
PUC	Public Utilities Commission
PV	Photovoltaic
PV	Present value
PV	Solar photovoltaics
PVC	Polyvinylchloride

QA/QC	Quality assurance/quality control
qty	Quantity
R, RD	Radius
RA	Return air
Rc	Refrigerant charge
RD&D	Research, development, and demonstration
REC	Renewable Energy Certificate
Ref	Reference
Reinf	Reinforcement
Req	Requirement, required
Rev	Revision
RFP	Request for proposals
RFQ	Request for qualifications
RH	Relative humidity
RTU	Rooftop package unit
RUL	Remaining useful life
S, SW	Switch
SA	Supply air
San	Sanitary
SBS	Sick building syndrome
SBTF	Sustainable Building Task Force (CA)
Sch, Sched	Schedule
SD	Smoke detector; shop drawings; storm drain: supply duct
SEER	Seasonal energy efficiency ratio
Sftwd	Softwood
SHGC	Solar heat gain coefficient
SIR	Savings-to-investment ratio
SOG	Slab-on-grade
SO_2	Sulfur dioxide
SO_x	Oxides of sulfur (sulfur oxides)
SOW	Scope of work
SPB	Simple payback
SPIRIT	Army-developed point/credit-based system for measuring sustainability of buildings/development (modified version of LEED version 2.0)
SRI	Solar reflectance index
STC	Sound transmission coefficient
Std	Standard
Sym	Symbol; symmetry (-ical)
Sys	System
T	Ton
TAC	Toxic air contaminant
Thk	Thick(ness)
TL	Total losses
TOG	Total organic gases
Topo	Topography
TP	Total phosphorous
tpd	Tons per day
tpy	Tons per year

TS	Tensile strength
TSP	Total suspended particulate matter
TSS	Total suspended solids
TVOC	Total volatile organic compounds, see VOCs
U, UR	Urinal
UFGS	Unified Facilities Guide Specifications
UH	Unit heater
UMCS	Utility monitoring and control system
USGS	United States Geological Survey
UST	Underground storage tank
Util	Utility
UV	Ultraviolet
V	Volt(s)
VAR, VAV	Variable air volume
VB	Vapor barrier
VENT	Ventilation; ventilator
VMT	Vehicle miles traveled
VOC	Volatile organic compound
VOL	Volume
VP	Vent pipe
W	Watt; width; wide; west; wire
WB	Wet bulb; wood base
WBDG	Whole Building Design Guide
WC	Water closet
WD	Wood
WH, DHWH	Water heater; domestic hot water heater
WP	Waterproof; weatherproof
WPM	Waterproof membrane
WW, WTW	Wall to wall
Y, yd	Yard
YR	Year
ZEV	Zero emissions vehicle (minimum energy star rating of 40)

GLOSSARY

Abatement Reducing the degree or intensity of or eliminating pollution.

Absorption The process by which incident light energy is converted to another form of energy, usually heat.

Accessible Describes a site, building, facility, or portion thereof that complies with these guidelines.

Accessible route A continuous unobstructed path connecting all accessible elements and spaces of a building or facility. Interior accessible routes may include corridors, floors, ramps, elevators, lifts, and clear floor space at fixtures. Exterior accessible routes may include parking access aisles, curb ramps, crosswalks at vehicular ways, walks, ramps, and lifts.

Accessory dwelling unit A subordinate dwelling unit that is attached to a principal building or contained in a separate structure on the same property as the principal unit.

Acid rain Precipitation of dilute solutions of strong mineral acids, which are formed by the mixing of the atmosphere of various industrial pollutants – mainly sulfur dioxide and nitrogen oxide, with naturally occurring oxygen and water vapor.

Acrylics A family of plastics used for fibers, rigid sheets, and paints.

Active ventilation See Mechanical Ventilation.

Adapted plants Plants that reliably grow well in a given habitat with minimal attention from humans in the form of winter protection, pest protection, water irrigation, or fertilization once root systems are established in the soil. Adapted plants are considered to be low maintenance and noninvasive.

Adaptive reuse Renovating a space to allow it to be used for a different purpose.

Addendum A written or graphic instruction issued by the architect prior to the execution of the contract, which modifies or interprets the bidding documents by additions, deletions, clarifications, or corrections. An addendum becomes part of the contract documents when the contract is executed.

Adhesive A bonding material used to bond two materials together.

Adjacent site A site that has at least 25% of its boundary bordering parcels that are each at least 75% previously developed; a street or other right-of-way does not constitute previously developed land; rather, it is the status of the property on the opposite side of the street or right-of-way that matters; also, any fraction of the boundary that borders a waterfront other than a stream is excluded from the calculation; however, a site is still deemed to be adjacent if 25% of the portion adjacent to its boundary is separated from previously developed parcels by underdeveloped, permanently protected land averaging no more than 400 ft. in width and no more than 500 ft. in any one place; the undeveloped land must be permanently preserved as natural area, riparian corridor, park, greenway, agricultural land, or designated cultural landscape; permanent pedestrian paths connecting the project through the protected parcels to the bordering site may also be deemed to meet the requirement of SLL Prerequisite 1, Option 2 (that the project be connected to the adjacent parcel by a through-street or nonmotorized right-of-way every 600 ft. on average, provided the path or paths traverse the undeveloped land at no more than a 10% grade for walking by persons of all ages and physical abilities).

Adobe A heavy clay soil used in many southwestern states to make sun-dried bricks.

Aggregate Fine, lightweight, coarse, or heavyweight grades of sand, vermiculite, perlite, or gravel added to cement for concrete or plaster.

Agrifiber Any fibrous material generated from agricultural/bio-based products such as wheat, straw, and cereal straw, sunflower husks, hemp and bagasse (sugar cane).

Agrifiber board A composite panel used in construction that is made out of agricultural waste fiber. Material for agrifiber boards can come from cereal straw, sugarcane bagasse, sunflower seed husks, walnut shells, coconut husks, and agricultural pruning. The raw fibers are processed with resins to produce panel products that are similar to conventional panels made from wood fiber.

Air-conditioning A process that simultaneously controls the temperature, moisture content, distribution, and quality of air.

Air filter A device designed to remove contaminants and pollutants from air passing through the device.

Air-handling unit (AHU) A mechanical unit used for air-conditioning or movement of air as in direct supply or exhaust of air within a structure.

Air pollution The presence of contaminants or pollutant substances in the air that may be hazardous to human health or welfare or produce other harmful environmental effects.

Albedo (aka solar reflectance) The reflectance of sunlight; higher values equal higher reflectivity, and lower values equal lower reflectivity.

Aligned section A section view in which some internal features are revolved into or out of the plane of the view.

Allergen A substance capable of causing an allergic reaction because of an individual's sensitivity to that substance.

Alligatoring A pattern of rough cracking on a coated surface, similar in appearance to alligator skin.

Alternative fuel vehicle Vehicles that run on a fuel other than traditional petroleum fuels (petrol or diesel); also refers to any technology of powering an engine that does not involve solely petroleum (e.g., electric car, hybrid electric vehicles, and solar powered vehicles).

Alternating current (AC) Electrical current that continually reverses direction of flow. The frequency at which it reverses is measured in cycles per second, or Hertz (Hz). The magnitude of the current itself is measured in amps (A).

Alternator A device for producing alternating current (AC) electricity. Usually driven by a motor, but can also be driven by other means, including water and wind power.

Ambient lighting Lighting in an area from any source that produces general illumination, as opposed to task lighting.

Ambient temperature The temperature of the surrounding air.

American bond Brickwork pattern consisting of five courses of stretchers followed by one bonding course of headers.

Ammeter A device used for measuring current flow at any point in an electrical circuit.

Ampere (A) or Amp The unit for electric current; the flow of electrons. One amp is 1 C passing in 1 s. One amp is produced by an electric force of 1 V acting across a resistance of 1 Ω.

Analog The processing of data by continuously variable values.

Anemometer A device used to measure wind speed.

Angle of incidence Angle between the normal to a surface and the direction of incident radiation; applies to the aperture plane of a solar panel. Only minor reductions in power output within plus/minus 15°.

Animal dander Tiny scales of animal skin.

ANSI The American National Standards Institute. ANSI is an umbrella organization that administers and coordinates the national voluntary consensus standards system. http://www.ansi.org/

Appeal A formal written request to review the content of an exam question for accuracy, validity, or errors in content and grammar. Appeals must be specific to an exam question and must be submitted by the candidate to GBCI's Accreditation Department within 10 days of the exam appointment. The appeal must describe the content of the exam question and, if possible, the nature of error. Exam scores are not modified under any conditions.

Aquifer A subsurface water-bearing layer that will yield water in a usable quantity to a well or spring.

Arc A portion of the circumference of a circle.

Architect's scale The scale used when dimensions or measurements are to be expressed in feet and inches.

Array A number of solar modules connected together in a single structure.

As-built drawings Record drawings completed by the contractor and turned over to the owner at the completion of a project, identifying any change or adjustments made to the conditions and dimensions of the work relative to the original plans and specifications.

ASHRAE American Society of Heating, Refrigerating and Air-Conditioning Engineers.

Asphalt shingles They are shingles made of asphalt or tar-impregnated paper with a mineral material embedded; very fire resistant.

Assumed liability It is the liability, which arises from an agreement between people, as opposed to the liability, which arises from common or statutory law. See also Contractual Liability.

ASTM International Formerly the American Society for Testing Materials. They develop and publish testing standards for materials and specifications used by industry. http://www.astm.org/

Authority having jurisdiction (AHJ) Means the governmental body responsible for the enforcement of any part of the Standard codes or the official or agency designated to exercise such a function and/or the architect.

Axial load A weight that is distributed symmetrically to a supporting member, such as a column.

Axonometric projection A set of three or more views in which the object appears to be rotated at an angle, so that more than one side is seen.

Backfill Any deleterious material (sand, gravel, etc.) used to fill an excavation.

Baffle A single opaque or translucent element used to diffuse or shield a surface from direct or unwanted light.

Bake out Process of removing VOCs from a building by elevating the temperature in order to accelerate off-gassing.

Ballasts Electrical "starters" required by certain lamp types, especially fluorescents.

Balloon framing A system in wood framing in which the studs are continuous, without an intermediate plate for the support of second-floor joists.

Baluster A vertical member that supports handrails or guardrails.

Balustrades A horizontal rail held up by a series of balusters.

Banister That part of the staircase, which fits on top of the balusters.

Bar chart A calendar that graphically illustrates a projected time allotment to achieve a specific function.

Base A trim or molding piece found at the interior intersection of the floor and the wall.

Base building The core (common areas) and shell of the building and its systems that typically are not subject to improvements to suit tenant requirements.

Base flashing Consists of flashing that covers the edges of a membrane.

Baseline case versus design case Amount of design case water saved over the baseline case amount; the baseline case is based on the Energy Policy Act or 1992 (EPACT 1992) for flush and flow rates.

Basis of design (BOD) Information gathered to document the owner's project requirements.

Batt insulation An insulating material formed into sheets or rolls with a foil or paper backing to be installed between framing members.

Batten A narrow strip of wood used to cover a joint.

Beam A weight-supporting horizontal member.

Bearing wall A wall that supports any vertical loads in addition to its own weight.

Benchmark A point of known elevation from which the surveyors can establish all their grades.

Best management practice (BMP) A method, activity, maintenance procedure, or other management practice for reducing the amount of pollution entering a water body; the term originated from the rules and regulations developed pursuant to the federal Clean Water Act.

Bill of material A list of standard parts or raw materials needed to fabricate an item.

Bio-based Materials derived from natural renewable resources such as corn, rice or beets.

Biodegradable Capable of being decomposed by bacteria.

Biodiversity The tendency in ecosystems, when undisturbed to have a great variety of species forming a complex web of interactions. Human population pressure and resource consumption tend to reduce biodiversity dangerously; diverse communities are less subject to catastrophic disruption.

Biomass A renewable energy source; is a biological material derived from living, or recently living organisms, such as wood, waste, and alcohol fuels; commonly plant matter grown to generate electricity or produce heat.

Blackwater Wastewater generated from toilet flushing. Blackwater has a higher nitrogen and fecal coliform level than graywater. Some jurisdictions include water from kitchen sinks or laundry facilities in the definition of blackwater.

Blistering The condition that paint presents when air or moisture is trapped underneath and makes bubbles that break into flaky particles and ragged edges.

Blocking The use of internal members to provide rigidity in floor and wall systems. Also used for fire draft stops.

Blueprints Documents containing all the instructions necessary to manufacture a part. The key sections of a blueprint are the drawing, dimensions, and notes. Although blueprints used to be blue, modern reproduction techniques now permit printing of black-on-white as well as colors.

Board foot A unit of lumber of measure equaling 144 in.^{3}; the base unit (B.F.) is 1 in. thick and 12 in.^{2} or $1 \times 12 \times 12 = 144$ in.^{3}.

Boiler Equipment designed to heat water or generate steam.

Bond In masonry, the interlocking system of brick or block to be installed.

Boundary survey A mathematically closed diagram of the complete peripheral boundary of a site, reflecting dimensions, compass bearings and angles. It should bear a licensed land surveyor's signed certification, and may include a metes and bounds or other written description.

Breezeway A covered walkway with open sides between two different parts of a structure.

Brick pavers A term used to describe special bricks to be used on the floor surface.

British Thermal Unit (BTU) The amount of heat energy required to raise one pound of water from a temperature of 60°F to 61°F at 1 atm. One watt-hour equals 3413 BTU.

Brownfield Generally, these are sites that have previously been developed or used for some purpose that has ceased and whose former use resulted in potential pollution or the presence of hazardous substances.

Building codes Rules and regulations adopted by the governmental authority having jurisdiction over the commercial real estate, which govern the design, construction, alteration, and repair of such commercial real estate. In some jurisdictions, trade or industry standards may have been incorporated into, and made a part of, such building codes by the governmental authority. Building codes are interpreted to include structural, HVAC, plumbing, electrical, life-safety, and vertical transportation codes.

Building density The total floor area of a building divided by the total area of the site (square feet per acre).

Building envelope The enclosure of the building that protects the building's interior from outside elements, namely, the exterior walls, roof, and soffit areas. Also referred to as the building shell.

Building footprint The area of the building structure as determined by the perimeter of the building plan, which is typically the foundation walls; hardscapes, landscaping, and other nonbuilding facilities are not included in the building footprint.

Building inspector A representative of a governmental authority employed to inspect construction for compliance codes, regulations, and ordinances.

Building line An imaginary line determined by zoning departments to specify on which area of a lot a structure may be built (also known as a setback).

Building permit A permit issued by appropriate governmental authority allowing construction of a project in accordance with approved drawing and specifications.

Building-related illness A discrete, identifiable disease or illness that can be traced to a specific pollutant or source within a building. (Contrast with "Sick building syndrome.")

Building systems Interacting or independent components or assemblies, which form single integrated units, that comprise a building and its site work, such as pavement and flatwork, structural frame, roofing, exterior walls, plumbing, HVAC, electrical, etc.

Build-out The interior construction and customization of a space (including services, space, and stuff) to meet the tenant's requirements; either new construction or renovation (also referred to as fit-out or fit-up).

By-product Material, other than the principal material, that is generated as a consequence of an industrial process or as a breakdown product in a living system.

Caisson A below-grade concrete column for the support of beams or columns.

Callback A request by a project owner to the contractor to return to the job site to correct or redo some item of work.

Candela A common unit of light output from a source.

Cantilever A horizontal structural condition where a member extends beyond a support, such as a roof overhang.

Capillary The action by which the surface of a liquid, where it is in contact with a solid, is elevated or depressed.

Carbon dioxide levels (CO_2) CO_2 levels that indicate indoor ventilation effectiveness; compared to outdoor CO_2 levels, concentrations above 530 ppm indicate inadequate ventilation, whereas concentrations above 800 ppm indicate poor air quality.

Carbon footprint The measure of greenhouse gas emissions associated with an activity. It takes into account energy use, transportation methods, and other means of emitting carbon. A number of carbon calculators have been created to estimate carbon footprints, including one from the US Environmental Protection Agency (EPA).

Casement A type of window hinged to swing outward.

Catch basin A complete drain box made in various depths and sizes; water drains into a pit, then from it through a pipe connected to the box.

Caulk Any type of material used to seal walls, windows, and doors to keep out the weather.

Cavity wall A masonry wall formed with an air space between each exterior face.

Cement plaster A plaster that is composed of cement rather than gypsum.

Central HVAC system A system that produces a heating or cooling effect in a central location for subsequent distribution to satellite spaces that require conditioning; see also all-air, all-water, and air-water HVAC systems.

Centrifugal A particular type of fluid-moving device that imparts energy to the fluid by high-velocity rotary motion through a channel; fluids enter the device along one axis and exit along another axis.

Certificate for payment A statement from the architect to the owner confirming the amount of money due the contractor for work accomplished or materials and equipment suitably stored, or both.

Certificate of insurance A document issued by an authorized representative of an insurance company stating the types, amounts, and effective dates of insurance in force for a designated insured.

Certificate of occupancy Document issued by governmental authority certifying that all or a designated portion of a building complies with the provisions of applicable statutes and regulations, and permitting occupancy for its designated use.

Certificate of substantial completion A certificate prepared by the architect on the basis of an inspection stating that the work or a designated portion thereof is substantially complete, which established the date of substantial completion; states the responsibilities of the owner and the contractor for security, maintenance, heat, utilities, damage to the work, and insurance; and taxes the time within which the contractor shall complete the items listed therein.

Certified wood Wood-based materials used in building construction that are supplied from sources that comply with sustainable forestry practices, protecting trees, wildlife habitat, streams, and soil as determined by the Forest Stewardship Council or other recognized certifiable organizations.

Cesspool An underground catch basin for the collection and dispersal of sewage.

Chain of custody A document that tracks the movement of a product from the point of harvest or extraction to the end user.

Change order A written and signed document between the owner and the contractor authorizing a change in the work or an adjustment in the contract sum or time. The contract sum and time may be changed only by change order. A change order may be in the form of additional compensation or time; or less compensation or time known as a "deduction."

Checklist List of items used to check drawings.

Chiller Equipment designed to produce chilled water; see also vapor compression chiller (centrifugal, reciprocating) and absorption chiller.

Circuit A continuous system of conductors providing a path for electricity.

Circuit breaker A circuit breaker acts like an automatic switch that can shut the power off when it senses too much current.

Circumference The length of a line that forms a circle.

Clear floor space The minimum unobstructed floor or ground space required to accommodate a single, stationary wheelchair and occupant.

Clerestory A window or group of windows that are placed above the normal window height, often between two roof levels.

Coefficient of utilization (CU) The ratio of light energy (lumens) from a source, calculated as received on the workplane, to the light energy emitted by the source alone.

Column A vertical weight-supporting member.

Combustion An oxidation process that releases heat; on-site combustion is a common heat source for buildings.

Commissioning (Cx) A systematic process to verify that building components and systems function as intended and required; systems may need to be recommissioned at intervals during a building's life cycle.

Commissioning plan Document that describes the organization, schedule distribution of resources and documentation requirements of the commissioning process.

Common use Refers to those interior and exterior rooms, spaces, or elements that are made available for the use of a restricted group of people (e.g., occupants of a homeless shelter, the occupants of an office building, or the guests of such occupants).

Community connectivity Amount of connection between a site and the surrounding community; the physical location of the site relative to homes, schools, retail, restaurants, medical, and other services.

Compact fluorescent lamp (CFL) Small fluorescent lamp used as a more efficient alternative to incandescent lamps.

Component A fully functional portion of a building system, piece of equipment, or building element.

Composite wood A product consisting of wood or plant particles or fibers bonded together by a synthetic resin or binder. Examples include plywood, particle-board, OSB, MDF, composite door cores.

Composting toilet A dry plumbing fixture that contains and treats human waste via a microbiological process.

Compressor A device designed to compress (increase the density of) a compressible fluid; a component used to compress refrigerant; a component used to compress air.

Computer-aided drafting (CAD) A method by which engineering drawings may be developed on a computer.

Computer-aided manufacturing (CAM) A method by which a computer uses a design to guide a machine that produces parts.

Concrete block A rectangular concrete form with cells in them.

Condensation The process by which moisture in the air becomes water or ice on a surface (such as a window) whose temperature is colder than the air's temperature.

Condenser A device designed to condense a refrigerant; an air-to-refrigerant or water-to-refrigerant heat exchanger; part of a vapor compression or absorption refrigeration cycle.

Conductor A material used to transfer, or conduct, electricity, often in the form of wires.

Conduit A pipe or elongated box used to house and protect electrical cables.

Conservation Act of preserving and renewing, when possible, human and natural resources and the use, protection, and improvement of natural resources according to recognized principles that ensure their highest economic or social benefits.

Construction and demolition debris (C&D) Waste and recyclable materials from construction, demolition, deconstruction or renovation of existing buildings; excludes land clearing debris.

Construction documents A term used to represent all drawings, specifications, addenda, and other pertinent construction information associated with the construction of a specific project.

Construction waste management plan A plan that diverts construction debris from landfills or incinerators through recycling, salvaging, and reusing.

Contaminant Unwanted airborne element that may reduce indoor air quality (IAQ).

Contingency allowance A sum included in the Project budget designated to cover unpredictable or unforeseen items of work, or changes in the work subsequently required by the owner. See Budget, Project.

Contour line A line that represents the change in level from a given datum point.

Contract A legally enforceable promise or agreement between two or among several person. See also Agreement.

Controllability of systems Providing occupants direct control over temperature, airflow, and lighting in their spaces.

Convection Transfer of heat through the movement of a liquid or gas.

Cooling tower Equipment designed to reject heat from a refrigeration cycle to the outside environment through an open cycle evaporative process; an exterior heat rejection unit in a water-cooled refrigeration system.

Cornice The projecting or overhanging structural section of a roof.

Cost appraisal Evaluation or estimate (preferably by a qualified professional appraiser) of the market or other value, cost, utility, or other attribute of land or other facility.

Cost estimate A preliminary statement of approximate cost, determined by one of the following methods: (1) Area and volume method; cost per square foot or cubic foot of the building. (2) Unit cost method; cost of one unit multiplied by the number of units in the project; for example, in a hospital, the cost of one patient unit multiplied by the number of patient units in the project. (3) In-place unit method; cost in-place of a unit, such as doors, cubic yards of concrete, and squares of roofing.

Coving The curving of the floor material against the wall to eliminate the open seam between floor and wall.

Cradle-to-grave analysis Analysis of the impact of a product from the beginning of its source gathering processes, through the end of its useful life, to disposal of all waste products. Cradle-to-cradle is a related term signifying the recycling or reuse of materials at the end of their first useful life.

Crawl space The area under a floor that is not fully excavated; only excavated sufficiently to allow one to crawl under it to get at the electrical or plumbing devices.

Cross-section A slice through a portion of a building or member that depicts the various internal conditions of that area.

CSI Construction Specifications Institute. Membership organization of design professionals, construction professionals, product manufacturers, and building owners. Develops and promotes industry communication standards and certification programs. http://www.csinet.org/.

Current It is the flow of electric charge in a conductor between two points having a difference in electrical potential (voltage) and is measured in Amps.

Curtain wall An exterior wall that provides no structural support.

Cut-off voltage The voltage levels at which the charge controller (regulator) disconnects the PV array from the battery, or the load from the battery.

Damper A device designed to regulate the flow of air in a distribution system.

Dangerous or adverse conditions These are essentially conditions that may pose a threat or possible injury to the field observer, and that may require the use of special protective clothing, safety equipment, access equipment, or any other precautionary measures.

Date of agreement The date stated in the agreement. If no date is stated, it could be the date on which the agreement is actually signed, if this is recorded, or it may be the date established by the award.

Date of commencement of the work The date established in a notice to the contractor to proceed or, in the absence of such notice, the date of the owner contractor agreement or such other date and may be established therein.

Date of substantial completion The date certified by the architect when the Work or a designated portion thereof is sufficiently complete, in accordance with the contract documents, so the owner can occupy the work or designated portion thereof for use for which it is intended.

Datum point Reference point.

Daylight factor (DF) The ratio of daylight illumination at a given point on a given plane, from an obstructed sky of assumed or known illuminance distribution, to the light received on a horizontal plane from an unobstructed hemisphere of this sky, expressed as a percentage. Direct sunlight is excluded for both values of illumination. The daylight factor is the sum of the sky component, the external reflected component, and the internal reflected component. The interior plane is usually a horizontal work plane. If the sky condition is the CIE standard overcast condition, then the DF will remain constant regardless of absolute exterior illuminance.

Daylighting The controlled admission of natural light into a space through glazing with the intent of reducing or eliminating electric lighting. By utilizing solar light, daylighting creates a stimulating and productive environment for building occupants.

Dead load The weight of a structure and all its fixed components.

Decibel (dB) Unit of sound level or sound-pressure level. It is 10 times the logarithm of the square of the sound pressure divided by the square of reference pressure, 20 µPa.

Defective work The work not conforming with the contract requirements.

Deferred maintenance Physical deficiencies that cannot be remedied with routine maintenance, normal operating maintenance, etc., excluding de minimis conditions that generally do not present a material physical deficiency to the subject property.

Design–build construction When an owner contract with a prime or main contractor to provide both design and construction services for the entire construction project. Use of the design–build project delivery system has grown from 5% of US construction in 1985 to 33% in 1999 and is projected to surpass low-bid construction in 2005. If a design–build contract is extended further to include the selection, procurement, and installation of all furnishings, furniture, and equipment, it is called a "turnkey" contract.

Details An enlarged drawing to show a structural aspect, an aesthetic consideration, a solution to an environmental condition, or to express the relationship among materials or building components.

Development density The total area of all buildings within a particular area and expressed in square feet per acre or residential units per acre.

Diffuser A device designed to supply air to a space while providing good mixing of supply and room air and avoiding drafts; normally ceiling installed.

Digital The processing of data by numerical or discrete units.

Dimension line A thin unbroken line (except in the case of structural drafting) with each end terminating with an arrowhead; used to define the dimensions of an object.

Dimensions are placed above the line, except in structural drawing where the line is broken and the dimension placed in the break.

Direct costs (hard costs) The aggregate costs of all labor, materials, equipment and fixtures necessary for the completion of construction of the improvements.

Direct costs loan; indirect costs loan That portion of the loan amount applicable and equal to the sum of the loan budget amounts for direct costs and indirect costs, respectively, shown on the borrower's project cost statement.

Direct current (DC) Electrical current that flows only in one direction, although it may vary in magnitude. Contrasts with alternating current.

Discount factor The factor that translates expected benefits or costs in any given future year into present value terms. The discount factor is equal to $1/(1 + i)t$, where i is the interest rate and t is the number of years from the date of initiation for the program or policy until the given future year.

Discount rate The interest rate used in calculating the present value of expected yearly benefits and costs.

Diversity of uses or housing types The number of spaces or housing types, offices, homes, schools, parks, stores, per acre.

Dormer A structure that projects from a sloping roof to form another roofed area. This new area is typically used to provide a surface to install a window.

Downcycling Recycling a material in a manner that much of its inherent value is lost.

Downspouts Pipes connected to the gutter to conduct rainwater to the ground or sewer.

Drip irrigation system An irrigation system that slowly applies water to the root system of plants to maximize transpiration while minimizing wasted water and topsoil runoff. Drip irrigation usually involves a network of pipes and valves that rest on the soil or underground at the root zone.

Dry ponds Elevated areas that detain stormwater and slow runoff but are dry between rain events.

Drywall An interior wall covering installed in large sheets made from gypsum board.

Duct Usually sheet metal forms used for the distribution of cool or warm air throughout a structure.

Due diligence The process of conducting a walkthrough survey and appropriate inquiries into the physical condition of a commercial real estate's improvements, usually in connection with a commercial real estate transaction. The degree and type of such survey or other inquiry may vary for different properties and different purposes.

Dwelling unit A single unit that provides a kitchen or food preparation area, in addition to rooms and spaces for living, bathing, sleeping, and the like. Dwelling units include a single-family home or a townhouse used as a transient group home; an apartment building used as a shelter; guestrooms in a hotel that provide sleeping accommodations and food preparation areas; and other similar facilities used on a transient basis. For purposes of these guidelines, use of the term "Dwelling Unit" does not imply the unit is used as a residence.

Easement The right or privilege to have access to or through another piece of property such as a utility easement.

Eave That portion of the roof that extends beyond the outside wall.

Ecological/environmental sustainability Maintenance of ecosystem components and functions for future generations.

Ecological impact The impact that a human-caused or natural activity has on living organisms and their nonliving environment.

Ecosystem The interacting system of a biological community and its nonliving environmental surroundings. An ecological community together with its environment, functioning as a unit.

Egress A continuous and unobstructed way of exit travel from any point in a building or facility to a public way. A means of egress comprises vertical and horizontal travel and may include intervening room spaces, doorways, hallways, corridors, passageways, balconies, ramps, stairs, enclosures, lobbies, horizontal exits, courts, and yards. An accessible means of egress is one that complies with these guidelines and does not include stairs, steps, or escalators. Areas of rescue assistance or evacuation elevators may be included as part of accessible means of egress.

Electric current The flow of electrons measured in Amps.

Electrical grid A network for electricity distribution across a large area.

Electricity The movement of electrons (a subatomic particle), produced by a voltage, through a conductor.

Electrode An electrically conductive material, forming part of an electrical device, often used to lead current into or out of a liquid or gas. In a battery, the electrodes are also known as plates.

Element An architectural or mechanical component of a building, facility, space, or site, for example, telephone, curb ramp, door, drinking fountain, seating, or water closet.

Embodied energy The total energy that a product may be said to "contain," including all energy used in growing, extracting, and manufacturing it and the energy used to transport it to the point of use. The embodied energy of a structure or system includes the embodied energy of its components plus the energy used in construction.

Emissivity Ratio of energy emitted from a material to the energy radiated from a black body at the same temperature.

Energy Power consumed multiplied by the duration of use. For example, 1000 W used for 4 h is 4000 Wh.

Energy efficient products and systems Building components and appliances that use less energy.

Energy or greenhouse gas emissions per capita Total greenhouse gas emissions of a community divided by the total resident count.

ENERGY STAR rating A designation given by the EPA and the US Department of Energy (DOE) to appliances and products that exceed federal energy efficiency standards. The label helps consumers identify products that are energy efficient and will save money.

Energy use intensity Energy consumption divided by the area in square feet in a building; energy consumption is usually expressed as British thermal units (BTUs) per square foot or as kilowatt-hours of electricity per square foot per year ($kWh/ft.^2/year$).

Engineer's scale The scale used whenever dimensions are in feet and decimal parts of a foot, or when the scale ratio is a multiple of 10.

Enhanced commissioning A set of best practices extending responsibility beyond fundamental commissioning such that the process requires the commissioning authority to be on the job earlier and stay later; includes designating a commissioning authority prior to the construction document phase, conducting commissioning design reviews, reviewing contractor submittals, developing a systems manual, verifying operator training, and performing a postoccupancy operations review.

Environmental sustainability Meeting the needs of the present generation without compromising the ability of future generations to meet their needs.

Environmental tobacco smoke (ETS) Mixture of smoke from the burning end of a cigarette, pipe, or cigar and smoke exhaled by the smoker (also secondhand smoke or passive smoking). See Smoke-free Homes Program at www.epa.gov/smokefree.

Environmentally friendly A term often used to refer to the degree to which a product may harm the environment, including the biosphere, soil, water, and air.

Epicenter The point of the earth's surface directly above the focus or hypocenter of an earthquake.

Expansion joint A joint often installed in concrete construction to reduce cracking and to provide workable areas.

Expected useful life (EUL) The average amount of time in years that an item, component, or system is estimated to function when installed new and assuming routine maintenance is practiced.

Exploded view A pictorial view of a device in a state of disassembly, showing the appearance and interrelationship of parts.

Extension line A line used to visually connect the ends of a dimension line to the relevant feature on the part. Extension lines are solid and are drawn perpendicular to the dimension line.

Façade The exterior covering of a structure.

Face of stud (F.O.S.) Outside surface of the stud. Term used most often in dimensioning or as a point of reference.

Facility All or any portion of buildings, structures, site improvements, complexes, equipment, roads, walks, passageways, parking lots, or other real or personal property located on a site.

Fascia A horizontal member located at the edge of a roof overhang.

Felt A tar-impregnated paper used for water protection under roofing and siding materials. Sometimes used under concrete slabs for moisture resistance.

Fiber optics Optical, clear strands that transmit light without electrical current; sometimes used for outdoor lighting.

Fillet A concave internal corner in a metal component, usually a casting.

Filter A device designed to remove impurities from a fluid passing through the device; see also air filter.

Final completion Term denoting that the work has been completed in accordance with the terms and conditions of the contract documents.

Final inspection Final review of the project by the architect to determine final completion, prior to issuance of the final certificate for payment.

Final payment The payment made by the owner to the contractor, upon issuance by the architect of the final certificate for payment, of the entire unpaid balance of the contract sum as adjusted by change orders.

Finish grade The soil elevation in its final state upon completion of construction.

Fire barrier A continuous membrane such as a wall, ceiling, or floor assembly that is designed and constructed to a specified fire-resistant rating to hinder the spread of fire and smoke. This resistant rating is based on a time factor. Only fire-rated doors may be used in these barriers.

Fire compartment of fire zone An enclosed space in a building that is separated from all other parts of the building by the construction of fire separations having fire resistance ratings.

Fire door A door used between different types of construction that has been rated as being able to withstand fire for a certain amount of time.

Fire resistance rating Sometimes called fire rating, fire resistance classification, or hourly rating. A term defined in building codes, usually based on fire endurance required. Fire resistance ratings are assigned by building codes for various types of construction and occupancies, and are usually given in half-hour increments.

Fire-stop Blocking placed between studs or other structural members to resist the spread of fire.

Firewall Means a type of fire separation of noncombustible construction that subdivides a building or separates adjoining buildings to resist the spread of fire and that has a fire-resistance rating as prescribed in the NBC and has structural ability to remain intact under fire conditions for the required fire-rated time.

Flashing A thin, impervious sheet of material placed in construction to prevent water penetration or to direct the flow of water. Flashing is used especially at roof hips and valleys, roof penetrations, joints between a roof and a vertical wall, and in masonry walls to direct the flow of water and moisture.

Floodplain A floodplain (or flood plain) is flat or nearly flat land generally adjacent to a stream or river that experiences occasional or periodic flooding; land that has a likelihood of being flooded within a given storm cycle such as a 100-year storm.

Floor joist Structural member for the support of floor loads.

Floor plan A horizontal section taken at approximately eye level.

Floor-to-area ratio It is the density of nonresidential land use, exclusive of parking, measured as the total nonresidential building floor area divided by the total buildable land area available for nonresidential structures; for example, on a site with 10,000 ft.^{2} of buildable land area, an FAR of 1.0 would be 10,000 ft.^{2} of building floor area; on the same site, an FAR of 0.5 would be 5,000 ft.^{2}, and an FAR of 1.5 would be 15,000 ft.^{2} of built floor area etc.

Flush Even, level, or aligned.

Flush-out The operation of mechanical systems for a minimum of 2 weeks using 100% outside air upon completion of construction and prior to building occupancy to ensure safe indoor air quality.

Fly ash The fine ash waste collected from the flue gases of coal combustion, smelting, or waste incineration.

Footcandle A common unit of illuminance used in the United States. The metric unit is the lux.

Footings Weight-bearing concrete construction elements poured in place in the earth to support a structure.

Footlambert The US unit for luminance. The metric unit is the nit.

Formaldehyde A colorless, pungent, and irritating gas mainly used as a disinfectant and preservative and in synthesizing other compounds such as resins.

Fossil fuels Fuel derived from ancient organic remains such as peat, coal, crude oil, and natural gas.

Foundation plan A drawing that graphically illustrates the location of various foundation members and conditions that are required for the support of a specific structure.

Frieze A decoration or ornament shaped to form a band around a structure.

Frost line The depth at which frost penetrates the soil.

Fungi Any of a group of parasitic lower plants that lack chlorophyll, including molds and mildews.

Fuse A fuse is a device used to protect electrical equipment from short circuits. Fuses are made with metals that are designed to melt, when the current passing through them is high enough. When the fuse melts, the electrical connection is broken, interrupting power to the circuit or device.

Gallons per flush Measurement of water used by flush fixtures (water closets and urinals); per EPAct 1992, baseline rates for water closets is 1.6 gpf and urinals is 1.0 gpf.

Gallons per minute Measurement of water used by flow fixtures (faucets, showerheads, aerators, sprinkler heads); per EPAct 1992, baseline rates for faucets, showerheads, and aerators is 2.5 gpm.

Galvanized Steel products that have had zinc applied to the exterior surface to provide protection from rusting.

Gauge The thickness of metal or glass sheet material.

General conditions (When used by contractors) construction project activities and their associated costs that are not usually assignable to a specific material installation or subcontract. Example: temporary electrical power. (When used by everyone else) the contract document (often a standard form) that spells out the relationships between the parties to the contract. Example: the AIA Document A201.

General contract Any contract (together with all riders, addenda, and other instruments referred to therein as "contractor or any other person, which requires the general contractor or such other person to provide, or supervise or manage the procurement of, substantially all labor and material needed for completion of the Improvements.

Generator A mechanical device used to produce DC electricity. Power is produced by coils of wire passing through magnetic fields inside the generator. Most alternating current generating sets are also referred to as generators.

Gigawatt (GW) A measurement of power equal to a thousand million watts.

Gigawatt-hour (GWh) A measurement of energy. One gigawatt-hour is equal to 1 GW being used for a period of 1 h, or 1 MW being used for 1000 h.

Girder A horizontal structural beam for the support of secondary members such as floor joists.

Glare The effect produced by luminance within one's field of vision that is sufficiently greater than the luminance to which one's eyes are adapted; it can cause annoyance, discomfort, or loss in visual performance and visibility.

Grading The moving of soil to effect the elevation of land at a construction site.

Graywater (or greywater) Wastewater that does not contain toilet wastes and can be reused for irrigation after simple filtration. Wastewater from kitchen sinks and dishwashers may not be considered graywater in all cases.

Greenfields Land not previously developed beyond agriculture or forestry use.

Greenhouse gas A gas in the atmosphere that traps some of the sun's heat and preventing it from escaping into space. Greenhouse gases are vital for making the earth habitable, but increasing greenhouse gases contribute to climate change. Greenhouse gases include water vapor, carbon dioxide, methane, nitrous oxide, and ozone.

Green power Electricity generated from renewable energy sources.

Grid An electrical utility distribution network.

Grid-connected An energy-producing system connected to the utility transmission grid. Also called "grid tied."

Grout A mixture of cement, sand, and water used to fill joints in masonry and tile construction.

Guardrail A horizontal protective railing used around stairwells, balconies, and changes of floor elevation greater than 30 in.

Halogen lamp A special type of incandescent globe made of quartz glass and a tungsten filament, enabling it to run at a much higher temperature than a conventional incandescent globe. Efficiency is better than a normal incandescent but not as good as a fluorescent light.

Halon Ozone damaging chemicals used in fire fighting systems and extinguishers.

Hard costs Project costs directly related to construction and development activities such as contractor costs, labor and material costs, and costs related to direct service and material for the project. Not included are costs such as architectural and engineering fees, legal fees, closing fees, interest costs, etc., which are considered "soft costs."

Harmonic content Frequencies in the output waveform in addition to the primary frequency (usually 50 or 60 Hz). Energy in these harmonics is lost and can cause undue heating of the load.

Harvested rainwater Captured rainwater used for indoor needs, irrigation, or both.

Hazardous waste By-products of society that can pose a substantial or potential hazard to human health or the environment when improperly managed. Possesses at least one of four characteristics – ignitable, corrosive, reactive, or toxic, or appears on special EPA lists.

Head The top of a window or door frame.

Header A horizontal structural member spinning over openings, such a doors and windows, for the support of weight above the openings.

Header course In masonry, a horizontal masonry course of brick laid perpendicular to the wall face; used to tie a double-wythe brick wall together.

Heat exchanger A device designed to efficiently transfer heat from one medium to another (for example, water-to-air, refrigerant-to-air, refrigerant-to-water, steam-to-water).

Heat island effect The incidence of higher air and surface temperatures caused by solar absorption and reemission from roads, buildings, and other structures.

Heat pump A device that uses a reversible cycle vapor compression refrigeration circuit to provide cooling and heating from the same unit (at different times).

Heat recovery A process whereby heat is extracted from exhaust air before the air is dumped to the outside environment, the recovered heat is normally used to preheat incoming outside air; may be accomplished by heat recovery wheels or heat exchanger loops.

Hertz (Hz) Unit of measurement for frequency. Home mains power is normally 50 Hz in Europe and 60 Hz in the USA. The magnitude of the current is measured in Amps.

High-performance green building Buildings that create healthy indoor environments, and include design features that conserve water and energy; efficient use of space, materials, and resources; and minimize construction waste.

HVAC systems Heating, ventilation, and air-conditioning systems equipment, distribution systems, and terminals that provide the processes of heating, ventilating, and air-conditioning inside a building.

Hydronic system A heating or cooling system that relies on the circulation of water as the heat-transfer medium. A typical example is a boiler with hot water circulated through radiators.

Illuminance The density of the luminous flux incident on a surface, expressed in footcandles or lux. This term should not be confused with illumination (i.e., the act of illuminating or state of being illuminated).

Imperviousness Resistance of a material to penetration by a liquid such as water.

Incandescent light An electric lamp that is evacuated or filled with an inert gas and contains a filament (commonly tungsten). The filament emits visible light when heated to extreme temperatures by passage of electric current through it.

Incident light Light that shines onto the surface of a PV cell or module.

Indirect cost statement A statement by the borrower in a form approved by the lender of indirect costs incurred and to be incurred.

Indoor air pollution Chemical, physical, or biological contaminants contained in indoor air.

Indoor air quality (IAQ) According to the US Environmental Protection Agency and National Institute of Occupational Safety and Health, the definition of good indoor air quality includes (1) introduction and distribution of adequate ventilation air; (2) control of airborne contaminants; and (3) maintenance of acceptable temperature and relative humidity. According to ASHRAE Standard 62 1989, indoor air quality is defined as "air in which there are no known contaminants at harmful concentrations as determined by cognizant authorities and with which a substantial majority (80% or more) of the people exposed do not express dissatisfaction."

Indoor environmental quality (IEQ) The evaluation of five primary elements – lighting, sound, thermal conditions, air pollutants, and surface pollutants, to provide an environment that is physically and psychologically healthy for its occupants.

Infill site A site that is largely located within an existing community. For the purposes of LEED for Homes credits, an infill site is defined as having at least 75% of its perimeter bordering land that has been previously developed.

Inscribed figure A figure that is completely enclosed by another figure.

Insolation The amount of sunlight reaching an area, usually expressed in watt-hours per square meter per day.

Inspection Examination of work completed or in progress to determine its conformance with the requirements of the contract documents. The architect ordinarily makes only two inspections of work, one to determine substantial completion and the other to determine final completion. These inspections should be distinguished from the more general observations made by the architect on visits to the site during the progress of the work. The term is also used to mean examination of the work by a public official, owner's representative, or others.

Insulation Any material capable of resisting thermal, sound, or electrical transmission.

Integrated design team The team of all individuals involved in a project from very early in the design process, including the design professionals, the owner's representatives, and the general contractor and subcontractors.

Internal rate of return The discount rate that sets the net present value of the stream of net benefits equal to zero. The internal rate of return may have multiple values when the stream of net benefits alternates from negative to positive more than once.

Inverter An inverter converts DC power from the PV array/battery to AC power. Used either for stand-alone systems or grid-connected systems.

Irradiance The solar power incident on a surface, usually expressed in kilowatts per square meter. Irradiance multiplied by time gives insolation.

Irrigation efficiency Percentage of water used by irrigation equipment that is effective for irrigation that does not evaporate, blow away or fall on hardscape surfaces.

Isometric drawing A form of a pictorial drawing in which the main lines are equal in dimension. Normally drawn using 30°, 90° angle.

Jamb The side portion of a door, window, or any opening.

Joist A horizontal beam used to support a ceiling.

Joule (J) The energy conveyed by 1 W of power for 1 s, unit of energy equal to 1/3600 kWh.

Junction box A PV junction box is a protective enclosure on a PV module where PV strings are electrically connected and where electrical protection devices such as diodes can be fitted.

Key plan A plan, reduced in scale used for orientation purposes.

Kilowatt (kW) A unit of electrical power, 1000 W.

Kilowatt-hour (kWh) The amount of energy that derives from a power of 1000 W acting over a period of 1 h. The kilowatt-hour is a unit of energy. 1 kWh = 3600 kJ.

Lattice A grille made by criss-crossing strips of material.

Ledger Structural framing member used to support ceiling and roof joists at the perimeter walls.

LEED Leadership in Energy and Environmental Design. A sustainable design building certification system promulgated by the United States Green Building Council. Also an accrediting program for professionals (LEED APs) who have mastered the certification system. http://www.usgbc.org/

LEED Accredited Professional (LEED AP) The credential earned by candidates who passed the exam between 2001 and June 2009.

LEED Credit An optional LEED Green Building Rating System component where achievement results in the earning of points toward project certification.

LEED Credit Interpretation Request (CIR) Formal USGBC process in which a project team experiencing difficulties in the application of a LEED prerequisite or credit can seek and receive clarification.

LEED Enhanced Commissioning See Enhanced Commissioning.

LEED Green Building Rating System Voluntary, consensus-based, market-driven green building rating system based on existing proven technology.

LEED intent Primary environmental goal of each LEED prerequisite or credit.

LEED prerequisite Required LEED Green Building Rating System component whose achievement is mandatory and which does not earn any points.

LEED project boundary The portion of the project site submitted for LEED certification and remains constant for all required credit calculations; for single building developments, this is the entire project scope and is limited to the site boundary which is usually the legal property description; for multiple building developments (e.g., campus settings, industrial complexes), the LEED project boundary may be a reasonable portion of the development as determined by the project team.

LEED Technical Advisory Group (TAG) Consists of a committee of industry experts who assist in interpreting credits and developing improvements to the LEED Green Building Rating System.

Legend A description of any special or unusual marks, symbols, or line connections used in the drawing.

Lien A monetary claim on a property.

Life-cycle assessment (LCA) A process of evaluating the effects that a product has on the environment over the period of its life (cradle-to-grave), thereby increasing resource-use efficiency and decreasing liabilities.

Life-cycle cost A sum of all costs of creation and operation of a facility over a period of time.

Life-cycle cost analysis (LCC) A technique used to evaluate the economic consequences over a period of time of mutually exclusive project alternatives.

Lighting power density (LPD) A measure of the amount of installed lighting in a given area; often used to set a limit on the brightness of external lights.

Light shelf A horizontal element positioned above eye level to reflect daylight onto the ceiling.

Limit of liability The maximum amount, which an insurance company agrees to pay in case of loss.

Lintel A load-bearing structural member supported at its ends. Usually located over a door or window.

Live load A temporary and changing load superimposed on structural components by the use and occupancy of the building, not including the wind load, earthquake load, or dead load.

Load The electrical power being consumed at any given moment or averaged over a specified period. The load that an electric generating system supplies varies greatly with time of day and to some extent season of year. Also, in an electrical circuit, the load is any device or appliance that is using power.

Load-bearing wall A support wall that holds floor or roof loads in addition to its own weight.

Lumen (lm) The luminous flux emitted by a point source having a uniform luminous intensity of 1 candela.

Luminaire A complete electric lighting unit, including housing, lamp, and focusing and/or diffusing elements; informally referred to as fixture.

Lux The International System (SI) unit of illumination. It is the illumination on a surface 1 m² in area on which there is a uniformly distributed flux of 1 lumen.

Manifold A fitting that has several inlets or outlets to carry liquids or gases.

Market transformation Systemic improvements in the performance of a market or market segment.

Masonry opening The actual distance between masonry units where an opening occurs. It does not include the wood or steel framing around the opening.

Master format Industry standard for organizing specifications and other construction information, published by CSI and Construction Specifications Canada. Formerly a 5-digit number system with 16 divisions. Now a 6- or 8-digit numbering system with 49 divisions.

Master specification A resource specification section containing options for selection, usually created by a design professional firm, which once edited for a specific project becomes a contract specification.

Masterspec® Subscription master guide specification library published by ARCOM and owned by the American Institute of Architects. http://www.specguy.com/www.masterspec.com

Mastic An adhesive used to hold tiles in place; also refers to adhesives used to glue many types of materials in the building process.

Mechanical drawing Applies to scale drawings of mechanical objects.

Mechanics' lien A lien on real property created by statute in all states in favor of person supplying labor or materials for a building or structure for the value of labor or materials supplied by them. In some jurisdictions, a mechanic's lien also exists for the value of professional services. Clear title to the property cannot be obtained until the claim for the labor, materials, or professional services is settled.

Megawatt (MW) A measurement of power equal to 1 million W.

Megawatt-hour (MWh) A measurement of power with respect to time (i.e., energy). One megawatt-hour is equal to 1 MW being used for a period of 1 h, or 1 kW being used for 1000 h.

Mesh A metal reinforcing material placed in concrete slabs and masonry walls to help resist cracking.

Mezzanine or mezzanine floor That portion of a story which is an intermediate floor level placed within the story and having occupiable space above and below its floor.

Minimum efficiency reporting value (MERV) Mechanical system air filter efficiency rating ranging from 7 to 16.

Modules A system based on a single unit of measure.

Modulus of elasticity (E) The degree of stiffness of a beam.

Moisture barrier Typically a plastic material used to restrict moisture vapor from penetrating into a structure.

Mortar The mixture of cement, sand, lime, and water that provides a bond for the joining of masonry units.

Multizone HVAC system A central air-all HVAC system that utilizes an individual supply air stream for each zone; warm and cool air are mixed at the air handling unit to provide supply air appropriate to the needs of each zone; a multizone system requires the use of several separate supply air ducts.

Native vegetation Plants native to the locale. A plant whose presence and survival in a specific region is not due to human intervention. Certain experts argue that plants imported to a region by prehistoric peoples should be considered native. The term for plants that are imported and then adapt to survive without human cultivation is naturalized.

Natural ventilation Natural exchange of air or movement of air through a building by thermal, wind, or diffusion effects through doors, windows, or other intentional openings in buildings.

NEC US National Electrical Code, which contains guidelines for all types of electrical installations, which should be followed when installing a PV system.

Negligence Failure to exercise due care under normal circumstances. Legal liability for the consequences of an act or omission frequently depends upon whether or not there has been negligence.

Net metering The practice of exporting surplus solar power during the day (to actual power needs) to the electricity grid, which either causes the home owner electric meter to (physically) go backwards and/or simply creates a financial credit on the home owner's electricity bill. (At night, the homeowner draws from the electricity grid in the normal way).

Net size The actual size of an object.

Noise reduction coefficient (NRC) Average of the sound absorption coefficient of the four octave bands 250, 500, 1000, and 2000 Hz rounded to the nearest 0.05.

Nominal discount rate A discount rate that includes the rate of inflation.

Nominal size The call-out size. May not be the actual size of the item.

Nonbearing wall A wall that supports no loads other than it own weight. Some building codes consider walls that support only ceiling loads as nonbearing.

Nonconforming work Implemented work that does not fulfill the requirements of the contract documents.

Nonferrous metal They are metals such as copper or brass that contain no iron.

Nonrenewable resource A resource that can be depleted over time.

Oblique drawing A type of pictorial drawing in which one view is an orthographic projection and the views of the sides have receding lines at an angle.

Occupiable A room or enclosed space designed for human occupancy in which individuals congregate for amusement, educational, or similar purposes, or in which occupants are engaged at labor, and which is equipped with means of egress, light, and ventilation.

Off-gassing A process of evaporation or chemical decomposition by which vapors are released from materials.

Ohm The resistance between two points of a conductor when a constant potential difference of 1 V applied between these points produces in the conductor a current of 1 A.

Ohm's Law A simple mathematical formula that allows either voltage, current, or resistance to be calculated when the other two values are known. The formula is $V = I \times R$, where V is the voltage, I is the current, and R is the resistance.

Opinions of probable costs Determination of probable costs, a preliminary budget, for a suggested remedy.

Operating cost Any cost of the daily function of a facility.

Organic compounds Chemicals that contain carbon. Volatile organic compounds vaporize at room temperature and pressure. They are found in many indoor sources, including many common household products and building materials.

Orientation Position with respect to the cardinal directions, N, S, E, W.

Orthographic projection A view produced when projectors are perpendicular to the plane of the object. It gives the effect of looking straight at one side.

Outlet An electrical receptacle that allows for current to be drawn from the system.

Packaged air-conditioner A self-contained unit designed to provide control of air temperature, humidity, distribution, and quality; see also unitary air conditioner.

Parapet A portion of wall extending above the roof level.

Partial occupancy The occupancy by the owner of a portion of a projection prior to final completion.

Particulates Alternatively referred to as particulate matter (PM) or fine particles, and are tiny subdivisions of solid or liquid matter suspended in a gas or liquid.

Partition An interior wall.

Party wall A wall dividing two adjoining spaces such as apartments or offices.

Passive solar home A house that utilizes part of the building as a solar collector, as opposed to active solar, such as PV.

Patent defect A defect in materials, equipment of completed work, which reasonably careful observation could have discovered; distinguished from a latent defect, which could not be discovered by reasonable observation.

Performance specifications The written material containing the minimum acceptable standards and actions, as may be necessary to complete a project.

Perviousness Percentage of a paved area that is open and allows water to soak into the ground.

Phase An impulse of alternating current. The number of phases depends on the generator windings. Most large generators produce a three-phase current that must be carried on at least three wires.

Photometer An instrument for measuring light.

Photovoltaic (PV) Refers to any device, which produces free electrons when exposed to light.

Photovoltaic (PV) energy or solar Energy from the sun converted by photovoltaic cells into electricity.

Photovoltaic (PV) panel A term often used interchangeably with PV module (especially in single module systems).

Photovoltaic system All the parts connected together that are required to produce solar electricity.

Pile A steel or wooden pole driven into the ground sufficiently to support the weight of a wall and building.

Pillar A pole or reinforced wall section used to support the floor and consequently the building.

Planking A term for wood members having a minimum rectangular section of 1.5–3.5 in. in thickness. Used for floor and roof systems.

Plans All final drawings, plans and specifications prepared by the borrower, borrower's architects, the general contractor or major subcontractors, and approved by lender and the construction consultant, which describe and show the labor, materials, equipment, fixtures and furnishings necessary for the construction of the improvements, including all amendments and modifications thereof made by approved change orders (and also showing minimum grade of finishes and furnishings for all areas of the Improvements to be leased or sold in ready-for-occupancy conditions).

Plat A map or plan view of a lot showing principal features, boundaries, and location of structures.

Plenum An air space (above the ceiling) for transporting air from the HVAC system.

Plug load Refers to all equipment that is plugged into the electrical system, such as task lights, computers, printers, and electrical appliances.

Polarity The direction of magnetism or direction of flow of current.

Pollutant Generally, any substance introduced into the environment that adversely affects the usefulness of a resource or the health of humans, animals, or ecosystems.

Poly-vinyl chloride (PVC) A plastic material commonly used for pipe and plumbing fixtures and as an insulator on electrical cables. A toxic material, which is being replaced with alternatives made from more benign chemicals.

Post A vertical wood structural member generally 4×4 (100 mm) or larger.

Post-and-beam construction A type of wood frame construction using timber for the structural support.

Postconsumer materials/waste Recovered materials that are diverted from municipal solid waste for the purposes of collection, recycling, and disposition.

Postconsumer recycled content A product composition that contains some percentage of material that has been reclaimed from the same or another end use of its former, useful life; includes construction and demolition debris, materials collected from recycling programs (e.g., decking, furniture, cabinets) and landscaping waste (e.g., leaves, grass clippings, tree trimmings).

Postconsumer recycling Use of materials generated from resident or consumer waste for new or similar purposes such as converting wastepaper from offices into corrugated boxes or newsprint.

Potable water This is water that is suitable for drinking, generally supplied by the municipal water systems.

Power Basic unit of electricity equal to the product of current and voltage (in DC circuits). It is the rate of doing work, expressed as watts (W). For example, a generator rated at 800 W can provide that amount of power continuously. $1 W = 1 J/s$

Precast A concrete component that has been cast in a location other than the one in which it will be used.

Preconsumer materials/waste Materials generated in manufacturing and converting processes such as manufacturing scraps and trimmings and cuttings. Includes print overruns, over issue publications, and obsolete inventories. Sometimes referred to as "postindustrial."

Present value The current value of a past or future sum of money as a function of an investor's time value of money.

Pressed wood products A group of materials used in building and furniture construction that are made from wood veneers, particles, or fibers bonded together with an adhesive under heat and pressure.

Primer The first coat of paint or glue when more than one coat will be applied.

Prime Farmland A designation assigned by the US Department of Agriculture and is described as land that has the best combination of physical and chemical characteristics for producing food, feed, forage, fiber, and oilseed crops and is also available for these uses.

Progress payment Partial payment made during progress of the work on account of work completed and/or materials suitably stored.

Progress schedule A diagram, graph or other pictorial or written schedule showing proposed and actual of starting and completion of the various elements of the work.

Project cost Total cost of the project, including construction cost, professional compensation, land costs, furnishings and equipment, financing and other charges.

Project manual The volume(s) prepared by the architect for a project, which may include the bidding requirements, sample forms and conditions of the contract and the specifications.

Projection A technique for showing one or more sides of an object to give the impression of a drawing of a solid object.

Purlin A horizontal roof member that is laid perpendicular to rafters to help limit deflections.

Quarry tile An unglazed, machine-made tile.

Quick set A fast-curing cement plaster.

R-factor A unit of thermal resistance applied to the insulating value of a specific building material.

R-value The unit that measures thermal resistance (the effectiveness of insulation); the higher the number, the better the insulation qualities.

Radius A straight line from the center of a circle or sphere to its circumference or surface.

Radon (Rn) and radon decay products Radon is a radioactive gas formed in the decay of uranium. The radon decay products (also called radon daughters or progeny) can be breathed into the lung, where they continue to release radiation as they further decay.

Rafter A sloping or horizontal beam used to support a roof.

Rain garden A depressed area of the ground planted with vegetation, allowing runoff from impervious surfaces such as parking lots and roofs the opportunity to be collected and infiltrated into the groundwater supply or returned to the atmosphere through evaporation and evapotranspiration; considered a stormwater management strategy.

Rainscreen A method of constructing walls in which the cladding is separated from a membrane by an airspace that allows pressure equalization to prevent rain from being forced in. Often used for high-rise buildings or for buildings in windy locations.

Rainwater harvesting The practice of collecting, storing, and using precipitation from a catchment area such as a roof.

Rapidly renewable materials Resources that can be rapidly replenished (within a 10-year cycle) as they are used. These materials are typically harvested from fast-growing sources and do not require unnecessary chemical support. Examples include bamboo, cork, flax, wheat, wool, and certain types of wood.

RAPS (remote area power supply) A power generation system used to provide electricity to remote and rural homes, usually incorporating power generated from renewable sources such as solar panels and wind generators, as well as nonrenewable sources such as petrol-powered generators.

Readily accessible Describes areas of the subject property that are promptly made available for observation by the field observer at the time of the walk-through survey and do not require the removal of materials or personal property, such as furniture, and that are safely accessible in the opinion of the field observer.

Record drawings Construction drawing revised to show significant changes made during the construction process, usually based on marked-up prints, drawing and other data furnished by the contracts to the architect. Preferable to *As Built Drawings*.

Rectifier A device that converts ac to dc, as in a battery charger or converter.

Recycled content The portion of a product that is made from materials, diverted from the waste stream, usually stated as a percentage by weight, and used to manufacture new materials; manufacturing waste stream (preconsumer) and/or the consumer waste stream (postconsumer).

Recycled material Material that would otherwise be destined for disposal but is diverted or separated from the waste stream, reintroduced as material feedstock, and processed into marketed end-products.

Reference numbers They consist of numbers used on a drawing to refer the reader to another drawing for more detail or other information.

Reflectance The ratio of energy (light) bouncing away from a surface to the amount striking it, expressed as a percentage.

Refrigerant A heat transfer fluid employed by a refrigerating process, selected for its beneficial properties (stability, low viscosity, high thermal capacity, appropriate state change points).

Regenerative design A process-oriented systems theory based approach to design; the term "regenerative" describes processes that restore, renew, or revitalize their own sources of energy and materials, creating sustainable systems that integrate the needs of society with the integrity of nature. Sometimes referred to as Cradle-to-Cradle.

Regional materials Percentage (total material costs of the building) of a building's materials that have been extracted, processed, and manufactured within a 500-mile radius of the project site.

Regionally harvested or extracted materials These are materials taken from within a 500-mile radius of the project site.

Register An opening in a duct for the supply of heated or cooled air.

Regulator A device used to limit the current and voltage in a circuit, normally to allow the correct charging of batteries from power sources such as solar panels and wind generators.

Relative humidity The amount of water vapor in the atmosphere compared to the maximum possible amount at the same temperature.

Release of lien Instrument executed by a person or entity supplying labor, materials, or professional services on a project which releases that person's or entity's mechanic's lien against the project property.

Remaining useful life (RUL) A subjective estimate based upon observations, or average estimates of similar items, components, or systems, or a combination thereof, of the number of remaining years that an item, component, or system is estimated to be able to function in accordance with its intended purpose before warranting replacement. Such period of time is affected by the initial quality of an item, component, or system, the quality of the initial installation, the quality and amount of preventive maintenance exercised, climatic conditions, extent of use, etc.

Renewable energy Alternative energy that is produced from a renewable source.

Requisition A statement prepared by the borrower in a form approved by the lender setting forth the amount of the loan advance requested in each instance and including, if requested by the lender.

Residual value The value of a building or building system at the end of the study period.

Resistance (R) The property of a material, which resists the flow of electric current when a potential difference is applied across it, measured in ohms.

Resistor An electronic component used to restrict the flow of current in a circuit. Sometimes used specifically to produce heat, such as in a water heater element.

Retainage A sum withheld from progress payments to the contractor in accordance with the terms of the owner–contractor agreement.

Retaining wall A masonry wall supported at the top and bottom designed to resist soil loads.

Return air The air that has circulated through a building as supply air and has been returned to the HVAC system for additional conditioning or release from the building.

Reuse Extends the life of materials by salvaging and reusing for the same or similar use.

Roof drain A receptacle for removal of roof water.

Roof pitch The ratio of total span to total rise expressed as a fraction.

Rotation A view in which the object is apparently rotated or turned to reveal a different plane or aspect, all shown within the view.

Rough in To prepare a room for plumbing or electrical additions by running wires or piping for a future fixture.

Rough opening A large opening made in a wall frame or roof frame to allow the insertion of a door or window.

Salvaged material (or reused materials) Construction materials or decorative items recovered from existing buildings or construction sites and reused.

Sanitary sewer A conduit or pipe carrying sanitary sewage.

Scale The relation between the measurement used on a drawing and the measurement of the object it represents. A measuring device, such as a ruler, having special graduations.

Schedule of values A statement furnished by the contractor to the architect reflecting the portions of the contract sum allocated to the work and used as the basis for reviewing the contractor's applications for payment.

Schematic diagram A diagram using graphic symbols to show how a circuit functions electrically.

Scratch coat The first coat of stucco, which is scratched to provide a good bond surface for the second coat.

Section A view showing internal features as if the viewed object has been cut or sectioned.

Sedimentation Sediment is naturally occurring materials that are broken down by processes of weathering and erosion and is subsequently transported by the action of fluids such as wind, water, or ice, and/or by the force of gravity acting on the particle itself; it generally decreases the quality of the water and can age streams, rivers, and lakes.

Seismicity The worldwide or local distribution of earthquakes in space and time; a general term for the number of earthquakes in a unit of time, or for relative earthquake activity.

Septic tank A tank in which sewage is decomposed by bacteria and dispersed by drain tiles.

Shear distribution The distribution of lateral forces along the height or width of a building.

Shear wall A wall construction designed to withstand shear pressure caused by wind or earthquake.

Sheet steel Flat steel weighing less than 5 psf.

Shoring Temporary support made of metal or wood used to support other components.

Short-term costs Opinions of probable costs to remedy physical deficiencies, such as deferred maintenance, that may not warrant immediate attention, but require repairs or replacements that should be undertaken on a priority basis in addition to routine preventive maintenance. Such opinions of probable costs may include costs for testing, exploratory probing, and further analysis should this be deemed warranted by the consultant. The performance of such additional services are beyond this guide. Generally, the time frame for such repairs is within one to two years.

Sick building syndrome (SBS) Term that refers to a set of symptoms that affect some number of building occupants during the time they spend in the building and diminish or go away during periods when they leave the building. Cannot be traced to specific pollutants or sources within the building. (Contrast with "Building related illness").

Sill A horizontal structural member supported by its ends.

Single-line diagram A diagram using single lines and graphic symbols to simplify a complex circuit or system.

Single prime contract This is the most common form of construction contracting. In this process, the bidding documents are prepared by the architect/engineer for the owner and made available to a number of qualified bidders. The winning contractor then enters into a series of subcontract agreements to complete the work. Increasingly, owners are opting for a "design-build contract" under which a single entity provides design and construction services.

Site A parcel of land bounded by a property line or a designated portion of a public right-of-way.

Site disturbance The areas of the site that have been disturbed because of the project's scope and requirements.

Site improvement Landscaping, paving for pedestrian and vehicular ways, outdoor lighting, recreational facilities, and the like, added to a site.

Skylight A relatively horizontal, glazed roof aperture for the admission of daylight.

Slab-on-grade The foundation construction for a structure with no basement or crawl space.

Smart growth Managing the growth of a community in such a way that land is developed according to ecological tenets that call for minimizing dependence on auto transportation, reducing air pollution, and increasing infrastructure investment efficiency.

Soft costs Cost items that are excluded from direct construction costs. Soft costs normally include architectural and engineering fees, legal fees, costs of permits, real estate commissions, financing fees, construction interest and operating expenses, leasing and advertising and promotion, etc.

Solar energy Energy from the sun.

Solar heat gain coefficient Solar heat gain through the total window system relative to the incident solar radiation.

Solar module A device used to convert light from the sun directly into DC electricity by using the photovoltaic effect. Usually consists of multiple solar cells bonded between glass and a backing material. A typical solar module would be 100 W of power output (but module powers can range from 1 W to 300 W) and have dimensions of 2 × 4 ft.

Solar panel A device that collects energy from the sun and converts it into electricity or heat.

Solar power Electricity generated by conversion of sunlight, either directly through the use of photovoltaic panels, or indirectly through solar–thermal processes.

Solar reflective index (SRI) It is a measure of material's ability to reject solar heat in which black = 0 and white = 100. The index ranges from 0 (black), being least reflective, to 100 (white), being the most reflective. Materials that have high SRI are cooler choices.

Special conditions A section of the conditions of the contract, other than general conditions and supplementary conditions, which may be prepared to describe conditions unique to a particular project.

Specific gravity The ratio of the weight of a solution to the weight of an equal volume of water at a specified temperature, used with reference to the sulfuric acid electrolyte solution in a lead acid battery as an indicator of battery state of charge, more recently called relative density.

Specifications A part of the contract documents contained in the project manual consisting of written requirements for material, equipment, construction systems, standards, and workmanship. Under the uniform construction index, the specifications comprise sixteen divisions.

Stakeholders All parties that might be affected by a company's policies and operations, including shareholders, customers, employees, suppliers, business partners, and surrounding communities.

Storm sewer A sewer used for conveying rainwater, surface water condensate, cooling water, or similar liquid wastes exclusive of sewage.

Stormwater runoff Precipitation (rain and snow) that does not infiltrate into the ground or evaporate due to impervious land surfaces but instead flows into stormsewer systems, waterways, or onto adjacent land.

Street grid density A measurement of circulation permeability and multimodel travel feasibility; it is expressed in centerlines miles per square mile, often within a 1-mile radius around a project boundary.

Stucco A type of plaster made from Portland cement, sand, water, and a coloring agent that is applied to exterior walls.

Structural frame The components or building system that supports the building's nonvariable forces or weights (dead loads) and variable forces or weights (live loads).

Stud A light vertical structure member, usually of wood or light structural steel, used as part of a wall and for supporting moderate loads.

Subcontract Agreement between a prime contractor and a subcontractor for a portion of the work at the site.

Subcontractor A person or entity who has a direct or indirect contract with a subcontractor to perform any of the work at the site.

Substitution A material, product, or item of equipment offered in lieu of that specified.

Superintendent Contractor's representative at the site who is responsible for continuous field supervision, coordination, completion of the work and, unless another person is designated in writing by the contractor to the owner and the architect, for the prevention of accidents.

Supervision Direction of the work by contractor's personnel. Supervision is neither a duty nor a responsibility of the architect as part of professional services.

Surety bond A legal instrument under which one party agrees to answer to another party for the debt, default, or failure to perform of a third party.

Surge An excessive amount of power drawn by an appliance when it is first switched on. An unexpected flow of excessive current, usually caused by excessive voltage that can damage appliances and other electrical equipment.

Survey Observations made by the field observer during a walk-through survey to obtain information concerning the subject property's readily accessible and easily visible components or systems.

Sustainability The condition that meets the needs of present generations without compromising the needs of future generations. Achieving a balance among extraction and renewal and environmental inputs and outputs, as to cause no overall net environmental burden or deficit.

Sustainable Forestry Sustainable forest management (SFM) is the management of forests according to the principles of sustainable development.

Sustained Yield Forestry Management of a forest to produce in perpetuity a high level annual or regular periodic output through a balance between increment and cutting.

Swale A low-lying, often depressed and swampy, area of land; an open-ended swale can be created for use as a land drainage device.

Symbol Stylized graphical representation of commonly used component parts shown in a drawing.

Synergy Action of two or more substances to achieve an effect of which each is individually incapable. As applied to toxicology, two exposures together (e.g., asbestos and smoking) are far more risky than the combined individual risks.

System (A process) A combination of interacting or interdependent components assembled to carry out one or more functions.

Task lighting Light provided for a specific task, versus general or ambient lighting.

Tee A fitting, either cast or wrought, that has one side outlet at right angles to the run.

Temper To harden steel by heating and sudden cooling by immersion in oil, water, or other coolant.

Template A piece of thin material used as a true-scale guide or as a model for reproducing various shapes.

Tensile strength The maximum stretching of a piece of metal (rebar, etc.) before breaking; calculated in kPa.

Termite shield Sheet metal placed in or on a foundation wall to prevent intrusion.

Terrazzo A mixture of concrete, crushed stone, calcium shells, and/or glass, polished to a tile-like finish.

Thermal comfort The appropriate combination of temperature combined with airflow and humidity that allows one to be comfortable within the confines of a building. Individually, an expression of satisfaction with the thermal environment; statistically, such expression of satisfaction from at least 80% of the occupants within a space.

Thermal resistance (R) A unit used to measure a material's resistance to heat transfer. The formula for thermal resistance is: R _ Thickness (in inches)/k.

Thermostat An automatic device controlling the operation of HVAC equipment.

Third-party certification An independent and objective assessment of an organization's practices or chain of custody system by an auditor who is independent of the party undergoing assessment.

Three-phase power A combination of three alternating currents in a circuit with their voltages displaced at 120°, or one-third of a cycle.

Timely access Entry provided to the consultant at the time of the site visit.

Timely completion The completion of the work or designated portion thereof on or before the date required.

Title insurer The issuer(s), approved by interim lender and permanent lenders, of the title insurance policy or policies insuring the mortgage.

Tolerance The amount that a manufactured part may vary from its specified size.

Topographic survey The configuration of a surface including its relief and the locations of its natural and man-made features, usually recorded on a drawing showing surface variations by means of contour lines indicating height above or below a fixed datum.

Toxicity A reflection of a material's ability to release poisonous particulate.

Transformer A transformer is a device that changes voltage from one level to another. A device used to transform voltage levels to facilitate the transfer of power from the generating plant to the customer.

Transient lodging A building, facility, or portion thereof, excluding inpatient medical care facilities and residential facilities that contains sleeping accommodations. Transient lodging may include, but is not limited to, resorts, group homes, hotels, motels, and dormitories.

Transistor A semiconductor device used to switch or otherwise control the flow of electricity.

Transportation demand management Also known as travel demand management (both TDM), it is the application of strategies and policies to reduce travel demand (specifically that of single-occupancy private vehicles), or to redistribute this demand in space or in time.

Trap A fitting designed to provide a liquid seal that will prevent the back-passage of air without significantly affecting the flow of wastewater through it.

Triangulation A technique for making developments of complex sheet metal forms using geometrical constructions to translate dimensions from the drawing to the pattern.

Trimmer A piece of lumber, usually a 2 × 4, that is shorter than the stud or rafter but is used to fill in where the longer piece would have been normally spaced except for the window or door opening or some other opening in the roof or floor or wall.

Truss A prefabricated sloped roof system incorporating a top chord, bottom chord, and bracing.

Turbulence Any deviation from parallel flow in a pipe due to rough inner wall surfaces, obstructions, etc.

UL Underwriters Laboratories, Inc. A private testing and labeling organization that develops test standards for product compliance. UL standards appear throughout specifications, often in roofing requirements, and always in equipment utilizing or delivering electrical power (http://www.ul.com/).

Unfaced insulation Insulation that does not have a facing or plastic membrane over one side of it.

Union joint A pipe coupling, usually threaded, that permits disconnection without disturbing other sections.

Urea formaldehyde A combination of urea and formaldehyde that is used in some glues and may emit formaldehyde at room temperature.

Utility plan A floor plan of a structure showing locations of heating, electrical, plumbing and other service system components.

Vacuum Any pressure less than that exerted by the atmosphere.

Valley The area of a roof where two sections come together to form a depression.

Valve A device designed to control water flow in a distribution system; common valve types include globe, gate, butterfly, and check.

Vapor barrier The same as a moisture barrier.

Vapor compression chiller Refrigeration equipment that generates chilled water via a mechanically driven process using a specialized heat transfer fluid as refrigerant; comprised of four major components: a compressor, condenser, expansion valve, and evaporator; operating energy is input as mechanical motion.

Variable air volume (VAV) HVAC system A central air-all HVAC system that utilizes a single supply air stream and a terminal device at each zone to provide appropriate thermal conditions through control of the quantity of air supplied to the zone.

Vegetated roof A roof that is partially or fully covered by vegetation. By creating roofs with a vegetated layer, the roof can counteract the heat island effect as well as provide additional insulation and cooling during the summer.

Vehicle miles traveled (VMT) The number of miles driven by a vehicle within a given time frame and geographic area; it is influenced by several factors, typically increasing with a higher number of car trips and distance traveled as well as poorly planned development.

Vehicular way A route intended for vehicular traffic, such as a street, driveway, or parking lot.

Veneer A thin layer or sheet of wood.

Veneered wall A single-thickness (one-wythe) masonry unit wall with a backup wall of frame or other masonry; tied but not bonded to the backup wall.

Vent Usually a hole in the eaves or soffit to allow the circulation of air over an insulated ceiling; usually covered with a piece of metal or screen.

Vent stack A system of pipes used for air circulation and prevent water from being suctioned from the traps in the waste disposal system.

Ventilation The exchange of air, or the movement of air through a building; may be done naturally through doors and windows or mechanically by motor-driven fans.

Ventilation rate The rate at which indoor air enters and leaves a building. Expressed in one of two ways: the number of changes of outdoor air per unit of time (air changes per hour or "ach") or the rate at which a volume of outdoor air enters per unit of time (cubic feet per minute or "cfm").

Vertical pipe Any pipe or fitting installed in a vertical position or which makes an angle of not more than 45° with the vertical.

View A drawing of a side or plane of an object as seen from one point.

Vision glazing That portion of exterior windows above 2 ft. 6 in. and below 7 ft. 6 in. that permits a view to the exterior.

Volatile organic compound (VOC) A highly evaporative, carbon-based chemical substance that produces noxious fumes; found in many paints, caulks, stains, and adhesives.

Volt (E) or (V) The potential difference across a resistance of 1 Ω when a current of 1 A is flowing. The amount of work done per unit charge in moving a charge from one place to another.

Voltage drop The voltage lost along a length of wire or conductor as a result of the resistance of that conductor. This also applies to resistors. The voltage drop is calculated by using Ohm's law.

Voltage protection A sensing circuit on an Inverter that will disconnect the unit from the battery if input voltage limits are exceeded.

Voltage regulator A device that controls the operating voltage of a photovoltaic array.

Waiver of lien An instrument by which a person or organization who has or may have a right of mechanic's lien against the property of another relinquishes such right.

Warranty Legally enforceable assurance of quality or performance of a product or work, or of the duration of satisfactory performance. *Warranty*, *Guarantee*, and *Guaranty* are substantially identical in meaning; nevertheless, confusion frequently arises from supposed distinctions attributed to guarantee (or guaranty) being exclusively indicative of duration of satisfactory performance or of a legally enforceable assurance furnished by a manufacturer or other third party. The Uniform Commercial code provisions on Sales (effective in all states except Louisiana) use Warranty but recognize the continuation of the use of Guarantee and Guaranty.

Waste diversion The process of diverting waste from landfill; a waste reduction strategy focused on the recycling or composting of materials, thereby recovering what would otherwise have been waste for use in new products.

Waste pipe Discharge pipe from any fixture, appliance, or appurtenance in connection with a plumbing system that does not contain fecal matter.

Wastewater Spent or used water from a home, farm, community, or industry that contains dissolved or suspended matter.

Water–cement ratio The ratio between the weight of water to cement.

Water hammer The noise and vibration that develops in a piping system when a column of noncompressible liquid flowing through a pipeline at a given pressure and velocity is abruptly stopped.

Water main The water supply pipe for public or community use.

Waterproofing Materials used to protect below- and on-grade construction from moisture penetration.

Watt (W) The unit of electrical power commonly used to define the electricity consumption of an appliance. The power developed when a current of 1 A flows through a potential difference of 1 V; 1/746 of a horsepower. $1 W = 1 J/s$.

Watt-hour (Wh) A unit of energy equal to 1 W of power being used for 1 h.

Wetland In stormwater management, a shallow, vegetated, ponded area that serves to improve water quality and provide wildlife habitat.

Wetland vegetation It is vegetation that is adapted to hydric soils and hydrologic conditions normally found in wetlands; plants that require saturated soils to survive or can tolerate prolonged wet soil conditions.

Wind lift (wind load) The force exerted by the wind against a structure.

Wiring (connection) diagram A diagram showing the individual connections within a unit and the physical arrangement of the components.

Working drawings A set of drawings, which provide the necessary details and dimensions to construct the object. May include specifications.

Wythe A continuous masonry wall width.

Xeriscape™ A trademarked term referring to water-efficient choices in planting and irrigation design. It refers to seven basic principles for conserving water and protecting the environment. These include (1) planning and design; (2) use of well-adapted plants;

(3) soil analysis; (4) practical turf areas; (5) use of mulches; (6) appropriate maintenance; and (7) efficient irrigation.

Zenith angle The angle between directly overhead and a line through the sun. The elevation angle of the sun above the horizon is 90° minus the zenith angle.

Zinc Noncorrosive metal used for galvanizing other metals.

Zone numbers They consist of numbers and letters on the border of a drawing to provide reference points to aid in indicating or locating specific points on the drawing.

Zoning The legal restriction that deems that parts of cities be for particular uses, such as residential, commercial, industrial, and so forth.

Zoning permit A permit issued by appropriate governmental authority authorizing land to be used for a specific purpose.

SAMPLE EXAM QUESTIONS

According to the Green Building Certification Institute (GBCI),

The LEED™ Green Associate exam is designed to measure your skills and knowledge against criteria developed by Subject Matter Experts and to assess your knowledge and skill to understand and support green design, construction, and operations. The LEED™ Green Associate exam is comprised of 100 randomly delivered multiple choice questions and must be completed in 2 hours; total seat time for the LEED™ Green Associate exam will be 2 hours and 20 minutes including a tutorial and short satisfaction survey.

In order to pass any given LEED™ professional credentialing exam, you are required to achieve a minimum competency score of 170 out of 200. It is always advisable to check the GBCI website for the latest updates (www.gbci.org).

The questions shown below are provided for your convenience to assist you to better familiarize yourself with the format and general content of the exam. However, although the content of this exam is believed to be representative of the type of questions on the LEED Green Associate exam, it does not necessarily reflect the content that will appear on the actual exam. Moreover, your ability to correctly answer these sample questions does not necessarily guarantee your ability to successfully answer questions on the actual LEED Green Associate exam. It is primarily intended to provide an indication of areas requiring more attention prior to proceeding to taking the LEED™ AP Exam.

1 LEED GREEN ASSOCIATE SAMPLE EXAM QUESTIONS

Answers are followed by an asterisk.

1.1 Sustainable Sites

1. What are the requirements of SS Credit 8, Light Pollution Reduction? (Choose two)
    a. Interior lighting should be automatically controlled to shut off during nonbusiness hours*
    b. Full-cutoff luminaires and shades must be used for all exterior lamps

    c. Conformance with zoned requirements as given in IESNA RP-33*

    d. Exterior lamps will meet a maximum 75% of the lighting power densities for exterior areas and building facades as defined by IESNA

2. A project that earns SS2 Development Density and looks to utilize a vegetated roof can count toward which of the following credits? (Choose two).

    a. SS5.1 Protect and restore habitat*

    b. SS7.1 Heat island effect*

    c. MR5.1 Regional materials

    d. WE1.1 Water efficient Landscaping

3. A large food manufacturer in White Plains, NY, has the following staff: 2 full-time managers, 58 full-time laborers, 16 part-time student workers who average 20 h/week and a cleaning staff of 3 who work an 8-h shift at night. The LEED team is pursuing SS4.2 Alternate transportation – Bikes. How many bike storage locations will need to be available and how many showers will the food manufacturer need to add to its site?

    a. 4 bike storage locations and 3 showers*

    b. 2 bike storage locations and 2 showers

    c. 2 bike storage locations and 1 shower

    d. 3 bike storage locations and 3 showers

4. An open grid pavement system with an SRI of 39 could count toward which of the following credits?

    a. SS5.2 Reduced Site Disturbance – Maximize Open Space

    b. SS7.1 Heat Island Effect Non Roof*

    c. SS6.2 Stormwater Design Quality Control*

    d. MR6 Rapidly Renewable Materials

5. A 20,000-ft.^{2} structure is being built on an open 50,000-ft.^{2} site. The applicable zoning ordinances have no open space requirements. The project should provide how many square feet of open space at a minimum to meet SS Credit 5.2, Site Development: Maximize Open Space?

    a. 0

    b. 5,000

    c. 10,000*

    d. 20,000

    e. 30,000

6. What is not a requirement of SS Prerequisite 1, Construction Activity Pollution Prevention? (Choose one)
   a. Create and implement an Erosion and Sedimentation Control (ESC) Plan for all construction activities associated with the project
   b. Conform to the 2003 EPA Construction General Permit or local erosion and sedimentation control standards, whichever is more stringent
   c. Prevent sedimentation of storm sewer or receiving streams
   d. Prevent polluting the air with dust and particulate matter
   e. Protect receiving stream channels from excessive erosion by implementing a stream channel protection strategy and quantity control strategies*

7. What is the intent of SS Credit 5.2, Site Development: Maximize Open Space?
   a. Conserve existing natural areas and restore damaged areas to provide habitat and promote biodiversity.
   b. Maintain open areas amid development to protect ecosystems and provide recreational space.
   c. Provide a high ratio of open space to development footprint to promote biodiversity.*
   d. Channel development to urban areas with existing infrastructure, protect greenfields and preserve habitat and natural resources.

8. The requirements for SS Credit 2.1, Development Density, are based on what? (Choose two)
   a. A typical 2-story downtown development*
   b. A typical urban mixed-use development density with retail, high-rise office and residential buildings, and city streets and parks
   c. A 60,000 ft.^{2}/acre net density*
   d. A 45,000 ft.^{2}/acre net density
   e. Density required for provision of basic community services

9. A large downtown industrial building renovation project has met the requirements for SS Credit 2, Development Density. The owner wants to achieve additional credits and is asking for your advice. Installation of a vegetated roof with native and adapted plants would help the project achieve which credits? (Choose three)
   a. SS Credit 3, Brownfield Redevelopment
   b. SS Credits 5.1 and 5.2, Site Development*
   c. SS Credits 6.1 and 6.2, Stormwater Design*
   d. SS Credit 7.1, Heat Island Effect
   e. SS Credit 7.2, Heat Island Effect*

10. Your team is designing an office building with 200 full-time staff, 200 half-time staff, and 20 peak-use visitors. There will be a parking lot with 100 spaces. You determine that you could meet the requirements for SS Credit 4.3, Alternative Transportation, by doing the following: (Choose three)

    a. Providing ZEV-classified vehicles for nine occupants, and providing them with nine designated spaces close to the main entrance*

    b. Providing five designated parking spaces for low-emitting or fuel-efficient vehicles*

    c. Providing three plug-in stations for electric vehicles anywhere in the parking lot*

    d. Ensuring that the company's fleet of vehicles achieves a score of 40 or more in the ACEEE guide

    e. Providing ZEV-classified vehicles for five occupants, and providing them with five designated spaces close to the main entrance

    f. Provide 21 designated parking spaces for low-emitting or fuel-efficient vehicles

11. What is a potential advantage of purchasing an office building located on a Brownfield site?

    a. Lower property costs*

    b. Treating contaminants on-site

    c. Limited liability

    d. Using solar detoxification technologies

    e. No previous development

12. Which of the following can help a project achieve SS Credit 7.2 Heat Island Reduction: Roof?

    a. 100% ENERGY STAR roof

    b. Low-sloped roof covered in SRI 29 material for 95% of the roof surface

    c. Low-sloped roof covered in SRI 29 material for 70% of the roof surface

    d. 100% green roof*

13. A commercial office project requires 500 parking spaces. There are 1000 available spaces for the project. What site feature would most qualify for SS Credit 7.1 Heat Island Effect: Non-Roof?

    a. Above-ground parking that uses natural asphalt

    b. Reduce the parking to 250 spaces

    c. An underground parking deck for 250 parking spaces*

    d. 250 of the parking spaces are shaded

14. What is the maximum distance a building's entrance needs to be from a rail or bus line to qualify for SS Credit 4.1 Public Transportation Access?

    a. 1/4 mile walking distance from one bus line

    b. 1/4 mile walking distance from two or more separate bus lines*

    c. 1/2 mile walking distance from one rail line*

    d. 1/2 mile walking distance from two or more separate rail lines

    e. 1/2 mile walking distance from two or more separate bus lines

15. Which of the following best defines light pollution? (Choose one)

    a. Glare from sunlight reflection off building windows in urban areas

    b. Pollution of an ecosystem with light, limiting visibility of night sky views*

    c. Pollution caused by chemicals and gasses given off by artificial lighting

    d. Pollution by light-weight materials carried by wind

1.2 Water Efficiency

1. When targeting WE2 Innovative wastewater technologies, which of the following strategies can be used to treat 50% of wastewater on site to tertiary standards?

    a. Packaged biological nutrient systems*

    b. Constructed wetlands*

    c. Biological treatment

    d. High efficiency filtration systems*

2. According to LEED, Potable Water is defined as:

    a. Water from kitchen sinks.

    b. Water from irrigation systems.

    c. Untreated household wastewater that has not come into contact with toilet waste.

    d. Water suitable for dinking and supplied from well or municipal water systems*

3. What standard mandated the use of water conserving-plumbing fixtures to reduce water usage?

    a. ASHRAE/IESNA 90.1

    b. Montreal Protocol

    c. Energy Policy Act of 1992*

    d. 2003 EPA Construction General Permit

4. You are required to provide what data for compliance with WE Credit 3, Water Use Reduction? (Choose three)
   a. Dishwasher flow rates
   b. Number of fixtures
   c. Lavatory flow rates*
   d. Fixture model and manufacturer*
   e. Default baseline reduction of 20%
   f. Male-to-female ratio, in special occupancy situations*

5. A residential high-rise building uses all of the following strategies or fixtures to reduce potable water use. Which do not contribute to compliance with WE Credit 3.1, Water Use Reduction? (Choose three)
   a. Efficient horizontal-axis clothes washers*
   b. Waterless urinals
   c. Efficient janitors sink
   d. Efficient hand wash fountains in the gym
   e. Reduction of water used during cooling tower blowdown*
   f. Use of rainwater for toilet flushing
   g. Efficient lavatory sinks
   h. Efficient dishwasher in the commercial kitchen*

6. The project team on a renovation project wants to reuse some of the water fixtures. To earn WE Credit 3 Water Use Reduction, the existing faucets would need a maximum gpm rate of _____ and the water closets would need a maximum gpf rate of _____.
   a. 1.6, 1.0
   b. 2.2, 1.6*
   c. 2.5, 1.0
   d. 1.0, 1.6
   e. 2.2, 2.2

7. Graywater is most often used for:
   a. Dishwashers.
   b. Drinking.
   c. Irrigation.*
   d. Swimming pools.

8. By the US Green Building Council estimates, how many more billions of gallons of water do Americans extract than they return to the natural water system each year? (Choose one)
   a. 2100
   b. 4600

    c. 3700*

    d. 3300

9. The highest form of wastewater treatment, which removes of organics, solids, and nutrients in addition to providing biological or chemical polishing, is which of the following? (Choose one)

    a. Underground effluence treatment

    b. Potable system filtering

    c. Biological treatment

    d. Tertiary treatment*

1.3 Materials and Resources

1. Select four of the following materials below that contribute to the rapidly renewable content calculation under MR Credit 6, Rapidly Renewable Content? (Choose four)

    a. Straw-plastic composite decking*

    b. Cotton-insulated ductwork

    c. Landscaping plants with a less-than–10-year life cycle

    d. Cork flooring*

    e. Linoleum*

    f. Composite wood panels

    g. Bamboo cabinets*

    h. Oak plywood flooring

2. Construction waste is estimated to make up ____% of the total solid waste stream in the United States.

    a. 50

    b. 40*

    c. 25

    d. 20

3. A project team for a 18,000-ft.^{2} commercial building in a suburban office park wants to earn as many anticipated credits as possible during the LEED design submittal. At the same time, it also wants to coordinate the entire construction process as much as possible in advance to help ensure compliance with credits that can only be submitted for the construction submittal. In which area of LEED does the latter concern predominate? (Choose one)

    a. Sustainable Sites

    b. Water Efficiency

    c. Energy and Atmosphere

    d. Materials and Resources*

    e. Environmental Quality

4. When pursuing MR1.1 Building Reuse, 55% (based on surface area) of the building structure must be maintained. Which three items can they include?

    a. Exterior skin and framing systems*

    b. Roof decking*

    c. Window assemblies

    d. Structural floor*

5. Requirements of MR Prerequisite 1, Storage and Collection of Recyclables, include which of the following? (Choose two)

    a. Collection of organic waste

    b. Specific sizing of recycling collection areas

    c. Recycling areas should be within 200 yard of entrances

    d. Collection of corrugated cardboard and paper*

    e. Collection of metal*

6. You are renovating and adding to a 15,000-ft.^{2} historic building and seeking LEED certification for the resultant 50,000-ft.^{2} building. You will reuse 95% of the existing building structure. Many of the floors, trim, and doors in the existing building are not suitable for its new use, but will be used in the addition, defraying 6% of the total materials cost. Which of the following credits would apply? (Choose one)

    a. MR Credits 1.1, 1.2, and 1.3, Building Reuse

    b. MR Credits 1.2 and 1.3, Building Reuse

    c. MR Credit 1.1, Building Reuse*

    d. MR Credit 1.3, Building Reuse

    e. None of the above

7. You are working on a project and the owner decided to pursue LEED certification after construction has already started. You are trying to document MR Credit 4, Recycled Content, and have asked a representative of your supplier about the recycled content in the steel-framing members. She tells you that all of her firm's suppliers attempt to provide steel with 80% postconsumer recycled content or better. Using this information, you may count what percentage of recycled content in the steel leads toward the credit?

    a. 0%

    b. 5%

    c. 25%*

d. 40%

e. 80%

8. A renovation project is attempting MR Credit 3, Materials Reuse. The materials used to achieve this credit may also be applied to which other credit?

   a. MR Credit 5, Regional Materials*

   b. MR Credit 2, Construction Waste Management

   c. MR Credit 4, Recycled Content

   d. MR Credit 3, Materials Reuse

   e. MR Credit 1, Building Reuse

9. In order to receive a credit point for MR credit 3.2 – Material Reuse, what must be achieved?

   a. Use salvaged, refurbished, or reused materials for an additional 10% beyond MR Credit 3.1 (15% total, based on cost).

   b. Use salvaged, refurbished, or reused materials for an additional 3% beyond MR Credit 3.1 (10% total, based on cost).

   c. Use salvaged, refurbished, or reused materials for an additional 5% beyond MR Credit 3.1 (10% total, based on cost).*

   d. Use salvaged, refurbished, or reused materials for an additional 8% beyond MR Credit 3.1 (10% total, based on cost).

10. MR Credit 1.1, Building Reuse, stipulates that a minimum surface area (Walls, Roofs, and Floors) must be maintained. This total percentage is:

    a. 25%

    b. 50%

    c. 75%*

    d. 100%

11. A project team has included regionally purchased new plumbing products in the project plan to achieve MR Credit 5.1 Regional Materials. The plumbing products will need to be included in which other credit?

    a. MR Credit 4 Recycled Content*

    b. MR Credit 1.2 Building Reuse

    c. MR Credit 2.1 Construction Waste Management

    d. MR Credit 3.1 Resource Reuse

1.4 Energy and Atmosphere

1. Which four commissioning activities should, at a minimum, be completed of the following systems in order to fulfill EAp1, Fundamental Commissioning of the Building Energy Systems? (Choose four)

    a. Renewable energy systems*
    b. CO_2 monitoring controls
    c. Lighting and daylighting controls*
    d. Domestic hot water systems*
    e. Building envelope
    f. HVAC&R systems*

2. How many points would a building that engages in renewable energy contracts for 80% of its electricity receive based on EA6 Green Power?
    a. 1
    b. 2*
    c. 3
    d. None

3. In EA1 – Optimize Energy Performance, Process Energy is considered to include but is not limited to which of the following items? (Choose three)
    a. Office equipment*
    b. Parking garage ventilation
    c. Laundry washing*
    d. Elevators*
    e. Lights for the building grounds

4. Which of the following are requirements for EA Prerequisite 1, Fundamental Commissioning of the Building Energy Systems? (Choose three)
    a. Review contractor submittals applicable to systems being commissioned
    b. Develop Basis of Design*
    c. Incorporate commissioning requirements into the construction documents*
    d. Review building operation within 10 months after substantial Commissioning
    e. Verify the installation and performance of commissioned systems*

5. Using rooftop photovoltaic panels supplying 17.5% of a building's energy needs would help a project comply with EA Credit 2, On-Site Renewable Energy, and which other credit?
    a. EA Credit 6, Green Power
    b. EA Credit 1, Optimize Energy Performance*
    c. ID Credit 1, Innovation in Design, for exemplary performance under EA Credit 2
    d. SS Credit 7, Heat Island Effect

6. A manufacturing facility in Colorado is seeking LEED certification and installs backup generators powered by corn-based biodiesel, with capacity to provide 50% of the facility's electricity needs. According to EPA's Emissions and Generation Resource Integrated Database, an average of 60% of the facility's utility electricity comes from hydroelectric generation. How many points in EA Credit 2, On-Site Renewable Energy, is the facility eligible for?

   a. 0*
   b. 1
   c. 2
   d. 3

7. An owner of a 20,000-ft.^{2} office building wants to earn as many points as possible under EA Credit 1, Optimize Energy Performance, but cannot afford an energy modeling process. A compliance path using which standard would be most appropriate for this situation?

   a. ASHRAE 90.1-2007
   b. Advanced Buildings Core Performance Guide*
   c. Building Performance Rating Method
   d. Advanced Energy Design Guide

8. Achieving EA Credit 4, Enhanced Refrigerant Management, requires that the project team calculate what? (Choose two)

   a. Ozone-depletion potential of CFC refrigerants used
   b. LCODP and LCGWP*
   c. Ozone-depletion potential of Halons used
   d. Pounds of refrigerant and tons of cooling required by the project*
   e. Refrigerant atmospheric impact

9. Which item that must be performed by an independent commissioning agent to receive the credit for Enhanced commissioning (EA 3):

   a. Perform one commissioning design review of the project requirements, basis of design and design documents prior to 50% construction document phase.
   b. Review of contractor submittals applicable to the systems being commissioned.*
   c. Develop a systems manual that provides the necessary information to operate the commissioned systems.
   d. Verify that the requirements for training operating personnel and building occupants are completed.
   e. Review building operation within 10 months of substantial completion and provide a plan for resolving and commissioning related issues.

10. Which of the following systems are not eligible for EA Credit 2 –
    On-site Renewable Energy? (Choose two)
    a. Geo-exchange System*
    b. Geothermal Energy Systems
    c. Architectural Features such as passive solar and daylighting
       strategies*
    d. Solar Thermal Systems
11. To earn points for EA Credit 5: Measurement and Verification, the
    M&V period must cover a length of time not less than:
    a. 6 months following construction completion
    b. 1 year of postconstruction occupancy*
    c. 18 months following construction completion
    d. 2 years of postconstruction occupancy
12. EA Credit 4, Enhanced Refrigerant Management, sets a maximum
    threshold for what environmental impact of the built environment?
    (Choose one)
    a. Ozone depletion
    b. The combination of chlorofluorocarbons and ozone depletion
    c. The combination of ozone depletion and global warming potential*
    d. Global warming potential
13. For the purposes of EA Credit 2, On-Site Renewable Energy, which
    of the following are considered on-site renewable energy systems?
    (Choose three)
    a. Renewable energy credits
    b. Wave and tidal power systems*
    c. Photovoltaic systems*
    d. Geothermal heating systems*
    e. Ground source heat pumps
    f. Passive solar features
14. Buildings in the United States consume roughly what percentage of
    energy annually?
    a. 52%
    b. 60%
    c. 37%*
    d. 25%
15. Which of the following refrigerants has the lowest ozone depletion
    potential (ODP)?
    a. HCFC
    b. CFC

    c. HFC*

    d. Halon

16. Which strategy can earn EA Credit 1.1 Optimize Energy Performance: Lighting Power?

    a. Comply with ASHRAE 90.1-2007

    b. Reduce lighting power density to 16% below the standard*

    c. Provide lighting controls to at least 90% of occupants

    d. For at least 90% of regularly occupied areas, achieve a minimum daylight factor of 2% and provide glare control

1.5 Indoor Environmental Quality

1. To meet the requirements of EQ4.3 – Low emitting materials – Carpet systems, the carpet will need to meet the testing and product requirements of:

    a. Carbon Trust Good Practice Guide

    b. Green Label Plus Program*

    c. South Coast Air Quality Management District Rule 1168

    d. Green Seal Standard GS11

2. Which of the following items would not be required for a project to achieve EQ3.1 Construction IAQ Management Plan?

    a. An HVAC subcontractor on a high-rise building ensures that all return duct is covered with plastic so not to allow contaminants to enter the duct system.

    b. If HVAC equipment is used prior to startup, the same subcontractor ensures that MERV8 filters are installed on all return ducts.

    c. Construction processes that create large amounts of dust including sanding drywall or sawing wood must occur only on days when the ventilation system is not running.*

    d. During construction, the subcontractor ensures that they meet or exceed the control measures of SMACNA IAQ Guidelines for Occupied Building under Construction, 1995, Chapter 3.

3. Which of the following two actions taken during construction of a 3-story office building would allow the project to achieve EQ Prereq2 Environmental Tobacco Smoke Control?

    a. Interior smoking areas should be exhausted directly out the building*

    b. Smoking will not be allowed except in designated smoking areas*

    c. Posting a "No Smoking" sign on the exterior of the building

    d. Exterior smoking areas are located 10 ft. from the building

4. To meet EQ Credit 8.1 Daylight and Views, the calculation option requires that you achieve a minimum glazing factor of ___ in a minimum of ___ of regularly occupied spaces. The measurement option requires demonstrating illumination of ___ footcandles throughout ___ of regularly occupied spaces.
   a. 5%, 75%, 20, 75%
   b. 5%, 90%, 20, 90%
   c. 2%, 75%, 25, 75%*
   d. 2%, 75%, 25, 90%
   e. 2%, 75%, 20, 80%

5. Products certified under which two programs can help a project earn EQ Credit 4.3, Low-Emitting Materials? (Choose two)
   a. GreenGuard
   b. GreenSpec
   c. Green Label Plus*
   d. Green Seal
   e. FloorScore*

6. Which of the following credits requires submission of photos demonstrating implementation of requirements? (Choose one)
   a. EQ Credit 8.2, Daylight and Views
   b. SS Prerequisite 1, Construction Activity Pollution Prevention
   c. SS Credit 5.1, Site Development
   d. EQ Credit 3.1, Construction IAQ Management Plan*
   e. SS Credit 1, Site Selection

7. Which standard does EQ Credit 1, Outdoor Air Delivery Monitoring, reference?
   a. ASHRAE 90.1-2007
   b. ASHRAE 55-2004
   c. ASHRAE 62.1-2004*
   d. CIBSE Applications Manual 10: 2005

8. You are an advisor on a LEED application for a multistory office building in Virginia. Internally generated cooling loads dominate design considerations. Which credit should the project focus on to reduce the cooling load? (Choose one)
   a. EQ Credit 8.1, Daylighting and Views*
   b. WE Credit 1, Water-Efficient Landscaping
   c. EA Credit 4, Enhanced Refrigerant Management
   d. EQ Credit 1, Outdoor Air Delivery Monitoring

9. On a project seeking LEED certification, a subcontractor insists on assembling the doors using adhesives that do not comply with SCAQMD Rule #1168. What is the best approach relative to EQ Credit 4.1, Low-Emitting Materials?
   a. Assemble the doors in a ventilated area of the building using MERV 8 filters
   b. Complying instead with GS-11
   c. To assemble the doors off-site*
   d. Not attempting compliance with the credit

10. When calculating the glazing factor of a design to verify compliance with EQ Credit 8.1, Daylight and Views, which of the following should be taken into consideration in your calculation? (Choose four)
    a. Window areas*
    b. Effect of light shelves
    c. Square footage of all regularly occupied spaces*
    d. Visible transmittance of windows*
    e. Effect of glare control
    f. Geometry factor by glazing type*

11. According to EQ 7.2, Thermal Comfort, corrective action must be taken if the percentage of occupants that are dissatisfied with the thermal comfort level exceeds:
    a. 10%
    b. 15%
    c. 20%*
    d. 25%
    e. 30%

12. In using the "Flush-Out" method (Option 1) to earn EQ 3.2: Construction IAQ Management Plan (before occupancy), what is the total volume of outdoor air that must be supplied to the building before occupancy?
    a. 7,000 cubic feet per square foot of floor area
    b. 10,000 cubic feet per cubic foot of interior space
    c. 14,000 cubic feet per cubic foot of interior space
    d. 14,000 cubic feet per square foot of floor area*

13. What is the minimum distance a designated smoking area must be away from any outdoor air intakes to achieve EQ Prerequisite 2 ETS Control?
    a. 15 ft.
    b. 20 ft.

    c. 25 ft.*

    d. 30 ft.

    e. 50 ft.

14. Which standard sets VOC limits for commercial interior nonflat paints?

    a. ASHRAE Standard 62.1-2004

    b. SCAQMD Rule 1113

    c. GC-03

    d. GC-11

    e. GS-11*

    f. GS-03

15. A tenant wants to minimize energy use on their project. What is the best strategy to likely achieve this?

    a. Increase ventilation

    b. Signing a 2-year contract to purchase green power

    c. Task lighting with occupancy sensors*

    d. Purchasing tradable renewable certificates

    e. Daylighting 90% of spaces

1.6 Innovative Design and Miscellaneous

1. Name one aspect of the LEED Accredited Professional's role on a LEED project as defined in the intent of ID Credit 2, LEED Accredited Professional?

    a. To ensure the integrity of the LEED process

    b. To efficiently provide interpretations of LEED requirements when needed

    c. To support an integrated design process*

    d. To advocate for the best environmental choices possible on a project

2. Using public transportation and promised access to a municipal parking garage, a multiuse project is able to reduce its parking capacity by 55% compared with the original design, while still offering 10% more parking spaces than required by city ordinances. Through the implementation of multiple transport and parking options, the project is able to demonstrate a quantitative reduction in automobile use. Which credit does the project comply with? (Choose one)

    a. SS Credit 4.4 Alternative Transportation

    b. ID Credit 1, Innovation in Design*

    c. SS Credit 7.1, Heat Island Effect: Non-Roof

    d. SS Credit 7.2, Heat Island Effect: Roof

    e. SS Credit 2, Development Density

3. Which of the following are valid criteria for achieving an innovation credit in a category not otherwise specifically addressed by LEED? (Choose three)
   a. The innovation includes an educational aspect
   b. A point was awarded for another project for the same innovation
   c. The project demonstrates quantitative performance improvements for environmental benefit*
   d. The process or specification is comprehensive to the project being certified*
   e. The formula developed for the credit is applicable to other projects*
4. Which of the following are determining factors in a project's cost for LEED certification? (Choose two)
   a. USGBC membership, project size, and project type (office, mixed use, education, government)
   b. LEED rating system (LEED-NC, LEED-EB, etc.) and project type
   c. Project size and LEED rating system*
   d. USGBC membership and certification level being sought
   e. Certification level earned*
5. For which credits do you need to calculate FTE building occupants? (Choose three)
   a. EQ Credit 8, Daylight and Views
   b. WE Credit 3, Water Use Reduction*
   c. EQ Credit 2, Increased Ventilation
   d. SS Credit 4, Alternative Transportation*
   e. EA Credit 1, Optimize Energy Performance
   f. WE Credit 2, Innovative Wastewater Technologies*
6. Which are conditions for submitting an appeal to LEED? (Choose three)
   a. After a credit has been denied in the Preliminary LEED Review
   b. After a credit has been denied in the Final LEED Review*
   c. Upon paying $500 to USGBC per credit appealed*
   d. Within 60 business days of denial
   e. Upon submitting a project narrative and three project highlights*
7. When can a project be referred to as LEED certified?
   a. When a project has been registered with LEED
   b. When the Final LEED Review is delivered
   c. When the project team accepts the Final LEED Review*
   d. When the LEED certificate and plaque are delivered

8. Which of the following should be included with a project's LEED application? (Choose three)
   a. Building measurements with metric conversions
   b. A check or credit card payment to the U.S. Green Building Council*
   c. A list of all credits approved through CIRs
   d. Project narrative including three highlights*
   e. Photos or renderings of the project*

9. Before submitting a CIR, a project team should first do all of the following except:
   a. Consult the LEED *Reference Guide*
   b. Self-evaluate whether the project meets the credit intent
   c. Prior to submitting a new CIR, see if there is an existing one very similar to your situation
   d. Prepare a full project narrative for submittal with the CIR*

10. An existing 3-story building undergoing major renovations is most likely a candidate for certification under which LEED rating system?
   a. LEED for Existing Buildings
   b. LEED for New Construction*
   c. LEED for Core and Shell
   d. LEED for Commercial Interiors

11. Which of the following are requirements for submittal documentation for ID Credit 2, LEED Accredited Professional? (Choose two)
   a. List number of LEED APs on the project
   b. Provide a copy of the LEED AP certificate*
   c. Provide list of project experience during last 2 years, including LEED certification levels achieved or pending
   d. Provide documentation with the design submittal
   e. Give the name of the LEED AP's company*

12. Environmental Building News is best described as a:
   a. Monthly periodical on new EPA regulations*
   b. Monthly periodical on sustainable design and construction
   c. Monthly periodical produced by Building Green
   d. Monthly periodical produced by USGBC and GBCI

13. Designers of a new manufacturing facility seek to earn an Innovation in Design credit in the LEED NC Rating system by complying with building industry acoustical standards. Which of the following organizations provide these standards?

a. American Society of Heating, Refrigeration, and Air-Conditioning Engineers

b. Architectural Nation Standards Institute

c. American National Standards Institute*

d. Environmental Protection Agency

14. What are the typical costs for registering and certifying a building (for LEED)? The correct response is that these costs are based on:

a. Location and lot size

b. Building square footage*

c. Project type

d. Type of construction

15. A high-rise building project has a $10 million materials budget. What estimated value of materials is needed to earn an Innovation in Design point for regional extracted materials?

a. $3.6 million

b. $1 million

c. $1.8 million

d. $2 million*

e. $3 million

f. $4 million

16. What factors impact the cost of building certification?

a. Number of credits applied for

b. Square footage*

c. Number of credits earned*

d. Having more than one LEED AP on the project

e. Number of CIRs submitted*

17. What does a LEED rating reflect?

a. The cost of a building

b. How green a building is*

c. The carbon footprint of a building's occupants

d. The location of a building

18. What is the purpose of Sustainable Sites Credit 2?

a. Reducing energy use in residences

b. Rehabilitation of Superfund sites

c. Encouraging development in areas that already have some infrastructure, thus protecting natural resources*

d. Ensuring that the project has a reasonable amount of parking spaces

19. Using the payback period to make a decision, which project is the most likely to be chosen?

   a. A project that costs $1 million and is expected to generate $15,000 per year.

   b. A project that costs $90 million and is expected to generate $9 million per year.

   c. A project that costs $6.5 million and is expected to generate $65,000 per year.

   d. A project that costs $30 million and is expected to generate $5 million per year.*

20. What rate of fresh air distribution per occupant should be enough to prevent "sick building syndrome"?

   a. 100 ppm

   b. 100–200 ft.^{3}/h

   c. 900–1200 ft.^{3}/h*

   d. 3000–4000 ft.^{3}/h

21. Which is the term for an evaluation of a product's environmental impact over the entire time of its use?

   a. Sustainability

   b. Durability

   c. Life-cycle assessment*

   d. Chain-of-custody

22. What is the goal of meeting ADA standards?

   a. Indoor air quality

   b. Accessibility of buildings to people with disabilities*

   c. Reduction of carbon emissions

   d. Durability of a structure

23. Which LEED rating system requires durability planning?

   a. LEED for Schools

   b. LEED for Commercial Interiors

   c. LEED for Homes*

   d. LEED for Existing Buildings: Operation and Maintenance

24. What is the name for the procedure used to clear buildings of contaminants before they are occupied?

   a. Flush–out*

   b. Infiltration

   c. Ventilation

   d. Exfiltration

1.6.1 LEED Green Associate Answers Key and Explanations to Questions 17–24

17. *The answer is B.* LEED stands for Leadership in Energy and Environmental Design, and is a system for rating how green a building is. Although LEED ratings can be affected by factors such as carbon footprints and how a building relates to the site, a LEED score takes many other elements of a building into account. A LEED rating can affect a building's value, but LEED scores do not reflect a building's cost.

18. *The answer is C.* Sustainable Sites Credit 2 is intended to encourage urban development in areas where infrastructure already exists, which in turn helps protect natural resources. This is usually defined as using a building site that has been previously developed, is within at least .5 miles of both a residential area and at least ten basic services, and offers pedestrian access.

19. *The answer is D.* The formula for the payback period is to divide the cost of the project by the annual revenue expected. It is usually only one of the methods used in selecting a project. Using this system for these four choices, Answer A will take about 66 years, answer B will take 10 years, answer C will take 100 years, and answer D will take 6 years. Therefore, D has the shortest payback period.

20. *The answer is C.* Indoor air quality is very important because poor indoor air quality and poor air circulation can lead people who work or live in the building to become sick from indoor pollutants and irritants: "sick building syndrome." The American Society of Heating, Refrigeration and Air Conditioning Engineers (ASHRAE) recommend that fresh air pumps at a rate between 900 and 1200 cubic feet per hour per occupant in order to keep occupants productive and healthy.

21. *The answer is C.* A life-cycle assessment provides an evaluation of a product's impact over the entire life-cycle of its use, including from its manufacturing to its eventual disposal. This kind of analysis offers a way to choose materials, which are most beneficial in both the long and short term. Sustainability and durability are important characteristics of green products, and chain-of-custody is a term that only applies to the life cycle of wood products.

22. *The answer is B.* ADA stands for Americans with Disabilities Act; the ADA Standards for Accessible Design are intended to lay out the structural requirements for access of buildings by people with disabilities. These guidelines apply during all phases of a building's life cycle and are published by the United States Department of Justice.

23. *The answer is C.* Durability is important because a durable structure lasts longer and prevents waste. This is of particular concern for homes. LEED for Homes rating system requires a durability assessment that includes potential concerns and plans for dealing with issues that can threaten the integrity of the structure.

24. *The answer is A.* After construction is completed, a building must be flushed with outside air – either via the HVAC system in place or with windows – before it may be occupied. This process is called flush-out. Infiltration and exfiltration both refer to unintended leakages of air into or out of buildings.

62. *The answer is D.* TVOC refers to total volatile organic compounds; a volatile organic compound (VOC) is a carbon-based chemical that can be carcinogenic. Some household products emit VOCs. The TVOC of a room or indoor space is the total amount of VOCs in the air of that space.

1.6.2 Two Sample Questions from the v4 Green Associate Exam (Courtesy USGBC)

1. When applying for innovation credits, a project team
   a. Cannot submit any previously awarded innovation credit.
   b. May receive credit for performance that doubles a credit requirement.*
   c. May submit a product or strategy that is being used in an existing LEED credit.
   d. May receive a credit for each LEED Accredited Professional that is on the project team.

This question represents Knowledge Domain A. LEED Process, credit categories and Task Domain A. LEED Green Associate Tasks A developer wants to make a profit by building a new office that maximizes daylighting and views.

2. A developer wants to make a profit by building a new office that maximizes daylighting and views. What actions might the developer take to fulfill all parts of the triple bottom line?
   a. Restore habitat onsite
   b. Purchase ergonomic furniture
   c. Pursue local grants and incentives
   d. Provide lighting controllability for occupants*

This question represents: Knowledge Domain I. Project Surroundings and Outreach, environmental impacts of the built environment and Task Domain A. LEED Green Associate Tasks, assist others with sustainability goals.

2 LEED AP WITH SPECIALTY SAMPLE EXAM QUESTIONS

Answers are followed by an asterisk.

1. Which of the following are possible consequences of lowering the lighting power density on a project? (Choose two)
   a. Reduce heating loads
   b. More controllability over indoor lighting
   c. Reduce cooling loads*
   d. Energy savings associated with less energy required for lighting*

2. Which refrigerant is environmentally preferable?
   a. CFC
   b. HCFC
   c. HFC
   d. All of the above*

3. Which of the following factors is NOT included in the calculations for indoor water use reduction? (Choose two)
   a. Whether or not lavatory sinks have automatic controls.
   b. The total number of water-efficient water closets and urinals in the building.*
   c. The total number of full-time equivalents.
   d. The flow rate of the showerheads.
   e. The average gallons per use (AGU) of dishwashers.*

4. When submitting Design phase credits, which of the following represent possible review responses issued by the USGBC? (Choose two)
   a. Accepted
   b. Denied*
   c. Anticipated*
   d. Not approved
   e. Earned
   f. Achieved

5. Which of the following standards relates specifically to carpets?
   a. Green Seal Standard 37
   b. South Coast Air Quality Management District, Rule 1113
   c. Green Label Plus Program*
   d. ASHRAE 55-2004
   e. Green Label Program

6. Which of the following strategies might be applied toward achieving construction waste management? (Choose two)

   a. Use materials with recycled content (RC) so the sum of postconsumer RC + 1/2 of the preconsumer RC is ≥10% (based on cost) of all materials.
   b. Crushing existing concrete foundations on-site to reuse as baserock.*
   c. Donating excavated soil to an adjacent Habitat for Humanity building undergoing construction.
   d. Diverting wood scraps from the landfill by selling them to a nearby facility that produces particleboard.*
   e. Diverting lead-based paint found on-site from the landfill by taking it to a hazardous materials facility.

7. A new 6-story apartment complex is being constructed in an urban setting. Because of budget concerns, the project team has chosen to replace bamboo cabinets with cabinets made from particleboard produced mainly from discarded wood chips. Based on the information above, which of the following credits would most likely be affected (either positively or negatively) by this decision? (Choose three)
   a. Certified wood
   b. Low emitting materials, composite wood, and agrifiber products*
   c. Building reuse, nonstructural elements
   d. Recycled content*
   e. Rapidly renewable materials*

8. Which of the following are covered under low-emitting materials with respect to LEED? (Choose three)
   a. Clear finishes on interior elements*
   b. Carpet*
   c. Insulation
   d. Medium-density fiberboard used on interior doors*
   e. Roofing adhesive

9. Which of the following represents xeriscaping?
   a. Planting trees that can provide food for the building occupants such as apple or orange trees.
   b. An advanced drip irrigation system that responds to water demand by sensing soil moisture content.
   c. A landscape design that directs the flow of rainwater on-site to the area that requires the most irrigation.
   d. A landscape design that requires little or no irrigation or maintenance.*
   e. Using artificial turf in place of real grass.

10. A 5-story multifamily residential building built in the 1950s is undergoing a major renovation. The existing air-conditioning equipment contains CFC-based refrigerants that are known to be harmful to the environment. In order to comply with Fundamental Refrigerant Management, what is the required timeline for phasing out CFCs in all air-conditioning equipment?
    a. CFCs must be phased out no later than 10 years after project completion.
    b. CFCs must be phased out prior to project completion.*
    c. CFCs must be phased out no later than 1 year after project completion.
    d. CFCs must be phased out before construction breaks ground.
    e. None of the above. There is no required timeline for phasing out CFCs in buildings built before 1973.

11. "Tertiary treatment" is a term that refers to which of the following?
    a. FSC-certified wood.
    b. Water-efficient landscaping.
    c. Wastewater*
    d. Construction indoor air quality management
    e. Low-emitting materials

12. A reduction in overall water quality due to an increase in the concentration of chemical nutrients would be an example of which of the following?
    a. Xeriscaping
    b. Eutrophication*
    c. Denitrification
    d. Osmosis
    e. Hypoalimentation

13. Which of the following account for the most energy usage in buildings according to the USGBC? (Choose two)
    a. Space heating*
    b. Cooling*
    c. Domestic hot water
    d. Lighting*
    e. Plug-in electronics such as computers, copiers, and fax machines.

14. Which of the following constitutes preconsumer recycled content?
    a. A tile facility that reuses shards of tile broken during the manufacturing process to make new tile.
    b. A steel manufacturing facility that uses recycled steel melted down from old cars.

    c. A steel manufacturing facility that recycles rebar from old construction projects to make steel beams.

    d. A curtain manufacturer that purchases scrap trimmings from a carpet manufacturing facility.*

    e. All of the above.

15. Which pieces of information are required to complete FSC-certified wood? (Choose three)

    a. The volume of wood used on the project and the exact location where the wood was harvested.*

    b. The appropriate Chain of Custody (COC) Certification number.*

    c. The name of the product manufacturer.*

    d. The cost of each wood product.*

    e. A list of paints and/or coatings used on each FSC-certified product.

16. Which of the following standards listed in the LEED *Reference Manual* relate to low emitting materials, paints and coatings? (Choose two)

    a. Green Seal Standard 11.*

    b. South Coast Air Quality Management District (SCAQMD) Rule 1113.*

    c. Green Label Plus Program

    d. Green Seal Standard 35*

    e. 2003 EPA Construction General Permit

17. Which of the following refrigerants generally has a lower global warming potential?

    a. PVCs

    b. HFCs

    c. HCFCs*

    d. DDT

18. Which of the following will help allow you to size smaller cooling systems? (Choose three)

    a. Larger R values*

    b. Larger U values

    c. Higher solar heat gain coefficient

    d. Lower lighting power density*

    e. Energy recovery systems on equipment*

19. Which of the following does *not* constitute "process energy" according to the LEED *Reference Manual*?

    a. Computers

    b. Miscellaneous office equipment

    c. Cooling towers*

    d. Elevators

    e. Kitchen refrigeration

20. Which of the following statements is TRUE regarding Credit Interpretation Requests? (Choose two)

    a. They must be no more than 400 words.*

    b. They may only reference one credit.*

    c. They may not include plans or drawings.*

    d. They must be submitted by the project owner.

    e. They must be no more than 6000 characters.

21. In general, what is the advantage of accomplishing the LEED goal-setting meeting early on in the project? (Choose two)

    a. To evaluate costs of pursuing various credits and strategies.*

    b. To begin testing Indoor Air Quality.

    c. To determine feasibility of design strategies and associated credits.*

    d. To earn an ID Credit for completing the Integrated Design Process.

    e. To receive discounted project registration fees.

22. Why would it be important to start working on recycled content materials, early in the process?

    a. To determine if recycled wood may contain phenol formaldehyde.

    b. To select mechanical equipment with a high recycled content.

    c. To minimize the ozone depletion potential of certain materials that contain high amount of recycled content.

    d. To submit recycled content materials during the design phase review.

    e. To be able to specify materials with recycled content during schematic design.*

23. Which of the following strategies would apply toward a reduction in potable water usage for water-efficient landscaping with respect to LEED? (Choose two)

    a. Use of municipally supplied, nonpotable water.*

    b. Elect to build the project in a location that has a sufficient amount of rainfall.

    c. Minimize the amount of overall landscaped area.

    d. Select plants with low microclimate factors.

    e. Use of a rainwater catchment system to irrigate landscaping.*

24. Which site would most likely qualify for Community Connectivity?

    a. A parking lot located in a dense urban area.*

    b. A suburban Brownfield site located within 1/4 mile of light rail public transportation.

    c. An empty lot located near several onramps to a large freeway intersection.

    d. A 7-story food processing facility that includes commercial office space located near the center of a 15-acre farm.

    e. A Greenfield site located within a short walking distance of many common amenities.

25. What is the relationship between Credit Interpretation Rulings and Innovation in Design credits?

    a. All ID Credits must be established via the CIR process.

    b. Some ID credit strategies have been established via the CIR process.*

    c. An ID credit is not guaranteed unless it has been established via the CIR process.

    d. Any product being applied to an ID credit, must first be submitted and approved via a CIR.

    e. The primary purpose of CIRs is to establish acceptable strategies for ID credits.

26. Renewable Energy Certificates contribute to which of the following?

    a. Offset for water usage.

    b. Reduced energy costs for the building.

    c. Increased building energy efficiency.

    d. A reduction in overall global emissions.*

    e. Utility-based incentive programs.

27. Which of the following is true regarding regional materials with respect to LEED?

    a. They must be manufactured locally.*

    b. They must be transported across no more than 1500 miles.

    c. They must be FSC-certified.

    d. They must not have traveled via plane or boat.

    e. They must be assembled onsite.

28. Which of the following water sources must be contained and treated before it may be used to irrigate landscaping? (Choose two)

    a. Approved graywater from lavatory sinks and showers where nontoxic soaps were used.

    b. Blackwater from urinals.*

    c. Rainwater collected on site.

    d. Cooling tower water.*

    e. Nonpotable water supplied by a public agency.

29. Which of the following is true regarding exemplary performance thresholds?

a. Points for exemplary performance are always counted under the category where the credit is listed.

b. All exemplary performance points should always be counted under the Innovation in Design category.*

c. All Innovation in Design points are achieved via an exemplary performance.

d. The exemplary performance thresholds are always double the base-level thresholds for achieving a credit.

30. Which of the following are true regarding costs? (Choose two)
   a. Certification fees are waived for Platinum projects.*
   b. Recertification fees are the same as initial certification fees for Existing buildings.
   c. Certification fees changed in 2002 and again in 2005.*
   d. Certification fees are based on project location.
   e. Certification fees for a LEED Silver project differ from those of a LEED Gold project.

31. Which of the following represents an example of heat island effect? (Choose two)
   a. Increased local temperatures caused by greenhouse effect from air pollution.
   b. Heat radiated from an urban area with impervious surfaces.*
   c. Incorporating water into building design to cool local temperature.
   d. Increased local temperatures due to highly reflective materials.
   e. Increased cooling loads during summer months due to dark-colored roof surfaces.*

32. Which would most likely help the project team earn points for Green Power?
   a. Solar panels on the roof.
   b. Wind towers located in the parking lot.
   c. Renewable Energy Certificates.*
   d. Geoexchange systems located under the parking lot.
   e. Geothermal systems located under the building.

33. A concrete building is completely deconstructed on-site and the concrete is crushed and reused as base rock for the new foundation. This strategy might apply to which credits? (Choose three)
   a. Building Reuse.
   b. Regional Materials.*
   c. Construction IAQ Management.

    d. Construction Waste Diversion.*

    e. Materials Reuse.*

34. Which of the following represents a benefit of using graywater for landscape irrigation? (Choose two)

    a. Less water sent to the wastewater treatment facility.*

    b. Reduces stormwater runoff onsite.

    c. Reduces soil erosion from vegetation areas.

    d. Reduced demand on potable water for irrigation.*

    e. Reduces potable water used for flush fixtures.

35. Which reduces outdoor air pollution the most? (Choose two)

    a. Eliminating CFCs from refrigerants.

    b. Incentivizing employees to carpool or take public transportation.*

    c. Installing CO_2 sensors throughout project space.

    d. Selecting a site within walking distance of amenities.*

    e. Implementing a thermal comfort survey.

36. In LEED v4, a green (vegetated) roof can contribute to which 3 credits? (Choose one)

    a. Heat Island Reduction, Optimize Energy Performance, Site Development – Protect or Restore Habitat*

    b. Optimize Energy Performance, Green Power and Carbon Offsets, Neighborhood Development*

    c. Heat Island Reduction, Optimize Energy Performance, Low Emitting Materials*

    d. Heat Island Reduction, Optimize Energy Performance, Green Power and Carbon Offsets

2.1 Two Sample Questions from the v4 LEED AP with Specialty Exam (Courtesy USGBC)

1. The city is building a new botanical garden and is attempting LEED® certification. What could the educational program include to earn an Innovation in Design Credit?

    a. Present the building's sustainable features at the grand opening

    b. Present the building's sustainable features at a town hall meeting

    c. Provide ongoing weekly tours highlighting the building's sustainable features*

    d. Publish a press release to the local newspaper outlining the building's sustainable features

This question represents: Knowledge Domain: LEED system synergies (e.g., energy and IEQ; waste management).

2. How should athletic fields be treated in the calculations for WE Credit, Outdoor Water Use Reduction?
   a. Must be calculated using 100% potable water
   b. May be included or excluded from the calculations*
   c. May be calculated using a standard 20% reduction from baseline
   d. Must be calculated using at least 20% from an alternative water source

This question represents: Knowledge Domain: Outdoor water use reduction: irrigation demand (e.g., landscape water requirement; irrigation system efficiency; native and adaptive species) and Task Domain(s): be a resource for LEED credit achievement (e.g., provide resources, training, tools, demonstrations of sample credits), manage LEED template(s)/certification process in LEED Online (e.g., review for completion), identify project-specific strategies, educate others (and self).

BIBLIOGRAPHY

Allen, E., Iano, J., 2013. Fundamentals of Building Construction: Materials and Methods, sixth ed. John Wiley, New York.

Andreasen, C., Lords, M., Miller, K.R., 2009. Contrasting bid shopping and project buyout. In: Proceedings of the Forty-Fifth Annual ASC Conference, Gainesville, Florida, April 1–4.

Apgar, M., 1998. The alternative workplace: changing where and how people work. Harvard Bus. Rev. May–June, 121–135.

Armer, G.S.T., 2001. Monitoring and Assessment of Structures. Spon Press, New York.

Arnold, C., 2006. Green movement sweeps U.S. construction industry. NPR July.

ASHRAE, 2008. Proposed Standard 189P: Standard for the Design of High-Performance Green Buildings, second public review. Atlanta, GA, July.

Askar, N., 2014. Waste Not, Want Not, Plumbing and Mechanical. February.

ASTM. Standard Guide for Property Condition Assessments: Baseline Property Condition Assessment, Designation: E2018-01.

Bady, S., 2008. Green building programs more about bias than science, expert argues. Professional Builder.

Bennett, F.L., 2003. The Management of Construction: A Project Life Cycle Approach. Butterworth-Heinemann.

Bezdek, R.H., 2008. In: Green Building: Balancing Fact and Fiction, Cannon, S.E. and Vyas, U.K., Moderators. Real Estate Issues 33 (2), 2–5.

Bonda, P., Sosnowchik, K., 2007. Sustainable Commercial Interiors. John Wiley & Sons, Hoboken, NJ.

Brundtland Commission, 1987. The Brundtland Report: Our Common Future. U.N. World Commission on Environment and Development.

Build it Green, 2007. Build It Green, New Home Construction – Green Building Guidelines, 2007 Edition.

Burr, A.C., 2008. CoStar Green Report: The Big Skodowski. CoStar Group, January 30.

Butters, F., 2008. In: Green Building: Balancing Fact and Fiction, Cannon, S.E., and Vyas, U.K., moderators. Real Estate Issues 33 (2), 10–11.

Calow, P., 1998. Handbook of Environmental Risk Assessment and Management. Blackwell Science Ltd, Oxford.

Ching, F.D.K., Shapiro, I.M., 2014. Green Building Illustrated. John Wiley & Sons, Hoboken, NJ.

Coggan, D.A., 2009. Intelligent Building Systems (Intelligent Buildings Simply Explained). Available from: http://www.coggan.com/intelligent-building-systems.html (accessed 06.01.2009).

Craiger, P., Shenoi, S. (Eds.), 2007. Advances in Digital Forensics III (IFIP International Federation for Information Processing). Springer, New York.

De Chiara, J., Panero, J., 2001. Time-Saver Standards for Interior Design and Space Planning. McGraw-Hill, New York.

Deasy, C.M., 1985. Designing Places for People: A Handbook for Architects, Designers, and Facility Managers. Whitney Library of Design, New York.

Del Percio, S., 2007. What's wrong with LEED™? American City, Issue 14, Green Building, Spring 2007. Available from: http://www.americancity.org/ (accessed April 2007).

DiLouie, C., 2006. Why do daylight harvesting projects succeed or fail?, Lighting Controls Association, March.

Dorgan, C., Cox, R., Dorgan, C. The value of the commissioning process: costs and benefits. Farnsworth Group, Madison WI, Paper presented at the 2002 US Green Building Council Conference, Austin, Texas.

Dunlap, R.E., Mertig, A.G., 1992. American Environmentalism: The U.S. Environmental Movement, 1970–1990. Taylor and Francis, Philadelphia.

Dworkin, J.F., 1990. Waterproofing below grade. The Construction Specifier, March.

Edwards, S., 1986. Office Systems – Designs for the Contemporary Workspace. PBC International Inc, New York.

EIA, 2008. Assumptions to the Annual Energy Outlook, Energy Information Administration, 2008; cited in Green Building Facts, U.S. Green Building Council, November 2008.

Federal Emergency Management Agency, 2004. FEMA 426, Reference Manual to Mitigate Potential Terrorist Attacks in High Occupancy Buildings.

Fisk, W.J., Rosenfeld, A.H., 1997. Estimates of improved productivity and health from better indoor environments. Indoor Air 7, 158–172.

Fisk, W.J., Rosenfeld, A.H., 1998. The indoor environment – productivity and health – and $$$. Strategic Planning Energy Environ. 17 (4), 53–57.

Friedman, A., 2015. Fundamentals of Sustainable Neighbourhoods. Springer, New York.

Gavin, J., 2015. Construction outlook: an economic recovery finds its footing. Electrical Contractor Magazine, January.

Global Green Building Trends, 2008. SmartMarket Report. McGraw-Hill Construction, New York.

Goldsmith, S., 1999. Designing for the Disabled – The New Paradigm. Architectural Press, An imprint of Butterworth-Heinemann, Oxford.

Gore, A., 2006. The future is green. Vanity Fair, Special Green Issue, May.

Gottfried, D., 2000. Sustainable building technical manual, U.S. Green Building Council. U.S. Department of Energy, U.S. Environmental Protection Agency, Washington, DC.

Graham, Jr., O.L. (Ed.), 2000. Environmental Politics and Policy, 1960s–1990s. Pennsylvania State University Press, University Park.

Green Building Certification Institute, 2009a. LEED® Green Associate Candidate Handbook. Green Building Certification Institute.

Green Building Certification Institute, 2009b. LEED® Professional Accreditation Handbook. Green Building Certification Institute.

GSA, 2004. LEED Cost Study – Final Report, Submitted to the U.S. General Services Administration. Steven Winter Associates, Inc., October.

Gunn, R.A., Burroughs, M.S., 1996. Work spaces that work: designing high-performance offices. The Futurist, March–April, pp. 19–24.

Haasl, T.P., Claridge, D., 2003. The cost effectiveness of commissioning. HPAC Engineering October, 20–24.

Hall, P., 2014. Green building accounts for nearly 1/4 of U.S. CRE construction. National Mortgage Professional Magazine, October 30.

Harmon, S.K., Kennon, K.E., 2001. The Codes Guidebook for Interiors, second ed. John Wiley & Sons, New York.

Hellier, C.J., 2001. Handbook of Nondestructive Evaluation. McGraw-Hill, New York.

Hess-Kosa, K., 1997. Environmental Site Assessment Phase I, second ed. CRC Press, Boca Raton, FL.

Horman, M.J., Riley, D.R., Pulaski, M.H., Magent, C., Dahl, P., Lapinski, A.R., Korkmaz, S., Luo, L., Harding, N.G., 2006. Delivering green buildings: process improvements for sustainable construction. J. Green Building 1 (1), 123–140.

Hosey, L., 2013. Why architects must lead on sustainable design. GreenBiz, March.

Houghton-Evans, R.W., 2005. Well Built?: A Forensic Approach to the Prevention, Diagnosis and Cure of Building Defects. RIBA Enterprises.

Howe, J.C., 2010. Overview of green buildings. National Wetlands Newsletter 33 (1).

International Institute for Sustainable Development (IISD), 2013. Telling Our Stories, Writing Our Future. Annual report, 2013–2014.

Kats, G., 2007. The costs and benefits of green. Capital E Analytics.

Keeler, M., Burke, B., 2009. Fundamentals of Integrated Design for Sustainable Building. John Wiley & Sons, New York.

Kennett, S., 2008. Making BREEAM robust, Building. Available from: http://www.building.co.uk/story (accessed March 2008).

Kibert, C.J., 2012. Sustainable Construction – Green Building Design and Delivery, third ed. John Wiley, Hoboken, NJ.

Kriss, J., 2014. USGBC Professional Exams Evolving with Launch of LEED v4, vol. 17. BNP Media.

Kubal, M.T., 2000. Construction Waterproofing Handbook. McGraw-Hill, New York.

Kubba, S.A.A., 2003. Space Planning for Commercial & Residential Interiors. McGraw-Hill, New York.

Kubba, S.A.A., 2007. Property Condition Assessments. McGraw-Hill, New York.

Kubba, S.A.A., 2008. Architectural Forensics. McGraw-Hill, New York.

Kubba, S., 2010. Green Construction, Project Management and Cost Oversight. Elsevier, Philadelphia, PA.

Kubba, S., 2012. Handbook of Green Building Design and Construction: LEED®, BREEAM® and GREEN GLOBES®. Elsevier, Philadelphia, PA.

Langdon, D., 2004. Costing Green: A Comprehensive Cost Database and Budgeting Methodology.

LEED®, 2008. LEED™ – NC Green Building Rating System for New Construction & Major Renovations, Version 2.2, USGBC. Available from: www.usgbc.org. (accessed August 2008).

LEED, 2014a. LEED™ v4 for Building Operations and Maintenance, (USGBC). Updated October 1, 2014.

LEED, 2014b. LEED™ v4 Rating System Selection Guidance (posted in LEED), July 29, 2014.

Levy, M., Salvadori, M.G., Woest, K. (Eds.), 1994. Why Buildings Fall Down: How Structures Fail. W.W. Norton & Co Inc, New York.

Lewis, B.T., Payant, R., 2001. Facility Inspection Field Manual: A Complete Condition Assessment Guide. McGraw-Hill, New York.

Lippiatt, B.C., Norris, G.A., 1998. Selecting environmentally and economically balanced building materials. Sponsored by: National Institute of Standards and Technology.

Loveland, J., 2003. Daylight by design. LD + A magazine, October.

Luhmann, T., 2007. Close Range Photogrammetry: Principles, Techniques and Applications. John Wiley, Hoboken, NJ.

Lupton, M., Croly, C., 2004. Designing for daylight. Building Sustainable Design magazine, February.

Lusk, J., Kaplow, S., 2015. Green code mandatory on April 1 for all building in Baltimore. Green Building Law Update, March 17.

Macaluso, J., 2009. An Overview of the LEED™ Rating System, Empire State Development. Green Construction Data, May.

Macdonald, S. (Ed.), 2002. Concrete Building Pathology. Blackwell Publishing Ltd, Oxford.

Madsen, J.J., 2008. The Realization of Intelligent Buildings. Buildings magazine, March.

Mago, S., Syal, M., 2007. Impact of LEED™–NC projects on construction management practices. MS thesis, Michigan State University.

Markovitz, M., 2008. The differences between Green Globes and LEED. ProSales Magazine, August 27.

Martín, C., Foss, A., 2006. All that glitters isn't green and other thoughts on sustainable design for the PATH partners. National Building Museum, Fall.

May, S., 2009. Do green design strategies really cost more? DC&D Technologies, Inc., DCD Construction Magazine.

Mendler, S., Odell, W., 2000. The HOK Guidebook to Sustainable Design. John Wiley & Sons, New York.

Moloney, C., 2014. LEED v4 changes: New Credit categories? Poplar Network, June 26.

Montoya, M., January 2010. Green Building Fundamentals, second ed. Prentice Hall.

Morris, P., Langdon, D., 2009. 2009 Market Update – A Guide to Working in a Recession, January.

Morris, P., Matthiessen, L.F., Langdon, D., 2007. Cost of Green Revisited, July.

Muldavin, S., 2007. A strategic response to sustainable property investing. PREA Quarterly, Summer, pp. 33–37. Available from: www.muldavin.com.

Murphy, B.L., Morrison, R.D., 2007. Introduction to Environmental Forensics, second ed. Academic Press.

Nadel, B.A., 2004. Building Security – Handbook for Architectural Planning and Design. McGraw-Hill, New York.

National Institute of Building Sciences, The Whole Building Design Guide. Available from: www.wbdg.org/sustainable.php.

Needy, K.L., Ries, R., Gokhan, N.M., Bilec, M., Rettura. B., 2004. Creating a framework to examine benefits of green building construction. In Proceedings from the American Society for Engineering Management Conference, October 20–23, Alexandria, Virginia, pp. 719–724.

Ochsendorf, J., 2012. Challenges and opportunities for low-carbon buildings, the National Academy of Engineering (NAE).

Piper, J.E., 2004. Handbook of Facility Assessment. Fairmont Press, Liburn, GA.

Poynter, D., 1997. Expert Witness Handbook: Tips and Techniques for the Litigation Consultant, second ed. Para Publishing, Santa Barbara, CA.

Propst, R., 1968. The Office – A Facility Based on Change. The Business Press, Elmhurst, Illinois.

Prowler, D., 2008. Whole Building Design. Donald Prowler & Associates, Revised and updated by Vierra, S. – Steven Winter Associates, Inc. Last updated: 08.07.2008.

Pulaski, M.H., Horman, M.J., Riley, D.R., 2006. Constructability practices to manage sustainable building knowledge. J. Architect. Eng. 12 (2).

Ratay, R.T., 2005. Structural Condition Assessment. John Wiley & Sons, New York.

Reznikoff, S.C., 1989. Specifications for Commercial Interiors. Whitney Library of Design, New York.

Rogers, E., Kostigen, T.M., 2007. The Green Book. Three Rivers Press, New York.

Sale, K., 1993. The Green Revolution: The American Environmental Movement, 1962–1992. Hill and Wang, New York.

Samaras, C., 2004. Sustainable Development and the Construction Industry: Status and Implications. Carnegie Mellon University, Rwanda, website: http://www.andrew.cmu.edu/user/csamaras

Samarasekera, R., 2014. Introducing the New LEED Online (posted in LEED), 27 January.

Sampson, C.A., 2001. Estimating for Interior Designers. Watson-Guptill Publications, New York.

Sara, M.N., 2003. Site Assessment and Remediation Handbook, second ed. CRC Press, Boca Raton, FL.

Scheer, R., Woods, R., 2007. Is There Green in Going Green? SBM, April.

Smart Market Report, 2013. World Green Building Trends: Business Benefits Driving New and Retrofit Market Opportunities in over 60 Countries. McGraw-Hill, New York.

Smith, W.D., Smith, L.H., 2001. McGraw-Hill On-Site Guide to Building Codes 2000: Commercial and Interiors. McGraw-Hill Professional Publishing, New York.

Starr, J. Nicolow, J., 2007. How water works for LEED, BNET, CBS Interactive Inc., October.

Steiner, A., 2014. Climate change threatens global health security: UNEP. IBNLive, November.

Steven Winter Associates, 1997. Accessible Housing by Design. McGraw-Hill, New York.

Stodghill, A., 2008. It's the Environment Stupid, LEED™ vs. Green Globes, August 27.

Sullivan, P.J., Agardy, F.J., Traub, R.K., 2000. Practical Environmental Forensics: Process and Case Histories. Wiley, Hoboken, NJ.

Syal, M., 2005. Impact of LEED™ projects on constructors. AGC Klinger Award proposal. Michigan State University, Construction Management Program, School of Planning Design and Construction, E. Lansing, MI.

Tilton, R., Jackson, H.J., Rigby, S.C., 1996. The Electronic Office: Procedures & Administration, eleventh ed. South-Western Publishing Co, Cincinnati, OH.

Turner, C., Frankel, M., 2008. Energy performance of LEED for new construction buildings. New Buildings Institute, March 4.

U.S. Green building Council, 2008. U.S. Green building Council, LEED™ 2009 for New Construction and Major Renovations, November.

U.S. Green building Council, 2009. U.S. Green building Council, LEED™ Reference Guide for Green Building Design and Construction, For the Design, Construction and Major Renovations of Commercial and Institutional Buildings, Including Core & Shell, and K-12 School Projects, 2009 Edition.

United Nations, 2007. Report of the World Commission on Environment and Development. General Assembly Resolution 42/187, 11 December 1987. Retrieved: November 2007.

USACE, 2008. USACE Army LEED™ Implementation Guide. Headquarters, U.S. Army Corps of Engineers, January 15.

USGBC, 2008. USGBC, United States Green building Council. Available from: www.usgbc.org (accessed 2008–2009).

USGBC, 2013. USGBC LEED v4 User Guide, June. Available from: http://www.cabrillo.edu/~msoik/3/LEED%20v4%20guide.pdf

Vermont Green Building Network, 2009. Green building rating systems. Available from: www.vgbn.org/gbrs.php (accessed June 2009).

Watson, R., 2012. Attention USA Today: Green buildings need to make financial sense. GreenBiz, November.

Wilson, S.R., Dunn, S., 2007. Green downtown office markets: A future reality. CB Richard Ellis.

Woods, J.E., 2008. In: Green Building: Balancing Fact and Fiction, Cannon, S.E. and Vyas, U.K., moderators. Real Estate Issues 33 (2), 10.

Woods, J. E., 2008. Expanding the principles of performance to sustainable buildings, focus on green building. Entrepreneur Media, Inc., Fall.

Yoders, J., 2008. Integrated Project Delivery Using BIM, Building Design and Construction, April 1.

Yudelson, J., 2008a. The Green Building Revolution. US Green Building Council.

Yudelson, J., 2008b. The Business Case for Green. Yudelson Associates, March.

Zelinsky, M., 1998. New Workplaces for New Workstyles. McGraw-Hill, New York.

Zwick, D.C., Miller, K.R., 2004. Project Buyout. Journal of Construction Engineering and Management March/April, 245–248.

SUBJECT INDEX

A

AAA. *See* American Arbitration Association (AAA)

AAC. *See* Autoclaved aerated concrete (AAC)

AC breaker panel, 482

Accreditation process, 72

ACEEE. *See* American Council for an Energy Efficient Economy (ACEEE)

Acidification, 29

Acoustical control, 347

ACQ. *See* Alkaline copper quaternary (ACQ)

Active solar energy systems, 477–485
photovoltaic (PV), 477–479
solar electric systems, 479–483
types of, 483–485

Adaptable structural elements, 180

Adhesives, primary types of, 254

Adobe construction, 236

Adobe floor. *See* Earthen flooring

Adobe LiveCycle technology, 95

ADR. *See* Alternative dispute resolution (ADR) techniques

Advanced energy design guide, 416

Advanced framing techniques, 257, 258, 285
benefits from, 257

Advanced metering credit, 418

AECB. *See* Association for Environment Conscious Building (AECB)

Aeration, 250

AFUE. *See* Annual fuel utilization efficiency (AFUE)

Agency for Toxic Substances and Disease Registry (ATSDR), 148

Aggregates, 109

Agricultural residue (ag-res)
boards, 256
panels, 256

Agrifiber boards, 290

AHERA. *See* Asbestos Hazard Emergency Response Act (AHERA)

AHP. *See* Analytical hierarchy process (AHP)

AIA. *See* American Institute of Architects (AIA)

Air-based systems, 449

Airborne asbestos fibers, 314

Airborne bacteria, 332

Air cleaners, 368, 369

Air cleaning filters, 365

Air conditioner systems, 426, 435, 439, 442, 456
component of, 454
packaged. *See* Packaged air conditioner models
split-system. *See* Split-system air conditioners
types of, 440

Air-conditioning filters, 458

Air-cooled chillers, 454

Air duct distribution system, 444

Air filters, 420
function of, 456

Airflow device, 451

Air handling systems, 366

Air-handling unit, 434

Air movement, 366

Air pollutants, 29

Air pollution, 29

Air-pressurized water extinguishers, 506

Air sampling techniques, 307

Air-to-air heat exchangers, 367

Air-vent systems, 435

Air–water systems, 445

Alameda County Waste Management Authority, 269

Aldrin, 328

Alkaline copper quaternary (ACQ), 252

All-air systems, 445

All-water cooling-only systems, 445

All-water systems, 445

Alternating current (A/C) circuit, 467

Alternative dispute resolution (ADR) techniques, 563, 564

Alternative transportation, 196–200

Aluminum wire, 465

Aluminum wiring, 464

Printed in the United States
By Bookmasters